Chemical Physics of Intercalation

NATO ASI Series

Advanced Science Institutes Series

A series presenting the results of activities sponsored by the NATO Science Committee, which aims at the dissemination of advanced scientific and technological knowledge, with a view to strengthening links between scientific communities.

The series is published by an international board of publishers in conjunction with the NATO Scientific Affairs Division

A	**Life Sciences**	Plenum Publishing Corporation
B	**Physics**	New York and London
C	**Mathematical and Physical Sciences**	Kluwer Academic Publishers Dordrecht, Boston, and London
D	**Behavioral and Social Sciences**	
E	**Applied Sciences**	
F	**Computer and Systems Sciences**	Springer-Verlag
G	**Ecological Sciences**	Berlin, Heidelberg, New York, London,
H	**Cell Biology**	Paris, and Tokyo

Recent Volumes in this Series

Volume 165—Relativistic Channeling
edited by Richard A. Carrigan, Jr., and James A. Ellison

Volume 166—Incommensurate Crystals, Liquid Crystals, and Quasi-Crystals
edited by J. F. Scott and N. A. Clark

Volume 167—Time-Dependent Effects in Disordered Materials
edited by Roger Pynn and Tormod Riste

Volume 168—Organic and Inorganic Low-Dimensional Crystalline Materials
edited by Pierre Delhaes and Marc Drillon

Volume 169—Atomic Physics with Positrons
edited by J. W. Humberston and E. A. G. Armour

Volume 170—Physics and Applications of Quantum Wells and Superlattices
edited by E. E. Mendez and K. von Klitzing

Volume 171—Atomic and Molecular Processes with Short Intense Laser Pulses
edited by André D. Bandrauk

Volume 172—Chemical Physics of Intercalation
edited by A. P. Legrand and S. Flandrois

Series B: Physics

Chemical Physics of Intercalation

Edited by

A. P. Legrand

ESPCI
Paris, France

and

S. Flandrois

Domaine Universitaire
Talence, France

Plenum Press
New York and London
Published in cooperation with NATO Scientific Affairs Division

Proceedings of a NATO Advanced Study Institute on
Chemical Physics of Intercalation,
held June 10-19, 1987,
in Castera Verduzan, France

Library of Congress Cataloging in Publication Data

NATO Advanced Study Institute on Chemical Physics of Intercalation (1987: Castéra-Verduzan, France)
Chemical physics of intercalation / edited by A. P. Legrand and S. Flandrois.
p. cm.—(NATO ASI series. Series B, Physics; v. 172)
"Proceedings of a NATO Advanced Study Institute on Chemical Physics of Intercalation, held June 10-19, 1987, in Castera Verduzan, France"—T.p. verso.
"Published in cooperation with NATO Scientific Affairs Division.
Includes bibliographies and index.
ISBN 0-306-42831-8
1. Clathrate compounds—Congresses. I. Legrand, A. P. II. Flandrois, S. III. North Atlantic Treaty Organization. Scientific Affairs Division. IV. Title. V. Series.
QD474.N39 1987
541.2—dc19 88-2468
CIP

A Division of Plenum Publishing Corporation
233 Spring Street, New York, N.Y. 10013

Printed in the United States of America

SPECIAL PROGRAM ON CONDENSED SYSTEMS OF LOW DIMENSIONALITY

This book contains the proceedings of a NATO Advanced Reserach Workshop held within the program of activities of the NATO Special Program on Condensed Systems of Low Dimensionality, running from 1983 to 1988 as part of the activities of the NATO Science Committee.

Other books previously published as a result of the activities of the Special Program are:

Volume 148 INTERCALATION IN LAYERED MATERIALS
edited by M. S. Dresselhaus

Volume 152 OPTICAL PROPERTIES OF NARROW-GAP LOW-DIMENSIONAL STRUCTURES
edited by C. M. Sotomayor Torres, J. C. Portal, J. C. Maan, and R. A. Stradling

Volume 163 THIN FILM GROWTH TECHNIQUES FOR LOW-DIMENSIONAL STRUCTURES
edited by R. F. C. Farrow, S. S. P. Parkin, P. J. Dobson, J. H. Neave, and A. S. Arrott

Volume 168 ORGANIC AND INORGANIC LOW-DIMENSIONAL CRYSTALLINE MATERIALS
edited by Pierre Delhaes and Marc Drillon

PREFACE

The number and diversity of intercalation compounds have grown enormously in recent years. The unifying theme is the tailoring of properties by inserting guest atoms or molecules into more-or-less well defined interstitial sites defined by a host lattice. The properties of interest may be primarily electronic (synthetic metals such as doped polymers, intercalated graphite, cuprate and perovskite superconductors...), or ionic (transport in solid electrolytes, superionic conductors), magnetic or catalytic. In all cases an important step is to understand the rich and complex structural chemistry which results from competing microscopic interactions in low-dimensional, multicomponent systems. In particular, the most novel physical properties often depend sensitively on the effective dimensionality. We felt that it was time to bring together people working on different kinds of intercalation compounds, to share state-of-the-art knowledge and encourage cross-fertilization of ideas.

This Advanced Study Institute was concerned with layered materials like graphite intercalation compounds, transition metal chalcogenides and clays, and equally with conducting polymers and oxides. A more general approach to intercalation features has been sought through a discussion of the reactivity of the host structure towards the phenomena of intercalation, the nature of the redox centers, modification of the physical properties and the occurrence of phase transitions.

Finally, several contributions were devoted to practical applications of these materials, especially in the field of electrochemistry. In particular, the use of layered compounds is of great interest in rechargeable lithium batteries.

The success of this NATO ASI was due in large part to the efforts of the Association Scientifique Culturelle et Educative de Bonas to welcome all the participants. We thank greatly the Local Organizing Committee, namely, Dr. D. Bonnin and Dr. M.F. Quinton for their efficiency and the Scientific Committee (J.E. Fischer, A. Hérold, R. Schöllhorn) for their recommendations.

Although mainly supported by the NATO Scientific Affairs Division, this ASI was helped by the National Science Foundation (USA). We thank the two french companies Le Carbone Lorraine and la SAFT, for their cooperation.

A.P. Legrand
S. Flandrois

July 1987

CONTENTS

PART A. THE LECTURES

PART B. SEMINARS AND COMMUNICATIONS

PART A. THE LECTURES

SYNTHESIS OF GRAPHITE INTERCALATION COMPOUNDS

Albert Hérold

Laboratoire de chimie Minérale Appliquée
CNRS 158 - Université de Nancy I
BP 239 54506 Vandoeuvre lès-Nancy cedex France

INTRODUCTION

Graphite is one of the most important 2D host lattice by the number and the scientific and technical interest of its intercalation compounds. This paper is not a comprehensive review of GIC synthesis but a description of the thermodynamical and chemical conditions of intercalation into graphite, for a comparison with intercalation into other host lattices and, if possible, as a starting point of new syntheses. It partly develops ideas which were put forward at the 1st Conference on GIC, in La Napoule, 1977. I apologize to those uncited authors whose work merits being cited in a true review of GIC.

I - GRAPHITE AS A HOST LATTICE

1. Structure, aromatic character, and amphoteric behaviour

Graphite is the most simple 2D host lattice. Its stable hexagonal variety is formed by an ABAB stacking of plane monoatomic sheets, with an interplanar distance d_o = 335 pm (Figure 1)(The ABC stacking of rhomboedral graphite is metastable). The in-plane honeycomb disposition of carbon atoms corresponds to an sp_2 hybridisation : graphite sheets can be considered as aromatic macromolecules. Each carbon atom forms 3σ bonds with its closest neighbours. The 4th valence electron is used in a Π bond, which is in resonance between the 3σ directions. The in-plane carbon-carbon distance d_{C-C} of 142 pm is much less than the distance d_{C-C} in diamond (154 pm corresponding to the carbon sp_3 hybridisation), but it is a little higher than in benzene (139 pm), in which the π bond is distributed between only two σ directions. This description justifies the term "graphene" which was proposed to the IUPAC nomenclature commission for an isolated graphitic sheet [I, 1]. The binding π orbitals form the valence band and the antibinding orbitals the conduction band. Whereas the small aromatic molecules are insulators or semi-conductors [G], in an isolated graphene sheet, the "gap" is exactly zero. In graphite, interactions between the neibouring sheets induces a small overlapping of the two bands ; therefore, graphite is a semi-metal containing about 2.9 x 10^{18} positive and negative carriers per cubic cm at 4.2 K.

Because of its aromatic character, graphite has an amphoteric behaviour: it is able to accept electrons from electropositive species (electron donors)

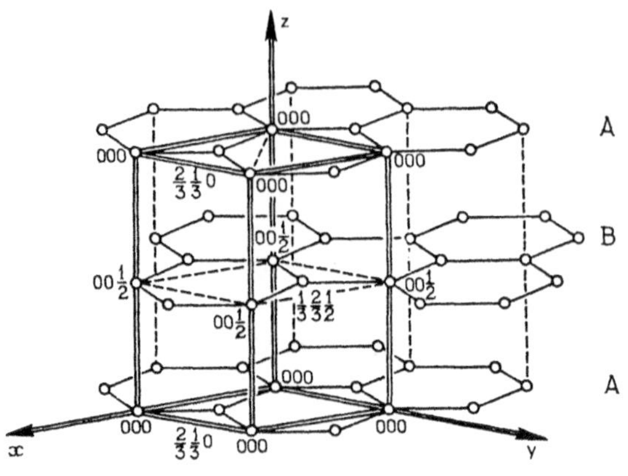

Fig. 1. Structure of hexagonal graphite
$a = d_{C-C}\sqrt{3} = 246$ pm. $c = 2 \cdot d_o = 670$ pm.

and to yield electrons to electronegative species (electron acceptors) leading to "synthetic metals" of high electrical conductivity.

According to the "ionic model" these electronic exchanges lead to cations or anions, which intercalate and are linked to the graphene sheets by a Coulomb attraction. One can suppose that the band structure is not modified for a low concentration of intercalate (rigid band model). Then it is of interest to compare the value of the graphite work function (which is close to 4.8 eV) to the first ionisation potential I of the most electropositive elements and to the electron affinity A of the most electronegative elements. (A is corresponding to the reaction $\frac{1}{2} X_2 + e^- \rightarrow X^-$)

Table 1. First ionisation potential I of alcali metals and electron affinity A of halogens in electron-volts.

	Li	Na	K	Rb	Cs
I	5.39	5.13	4.34	4.17	3.89
		F	Cl	Br	I
A		2.77	2.49	2.53	2.43

One can see that the formation of a cation M^+ through the transfer of an electron from Cs, Rb, or K to the graphite conduction band is exothermal. It is slightly endothermal for Na and Li. On the contrary the formation of an anion X^- by transferring an electron from the graphite valence band to a halogen is strongly endothermal. Consequently graphite is more an electron acceptor than an electron donor.

On the other hand, because of its amphoteric behaviour, graphite cannot be a very strong acceptor (nor, of course a strong donor). For instance the 2D and 3D ionocovalent host lattices of oxides of chalcogenides which do not function as donors, are much stronger acceptors than graphite [2] [3] [4].

An interesting evaluation of the acceptor or donor power is given by measuring the electrochemical oxidation of reduction potentials (see IV, 1, electrochemical intercalation).

2. The ionic model

The electrostatic nature of the bond in GIC was firstly proposed by Hofmann and coworkers (1934-38) [5] [6] for the sulfuric acid-graphite compounds and other "graphite salts". It is the basis of the ionic model developped since 1951 by Hennig [7] [8] and by Ubbelohde and Coll. [9] [A]. According to this model the free enthalpy of intercalation ΔG_i can be calculated as a sum of energy terms which may or may not depend on the charge transfer.

Figure 2 gives a schematic view of these energy terms. Cycles A and B respectively correspond to a metal M and an acceptor X. In both cases the free enthalpy ΔG_i is negative : intercalation spontaneously occurs.

The cycle C corresponds to the intercalation of a solvated anion X^-. Because X^- is not an oxidising agent, ΔG_i is positive : intercalation absorbs energy, we propose the term of "assisted intercalation" for this case.

Several energy terms merit comment :

a) "Spacing energy" W_s is required for increasing the interplanar distance from 335 pm to a value d_i which depends on the intercalate layer thickness. Values between 2.5 and 5 KJ per mole of carbon have been proposed for the energy $W_s(\infty)$ required for the separation to infinity of two graphene layers [10] [11]. W_s strongly varies with the interplanar distance d_i. Armand [12] following Crowell [13] calculated it by integration on the graphene planes of the attractive and repulsive interatomic forces :

$$W_s = N \cdot s \left[-\frac{A}{4d^4} + \frac{B}{10d^{10}} \right]_{d_o}^{d_i} \quad \text{with } A = \frac{d_o^5}{6k} \quad ; \quad B = \frac{d_o^{11}}{6k}$$

in which : N = Avogadro's number, s = surface area of a carbon atom, k = coefficient of compressibility along the $\vec{c}$ axis ($k = 5 \cdot 10^{-12}\ cm^2 \cdot dyne^{-1}$).
For $d_i = 2d_o$, W_s is more than 90 % of $W_s(\infty)$.

Although W_s is rather small, it plays an important role by limiting the number of intercalates : it is well known that neutral molecules such as noble gases and hydrocarbons are physisorbed at the external surface of graphite samples, especially on 001 planes. But they cannot intercalate, because they energy of physisorption is lower than the spacing energy.

In GIC, W_s is partially compensated by the Van der Waals energy of the intercalate adsorption on graphene layers.

b) When intercalation starts (also when the intercalate concentration in graphite is very low) one can suppose that the graphite band structure is not modified and write :
$|W(\text{capture of an electron})| = |W(\text{extraction of an electron})|$ = Work function = 4.8 eV. At higher intercalate concentration $|W(\text{capture})|$ decreases and W(extraction) increases. The corresponding variations depend on the change in the band structure and cannot be calculated a priori.

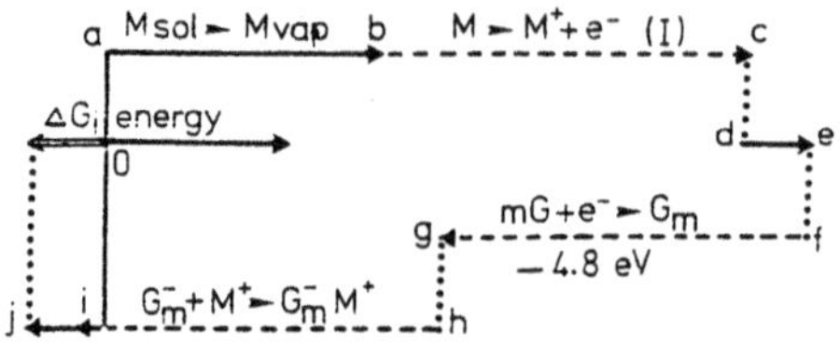

a b : metal sublimation

b c : metal ionisation

d e : spacing energy

f g : electron capture by G.

h i : Coulomb energy

i j : Van der Waals energy

Fig. 2. A. Free enthalpy of intercalation ΔG_i of a metal M into graphite as a sum of energies depending (----) or not depending (——) on the charge transfer.

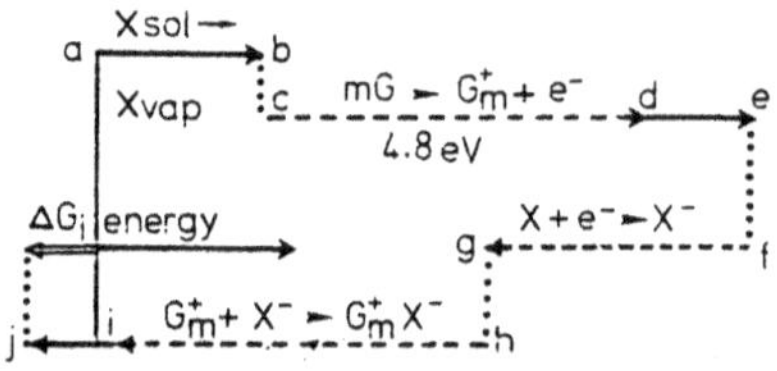

a b : acceptor sublimation

c d : graphite oxidation

d e : spacing energy

f g : electron capture by X

h i : Coulomb energy

i j : Van der Waals energy

Fig. 2. B. Free enthalpy of intercalation of an electron acceptor X.

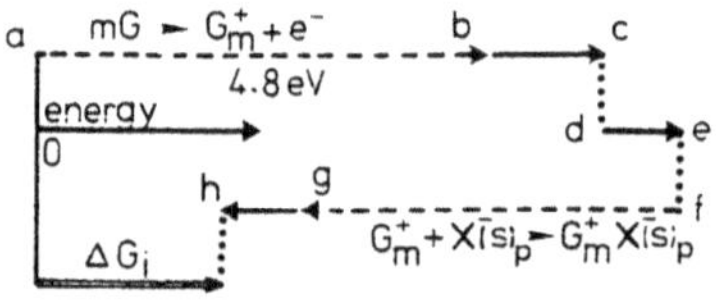

a b : graphite oxidation

b c : spacing energy

d e : 3D → 2D energy of solvated ions

f g : Coulomb energy

h g : Van der Waals energy

Fig. 2. C. Free enthalpy of intercalation of an anion X^- solvated by p molecules of electrolyte S.

c) Calculation of the electrostatic energy is uncertain by the following reasons :

- it is proportional to the in-plane concentration of ions, which is unknown a priori
- ionisation seems to be incomplete in the donor compounds, in which the bond is partly covalent (or metallic)
- the charge distribution in the graphene layers is uncertain

Although the calculation of Salzano and Aronson [11] for the CsC_8 compound led to reasonable energetical values, the ionic model has to be used carefully for predicting intercalation. Nevertheless, an evaluation of the energy terms which are taken into account in this model is of interest for understanding the experimental data.

3. Structure and stoichiometry

a) Layer distribution along the $\vec{c}$ axis - Stages -

Each intercalate can form GIC of different ordered stages. In a stage s, two layers of intercalate are separated by s graphene sheets. The repeat distance along the $\vec{c}$ axis perpendicular to these sheets is I_c(nm) : d_i (interplanar distance) + $(s-1)d_o$. The coexistence of free intervals between the graphene sheets, and of intervals occupied by layers of high in-plane density corresponds to the minimum of the spacing energy W_s, and their regular distribution along the $\vec{c}$ axis to the minimum of the electrostatic repulsive energy. Figure 3 shows a schematic view of a third stage compound in the classical model of stages and the pleated layer model. According to

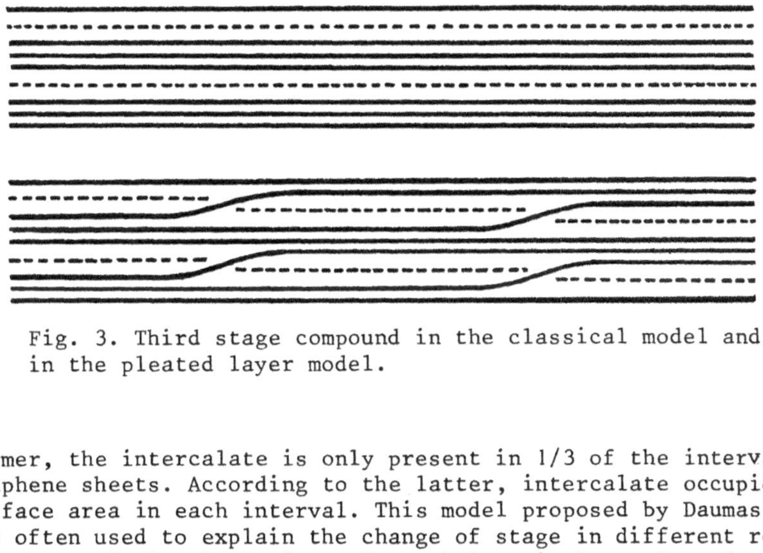

Fig. 3. Third stage compound in the classical model and in the pleated layer model.

the former, the intercalate is only present in 1/3 of the intervals between the graphene sheets. According to the latter, intercalate occupies 1/3 of the surface area in each interval. This model proposed by Daumas and Hérold [14] is often used to explain the change of stage in different reactions, and for theoretical calculations. Nevertheless it has only undergone partial crystallographic verification. Observations of domains surrounded by dislocations were made by Eeles and Turnbull [15] in bromine-graphite compounds and by Heerschap and Delavignette [16] in ferric chloride-GIC using transmission electron microscopy. High resolution TEM allowed Thomas et al. [17] a direct observation of the 001 graphene and $FeCl_3$ sheets, showing the "folds". More recently Levi-Setti et al. [18] measured the lateral extent of the domains by direct imaging of the surface distribution of the intercalate in $SbCl_5$ - G compounds, using secondary ion mass spectrometric (SIMS) mapping of $^{35}Cl^-$. The average domain size is $\sim$ 200 nm.

The small thickness of the graphene layers favourizes the electrostatic and strain interactions along the $\vec{c}$ axis, leading to ordered stages up to the 10th one. From this point of view, graphite differs from the other 2D host lattices, which are made of thicker and more rigid sheets, so that it is not generally possible to observe regular sequences beyond stage 4. Nevertheless, a recent precise study shows that GIC of stage higher than 2 contain stacking defects [19].

b) Charge distribution along c

Intercalated layers of donors M and acceptors X can be monatomic or monomolecular. Then the sequence along the $\vec{c}$ axis is $G^- M^+ G^- M^+$... or $G^+ X^- G^+ X^-$. But the sandwich unit can be more complex : for instance dichlorides of Mn, Fe, Co, Ni, Cd, trichlorides of Fe, Cr, form triple ionocovalent sheets, in which two planes of Cl^- ions surround a plane of cations M^{2+} or M^{3+} in octahedral sites. The signs of the successive planes are ... $G^+ Cl^- M^+ Cl^- G^+$... A symetrical arrangement exists in multilayers of donors, for instance $G^- K^+ H^- K^+ G^-$ ... or $G^- K^+ Hg^- K^+ G^-$.

Alternate planes of positive and negative charges can also be found in layers containing complex anions such as ClO_4^- , BF_4^- or PF_6^-, SbF_6^-, $SbCl_6^-$. The probable disposition of these anions which is shown in figure 4 also leads to a stacking of alternate positive and negative planes... $G^+ O^- Cl^+ O^- G^+$... or ...$G^+ F^- P^+ F^- G^+$..

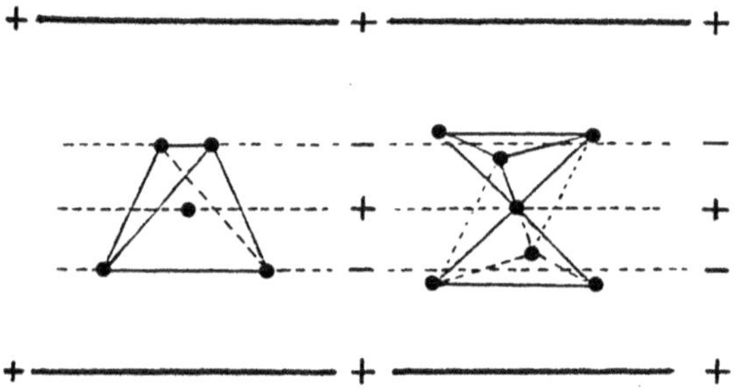

Fig. 4. Tetrahedral and octahedral anions between two graphene sheets.

c) In-plane organisation of the intercalate

GIC are topotactic compounds, their formula depend on the structural relations between the intercalated layers and the adjacent graphene sheets. The potential wells of these sheets are localised over the centers of the carbon aromatic cycles. Simultaneous occupation of the potential wells of the both graphene sheets on either side of the intercalate often occurs by shifting these planes from the graphite ABAB.. stacking sequence to the AAA.. one. Then, the coordination number of an intercalated ion is 12, an exceptional number in 2D intercalation chemistry.

Nevertheless, the geometrical structure of the graphene layers and their small acceptor - donor power lead to shallow potential wells, which favourize the formation of ternary compounds, and allow a large variety of structures. Three main kinds of layers can be distinguished :

α) Commensurate layers

Commensurability corresponds to a fitting of the intercalate lattice to that of the graphene sheets. It requires an in-plane extension or compression of the intercalate. Therefore it depends on the in-plane cohesion forces. In layers of single ionised atoms (such as alkali metals ...) the inplane cohesion is mainly due to the elastic energy : the spacing work and the strain energy around each atom decrease when the in-plane density increases. This kind of layers can be easely extended or compressed, so that they are often commensurate.

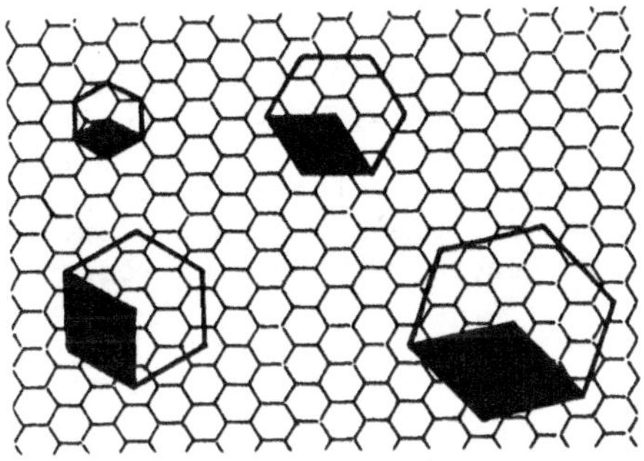

Fig. 5. Commensurate hexagonal lattices

Figure 5 shows the most common commensurate lattices, which are hexagonal ones. The parameter a equals the graphite parameter multiplied by $\sqrt{3}$, 2 or $\sqrt{7}$ respectively, and the lattices are turned by angles of 30°, 0°, 19°1 with respect to the lattice vectors of graphite.

Commensurability leads to compounds in which the "ideal" carbon / intercalate ratio is an integer or a rational fraction, such as LiC_6, KC_8, $C_{14}Sb_{1.2}Cl_6$ [D 1].

β) Incommensurate ordered layers

Thick ionocovalent layers (see b) are very rigid and consequently are well ordered, oriented with respect to the graphene sheets, but not generally commensurate. The "ideal" carbon / intercalate ratio is not an integer, but can be calculated : $\frac{c}{i} = \frac{n_g}{n_i} \cdot \frac{s_i}{s_g}$ in which n_g and n_i are the number of atoms (or molecules) per unit cell ($n_g = 2$), and s_i and s_g the unit cell surface areas.

γ) Ionomolecular layers

These layers are formed by association ions with neutral molecules : cations of electropositive metals solvated by polar or aromatic molecules (see II, 4); chloro-anions such as $AlCl_4^-$ associated with neutral Al_2Cl_6 molecules (see III, 3); oxyanions such as HSO_4^-, NO_3^- solvated by neutral acid molecules or non aqueous electrolytes (see IV, 1). The cohesion of this kind of intercalated layers results from ion-dipole or other Van der Waals attractive forces which often are too weak to impose a long range in-plane ordering. The liquid like layers generally present transitions to more ordered structures at low temperatures which are often modulated by interaction with the graphite lattice periodic potential. An "ideal" crystallographic formula cannot be easely calculated. Good examples of these structures are given by the two kinds of graphite nitrates α [20] and β [21].

d) "Staging"

Intercalate distribution in the graphitic host lattice (in other words stage, in-plane density, and structure) as a function of its concentration and of temperature and pressure is often called "staging".

It results from an equilibrium between different kinds of energies (elastic, electrostatic), which has been studied by Safran [22] and Fisher [23].

4. Spontaneous intercalation : techniques

Synthesis of GIC by spontaneous intercalation can be carried out using an intercalate in the vapour, liquid, or solid phase.

a) Intercalation from a vapour phase leads to the highest quality products, if the reagent is volatile enough. In the two bulb tube (figure 6), graphite at temperature T_G reacts with the intercalate at temperature T_i,

Fig. 6. The two-bulb tube

by the intermediary of its vapour. The same principle is used in many other forms of two-temperatures reactors. In several of them, the weight uptake of the graphite is measured continuously during intercalation [24] [25]. Under equilibrium conditions, the ratio i/C is a function of T_G and T_i, or T_G and p (saturating vapour pressure of i at T_i). The two temperature technique can be used in three different manners :

α) T_G is constant
T_i is the independent variable. The i/C = f(p) or f (T_G - T_i) curves are isothermal ones. Isotherms are of great interest for calculating the thermodynamic data [D2]. Howhever the necessity of measurements over a large range of pressure often leads to technical difficulties.

β) T_i is constant
and therefore the pressure p in the tube; T_G is the independent variable. The curves i/C = f(T_G - T_i) are isobaric ones. Figure 7 shows examples of these two kinds of curves

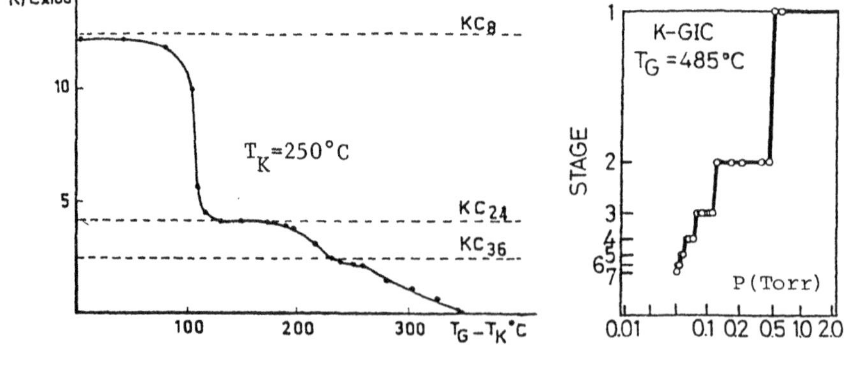

Fig. 7. Stages in potassium-graphite system
a) Isobaric curve determined by chemical analysis (Hérold) [26]
b) Isothermal curve determined by X ray technique (Suematsu) [27]

Comparison between these curves clearly shows that the phase rule is applicable to the stages which are true phases.

γ) $T_G - T_i$ is constant and close to zero : graphite is under the quasi saturating vapour of i at the temperature $T = T_G \sim T_i$. A "saturating curve" $i/C = f(T)$ corresponding to the graphite - bromine system is represented on figure 8 [28].

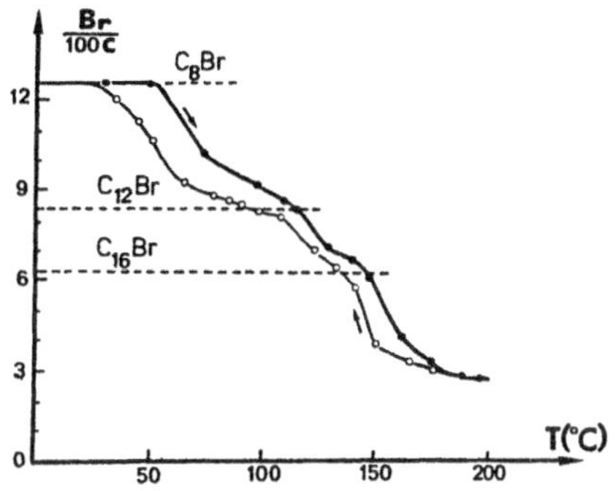

Fig. 8. "Saturation curve" of the bromine-graphite system

It shows that the concentration of intercalate decreases with increasing temperature. Above a limiting temperature intercalation becomes impossible. A good example of this is also given by chlorine, which cannot intercalate above - 20°C. Likewise, the 1rst stage HNO_3 - GIC can be obtained at - 20°, but not at room temperature; the 1st stage ICl - GIC decomposes above 50°C. The decrease of intercalate concentration with increasing temperature is a consequence of the exothernal character of intercalation, in agreement with Le Châtelier 's Principle of equilibrium displacement.

"Saturation curves" with $T_G \sim T_i$ have been also used for studying equilibria between graphite and mixtures of two intercalates, such as two alkali metals (See II, 2), two halogens (See III, 2).

b) Intercalation from a liquid phase can be very fast [28]. Therefore it is of interest to synthetise compounds of pure reagents such as lithium (See II, 1) or chlorine (See III, 2) which slowly reacts with graphite. It must be used carefully for other reagents, because fast intercalation can favourize the formation of defects.

Furthemore two kinds of reaction can be observed, when a liquid alloy or a solution is in contact with graphite

- intercalation of one reagent : lithium from a sodium bath (see I, 1) bromine from its solution in carbon tetrachlorid, SO_3 from its solution in SO_2 (See III, 7)
- intercalation of the two components, leading to ternary GIC : metallic alloys, metals solvated by polar or aromatic molecules (See II, 2 and 4), assisted intercalation of solvated anions and cations (See IV, 1).

c) Intercalation in the solid phase has been carried out by compression of a mixture of powdered graphite and divided lithium (See II, 1). Compounds of different stages can be synthetised according to the pristine Li/C ratio.

Applications of the ionic model and of spontaneous intercalation techniques will be exposed in part II (electron donors) and III (electron acceptors).

II - INTERCALATION OF DONORS

1. Pure metals

a) Experimental data

Binary compounds have been synthetised by the reaction of graphite with the following metals.

Li (Na) K Rb Cs/(Ca) Sr Ba / Eu (Sm) Yb.

The first observations concerning the reaction of alkali metals with graphite were made by Weintraub [1] and more precisely by Fredenhagen and Cadenbach [2]. The yellow colour of graphite after reaction with lithium vapour was first described by Hérold [3] and induced the studies of the group of Nancy (D. Guérard, A. Hérold, A. Métrot) which carried out intercalation of the other metals.

Intercalation of the haevy alkali metals is easy in two temperature reactors [3] [B] [D.2] [4] and leads to 1st stage compounds with an in-plane 2 · 2, R = 0 lattice and an MC_8 formula. At the higher stage, there is no long range ordering in the metal layers, and the in-plane density corresponds to a formula close to MC_{12s} [4] [E, p. 185].

Sodium hardly intercalates into graphite leading to compounds of stages 8 or 6 [5] [6] and more easely reacts with soft carbons [7].

The intercalation of lithium [8], alkaline earth metals [9] and bivalent, rather volatile lanthanides [10] is in competition with the formation of stable carbides. Therefore it must be carried out at temperatures as low as possible under very small vapour pressure. Intercalation remains superficial for Ca and Sm. Pure high stage compounds could only be synthetised for lithium using either the reaction of graphite with lithium - sodium liquid alloys [11] [12], or by compression of lithium and graphite powders [8].

b) Energetic aspects of metal intercalation

According to the ionic model, intercalation of metals requires their vaporisation and their ionisation. Table 2 contains a list of the first 14 metals arranged in order of the increasing values of the sum of their enthalpy of vaporisation (ΔH_V) and their first ionisation energy in eV.

Table 2. Classification of the metals as a function of values of the vaporisation enthalpy (ΔH_V) plus the first ionisation energy (I), in eV.

Cs	Rb	K	(Na)	Li
4.55	4.90	5.19	6.09	6.82
Ba	Sr	(Ca)		
6.65	7.06	7.57		
Eu	(Sm)	Yb		
7.37	7.48	7.86		
Tl	In	Ga		
7.72	8.04	8.57		

Discussion using the "ionic model"

- All already intercalated metals are in table 2.
- The heat of formation (-ΔH) of the MC_8 phases (table 3) increases from K to Cs [13] in agreement with table 2 data.

Table 3. Molar enthalpies of formation of the MC_8 compounds from liquid metals

Metal	K	Rb	Cs
- ΔH (KJ)	34.9	58.1	79.8
- ΔH (eV)	0.362	0.481	0.662

- The reasons underlying the stability of the sodium compounds with respect of those of lithium, and the change of the in-plane density from K to Li has not yet completely explained. Nevertheless, the small interplanar distance in lithium compounds leads to a small value of the spacing energy and a high absolute value of the electrostatic energy. It also favourizes the screening of the positive charge of the lithium atoms by the carbon ones, leading to a small value of the in-plane Li - Li repulsive energy.
- Intercalation of alkaline earth metals and lanthanides is only possible through the intervention of the 2nd (or 2nd and 3rd) valence electrons which stabilise the metal layers, probably by the formation of metal-metal in-plane bonds.
- In the III B metals the low symetry of the p orbital corresponding to the external electron is not favorable to the formation of a iono-metallic bond, so that these metals do not intercalate.

c) Criticism of the purely ionic model

Several experimental data cannot be explained using a purely ionic character of the graphite - metal bond.

- In 1st stage lithium graphite compounds, the sequence of the graphene and intercalated layers is Aα Aα Aα, leading to a Li - Li distance along the $\vec{c}$ axis of 370 pm, whereas the in-plane distance is 426 pm. These data are not compatible with the repulsive interactions of completely ionised lithium atoms, but, on the contrary, suggest the formation of iono-metallic Li - Li bonds along the $\vec{c}$ axis, in agreement with the good electrical conductivity of LiC_6 along this direction.
- The decrease of the in-plane density from the 1st to the other stages in heavy alkali metal compounds probably results from an increase of the Coulomb repulsive energy due to the change from an iono - metallic bond in stage 1 to an almost purely ionic bond in higher stages in agreement with electronic properties.

Although the metal graphite bond seems to be partly metallic - especially in the first stage compounds - the ionic model remains of interest in prediction of new metals which may intercalate. Several metals, such as lanthanum, cerium, thorium, uranium, have lower first ionisation energy than ytterbium (0.60 eV). But they lattice energy, measured by their enthalpy of vaporisation is much more higher. Therefore we believe that direct intercalation of new metals, leading to stable compounds is not possible. However, formation of metastable compounds by indirect intercalation cannot be excluded (See II, 5).

2. Alloys

Intercalation of alloys has been mainly studied by the group of Nancy (Billaud, Lagrange). Two kinds of compounds were synthetised, which contain :

a) Alloys in single layers

b) Alloys in multiple layers.

a) The most important ternary compounds containing single layers are the $M_{1-x} N_x C_8$ ones (M, N = K, Rb, Cs and $0 \leq x \leq 1$) [14] which can be synthetised by the successive or simultaneous action of two metals on the graphite. Under the equilibrium conditions between a liquid alloy $M_{1-y}N_y$ at T_i and a compound $M_{1-x} N_x C_8$ at $T_G = T_i + \sim 10°$ the ratio :

$$\frac{x}{1-x} \cdot \frac{1-y}{y} = K(T)$$ is independent of x and y.

Thus, the $M_{1-x} N_x C_8$ compounds can be considered as solid solutions of an MC_8 phase and an NC_8 phase. The constant K measures the enrichment of the intercalated alloy in the haevier metal with respect to the free alloy in equilibrium with it; the highest values of K are obtained for the potassium - cesium couple, in agreement with the enthalpies of intercalation given in table 3.

These data can be applied to the separation or the purification of alkali metals.

b) Alloys intercalated in multiple layers are made of a metal which easely intercalates in its pure state and a metal which hardly intercalates or not intercalate by itself.

The first compound of this type was obtained by Billaud in the reaction of a liquid sodium - barium alloy with pyrographite [15]. The other compounds discovered by Lagrange contain alloys of a haevy alkali metal and a less electropositive metal, such as mercury [16] thallium [17] and bismuth [18]. In these phases it can be admitted, as a first approximation, that the intercalated layers are triple (Fig. 9) :

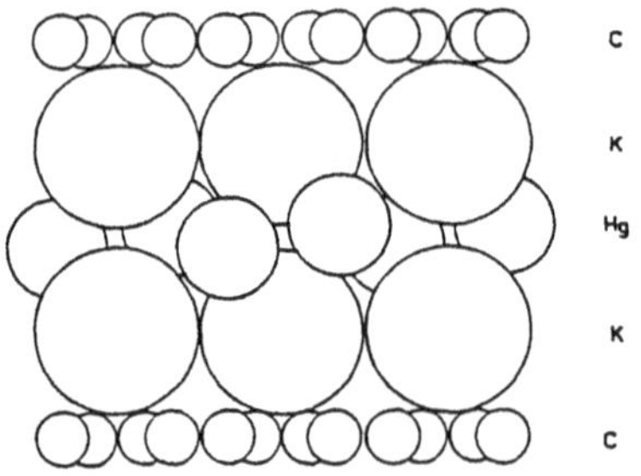

Fig. 9. Ideal structure of potassium - mercury - graphite compounds (Lagrange).

Two planes of potassium atoms of (2 · 2, R = 0) structure surround a plane of mercury atoms in prismatic sites. The formula for a first stage is $KHgC_4$. Compounds of thallium form more complex structures leading to the formula $MTl_{1.5}C_4$ for the first stage, whereas the bismuth compounds present a large variety of commensurate and incommensurate structures, leading to formula between $MBi_{0.5}C_{4s}$ to $MBiC_{4s}$ (s = stage).

Synthesis is carried out by liquid phase reaction; only the compounds of the volatile mercury can be prepared in the vapour phase. Table 4 gives a listing of the intercalated phases. Several of them exist under two allotropic forms α and β with small differences in the interplanar distances.

Table 4. List of GIC containing a heavy alkali metal associated with mercury, thallium, or bismuth.

Metal	K	Rb	Cs
Hg	s = 1 - 2	s = 1 - 2	
Tl	s = 1 - 2	s = 1 - 2	
Bi	s = 2 - 5	s = 1 - 7	s = 1 - 3

What is the stability of these phases, and the reasons for the different behaviour of Hg, Tl and Bi?

α) The iono-metallic bond between the alkali metal and the haevy metal leads to a strong in-plane cohesion of the triple layers. The sequence of alternating negative and positive charges along the $\vec{c}$ axis (for instance $(..\bar{C}\ \overset{+}{K}\ \overline{Hg}\ \overset{+}{K}\ \bar{C}\ ..)$ stabilises the triple layers with respect to the theoretically possible single layers.

β) Intercalation of an alkali metal associated with a less electropositive haevy metal results of a balance between two affinities :

- affinity of graphite for the alkali metal, which increases from potassium to cesium and also increases with the stage s.
- affinity of the haevy metal for the alkali metal, which increases with the haevy metal molar fraction in the alloy, and also depends on the nature of the two metals.

Figure 10 clearly illustrates the effect of the potassium amalgam composition on the possibility of intercalation and the nature of reaction products.

. If the potassium molar fraction is two low, no intercalation occurs
. If this fraction is to high, potassium intercalates alone.
. For intermediate values of concentration a cooperative intercalation of the two metals leads to ternary compounds.
. Only 1st and 2nd stage K - Hg - G compounds exist because, in a 3rd stage, the affinity of graphite for potassium would be too high to allow its association with mercury.

The different behaviour of Hg, Tl and Bi which appear in table 4, can be qualitatively explained using the hypothesis that the affinity of the haevy metals for the alkali ones increases along the series K Rb Cs and Hg Tl Bi in agreement with the values of the electronegativities. Another parameter which must be taken into account is the temperature : intercalation of alloys is exothermal and therefore they decompose above a certain temperature : synthesis of the K - Bi - G 1st stage compounds could not be carried out because of the high melting points of the alloys of convenient composition.

Alloy intercalation certainly offers wide unexplored possibilities for new synthesis : they are of special interest because of the superconductivity of certain phases.

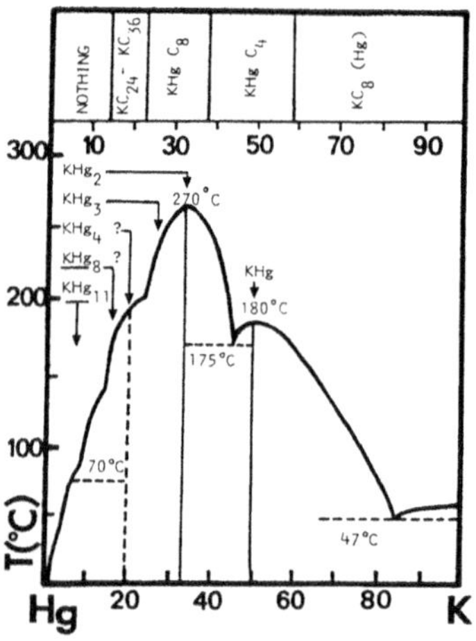

Fig. 10. Phase diagram of the K - Hg system and products of the potassium amalgam intercalation into graphite.

3. Hydrides

Hydrogen easely reacts with the yellow KC_8 leading to blue products. According to Saehr [19] and Colin [20] the reversible reaction can be written

$$KC_8 + x \frac{H_2}{2} \rightleftarrows K\,H_x\,C_8$$

$K\,H_x\,C_8$ is homogeneous for $0.5 \leqslant x \leqslant 0.67$. It belongs to the 2nd stage and contains triple layers made of 2 planes of potassium atoms in contact with the graphene layers, and an intermediate plane of hydrogen atoms. Of course the sequence of charges along the $\vec{c}$ axis is .. $C^- K^+ H^- K^+ C^-$... although hydrogen and potassium perhaps are not completely ionised [21].

RbC_8 can also be hydrogenated leading to Rb $H_{2/3}$ C_8, but only under pressure whereas Cs C_8 does not react. These results suggest that the formation of the K H_x C_8 ternaries corresponds to a balance between the affinity of graphite and hydrogen for the alkali metal. The former increases in the sery K Rb Cs (see the enthalpies of formation in table 3). Whereas the latter slightly decreases (the enthalpies of formation of KH, RbH and CsH from the liquid metals are respectively : 59.0, 52.2 and 56.4 KJ).

Direct reaction of graphite with KH, which was carried out by Guérard

[22] allowded to synthetise not only the blue $KH_x\ C_8$, but also the violet $KH_x\ C_4$ 1st stage ternary ($0.5 \leq x \leq 0.8$).

Certainly the affinity of the graphite matrix for the potassium atoms is higher in $KH_x\ C_8$ than in $KH_x\ C_4$. Therefore the dissociation pressure of the former should be higher than that of the latter, in agreement with the temperature ranges which were used for the syntheses : 250 - 300°C for $KH_x\ C_8$, 300 - 500°C for $KH_x\ C_4$.

No compounds of stage higher than 2 could be synthetised, probably because, above the stage 2, the graphite affinity for potassium is too strong, so that hydrogen is excluded : the second stage KC_{24} reacts with hydrogen leading to $KH_y\ C_{24}$ products ($0 \leq x \leq 0.5$) which contain the same kinds of layer than $KH_x\ C_8$ [21] with single layers of metal.

Sodium hydride NaH also reacts with graphite in scealed tubes to form ternary compounds of stages 1 to 8, containing triple intercalated layers [23]. The existence of a 1st stage whereas the metal forms only compounds of stages 6 - 8 (See 1) can be explained by the hydrogen - metal bonds which stabilise the triple Na - H - Na layers with respect to the single layers of metal.

Ternary Na - H - G compounds generally contain some included crystallised hydride so that their composition remains uncertain. In any case, the metal - hydrogen - graphite systems certainly offer large possibilities for new syntheses and present many unsolved thermodynamic problems.

4. Electropositive metals associated with molecules

a) Metals and Molecules

The list of metals intercalated with molecules (table 5) includes almost all the metals which form binary compounds (See 1) and several others (Be, Mg, Al, some lanthanides).

Table 5. Metals intercalated with molecules

Li	Na	K	Rb	Cs	/	Be	Mg	Ca	Sr	Ba
Al /	La	Ce	Sm	Gd		Tb	Dy	Ho	Er	Tm

The molecules are of two types :

- polar molecules : ammonia, amines, hexametlyphosphotriamide (HMPT) furan and tetrahydrofuran (THF), pyrydin, thiophene, dimethoxyéthane, etc..
- aromatic molecules : benzene, toluene, xylene.

b) Synthesis of metal-molecule-graphite ternaries can be carried out using three main techniques

α) Reaction of metals in solution with graphite.

Following the pioneering research of Rüdorff in 1954 [24] all metals of table 4 were intercalated in liquid ammonia or amines. The blue solutions contain solvated electrons $e^-(NH_3)_p$ which play the role of reducing agent and solvated cations $M^{n+}(NH_3)_q$, which intercalate [24 to 27].

All other molecules were only intercalated in association with alkali metals, mainly by the groups of Golé, Bonnetain, and Setton.

For instance, solutions in THF of orange-red [Naphtalene$^-$ M^+] complexes react with graphite by the transfer of an electron from naphtalene to graphite (which is a stronger acceptor) and intercalation of M^+ ions solvated by the THF polar molecules [28].

β) The reaction of polar or aromatic molecules in the liquid or vapour phase with binary metal graphite compounds provides a 2nd possibility of ternary compounds synthesis. For instance, C_6H_6 reacts with KC_{24} according to KC_{24} (s=2) + $xC_6H_6 \rightarrow K(C_6H_6)_x C_{24}$ (s=1) and forms a ternary compound with an increase in stage [29]. The same is true for THF [30] Curiously, C_6H_6 and THF do not react with KC_8. On the contrary ternary compounds are formed by action of THF on LiC_6 [31] [32] and of furan on KC_8 : a general study of the reactions of simple organic molecules with KC_8 and KC_{24} has been made by Setton [33].

γ) Electrochemical cathodic intercalation, (for instance in dimethyl sulfoxide as electrolyte) is a third technique for the synthesis of metal-molecule-graphite ternaries (See IV, 1).

c) Structure and stoichiometry of metal-molecule-graphite ternariesled to many discussions

Intercalated metals associated with molecules form ionomolecular layers (See I, 3), which generally do not have in-plane long range order. Therefore the carbon/metal and molecule/metal ratios are not generally integers. On the other hand, the interplanar distance of the alkali metal-ammonia-G ternary compounds is independant of the nature of the alkali metal (from Li to Cs) [24-26] and, therefore is mainly imposed by the thickness of the ammonia molecules. Likewise in the $M(C_6H_6)_x C_{24}$ compounds with $2.5 < x \leq 3.1$ d_i does not depend on the metal (M = K, Rb, Cs). This suggest that the interplanar distance is imposed by a perpendicular position of the benzene cycles with respect to the graphene sheets [29]. In the $M(THF)_x C_{24}$ ternaries the molecules cycles are parallel to the graphene sheets in the "poor" compounds ($x \sim 1.8$). They are perpendicular in the "rich" compounds $x \sim 2.6$) and turn around their symetry axis [36]. Oblique positions have been also proposed : see figure 11. Setton et al. tried with succes to calculate x

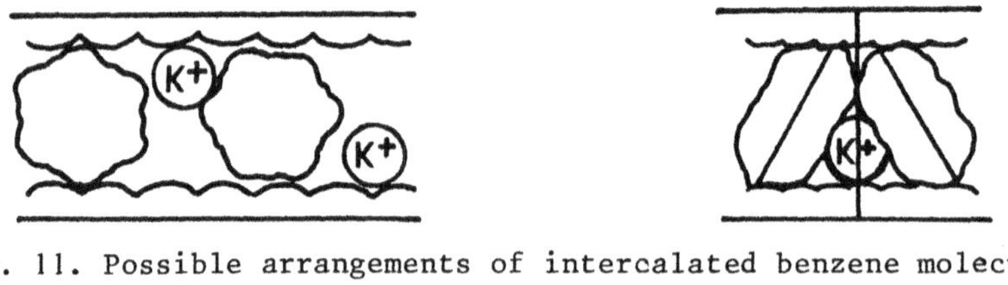

Fig. 11. Possible arrangements of intercalated benzene molecules.

in the ternaries KC_{24} (molecule)$_x$ from the interplanar distance and the volume of the components [34] and Solin et al. confirmed their hypothesis for the $M(NH_3)_x C_{24}$ ternaries [35]. According to Setton [38], better precision can be obtained by introducing in the calculation corrected molar volumes taking into account the charge distribution between intercalated species.

d) Energetic aspects of intercalation

Ternarisation of a binary metal-graphite compound corresponds to a complex balance of positive and negative energies (See I, 2)

- increase of $d_i \rightarrow$ increase of the spacing energy : $\Delta G_1 > 0$
- enlargement of d_i and diminution of the in-plane density of positive

ions → decrease in the Coulomb attraction between graphene sheets and intercalated layers : $\Delta G_2 > 0$

- screening of the positive charges by the molecules playing the role of spacers (Hennig) favourize intercalation : $\Delta G_3 < 0$

- association of ions and molecules corresponds to : $\Delta G_4 < 0$.

Interactions between metal atoms and polar molecules are mainly of the ion-dipole type : they correspond to a high energy : for instance Beguin measured an enthalpy of - 119KJ per mole of THF in the reaction :

$$KC_{24} + THF\ (vap) \rightarrow K\ (THF)_{2.5}\ C_{24} \quad [37]$$

On the other hand, ^{13}C NMR measurements indicate that the charge transfer from potassium atoms to the graphene sheets is not affected by THF absorption [36].

In the case of non polar molecules, interactions with the metal atoms are more complex : they probably imply a distribution of the charge left by these atoms between graphene sheets and intercalated molecules [38].

Many new synthesis in the field of metal-molecule-graphite ternaries may developp; they are of interest for catalysis.

5. Transition metals : inclusion products and intercalation complexes

According to the data of part II, the formation of or iono-metallic compounds between graphite and transition metals is very improbable. On the contrary one cannot exclude Π complexes of the ferrocene type. From an energy point of view, in situ reduction of intercalated transition metal halides is an intersting way to form these compounds : it leads to free metal atoms and, therefore it avoids the breaking of metal-metal bonds which is required if the synthesis is carried out from a bulk metal [F]. Many reducing agents have been used.

a) A priori, hydrogen seems to be the best one because the halide reduction forms gazeous HCl and therefore can lead to pure products. Unfortunately hydrogen does not intercalate into graphite : thus, halide reduction occurs at the edges on graphene layers, by migration of the intercalate, which requires rather high temperatures. According to Gross [39], $FeCl_3$ is first reduced into $FeCl_2$, and in a second step into α Fe, but the reduction remains incomplete, even at 1000°C. Particles of α-iron are very quickly dissolved in the dilute acids. Reduction of $CuCl_2$ into copper by hydrogen also remains incomplete. On the contrary intercalated $AuCl_3$ is completely reduced into metallic gold particles of 100 to 350 nm [40]. Part of the metal, especially the bigest particles are easely removed by a chlorine current, thus stay at the cristallites edges. Another portion does not react with chlorine, probably because it has been formed between graphene layers, according to the reactions : $AuCl_3 \rightleftarrows AuCl + Cl_2$; $3AuCl \rightleftarrows 2Au + AuCl_3$ Gold-graphite is probably the only pure TM - G inclusion product which has been prepared.

b) Alcali metals are stronger reducing agents than hydrogen, and are able to intercalate and, therefore, to reduce intercalated halides in situ. Reaction in the liquid or vapour phase generally leads to dispersions of small particles included between the graphene sheets. For instance, tantalum particles of 30-60 nm [40] were measured and more recently, Ni particles of 10 nm or less [41].

c) Liquid reagents allow reductions at or under room temperature. The first one was carried out by Klotz and Schneider, using solutions of alcali metals in liquid ammonia, [42]. Hexamethylphosphoric triamide has also been used as solvant [43]. A wide range of intercalated chlorides was reduced by Novikov, Vol'pin and their collaborators using solutions of alkali metals complexed by aromatic molecules (diphenyl, naphtalene) in solvants such as T.H.F. [44-45].

d) Electrochemical reduction which can be easely controlled by setting the current intensity, was used by Touzain et al. [46], and more recently by Yazami et al. [47]. Electrochemical cell is of the type

Li (metal) / non aqueous solvent / MX_n - G compound (MX_n = $FeCl_3$, $CoCl_2$).

e) Reduction of halide solutions by KC_8 or KC_{24} was also proposed for preparing TM - G compounds [48] [49], but a large part of the reduction probably occurs at the crystallites surface, because of the high mobility of intercalated potassium.

The in situ reduction products obtained by chemical or electrochemical means seem to be comparable. They consist in a mixture of pure graphite, alcali metal chloride, solvent, and transition metal. The chloride and solvent can be partly removed by washing and drying. X ray diagrams often contain reflexions corresponding to metal particles in its usual structure (for instance cubic centered α Fe). But they can be associated or completely replaced by new reflexions which have been attributed to transition metal-graphite complexes (TMGC). Such complex have been described for Cr Mo W Mn Fe Co Ni Ru Os Pd and Cu [44-47]. Intercalated metals cannot be removed from these complexes by inorganic acids. However the TMGC seem to be metastable with respect to biphasic metal-graphite mixtures : by heating at 400°C under an inert atmosphere (H_2, N_2 ..) they decompose into inclusion products.

Powder diagrams do not allow sure indexing of the X ray reflexions. Thus the structure proposed for TMGC vary. According to Touzain et al. [46], Fe-GC is formed of alternating layers of graphene and iron with d_i = 560 pm. The iron unit cell is tetragonal, with a = 282 pm, so that iron in-plane parameter is not commensurate with that of graphite.

Structures proposed by the Novikov group are based not only on X ray powder diagrams, but also on magnetic measurements, and spectrometric studies (X ray spectra, ESCA and Mössbauer spectra).

According to this group, Molybdenum-GC is comparable to dibenzene-molybdenum by its interplanar distance (370 pm for Mo-GC, 350 pm for the molecular complex), and by the valence zero of the metal. The same is true for chromium, whereas palladium would be intercalated with an oxidation number 1, as in its complexes of the olefinic and allylic type [50].

Intercalation complexes of iron and other metals with an interplanar distance of 580 to 600 pm would contain double layers of metals, with a irregular in-plane distribution (figure 12).

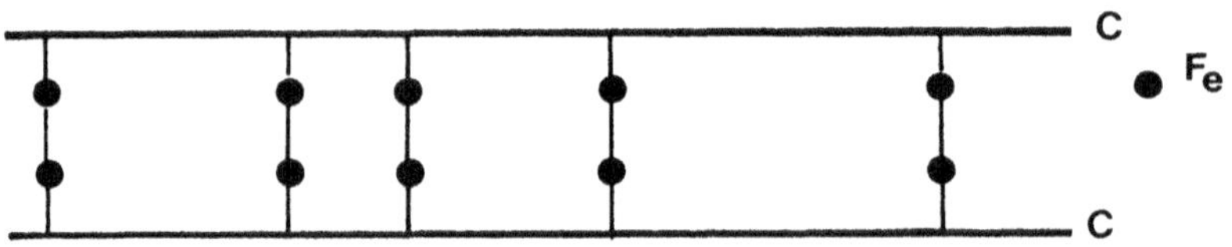

Fig. 12. Structure of iron GIC according to Novikov et al.

The metal-metal bonds observed by Mösbauer spectroscopy would also correspond to bonds perpendicular to the graphene sheets.

Although intercalated metals are inaccessible to liquid or gazeous reagents, TMGC have been successfully studied by the Vol'pin - Novikov group [12] for catalytic applications.

The catalysts available from Ventron Corporation under the name "Graphimet" [51] seem to be prepared by the reduction of intercalated T.M. chlorides by lithium diphenylide. They are biphasic products containing graphite and small particles of metal or metal oxide [52]

T.M. - G compounds and inclusion products are certainely interesting materials. Progress in elaboration of better defined samples and in crystallographic studies are desirable and seem to be possible.

III - INTERCALATION OF ELECTRON ACCEPTORS

1. Generalities

a) Overview of acceptor intercalates

The first observations relative to acceptor intercalation were reported by Schaffaütl in 1841 [1] and concerned H_2SO_4 - HNO_3 mixtures. Modern studies began with Thiele [2]. At present a large number of chemical species with very different physical and chemical properties have been intercalated. Table 6 contains a list of neutral intercalates classified according to the oxidation state of the central atom. Species such as acids which require an assisted intercalation will be classified in part IV. Brackets indicate uncertain intercalation.

In the following pages only several aspects of acceptor intercalation will be studied, using a small number of examples, with selected references. Forsman's review [F] is of particuliar interest.

Table 6. Intercalated acceptors
(I, II .. = oxidation state)

Halogens		F_2 Cl_2 Br_2 $Br_{1-x}Cl_x$ $I_{1-x}Cl_x$ $I_{1-x}Br_x$
Fluorides	II	Kr Xe
	III	B / Cl Br I
	IV	Si Ge / Ti / Xe
	V	As Sb Nb Ta / I
	VI	Mo W / U / Re /Os Ir Pt
Chlorides	II	Mg Zn Cd Hg / Mn / Fe Co Ni Pd / Cu
	III	B Al Ga In Tl / Cr / Fe Ru Os / Au
		Y Sm Eu Gd Tb Dy Ho Er Tm Yb Lu
	IV	(Si) Sn Zr Hf / Re / Pt
	V	Sb Nb Ta / Mo / U
	VI	W
Bromides	II	Cd Hg / Fe
	III	Al Ga Tl / Fe / Au
	V	U
Oxyfluorides	V	VOF_3
	VI	$XeOF_4$ / CrO_2F_2 $MoOF_4$ WOF_4
Oxychlorides	V	CrO_2Cl_2
Oxides	V	N_2O_5
	VI	SO_3 (SeO_3) / CrO_3
	VII	Cl_2O_7 (Re_2O_7)
Nitrates	II	Cu Zn
	III	Al Ga In / Fe CO
	IV	Ge Ti
Oxynitrates	V	MOX_3 M = V
(x = NO_3)	VI	MO_2X_2 M = Cr, Mo, W

b) <u>What is an electron acceptor?</u>

The word "acceptor" is ambiguous, as it is apparent in the following reactions :

Na	+	1/2 Cl_2	→	NaCl
reducing agent		oxidising agent		salt

Cl^-	+	$AlCl_3$	→	$AlCl_4^-$
Lewis base		Lewis acid		Complex anion

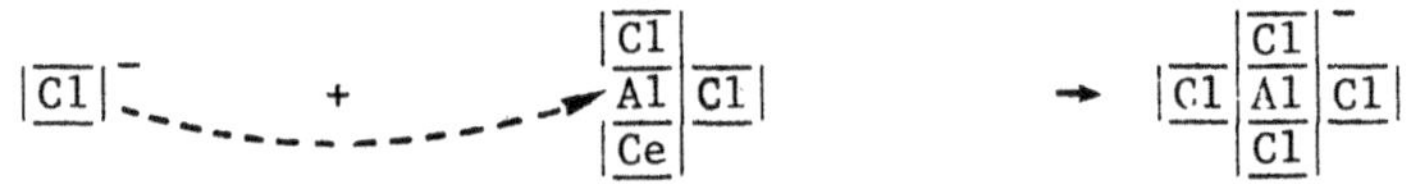

In the first reaction, each sodium atom (donor) yields an electron to a chlorine atom (acceptor) : the terms "acceptor" and "donor" are respectively equivalent to "oxidising" and "reducing agent".

In the second reaction, a chloride ion Cl^- (donor) yields an electron pair to an aluminium chloride molecule with an unoccupied p orbital (acceptor) : this electron pair becomes common to the two species, which form a complex anion $AlCl_4^-$. Although the first interpretation of the terms donor and acceptor corresponds to the general use in graphite intercalation chemistry, the acid-base reactions also play an important role in acceptor intercalation. For instance the stability of complex anions towards oxidation (see IV, 1) directly depends on the strength of the lewis acidity of the corresponding molecules (MX_n) and on the Lewis basicity of the simple anions X^{n-} (X = F, Cl, Br, O ...). Likewise, the formation of Brönsted acids such as $HSbF_6$ or H_2SO_4 corresponds to an acido-basic process according to Lewis :

$$SO_3 + H_2O \rightleftarrows H_2O \rightarrow SO_3 = H_2SO_4$$

$$SbF_5 + HF \rightleftarrows HF \rightarrow SbF_5 = HSbF_6$$

For a given anion X^{n-}, the strength of the Brönsted acid increases with the strength of the Lewis acidity of the neutral MX_n molecule. The stability of complex ions also depends on their charge : bivalent and a fortiori trivalent anions are less stable than monovalent ones because of the Coulomb repulsion between the negative charges. On the other hand, the most stable structures are the tetrahedral and octahedral ones. MX_4^- and MX_6^- ions result from the addition of an anion X^- on MX_3 and MX_5 molecules, respectively. On the contrary formation of an octahedral anion from an MX_4 molecule, requires the addition of 2 X^- ions, leading to a less stable bivalent MX_6^{2-} anion. This is the probable reason why the MX_4 halides intercalate with more difficulty than the MX_3 or MX_5 ones.

c) <u>Conditions of acceptor intercalation</u>

The data of table 1 show that the transfer of an electron from a haevy alkali metal to graphite corresponds to an emission of 0.75 to 1.3 eV, whereas the transfer of an electron from graphite to a halogen absorbs 2.1 to 2.4 eV. This electron transfer is favourable to the donor intercalation and unfavourable to that of an acceptor.

On the other hand, the interplanar distance in the acceptor compounds -

especially in element compounds - is higher than in the donor ones : the spacing energy which is a positive term, unfavourable to intercalation is higher for the acceptors than for the donors; and the electrostatic energy, which is a negative favourable term is lower in absolute value for the former than for the latter. Therefore acceptor intercalation is, a priori, more difficult than that of a donor. These remarks explain the significant role of the intercalate lattice energy and the role of the different kinds of addition reactions in acceptor intercalation.

α) Role of the lattice energy

As was shown in I, 2, intercalation is favourized by a low value of the intercalate lattice energy, in its free state (which is measured, to a first approximation, by the enthalpy of vaporisation ΔH_v). Fortunately, many electron acceptors form molecular lattices, so that ΔH_v corresponds to only a small Van der Waals energy. For instance, ΔH_v is 0.150eV per mole of Br_2, thus 0.075 per mole of Br, whereas it is 0.75eV per K mole.

The term ΔH_v is also small for 2D intercalates : in this case the sheets which exist in the free intercalate are recomposed between the graphene sheets after intercalation, so that ΔH_v is only the Van der Waals energy corresponding to the sheet separation, which is compensated by the similar energy relative to the interaction of intercalated sheets with the graphene ones. On the contrary intercalation of a 3D ionocovalent compound absorbs significant energy, because strong interatomic bonds are brocken by passing from the 3D pristine lattice to a 2D one.

Consequently most of the intercalated acceptors are molecular or lamellar species. Compounds with three-dimensional ionocovalent lattices, such as metallic M_2O, MO, M_2O_3, MO_2 oxides; MX, MX_2, MX_3 fluorides do not intercalate.

β) Role of addition mechanisms

In many cases, acceptor intercalation is favourized by addition reactions corresponding to a decrease of the free enthalpy. One of the most common is the formation of a complex anion by an acid-base reaction. Other kinds of addition processes will be shown in the following pages.

2. Halogens as intercalates

Bromine, chlorine and fluorine intercalate

The nature of the intercalated species and the energy aspects of intercalation will be discussed first for the haevy halogens and the complex behaviour of fluorine vis-à-vis graphite will be treated in 9.

Bromine easely intercalates in the vapour phase leading to a 2nd stage C_8Br at room temperature, as shown by isobaric curves [I, 26]. Above 49°C, the composition is $C_{14s}Br$ for the stage s [3]. Graphite sample in contact with liquid bromine at room temperature would form C_7Br (s = 2) [4].

The first in-plane crystallographic study [I, 15] was carried out on a 4th stage $C_{28}Br$ compound; it was concluded that Br_2 molecules were present lying between the graphene sheets. From an EXAFS study Heald and Stern [5] deduced that in most of them the Br-Br distance was 234 pm instead of 228 in free bromine. This increase in the Br-Br distance would correspond to a transfer of 0.16 e^- from the graphite to the * σ antibinding orbital of the bromine molecules. It is clear that the energy balance of this partial charge transfer is more favourable to the intercalation than a complete transfer to only some of the molecules leading to their breaking into Br^- ions. However, the conclusion of a recent work by Ghosh and Chung [6] is different : according to these authors who made use of X ray diffraction and Raman spectroscopy on C_7Br compounds prepared from graphite single crystals, bromine is intercalated in the form of chains containing polybromide anions such as Br_3^-, Br_4^-, Br_5^-. While oxidation of graphite by bromine with intercalation of Br^- ions is not possible, the addition of Br_2 molecules on the Br^- ions can evolve enough energy to allow intercalation. Nevertheless the affinity of bromine for graphite is small, so that the 1st stage cannot be synthetised. Therefore, bromine, as sodium, but in an opposite sense, is very sensitive to the degree of organisation of the carbon host lattice [7]. Chlorine slowly intercalates above - 20°C [8], leading to $C_{8s}Cl$ compounds (s ≥ 2) [9] which are probably similar to those of bromine. Their low stability is likely due to a decreased stability of the polyhalide anions in going from I to Cl.

The negative results concerning iodine intercalation can be explained using the ionic model by the high value of its atomic diameter, leading to a low absolute value of the electrostatic energy. Nevertheless one cannot exclude that some of the energy required for halogen intercalation is provided by a small overlapping of the p orbitals of the carbon and halogen atoms, which decreases from Cl to I with the atoms electronegativity.

Ternary compounds $C_{4s}\ X_{1-\alpha}\ Y_{\alpha}$ with X = I, Y = Br [10], X = I, Y = Cl [11], X = Br, Y = Cl [12] are more stable than the binary ones for certain values of α (especially $\alpha \approx 0.45$, for what reason ?). $C_{4s}\ I_{1-}\ Cl_{\alpha}$ exists for s = 1. The difference between the electronegativities of X and Y probably favourizes the formation of stable XY_2^- polyanions and stabilises the halogen layers by dipole-dipole interactions. Nevertheless, a partial charge transfer from the graphene sheets to the ICl molecules, similar to that proposed for the bromine-graphite compounds [5] cannot be excluded.

3. Chlorides

Chlorides form the most important group of intercalates. After the pioneering work of Croft [13] a large review of chloride intercalation was given by Stumpp [14]. Two main groups of chlorides are able to intercalate : volatile molecular ones such as $AlCl_3$ or $SbCl_5$ and less volatile lamellar ones, of the MCl_2 or MCl_3 type (see I, 3). For the reasons exposed in I, b, the 3D alkali metal and alkaline earth metal chlorides have not been intercalated.

In the 2D chlorides as in the molecular ones, two kinds of intercalation can be distinguished : cooperative intercalation with chlorine and sponteneous intercalation of the pure halide.

a) Cooperative intercalation of chlorine and a molecular chloride : example of $AlCl_3$

Cl_2 does not intercalate above -20°C, and, therefore cannot oxidise graphite by itself. Because aluminium compounds of oxidation number 1 or 2 are much less stable, Al_2Cl_6 molecules are not oxidising agents, and cannot intercalate by theirselves. Howhever a cooperative intercalation of Cl_2 and Al_2Cl_6 molecules is possible. From a thermodynamical point of view it can be described by the following steps :

$$(\alpha)\ 2mC\ (\text{graphite}) + Cl_2 \rightleftarrows 2C_m^+ + 2Cl^-$$

$$(\beta)\ 2Cl^- + Al_2Cl_6 \rightleftarrows 2AlCl_4^-$$

$$(\gamma)\ 2C_m^+ + 2AlCl_4^- + 2pAl_2Cl_6 \rightleftarrows 2C_m^+\ AlCl_4^-\ (Al_2Cl_6)_p \quad [15]$$

The energy of the acid-base reaction (β) allows intercalation. Accomodation of $AlCl_4^-$ ions and neutral molecules is easy in a semi-liquid ionomolecular layer. The ratios $m = C/AlCl_4^-$ and $p = Al_2Cl_6/AlCl_4^-$ depend on the synthesis conditions (chlorine pressure, temperature), so that two compounds of the same stage can differ in their structure [16] and electronic properties. This intercalation mechanism concerns all molecular chlorides which cannot oxidise graphite.

b) Sponteneous intercalation of $SbCl_5$

$SbCl_5$ intercalates by itself into graphite leading to compounds of an approximate formula $C_{12s}\ SbCl_5$ [17]. The crystallographic formula corresponding to the in-plane $\sqrt{7} \cdot \sqrt{7}$, R = 19.°1 lattice is $C_{14s}Sb_{1.2}Cl_6$ [18] Mössbauer spectra [19] [20] show the presence of Sb V and Sb III in the layers and suggest that $SbCl_5$ disproportionate into $SbCl_6^-$ and $SbCl_3$, playing both roles of an oxidising agent and of a Lewis acid. From a thermodynamic point of view the reaction steps can be written :

$$(\alpha) \quad SbCl_5 \rightleftarrows SbCl_3 + Cl_2$$

$$2m\ C + Cl_2 \rightleftarrows 2C_m^+ + 2\ Cl^-$$

$$(\beta) \quad 2Cl^- + 2SbCl_5 \rightleftarrows 2\ SbCl_6^-$$

$$(\gamma) \quad 2C_m^+ + 2SbCl_6^- + 2pSbCl_5 + SbCl_3 \rightleftarrows 2C_m^+SbCl_6^-(SbCl_5)_p(SbCl_3)_{0.5}$$

As for in (a) the values of m and p depend on the conditions (liquid or vapour phase, temperature). Accomodation of $SbCl_6^-$ ions and $SbCl_5$ orSb_2Cl_{10} molecules seems to be possible in the molecular layers. However the differences in structure and melting point (73.4°C for $SbCl_3$ and 2.8°C for $SbCl_5$) can lead to a segregation of the trichloride [20] which seems to be confirmed by NMR of ^{13}C [21].

Niobium and tantalum pentachlorides cannot oxidise graphite and therefore only intercalate in association with chlorine [19].

c) Cooperative intercalation of chlorine and a lamellar chloride : example of $NiCl_2$

Because of the low volatility of ionocovalent $NiCl_2$, intercalation requires a temperature of 500 - 600°C leading to compounds $C_m\ NiCl_{2+x}$ ($x \sim 0.13$).

The first step of the reaction is :

$m\ C$ graphite $+ 1/2\ Cl_2 \rightarrow C_m^+ + Cl^-$

$NiCl_3^-$ and $NiCl_4^{2-}$ ions exist in aqueous solutions, indicating a Lewis acidity of $NiCl_2$. But such ions cannot take place in the regular 2D intercalated lattice. Then two hypotheses have been proposed for introducing an excess of Cl^- ions in the intercalated layers.

α) It is well known that intercalated ionocovalent halides do not completely cover the graphene sheets : in a $NiCl_2$ 2nd stage compound, the covered surface area can be lower than 70 per cent of the theoretical value [23].

Flandrois et al. [24] have suggested that the intercalated chlorides form islands, and that Cl^- ions in excess are localised along the island boundaries.

β) An other hypothesis is the formation of vacancies in the metallic ion sublattice : then C_mNiCl_{2+x} should be written $C_n^+\ Cd_{1-y}^{2+}Cl_2^-$ with

$$p = \frac{m}{1+\frac{x}{2}} \quad \text{and} \quad 1-y = \frac{1}{1+\frac{x}{2}}$$

d) Spontaneous intercalation of lamellar chlorides: example of $FeCl_3$

The behaviour of iron trichloride with graphite is very complex. $FeCl_3$ can intercalate with a Cl/Fe ratio higher than 3, lower than 3 or exactly equal to 3. In this last case, the simplest hypothesis for explaining an excess of negative charges in the intercalated layers is the partial reduction of the Fe^{3+} ions into Fe^{2+} ones in the cationic sublattice, leading to a formula

$$C_m^{x+}\ Fe_x^{2+}\ Fe_{1-x}^{3+}\ Cl_3^- \quad \text{or} \quad C_p^+\ Fe^{2+}\ (Fe^{3+})_{q-1}\ Cl_{3q}^- \quad \text{with } p = \frac{m}{x} \text{ and } q = \frac{1}{x}$$

The formation of Fe^{2+} ions during intercalation is confirmed by the in-plane parameter increase from 606 pm in the free trichloride to 612 pm in the intercalated one [25] in agreement with the ion radii (64 pm for Fe^{3+} and 76 pm for Fe^{2+}).

4. Bromides

These compounds intercalate less easely than the chlorides : bromine is a less oxidising agent than chlorine and the increase in ionic radius from Cl^- to Br^- defavourizes intercalation by decreasing the complex ion stability and the absolute value of the Coulomb energy. The same reasons explain that no iodides could be intercalated.

5. Fluorides

Although UF_6 was already intercalated in 1958 by Maire [26] a methodic study of fluoride-graphite systems began in different countries between 1973 and 1975. SbF_5 intercalation was studied by Lalancette et al. [27], Onalovski et al. [28], Melin and Hérold [31] which also intercalated NbF_5 and TaF_5; $XeOF_4$, IF_5, Br F_3 by Selig and Gani [30] TiF_4 by Buscarlet et al. [32]. Fluoride-graphite compounds form now a vast field of research which was summarized in reference [F] and more recently by Watanabe et al. [33]. Only a few aspects of fluoride intercalation will be discussed here.

a) Intercalation of mono, di, and trifluorides has been only carried out for compounds of fluorine with non metallic elements such as halogens, noble gases and boron. Several of them (ClF, ClF_3, XeF_2) have been intercalated in liquid HF. Insertion can occur sponteneously (KrF_2, BrF_3, ClF in HF) or by using an oxidative agent (such as ClF for BF_3). The formula of intercalated species and the reaction mechanisms which often imply some halogen or noble gas evolution are not completely known. The same is true for halogen and noble gas fluorides in higher oxidation states. Metallic di- and trifluorides cannot not be intercalated, to the contrary of the corresponding chlorides. This difference of behaviour between fluorides and chlorides must be related to the higher fluorine electronegativity. Whereas di and trichlorides are molecular or lamellar compounds, the corresponding fluorides are 3D ionic compounds. Their high lattice energy leads to high melting and boiling points, and is unfavourable for intercalation (see 1c).

b) Among the pentafluorides only AsF_5 and SbF_5 spontaneously intercalate, leading to compounds of approximate formula $C_{6.5s}SbF_5$ [31] and $C_{8s}AsF_5$ [34] (s = stage).

These compounds contain a mixture of MF_5 and MF_3 molecules with MF_6^- ions produced by the equilibrium reaction :

$$3MF_5 + 2e^- \rightleftarrows 2MF_6^- + MF_3$$

This equilibrium has been studied in a large number of publications especially for the AsF_5 compounds [E] [35] which would reach an in-plane conductivity close to that of copper.

According to Bartlett et al. [36] AsF_5 and AsF_3 can be removed by pumping, leading to $C_m^+AsF_6^-$ compounds with K shall X ray spectrum similar to that of the compound $C_8^+AsF_6^-$ synthetised by the reaction of $O_2^+AsF_6^-$ with graphite.

Transition metal pentafluorides NbF_5 and TaF_5 were at first intercalated using chlorine as oxidising agent [31] with the formation of chlorofluoroanions such as MF_5Cl^- [37]. Intercalation is easier in a fluorine atmosphere [38].

c) All hexafluorides spontaneously intercalate except WF_6 [39] (33, p. 262-6). Magnetic measurements [40] have shown that the intercalated species are MF_6 molecules associated with MF_6^- ions containing the metal M at the oxidation state V. In the corresponding formula $C_n^{x+}(MF_6)_x^-(MF_6)_{1-x}$, the value of x linearly increases with the electronic affinity of the MF_6 molecules, which is given in the following table. The relatively small value of WF_6 affinity explains that this compound does not intercalate by itself.

Table 7. Electronic affinities of MF_6 molecules in eV according to Bartlett and Mc Quillan [41]

Mo				
1.36				
W	Re	Os	Ir	Pt
1.08	1.34	1.6	1.86	2.11
U				
1.38				

6. Oxyhalides

The conditions of oxyhalide intercalation are similar to those of halides, and have often been described in the same papers, for instance, $X_eO_2F_2$ in [30], $MoOF_4$ and WOF_4 in [38]. Therefore no particuliar discussion of oxyhalides is required here. Nevertheless, the negative attemps concerning intercalation of SO_2F_2 and SO_2Cl_2 merit attention. The oxidising power of these compounds is probably comparable to that of SO_3, which easely intercalates. The difference in behaviour between the oxyhalides and the trioxide is certainly a consequence of the absence of stable anions derived from the former whereas the latter forms stable sulfate and disulfate ions SO_4^{2-}, $S_2O_7^{2-}$.

7. Oxides

The first oxide-graphite compound was synthetised by Platzer [42] : reaction of a solution of CrO_3 in anhydrous acetic acid with natural graphite has led to a compound of approximate formula $C_{13}CrO_3$ of stage 2. Direct intercalation of pure N_2O_5, SO_3 and Cl_2O_7 in the vapour phase was carried out by Fuzellier [43] [44] whereas SeO_3 intercalation remains uncertain.

All intercalated oxides are strong oxidising agents. Partial reduction of CrO_3 during intercalation leads to solid oxides [45]. On the contrary intercalation of N_2O_5, Cl_2O_7 and SO_3 is respectively accompagnied by an emission of brown-red NO_2, red Cl_2O_6 [43] and colourless SO_2 [46]. Intercalation implies not only a redox process, but also an acid-base reaction : the strong Lewis acidity of N_2O_5, SO_3 and Cl_2O_7 allows the formation of very stable anions NO_3^-, SO_4^{2-}, and ClO_4^-.

For instance, analytical determinations of evolved SO_2 [47] allow to writing the intercalation of SO_3 :

$$21sC + 4.66SO_3 \rightarrow C_{21s}(SO_4^{2-})_{0.5}(SO_3)_{3.66} + 0.5\ SO_2 \quad (s = \text{stage})$$

It is well known that SO_3 easely polymerises in chains : an association of the SO_4^{2-} ion with SO_3 molecules in linear polysulfate $S_nO_{3n+1}^{2-}$ is probable and its exothermal formation favourizes intercalation.

Molecular oxides can also be intercalated from solutions, for instance SO_3 from its solutions in liquid SO_2 [46] [48].

8. Nitrates

Intercalation of molecular volatile nitrates and oxynitrates into graphite by Stumpp and coworkers [48] has lead to a new family of GIC.

Nitrates and oxynitrates are oxidising agents and therefore the reaction spontaneously occurs. Chromyl nitrate-graphite compounds can also be synthetised by the reaction of N_2O_5 with intercalated CrO_3.

In these new materials, the nature of intercalated species and consequently the chemical reactions implied in intercalation are not well known.

The reaction with graphite of $NO[(FeNO_3)_4]$ in nitromethane leading to the intercalation of $Fe(NO_3)_3$ molecules and $[Fe(NO_3)_4]^-$ anions [49] is better understood : the process of nitrosyl salts intercalation will be discussed in IV, 2.

9. Reaction of fluorine with graphite

a) "Covalent" compounds : graphite fluorides

Fluorine does not react with graphite at room temperature. But the formation of "graphite fluorides" CF_x was soon observed in the electrolytic cells used for fluorine elaboration, which operate at higher temperatures. According to Watanabe [33] [50] reaction between graphite and gazeous fluorine starts near 300°C, and the x value continuously increases with the temperature. Black C_2F is formed between 350-450°C, whereas white CF is obtained in pure form near 600-640°C. Mixtures of the two phases coexist between these temperatures. A covalent structure for graphite monofluoride CF was proposed by Rüdorff [51]and is shown in figure 13. The aromatic graphene layers corresponding to an sp_2 hybridisation of the carbon atoms are replaced by aliphatic pleated layers corresponding to the sp_3 hybridisation and in which the cycles take the form "chair". This structural change and the absence of free electrons explain the low electrical conductivity and the transparency of the colourless monofluoride.

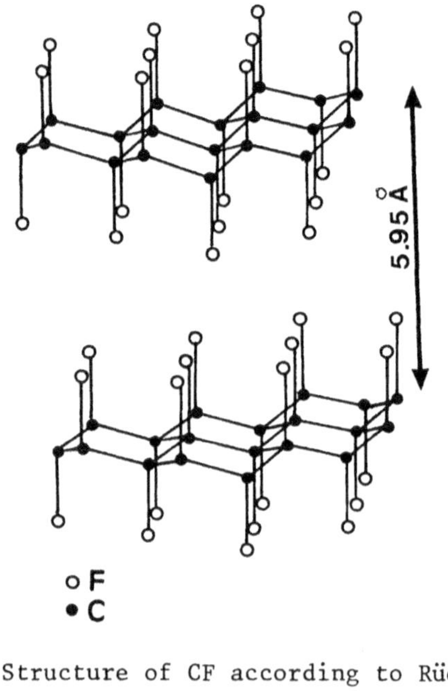

Fig. 13. Structure of CF according to Rüdorff

In C_2F, the fluorination concerns only one side of the graphene layers, leading to a regular sequence of free interspaces and of interspaces containing two layers of fluorine atoms (as in CF) [50].

Covalent CF bonds are often formed during fluoride intercalation leading to a decrease in electrical conductivity [33].

A comparison between the bonding energy in diamond, graphite and CF can explain the reason why fluorine, which is the most electronegative element (X = 4 in the Pauling's scale) form covalent compounds, whereas chlorine and bromine (X = 3 and 2.5 respectively) only give true GIC with ionic bonds:

Graphite is only a little more stable than diamond :

Diamond → graphite $\Delta H = -1.9$ KJ

In diamond each carbon atom for 4 aliphatic C-C bonds corresponding to an energy $\Delta H = -347$ KJ per bond. In CF, each carbon atom for 3 C-C bonds and 1 C-F bond with a ΔH of -485 KJ. Therefore, the bonding energy in CF is $-485 - (-347) + 1.9 \approx -130$ KJ. This calculation corresponds to the reaction of fluorine atoms with graphite. If fluorine reacts in molecular form the dissociation enthalpy of the F_2 molecule (74.4 KJ) must be added :

$$\text{C (graphite)} + \frac{1}{2} F_2 \rightarrow CF \qquad \Delta H \approx -52.5 \text{ KJ}$$

Similar calculations give the values of + 151 and + 176 KJ respectively for the enthalpies of formation of hypothetical CCl and CBr compounds, which therefore cannot be synthetised.

b) True fluorine GIC

In 1982, fluorine has been intercalated into graphite by Watanabe and his collaborators with small amounts of metallic fluorides (LiF, MgF_2, AlF_3, CuF_2) at temperatures lower than 300°C, so that there was no formation of covalent fluorides. For instance, 1st stage compounds of formula $C_{9-13}F(LiF)_{0.1-0.002}$ with interplanar distances between 938 and 944 pm have been obtained [52]. About at the same time Selig and coworkers have intercalated fluorine in the presense of HF, AsF_5, or BF_3. Compounds of stage 5 to 1 have been synthetised. The interplanar distance of 605 pm is close to that of 670 pm measured in the graphite-chlorine compounds except at the 1st stage compounds in which the value of 1100 pm corresponds to a double layer of fluorine atoms [53]. More recently intercalation of pure fluorine has been carried out by Touhara [54]. A period of 3-6 months is required for the synthesis of C_xF compounds ($6 < x \leq 14$). The repeat distance I_c is 943 pm for a C_6F 2nd stage compound : thus d_i is about 608 pm, a value close to that measured by Selig et al.

In the three cases, the in-plane electrical conductivity increases during fluorine intercalation. A maximum is observed for a F/C ratio close to 0.075 in the first kind of compounds, and a plateau between the ratios 0.05 to 0.20 in the second one. The decrease in conductivity at higher fluorine concentration probably corresponds to the formation of covalent carbon-fluorine bonds.

These recent studies have open a new dimension in the chemistry of fluorine-graphite compounds. Nevertheless, the exact state of the intercalated fluorine, the mechanism of the charge transfer, and the role of the impurities are not clearly understood. Graphite fluorides have found applications in lubrication and as cathode material in batteries [33]. The new fluorine-graphite compounds seem to be also of industrial interest.

10. Graphitic oxide

The C-O bond in aliphatic compounds has an enthalpy of formation of - 343 KJ which close to that of the C-C bond (- 347 KJ). On the other hand, the O-O bond in the O_2 molecule is very strong : $\Delta H = -571$ KJ : therefore the reaction of molecular oxygen with graphite leading to a solid covalent oxide appears to be not possible. Graphitic oxide often called graphitic acid, discovered by Brodie in 1855 [56] results from the oxidation of graphite in liquid phase, in the presence of powerful oxidising agents and strong electronegative radicals. Oxidation by the "Brodie mixture" of $KClO_3$ and HNO_3 is slow; it is faster using the Hummers reagent (MnO_4^- ions in a sulfuric solution). In these media, covalent C-O-R bonds are formed (R : $-NO_2$, $-SO_3H$ ...) which are hydrolysed by washing leading to a complex yellow-brown hydrophilic material.

The oxygen atoms are mainly bonded to the graphene sheets in form of epoxy functions (which would lead to an ideal formula C_2O) but also in form of - OH and - COOH functions, which give to the graphitic oxide its acid properties and its hydrophilic behaviour.

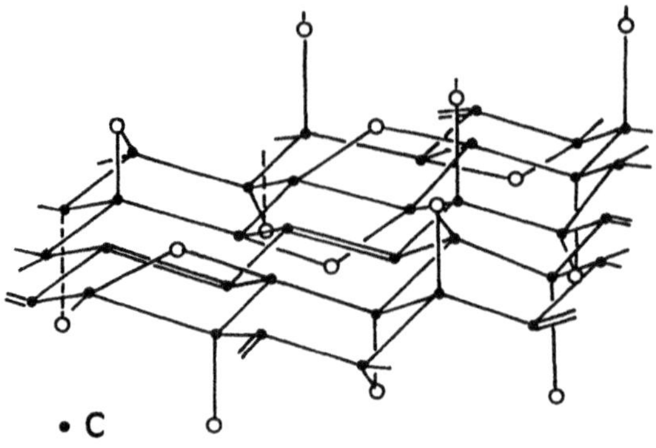

Fig. 14. Graphitic oxide layer according to Hofmann et al.

Recent analyses of Yazami and Touzain [56] have led to a composition close to the Hofmann's formula $C_8O_2(OH)_2$ [57].

IV - ASSISTED INTERCALATION (Intercalation with absorption of energy)

Several species which do not spontaneously react with graphite such as ions can be intercalated with energy absorption. This energy can be furnished either electrochemically or by chemical reactions. Other species can be intercalated under U.V. inadiation.

1. Electrochemical intercalation

a) Anodic intercalation : - Graphite salts -

Electrochemical intercalation of sulfuric acid observed by Thiele in 1932 [1] was studied by many authors: Rüdorff and Hoffmann (1938), Hennig (1951), Ubbelohde (1960), Aronson (1971) and more recently by Metrot (1979, 83, 87) Besenhard (1983), Eklund et al. (1986). See references in [2]

Intercalation occurs when graphite is used as anode in concentrated sulfuric acid. Successive stage (from 8 to 1) have been obtained. Complete intercalation leading to stage 1 can be written :

$$mC + 3.5\ H_2SO_4 \rightleftarrows C_m^+\ HSO_4^-\ (H_2SO_4)_{2.5} + H^+ + e^- \quad (m \approx 24)$$

Although intercalated HSO_4^- ions are solvated by neutral molecules of H_2SO_4, "graphite sulfat" is considered as a binary compound. The same is true for other graphite salts prepared by electrochemically or chemically assisted intercalation (table 8).

A large part of them were first studied by Rüdorff et al. [3] [4], whereas the organic acids have been recently intercalated by Boehm and coworkers [5] and HBF_4 by Metrot et al. [6].

On the other hand intercalation of oxo and fluoroanions has been carried out in non aqueous solvents such as propylene carbonate, nitromethane, methylene chloride by Besenhard et al [7, 8], Touzain et al. [9, 10] Billaud et al [11, 12]. The corresponding compounds are considered as ternary ones.

Table 8. Reagents intercalated with energy absorption

Electron acceptors						
Haloacids		HF / HBF_4 / $HSbF_6$		$HAsF_6$		
		$HAuCl_4$ / H_2PtCl_6				
Inorganic	V	* HNO_3	H_3PO_4	$H_4P_2O_7$	H_3AsO_4	
Oxoacids	VI	H_2SO_4	$H_2S_2O_7$	HSO_3F	HSO_3Cl	H_2SeO_4
	VII	* $HClO_4$	(H_5IO_6)			
Organic		CF_3COOH / CF_3SO_3H				
oxoacids		$R\text{-}SO_3H$ with $R = CH_3$ to C_4H_9				
		$R = C_4F_9$, C_6F_{13}, C_8E_{17}				
		$HO_3S - CH_2 - CH_2 - SO_3H$				
Haloanions		BF_4^- PF_6^- AsF_6^- SbF_6^- TaF_6^-				
		$AlCl_4^-$				
Oxoanions		NO_3^- ClO_4^-				
Electron donors						
Alkali metal ions		NH_4^+ NR_4^+				

* Spontaneous intercalation

b) Cathodic intercalation

This has been less studied than the anodic intercalation which seems to be of higher interest in batteries. The first was carried out by Dzurus and Hennig [13] using solutions of NH_4Cl or NH_4NO_3 in liquid ammonia at low temperature. Compounds of approximate formula NH_4^+ $(NH_3)_4$ C_{200}^- were synthetised. In these compounds NH_4^+ plays the same role as the alkali metal ions in the Metal-NH_3-graphite ternaries (See II, 4); in other words the NH_4 radical plays the same role as an alkali metal. Similar richer compounds $(NMe_4)^+(NH_3)_xC_{24}^-$ have been obtained using methyl-ammonium salts.

More recently alkali metal ions and NMe_4^- ions have been intercalated in dimethylsulfoxyde by Besenhard et al [14] leading to ternary compounds, with C/M^+ ratios of 18 for lithium and 24 for the others metals. According to Marcus and Touzain [15] formula of the potassium compounds is $K^+(DMSO)_6C_{24s}^-$ (s = stage).

These data show that cations as anions intercalate in association with solvent molecules. Nevertheless, pure alkali metals have been intercalated by using two kinds of solid electrolytes

- a potassic glass beyond 200°C [16]
- polyethlene oxide containing a solution of lithium or potassium salts ($LiClO_4$, KSCN) [17]; this electrolyte allows slow intercalation of alkali metals at room temperature. The use of solid electrolytes is of interest for measuring the free enthalpy of formation of the alkali metal-graphite compounds.

c) Organisation of intercalated layers

As indicated in I, 3, intercalated acids and solvated ions form iono-

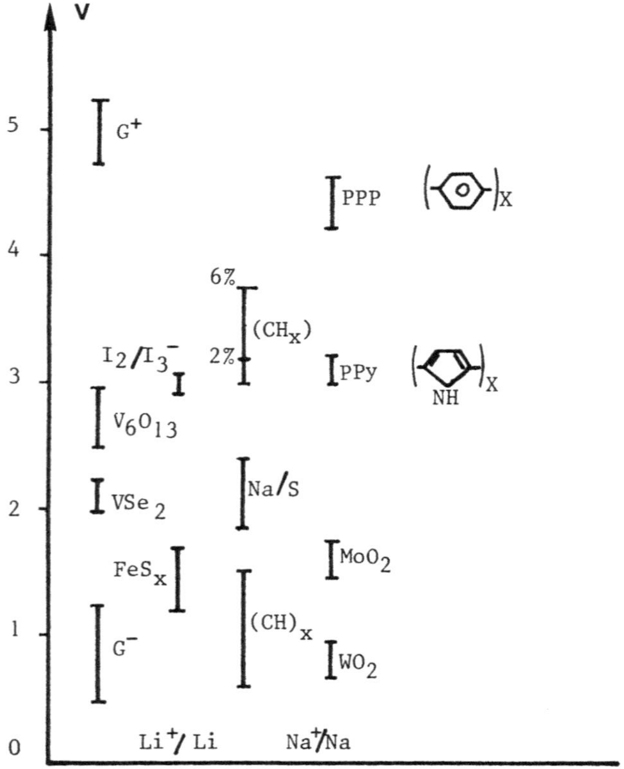

Fig. 15a. Intercalation Potentials (Armand)

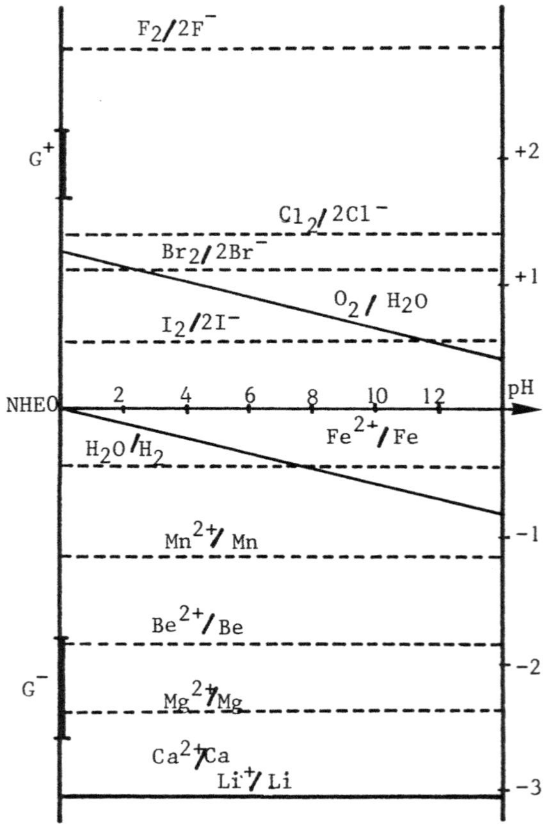

Fig. 15b. Potential scale

molecular layers which generally do not have long range in-plane order at room temperature.

The probable arrangement of the tetrahedral and octahedral anions between the adjacent graphene sheets is shown in figure 4. The triangular NO_3^- ions are approximately perpendicular to the carbon sheets in α graphite nitrate of formula $C_{5s}HNO_3$ [I, 20] and parallel to them in the β nitrate $C_{8s}HNO_3$ [I, 21].

The interplanar distance in the R-SO_3H and HSO_3-CH_2-CH_2-SO_3H GIC is about 795 pm: this value is close to that of the graphite sulfate indicating that the organic chain is parallel to the graphene sheets. On the contrary the high values of d_i measured for the perfluoroacid-GIC certainly corresponds to a bilayered structure : the occupied graphene interspaces contains two layers of intercalate;in each layer the $-SO_3^-$ functions are in contact with a graphene sheet; the fluorocarbonaceous chains in oblique position with respect to the graphene sheets are in contact in the middle of the interspace (Figure 14a).

In the $C_{24s}^+MF_6^-$ $(CH_3NO_2)_2$ ternaries [12] the interplanar distances (775 pm for PF_6^- and 805 for SbF_6^- ions) seem to be imposed by the anion diameter. On the contrary, in the $C_{24s}^+MF_6^-(PC)_4$ compounds the high values of d_i (1390 to 1420 pm) suggest a tetrahedral arrangement of the propylene carbonate molecules around the MF_6^- ion [9](See figure 14b).

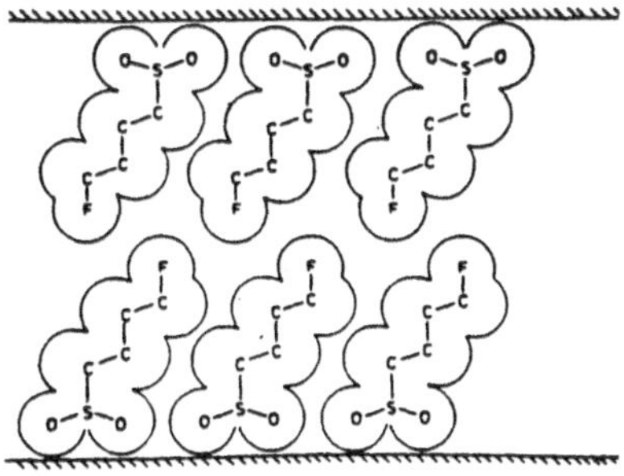

Fig. 14a
Structure of the graphite-$C_4F_9SO_3H$ 1st stage compound (Boehm et al.)

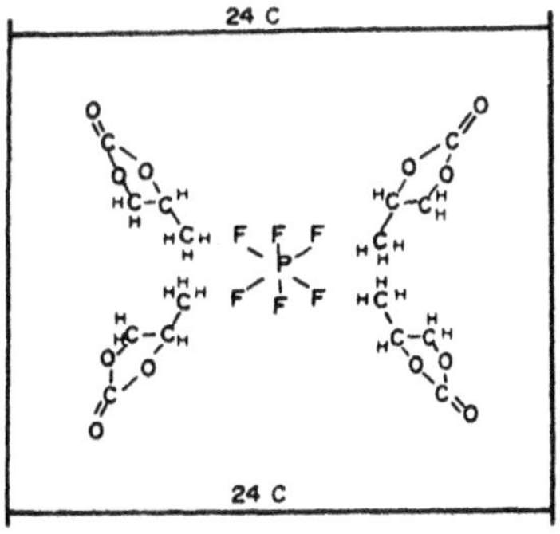

Fig. 14b
Tetrahedral arrangement of C.P molecules in $C_{24}PF_6(CP)_4$ (Touzain et al.)

d) Intercalation potentials

Anionic intercalation starts when the difference of potential between the graphite electrode and the electrolyte reach an initial value E_a^i, and it is terminated at a final potential E_a^f. Symetrically cationic intercalation occurs between negative potentials E_c^i and E_c^f. Direct measurements of anodic and cathodic potentials are not easy, and, in practice, intercalation potentials are determined using reference electrodes. Figure 15a shows intercalation domains $E_a^f - E_a^i$ and $E_c^i - E_c^f$ for graphite and several host lattices according to M. Armand [18]. Reference potential is Li^+/Li.

Inorganic compounds (V_6O_{13}, VSe_2, FeS_x, MoO_2, WO_2) are acceptor host lattices, which intercalate cations. Polypara-phenylene and polypyrrole are donors, which accept anions. Graphite and polyacetylene are amphoteric host lattices, which accept positive and negative ions.

It is clear that anionic intercalation into graphite necessitates very high potentials: in agreement with the data of table 1, graphite is a very weak acceptor (especially weaker than polyacetylene). Cathodic intercalation also requires low potentials : graphite is a rather weak electron acceptor, comparable to $(CH)_x$, but much weaker than inorganic acceptors such as V_6O_{13} or VSe_2. Generally the inorganic 2D and 3D host lattices intercalate the cations without solvent molecules [19]. On the contrary it was shown that the ions intercalated in graphite are solvated : this may be related to the shallow depth of the graphite potential wells, which is a consequence of their geometrical form and of the low donor and acceptor power of graphite.

A potential scale is useful in understanding intercalation conditions. Generally, redox potentials are given in water solutions, with respect to the normal hydrogen electrode (NHE). The theoretical potential of the Li^+/Li electrode in water is - 3.05V. In figure 15b. intercalation potentials into graphite are indicated by shifting Armand's values by + 3.05V. Water oxidation and reduction walls at 298K have also been reported versus p_H.

It clearly appears that common metal ions such as nickel, iron, manganese, or chromium do not intercalate because the corresponding Me^{2+}/Me potentials are less negative than that of graphite reduction ($E_c^{\frac{1}{2}}$). Only very electropositive metals can be intercalated leading to metal-molecule-graphite ternaries similar to those described in II, 4.

Similarly iodide bromide and chloride ions are oxidised above the anionic insertion potential. Intercalation of solvated fluoride ions is thermodynamically possible, but seems to have never been carried out.

The graphite cationic and anionic insertion potentials $E_c^{\frac{1}{2}}$ and $E_a^{\frac{1}{2}}$ are outside the range of water stability. Therefore intercalation from a water solution is in principle not possible. This is completely verified for cations because the $2H^+/H_2$ potential is much higher than $E_c^{\frac{1}{2}}$. It is also well known that NO_3^-, HSO_4^-, ClO_4^- anions cannot be intercalated into graphite from their dilute water solutions, although the high oxygen supertension allows the use of potentials higher than that of the oxygen electrode, such as the $Cl_2/2Cl^-$ one (1.36V). According to Fuzellier [20], intercalation of HNO_3^- ions solvated by HNO_3 and H_2O molecules starts beyond a ponderal concentration of 60% in HNO_3. The case of H_2SO_4 is discussed in [21].

e) Role of the acid strength and of the complex anion stability

It is well known that only strong Brönsted acids and the corresponding ions can be intercalated. Weak protonic acids such as CH_3COOH or H_3BO_3 do not intercalate.

The explanation of this fact, is contained in III, 1b : a complex anion results from the addition of a Lewis acid (BF_3, $AlCl_3$, SO_3..) and a Lewis base (F^-, Cl^-, O^{2-}...). Likewise, a Brönsted acid results from the addition of a Lewis acid and a molecule such as HF, HCl, H_2O, which can function as a Lewis base. The stability of the complex anion towards oxidation and the strength of the corresponding Brönsted acid are connected, so that an ion derived from a week Brönsted acid is oxidised and cannot intercalate.

f) Role of the anion charge

One can observe that only monovalent anions (BF_4^-, AsF_6^-... ClO_4^-...) have been intercalated in association with non aqueous solvants. Anions such as SO_4^{2-}, PO_4^{3-}... are very stable, and are certainely not easely oxidised. What is the reason why they have not been intercalated?

Several explanations can be proposed :

- For the same charge transfer per carbon atom, intercalation of polyvalent anions leads to a less uniform charge distribution than that of monovalent ones, and therefore absorbs more energy.

- The 3D → 2D transformation required by ion intercalation leads to their partial desolvatation, which absorbs more energy for strongly solvated polyvalent ions than for monovalent ones.

According to these remarks the intercalation potential of polyvalent anions should be higher than that of the monovalent ones, and perhaps cannot be reached in usual non aqueous solvents.

Nevertheless one cannot exclude that the non-intercalation of polyvalent anions and polyacids is only a consequence of the low solubility of the corresponding salts in usual non aqueous electrolytes.

g) Role of the solvent

As indicated in II, 2, the insertion potential is the ratio $\Delta G_i/nF$ of the free enthalpy of intercalation of an ion-gram to the corresponding charge (n = ion valence; F = 1 faraday). Thus this potential depends on the electrostatic energy, and therefore on the ion charge, as explained in the preceding discussion. It also depends on the solvent. For instance the $ACCl_4^-$ anion is oxidised in propylene carbonate, according to :

$$AlCl_4^- \rightleftarrows AlCl_2^+ + Cl_2 + 2e^- \quad [22]$$

On the contrary it can be intercalated from a nitromethane solution [23]. This difference in behaviour may be related to the difference of solvatation in the two electrolytes. The tetrahedral coordination in P.C. leads to a lower absolute value of the Coulomb energy than the planar coordination in nitromethane.

This example clearly show that the experimental values of insertion potentials of figure 15 are only valid for monovalent ions and certain electrolytes, and therefore must be used carefully for predicting intercalation.

h) Charge transfer and stoichiometry

Electrochemical intercalation allows direct measurements of charge transfer. Figure 16 shows the variation of potential during sulfuric acid intercalation versus charge [24]. The plateaus GB and FA correspond to the equilibria: stage 3 $\rightleftarrows$ stage 2 and stage 2 $\rightleftarrows$ stage 1 respectively. Oblique segments CG, BF, and AE indicate an increase of charge without change in stage : in the present case, an excess of positive charge in graphene layers is compensated by protons evolving from intercalated layers, leading to an increase in the concentration of HSO_4^- ions. This "overcharging" phenomenon seems to be a general process in electrochemical and chemical intercalation, so that the charge transfer and therefore the electronic properties can be different for two samples containing the same intercalate at the same stage.

i) The capacitor model and the origin of potentials

According to Metrot [24] all charge - potential curves may be similar to that of figure 16a. The capacitor model establishes a relation between

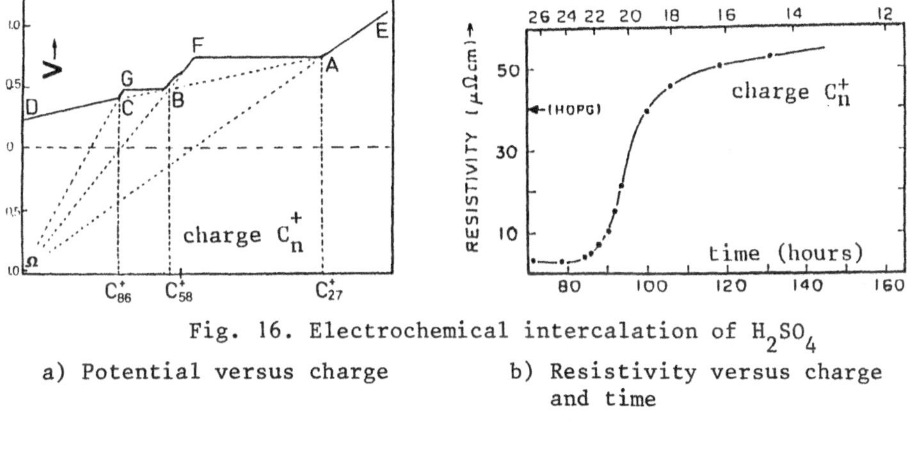

Fig. 16. Electrochemical intercalation of H_2SO_4

a) Potential versus charge b) Resistivity versus charge and time electroadsorption and electrochemical insertion. In particular the straight lines AE, BF and CG correspond to an electroadsorption phenomenon on different surface areas of the graphene sheets, and therefore converge to an Ω point which is the true potential origin. Thus E_a^1 is mesured by the distance ΩD.

j) "Overoxidation"

This phenomenon starts in sulfuric acid when the charge transfer increases above 1 elemental charge per 21 carbon atoms : then covalent carbon-oxygen (or carbon - HSO_4) bonds are formed. These conditions are comparable to those used in the synthesis of graphitic oxide (See III, 1c).

The created covalent "defects" lead to a strong increase in the in-plane electrical resistivity [25] (See figure 16b). Overoxidation is certainly a general phenomenon which occurs in chemical as in electrochemical insertion, and limits the conductivity of the graphite acceptor compounds.

2. Chemically assisted intercalation

a) Inorganic acids

Insertion of inorganic acids (table 8) by chemical oxidation preceded their electrochemical intercalation. The energy required for intercalation is furnisched by a redox chemical reaction. The stage of the graphite salt can be adjusted by using the exact amount of a convenient oxidising agent such as Cl_2, Br_2, CrO_3, $Cr_2O_7^{2-}$, CrO_4^{2-}, MnO_4^-. For instance intercalation of sulfuric acid in the presence of bichromate ions can be decomposed into an intercalation reaction (α) and a connected redox reaction (β) :

$$(\alpha)\ 24C + 3.5H_2SO_4 \rightleftarrows C_{24}^+\ HSO_4^-(H_2SO_4)_{2.5} + H^+ + e^-$$

$$(\beta)\ \frac{1}{6}(Cr_2O_7^{2-} + 14H^+ + 6e^- \rightleftarrows 2Cr^{3+} + 7H_2O)$$

The normal potential Cr VII/Cr III is 1.33 V at $p_H = 0$, a value which is much lower than the intercalation potential E_a^1 : intercalation is not possible in aqueous solution, but can be carried out in concentrated acid, because its strong affinity for the water produced in (β) (about 80 KJ or 0.9 eV) leads to an increase of the Cr VII/Cr III redox potential. The same is true for other acids.

b) Oxidising acids

Reagents such as HNO_3, $HClO_4$, $H_2S_2O_7$ and sulfuric oleums merit special comment : they are able to oxidise graphite and to solvate the corresponding ions and therefore intercalate spontaneously. Insertion of HNO_3 and $HClO_4$ can be carried out in the vapour phase, leading to NO_2 or ClO_3 evolution [26].

HNO_3 intercalation can be written

$$25sC + 6HNO_3 \rightarrow C_{25}^+\ NO_3^-(HNO_3)_4 + NO_2 + H_2O\ [27].$$

The domain of intercalation of sulfuric oleums is given in [21].

c) Hydrofluoric acid, fluoroacids and fluoroanions

Because of the large difference between the electronegativities of fluorine and hydrogen (4 - 2.1 = 1.9 in the Pauling scale), HF is a strongly polar molecule. It can react with F^- ions to form linear $(FHF)^-$ anions, which are present in graphite hydrofluoric acid compounds.

Intercalation of liquid HF either electrochemically or by using the preceding oxidising agents leads to compounds of formula $C_{24}^+HF_2^-(HF)_{3.6-5}$ for the 1st stage, which are true GIC with good electrical conductivity [4]. On the other hand, liquid HF reacts rapidly with graphite under a pressure of gaseous fluorine at room temperature and forms $C_{6s}^+HF_2^-$ compounds (s > 2). These phases contain HF_2^- ions lying between the graphene sheets. But their rather low electrical conductivity indicates the presence of covalent defects.

After a time of about one day $C_mF_{1-x}(HF)_x$ compounds are formed, with $2 < m < 6$ and C/H ≈ 11-12 [28]. These materials are similar to covalent fluorides (See III, 9).

In the HF molecule, the fluorine atom is close to an F^- ion and thus is able to yield an electron pair : HF can function as a Lewis base and react with Lewis acids such as BF_3 or SbF_5 to form Brönsted fluoroacids HBF_4 and $HSbF_6$. For instance, a solution of SbF_5 oxidising molecules in liquid HF intercalates as a mixture of SbF_6^- ions and $HSbF_6$ molecules. Intercalation of fluoroacids can also be carried out by the reaction of Lewis acids (GeF_4, PF_5, AsF_5) with hydrofluoric acid - graphite compounds : for instance, a 1st stage compound $C_{12}HAsF_6$ has been synthetised by this method [28].

An interesting technique for fluoroanion insertion was proposed by Forsman, in which the nitronium ion NO_2^+ plays the same role of oxidising agent as in HNO_3 intercalation [29] :

$$2\ HNO_3 \rightleftarrows NO_2^+ + NO_3^- + H_2O$$

$$mC(\text{graphite}) + NO_2^+ \rightarrow C_m^+ + NO_2$$

$$C_m^+ + NO_3^- + pHNO_3 \rightarrow C_m^+\ NO_3^-(HNO_3)p$$

Similarly the nitryle salts NO_2BF_4 or NO_2MF_6, (M = P, Sb), dissolved in non aqueous solvants such as DMSO [30] or nitromethane [31] reacts with graphite : ternary compounds resembling those obtained by the electrochemical technique are formed. The general equation for a hexafluoroanion intercalation in a solvent S can be written :

$$mC + NO_2^+MF_6^+ + pS \rightarrow C_m^+MF_6(S)_p + NO_2$$

Ternary compounds can also be obtained by the reaction of graphite with NO_2BF_4 and NO_2PF_6 salts in the vapour phase, leading to compounds of formula

$$C_{24s}^+BF_4^-\ (NO_2BF_4)_2 \text{ and } C_{24s}PF_6^-\ (NO_2PF_6)_2 \qquad [32]$$

c) Remarks relative to the difference between spontaneous and assisted intercalation

From a thermodynamic point of view these two kinds of insertion differ in the sign of the free enthalpy of reaction, which is negative in the former, and positive in the latter (See I,2.). However the distinction between these two kinds of intercalation is less clear from a chemical viewpoint :
- the action of nitryle salts in the vapour phase on graphite leads to a spontaneous intercalation.
- intercalation of fluoroanions from their solutions in non aqueous solvants absorbs energy, and therefore is an assisted intercalation. If nitryle salts are used for intercalating fluoroanions, the NO_2^+ ion does not intercalate, and plays only the role of an oxidising agent which could be replaced by another one. Therefore the reaction is closer to an assisted intercalation than to a spontaneous one.

3. Intercalation by U.V. irradiation

The first investigations concerning intercalation into graphite under U.V. radiations were carried out by Boehm and Schlögl [33] : the beam of a mercury high pressure burner was sent on solvents such as CCl_4, $C_2H_4Cl_2$, C_6H_{14} containing ferric chloride in the presence of powdered natural graphite. The final product was a stage 1 compound containing $FeCl_3$, but intercalation progresses by successive steps : during an induction period of 15 minutes d_{002} remains unchanged, whereas C-Cl bonds are formed by reaction of graphite with atomic chlorine. The first intercalate is $FeCl_2$, which is afterwards oxidised in $FeCl_3$.

The complex action of U.V. radiations on the mixture solvent-$FeCl_3$-graphite seems to be not completely understood.

However, new GIC containing tin IV compounds and iron carbonyls have been intercalated by the U.V. irriadiation technique.

- Reaction of $SnCl_4$ with graphite was carried out by U.V. irradiation in $ClCl_4$ or 1,2-dichloroethane [34, 35], compounds of stages 1, 2, 3, mixed with unreacted graphite were formed. According to analytical and spectroscopical data, intercalated species are $SnCl_6^{2-}$ ions associated with solvent molecules.

- Intercalation of $(CH_3)_3SnCl$ was carried out under U.V. radiations in different solvents the best being the mixture $Sn(CH_3)_4 + SnCl_4 + CCl_4$ [34, 36]. Intercalated species seem to be oligomeric ions such as

$$\{Cl-Sn(CH_3)_3-[Cl-(CH_3)_3]_n-Cl\}^-$$

Photochemical insertion of iron carbonyls [37, 38] $Fe(CO)_5$, $Fe_2(CO)_3$, $Fe_3(CO)_{12}$ in different solvents (THF, benzene, hexane) also led to mixtures of stages and intercalated layers containing carbonyls and solvent molecules. Mössbauer and Raman Spectra, and ^{13}C NMR indicate that graphene layers are oxidised and that intercalated layers contains ions such as $[Fe(CO)_4]^{2-}$, $[HFe(CO)_4]^-$, $[Fe_4(CO)_{13}H]^-$ besides neutral species and solvent molecules. Iron carbonyl-graphite compounds spontaneously decompose in a few weeks, except when THF is used as solvent.

In conclusion, energetical U.V. photons decompose solvent and intercalate and produce free radicals and ions which are able to oxidise graphite and induce intercalation.

Development of this way of synthesis will depend on a better control of reaction mechanisms, leading to more homogeneous materials.

A review intercalation under U.V. radiations was recently presented by Boehm [39].

V - BIINTERCALATION

1. Definition and nomenclature

Let us consider a GIC of stage 2 containing an intercalate i :

The stacking sequence of graphene G and intercalate i along the c axis can be written : ...GiGGiGGi... or more briefly G_2i. This compound contains free interlayer galleries : therefore it can play the role of host lattice for a second intercalate k leading to a compound with alternating layers according to :

...GiGGiG... + k → ...GiGkGiG...

or G_2i + k → GiGk

The term biintercalation has been proposed to the IUPAC for this kind of reaction. In the above example, the ratio k/i of the number of k and i layers equals 1, so that the two intercalates play the same role in the graphite biintercalation compound (GBC) GiGk. But more complicated reactions are possible in principle.

If the host lattice is a 4th stage, GiGGGGi... or G_4i, three types of compounds are the theoretically possible.

a) ...GiGGGGiG.... + k → ..GiGGkGGiG...

or G_4i + k → G_2iG_2k

In this case, i and k play a symmetrical role in the final compound, which remains unsaturated

b) ...GiGGGGiG.... + 2k → GiGkGGkGiG...

or G_4i + 2k → $GiGkG_2k$

k/i = 2, and the final compound is unsaturated

c) ...GiGGGGiG.... + 3k → GiGkGkGkGi...

or G_4i + 3k → $Gi(Gk)_3$

k/i = 3, and the final compound is saturated.

The number of theoretically possible new compounds increases with the stage of the host lattice.

2. Structural aspects

a) Repeat distance and stages

Crystallographic studies have shown that in a saturated GiGk type compound the repeat distance I_c along the c axis is very close to the sum of the interplanar distances d_i and d_k of the two intercalates, measured in binary compounds.

Saturated biintercalation compounds are often considered as 1st stage materials. However, in a GiGk compound, 2 layers of i are separated by 2 carbon sheets. In a $Gi(Gk)_3$ compound two layers of i are separated by 4 carbon sheets, so that the term "1st stage" is ambiguous and may be completed by the sequence along the c axis, using the symbols G, i, k ...

b) Macrostructure (texture)

Figure 17 shows a schematic representation of a GiGk compound according to the pleated layer model. The first intercalate forms triple layers (for instance $K^+ H^- K^+$ layers); the second forms single layers (for instance K, Rb, or Cs layers).

According to this model, each interlayer space contains the two intercalates, which are in contact along the folds of graphene sheets (dislocation loups). Thus, GBC are polysynthetic materials, and cannot be obtained in the form of true single crystals. Nevertheless, they are single phases, from a crystallographic point of view and in agreement with the phase rule. Therefore, the term GIG heterostructure, which is often used by physicists does not seem to be the best appropriate for this kind of material.

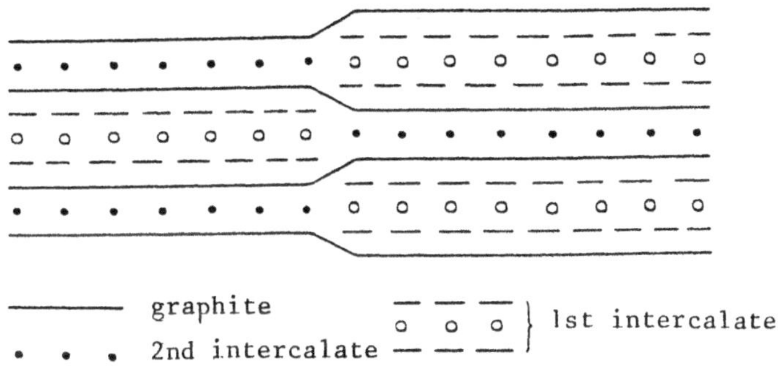

Fig. 17. Schematical representation of the limits of domains in a GIC containing one intercalate in triple layers and a second in simple layers.

3. Experimental data

The first well defined GBC seem to have been synthetised by Freeman who intercalated N_2O_5 into a G-$FeCl_3$ compound [1] and by Lagrange et al [2] who intercalated haevy alkali metals into the second stage compound $KC_8H_{2/3}$. More than 30 GBC have now been described, generally of GiGk type [3, 4]. Only a few examples will be given here, which can be classified into 4 categories :

a) i is an acceptor, k is an acceptor
$CoCo_2$, $CdCl_2$ — $AlCl_3$, $GaCl_3$

b) i is a donor, k is a donor
KH_x — K, Rb, Cs

c) i is an acceptor, k is a donor
$CoCl_2$, $CdCl_2$ — Li, Na, K, Rb, Cs

d) i is a donor, k is an acceptor
As yet no synthetised compounds.

Recently, a GIC containing 3 different layers of intercalate has been synthetized [5].

4. Discussion of biintercalation conditions

Unsaturated GIC which are used as host lattices in biintercalation are less stable than pristine graphite. Therefore the synthesis of compounds with alternating layers is in competition with several kinds of other reactions such as :

a) "Ternarisation" leads to compounds in which each layer contains the two intercalates in the form of a solid solution (2 haevy alkali metals, $AlCl_3$ and $GaCl_3$) or of addition compounds (FeCl3 and $NiCl_2$) [6].

b) Segregation leads to biphasic systems :

$$4LiC_{12}\ (s = 2) + 3k \rightarrow 3LiC_6\ (s = 1) + 3KC_8\ (s = 1)$$

$$3KC_{24}\ (s = 2) + 8Li \rightarrow 8LiC_6\ (s = 1) + 3KC_8\ (s = 1)$$

c) Displacement of i by k

$$C_8Br\ (s = 2) + 2ICl \rightarrow 2C_4ICl (s = 1) + \frac{1}{2}Br_2$$

d) Chemical reaction between i and k

$$CdCl_2 + 2K \rightarrow 2KCl + Cd$$

It is clear that, in many cases GBC are metastable species, so that their synthesis requires very mild conditions.

5. Conclusion

Although biintercalation is less easy than insertion into pure graphite, it can certainly lead to a wide range of new compounds. Taking into account the change in electronic properties of graphite produced by the first intercalate i, insertion of species k which are not able to react with pure graphite may be expected.

Research on biintercalation can lead to progress in our knowledge of the binary compounds used as host lattices. Moreover, structural, electronic, and magnetic interactions between two kinds of intercalate are certainly of scientific interest.

GENERAL CONCLUSION

Because of its amphoteric behaviour and the shallow depth of its potential wells, graphite is able to intercalate an exceptionally wide variaty of elements and compounds. One can expect in the near future the synthesis of a large number of new compounds, especially ternary and biintercalation compounds.

New progress in understanding the conditions of intercalation can be expected by the association of chemical, physical and thermodynamic studies.

AKNOLEDGMENTS

I would like to express my thanks to Dr. E. McRae for his careful reading of the manuscript.

REFERENCES

REVIEW PAPERS
A) A.R. Ubbelohde, F.A. Lewis : Graphite and its Crystal Compounds, Clarendon Press, Oxford, 1960
B) A. Hérold, R. Setton, N. Platzer, Composés d'Insertion du Graphite, in Les Carbones, A. Pacault editeur, Masson, 1964, Vol. 2, p458-683
C) L.W. Ebert, Ann. Rev. Mat. Science 6(1976) 182-211
D) 1. A.R. Ubbelohde (p1-31) ; 2. A. Hérold (323-421) ; 3. M.S. Dresselhaus and G. Dresselhaus (422-480) ; 4.J.E. Fischer (481-532) ;5. M.S. Whittingham and L.B. Ebert (533-562) in Intercalated Layered Materials, edited by F.A. Lévy,D. Reidel Publishing Company, 1979
E) M.S. Dresselhaus and G. Dresselhaus, Advances in Physics, Vol. 30, 2(1981) p139-326
F) A. Hérold, 1st Int. Conf. on GIC, 1977, in Materials Science and Engineering Vol. 31, 1977, 1-16
G) A.R. Ubbelohde, ibidem, 34-51
H) W.C. Forsmann, T. Dziemianowicz, K. Leong, D. Carl, Synth. Met. 5(1983) 77-100

The International Conferences on GIC of La Napoule 1977, Provincetown 1980, Pont-à-Mousson 1983, Jerusalem 1987 will be called ISGIC 1, 2, 3 and 4, respectively.

PART I
1. H.P. Boehm, R. Setton, E. Stumpp, Synth. Met. II (1985)363
2. C. Delmas, This Volume, plenary lecturce
3. R. Schöllhorn, idem
4. J. Rouxel, idem
5. W. Rüdorff, U. Hofmann, Z. electrochem. 40 (1934) 511
6. W. Rüdorff, U. Hofmann, Z. Anorg. Allg. Chem. 288 (1938)1
7. G. Hennig, J. Chem. Phys. 19 (1951)922
8. M.L. Dzurus, G. Hennig, J. Am. Chem. Soc. 79 (1957) 5897
9. F.R. McDonnel, R.C. Pink, A.R. Ubbelohde, J. Chem. Soc, London (1951)191
10. R. Setton, Bull. Soc. Chim. Fr. (1960)1758
11. F.J. Salzano, S. Aronson J. Chem. Phys. 45(1966)2221
12. M. Armand, Thèse, Grenoble (1978)
13. A.D. Crowell, J. Chem. Phys. 22.8 (1954)1397
14. N. Daumas, A. Hérold, C.R. Acad. Sci. C. 268 (1969) 373 and Bull. Soc. Chim. Fr. (1971)1598
15. W.T.Eeles,J.A. Turnbull, Porc. Roy. Soc. (London A. 283 (1963) 179
16. M.Heerschap, P. Delavignette, Carbon (1967) 5.383
17. R.Levi-Setti, G. Crow, Y.L. Wang, N.W. Parker, R. Mittleman, Phys. Rev. Lett.54, 24 (1985) 2615
18. J.M. Thomas, G.R. Millward, R.F. Schloögl, H.P. Boehm, Mater. Res. Bull. 15 (1980)671
19. R. Rousseaux, G. Besson, C. Tchoubar, M. R.S. Meeting, Symposium K, Boston 1986
20. E.J. Samuelsen, R. Moret, R. Comes, H. Fuzellier, M. Klatt, M. Lelaurain, A. Hérold, Synth. Met. 10 (1984/85) 13
21. C.Clinard, D. Tchoubar, C. Tchoubar, F. Rousseaux, H. Fuzellier, ISGIC 3, in Synth. Met. 7 (1983) 333 and Clinard, Diplôme Supérieur de Recherches, Orléans (1987)

22. A. Safran, this volume, plenary lecture
23. J.E. Fischer, idem
24. A. Hérold, Bull. Soc. Chim. Fr., 1960, 533
25. A. Métrot, in Progress in vacuum microbalance techniques Vol.3, p153, Ed. C. Eyraud, M. Escoubes, Heiden, 1975
26. A. Hérold, Bull. Soc. Chim. Fr. (1955)999
27. A. Suematsu, Colloque Franco-Japonais 1985 in Ann. Phys. Fr. 11(1986)75
28. A. Hérold, ISGIC 4, to be published in Synth. Met.

PART II

1. E. Weintraub, U.S. Patent 922, 645 (1909) and Brevet Français 5 85 185, (1924)
2. K. Fredenhagen, G. Cadenbach, Z. Anorg. Allg. Chem. 158 (1926), 249
3. A. Hérold, Bull. Soc. Chim. Fr. (1955)999
4. A. Hérold, G. Furdin, D. Guérard, L. Hachim, M. Lelaurain, N.E. Nadi, R. Vangelisti, Proc. Int. Symp. Tsukuba, in Physica 105 B (1981)253
5. R.C. Asher, J. Inorg. Nucl. Chem, 10 (1959)238
6. A. Métrot, D. Guérard, D. Billaud, A. Hérold, Synth. Met. 1 (1979/80)363
7. A. Métrot, A. Hérold, J. Chim. Phys. (1969)71
8. D. Guérard, A. Hérold, Carbon 13 (1975)337
9. D. Guérard, M. Chaabouni, P. Lagrange, M. El Makrini, A. Hérold, Carbon, 18 (1980)257
10. M. El Makrini, D. Guérard, P. Lagrange, A. Hérold, ibid p203
11. D. Billaud, E. McRae, J.F. Marêché, A. Hérold, Synth. Met. 3 (1981)21
12. S. Basu, C. Zeller, P. Flanders, C.D. Fuerst, W.D. Johnson, J.E. Fischer, Mater. Sci. Eng. 38 (1979)275
13. A. Hérold, D. Saehr, Bull. Soc. Chim. Fr. (1960) 1039
14. D. Billaud, A. Hérold, Bull. Soc. Chim. Fr. (1972) 103, and (1974) 2402 and 2407
15. D. Billaud, A. Hérold, Bull. Soc. Chim. Fr. (1978)131
16. M. El Makrini, P. Lagrange, D. Guérard, A. Hérold, Carbon 18 (1980), 211
17. M. El Makrini, P. Lagrange, A. Hérold, Carbon 18 (1980) 374
18. P. Lagrange, A. Bendriss-Rerhrhaye, J.F. Marêché, E. McRAE, Proc. Int. Symp. Tsukuba, in Synth. Met. 12 (1985) 201
19. D. Saehr, A. Hérold, Bull. Soc. Chim. Fr. 11(1965)3130
20. M. Colin, A. Hérold, Bull. Soc. Chim. Fr. 6(1971) 1982
21. A. Hérold, P. Lagrange, ISGIC 1, Mater. Sc. Eng. 31 (1977)29 and 33
22. D. Guérard, C. Takoudjou, F. Rousseaux, ISGIC 3, in Synth. Met. 7(1983)43
23. D. Guérard, N.E. Elalem. C. Takoudjou, Proc. Int. Symp. Tsukuba, in Synth. Met. 12 (1985)195
24. W. Rüdorff, E. Schulze, Angew. Chem. 66 (1954)305
25. W. Rüdorff, E. Schulze, O. Rubish, Z. Anorg. Allg. Chem, 282 (1955) 223
26. W. Rüdorff, Chimia, 19 (1965)489
27. W. Craven, W. Ostertag, Carbon (1966)223
28. C. Stein, L. Bonnetain, J. Golé, Bull. Soc. Chim. Fr. (1966)3166
29. L. Bonnetain, Ph. Touzain, A. Hamwi, ISGIC 1, in Mat. Sc. Eng. 31 (1977)45
30. F. Beguin, P. Marceau, R. Setton, M.F. Quinton, Proc. 16th Bienn. Conf. on Carbon, San Diego, USA (1983)252
31. F. Beguin, H. Estrade-Szwarckopf, J. Conard, P. Lauginie, P. Marceau, D. Guérard, L. Facchini, ISGIC 3, in Synth. Met. 7 (1983)77
32. F. Beguin, B. Gonzalez, J. Conard, H. Estrade-Szwarckopf, D. Guérard, Int. Symp. on GIC, Tsukuba (1985) in Synth. Met, 12 (1985) 187
33. J. Jegoudez, C. Mazières, R. Setton, ISGIC 3, in Synth. Met. 7 (1983)85
34. R. Setton, F. Beguin, J. Jegoudez, C. Mazières, Rev. Chim. Min. 19 (1982) 360
35. S.A. Solin, Y.B. Fan, X.W. Qian, Proc. Int. Symp. Tsukuba, in Synth. Met. 12 (1985)181

36. A.P. Legrand, L. Facchini, D. Bonnin, J. Bonat, M.F. Quinton, F. Beguin ibidem, p175
37. F. Beguin, "Carbon 86", Baden-Baden, Ext. Abstr. p424
38. R. Setton, ibidem, p420
39. R. Gross, Thèse de Docteur Ingénieur, Nancy (1962)
40. R. Vangelisti, A. Hérold, C.R. Acad. Sci. Paris, 280, C(1975)571 and 286, C(1978)289
41. R. Erre, F. Beguin, D. Guérard, S. Flandrois, "Carbon 86" Baden-Baden, extend. abstr. p516
42. H. Klotz and A. Schneider, Naturwiss 49 (1962)448
43. G. Bewer, N. Wichmann, H.P. Boehm, ISGIC 1 in Mat. Sc. Eng. 31 (1977) 73-76
44. M.E. Vol'pin, Yu. T.Struchnov, Yu. N. Novikov, V.A. Semion, Zh. Obshsb. Khim. 41(1971)2426
45. M.E. Vol'pin, Yu N. Novikov et al., J. Am. Chem. Soc. 97(1975)3366
46. Ph. Touzain, A. Chamberod, A. Briggs, ISGIC 1, in Mat. Sci. Eng, 31 (1977)77
47. R. Yazami, Ph. Touzain, G. Chouteau, A. Briggs, Synth. Met., 12(1985)485
48. D. Braga et al. , J. Chem. Soc. Chem. Comm.(1978)127
49. M. Inagaki, Y. Schiwachi, Y. Maeda, J. Chem. Phys., 81(1984)847

PART III
1. P. Schaffaütl,J. prakt. Chem. 21(1841)155
2. H. Thiele, Z. Anorg. Allg. Chem. 207(1932) 340-52
3. G. Furdin, ISGIC 3, in Synth. Met. 7(1983)101
4. J.S. Culik, D.D.L. Chung, Mater. Sci. Eng. 37(1979)213
5. S.M. Heald, E.A. Stern, ISGIC 2 in Synth. Met. 2(1980)87
6. D. Ghosh, D.D.L. Chung, Mat. Res. Bull. 18(1983)1179
7. M.C. Robert, M. Oberlin, J. Mering, Chemistry and Physics of Carbons, edited by P.L. Walker and P.A. Thrower, Marcel Dekker Inc., Vol. 10 p141-210
8. G. Hennig, J. Chem. Phys. 20(1952) 1443
9. G. Furdin, M. Lelaurain, E. McRae, J.F. Marêché, A. Hérold, Carbon, 17(1979)329
10. G. Colin, A. Hérold, Bull. Soc. Chim. Fr. 9(1972)2345
11. B. Bach, A. Hérold, ibid, 5(1968) 1978
12. G. Furdin, A. Hérold, ibid, 5(1972)1768
13. R.C. Croft, Austral. J. Chem. 9(1956)184
14. E. Stumpp, ISGIC 1, in Mater. Sc. Eng, 31(1977)53
15. M.L. Dzurus, G.R. Hennig, J. Am. Chem. Soc. 79(1957)1051
16. P. Behrens, U. Wiegand, W. Metz, "Carbon 86", Baden-Baden, Ext. Abstr. p502
17. J. Melin, A. Hérold, C.R. Acad. Sc. Paris, 269(1969)877
18. J. Melin, A. Hérold, Carbon 13(1975)357
19. J. Melin, Thèse Nancy (1976)
20. P. Boolchand, W.J. Bresser, D. McDaniel, K. Sisson, V. Yeh, P.C. Ecklund, Sol. State Com. 40(1981) 1049
21. J.M. Friedt, L. Soderholm, R. Poinsot, R. Vangelisti, ISGIC 3, in Synth. Met. 8(1983)99
22. J. Conard, ISGIC 4, Jerusalem 1987, to be published in Synth. Metals
23. C. Hauw, J. Gaultier, S. Flandrois, O. Gonzalez, D. Dorignac, R. Jagut, ISGIC 3, in Synth. Met. 7(1983)213
24. S. Flandrois, A.W. Hewat, C. Hauw, R.H. Bragg, ibid, 305
25. F. Rousseaux, R. Vangelisti, A. Plançon, D. Tchoubar, Revue de Chimie Minérale, 19(1982)572
26. J. Maire, Proc. 2nd U.N. Int. Conf. Peaceful Uses Atom. Energy, Geneva, 1958, p392
27. J.M. Lalancette, J. Lafontaine, J. Can. Soc. Chem. Comm., 21 (1973)815
28. A.A. Opalovskii, A.S. Nazarov, A.A. Uminskii, Zh. Neorg. Khim. 19, 6 (1974)1512

29. A.A. Opalovskii, Z.M. Kuznetzova, Y.V. Chichgov, A.S. Nazarov, A.A. Uminskii, ibid, 19, 8(1974)2071
30. H. Selig, O. Gani, Inorg. Nucl. Chem. Lett. 11(1975)75
31. J. Melin, A. Hérold, C.R. Acad. Sci. Paris, 280(1975)641
32. E. Buscarlet, P. Touzain, M. Armand, L. Bonnetain, ibid p1313
33. N. Watanabe, H. Toubara, T. Nakajima, H. Selig in Inorganic Solid fluorides, edited by P. Hagenmuller, Acad. Press. Inc. 1975, p332-366
34. E.R. Falardeau, L.R. Hanlon, T.R. Thompson, Inorg. Chem. 17(1978)301
35. M. Lelaurain, J.F. Marêché, G. Furdin, E. McRae, A. Hérold, ISGIC 4, Jerusalem 1987, to be published in Synthetic Metals
36. N. Bartlett, T. Mallouk, F. Okino, G. Rosenthal, J. Verniolle, J. Fluorine Chem. 23.(1983)221
37. D. Ravaine, J. Boyce, A. Hamwi, Ph. Touzain, ISGIC 2, in Synth. Met. 2(1980)249
38. A. Hamwi, Ph, Touzain, Rev. de Chim. Min. 19(1982)432
39. A. Hamwi, S. Mouras, D. Djurado, J.C. Cousseins, "Carbon 86" Baden-Baden, Ext. Abstr. p454
40. D. Vaknin, D. Davidov, H. Selig, J. Fluorine Chem., 32 (1986)345
41. N. Bartlett, B.W. McQuillan, in "Intercalation Chemistry", M.S. Whittingham, A.J. Jacobson editors, Academ. Press,(1982), 19
42. N. Platzer, B. de la Martinière, Bull. Soc. Chim. Fr. (1961)177
43. H. Fuzellier, Thèse Nancy, (1974)
44. H. Fuzellier, A. Hérold, Proceed. 4th London, Int. Carbon and Graphite Conf., (1974)322
45. L.B. Ebert, L. Matty, Synth. Met. 4(1982)345
46. A. Yaddaden, M. Lakjadj, P. Vast, Rev. Chim. Min. 19 (1981)602
47. G. Furdin, M. Klatt, M. Lelaurain, A. Hérold, ISGIC, Jérusalem 1987, to be published in Synth. Met.
48. E. Stumpp. G. Nietfeld, K. Steinwede, K.D. Wageringel, ISGIC 3, in Synth. Met. 7(1983)143
49. E. Stumpp, S. Gieschinger, G. Nietfeld, "Carbon 86", Baden-Baden, Ext. Abstr. p451
50. N. Watanabe, Proc. of the Yamada Conf. IV, Sendai 1980, in Physica 105 B (1981)17
51. W. and G. Rüdorff, Z. Anorg. Allg. Chem. 253 (1947)281
52. T. Nakajima, M. Kawaguchi, N. Watanabe, Carbon, 20 (1982)287 and ISGIC 3 in Synth. Met. 7 (1983)117
53. I. Palchan, D. Vaknin, D. Davidov, H. Selig, "Carbon 86" Baden-Baden, Ext. Abstr. p462
54. I. Toubara, ISGIC 4, Jerusalem, 1987 to be published in Synth. Metals
55. B.C. Brodie, Ann. Chim. Phys. 45(3)(1855)351
56. R. Yazami, Ph. Touzain, Synth. Met. 12(1985)351
57. A. Clauss, R. Plass, H.P. Boehm, U. Hofmann, Z. Anorg. Allgem. Chemie, 291(1957)205

PART IV
1. H. Thiele, Z. Anorg. Allg. Chem. 206(1932)407
2. P.C. Eklund, C.H. Olk, F.J. Holler, J.G. Spolar, E.T. Arakawa, J. Mater. Res. 1, 2(1986)361
3. W. Rüdorff, U. Hofmann, Z. Anorg. Allg. Chem. 238 (1938)1
4. W. Rüdorff, W.F. Siecke, Ber, 91 (1958)1348
5. W. Helle, H.P. Boehm, B. Buisinger, "Carbon 86", Baden-Baden, Ext. Abstr. p485
6. A. Métrot, P. Willman, A. Hérold, Carbon 19 (1981)119
7. J. Besenhard, H.P. Fritz, J. Electrochem. Soc. 119 (1972)1697
8. J. Besenhard, H.P. Fritz, H. Mohwald, J.J. Nickl, "Carbon 80" Baden-Baden, Ext. Abstr. p147
9. A. Jobert, Ph. Touzain, L. Bonnetain, Carbon 19 (1981) p193
10. P. Touzain, A. Jobert, Bull. Soc. Chim. Fr. I, (1983)110
11. D. Billaud, A. Chenite, A. Métrot, Carbon, 20 (1982)493

12. D. Billaud, A. Chenite, J. of Power Sources 13 (1984)1
13. M. Dzurus, G. Hennig, J. Chem. Phys. 27 (1957)275
14. J.O. Besenhard, H.P. Fritz, H. Möhwald, J.J. Nickl, "Carbon 80", Baden-Baden, Ext. Abstr. p147
15. B. Marcus, Ph. Touzain, "Carbon 86", Ext. Abstr., p430
16. S. Aronson, F.J. Salzano, D. Bellafiore, J. Chem. Phys. 49(1968)434
17. R. Yazami, Ph. Touzain, Solid State Ionics, 9-10 (1983)489
18. M. Armand, private communication
19. D.W. Murphy, S.A. Sunshine, S.M. Zahurak, this volume, plenary lecture
20. H. Fuzellier, A. Métrot, B. Bouayad, A. Hérold, 15th Biennal Carbon Conf. Univ. of Pennsylvania P.A., Ext. Abstr. (1981)391
21. B. Bouayad, A. Marrouche, M. Thili, H. Fuzellier, A. Métrot, ISGIC 3, in Synth. Met. 7(1983)159
22. P. Touzain, A. Jobert, Electrochimica Acta, 26(1968)1133
23. D. Billaud, A. El Haouari, private communication
24. A. Métrot, ISGIC 3, in Synth Met. 7(1983)177
25. A. Métrot, J. Fisher, Synth. Met. 3 (1981)201
26. H. Fuzellier, Thèse Nancy 1974
27. S. Longhin, R. Grayeski, J.E. Fischer, J. Chem. Phys, 69, 8 (1978)3740
28. See (III, 33 p344)
29. W.C. Forsman, 13th Biennal Conf. on Carbon, Irvine C.A. 1977, Ext. Abst. p153
30. W.C. Forsman, H. Mertwoy, Synth. Met, 2 (1980)171
31. D. Billaud, A. Pron, F. Lincoln Vogel, ibid, 177
32. D. Billaud, A. Chenite, Mat. Res. Bull. 18 (1983)1001
33. R. Schlögl and A.P. Boehm, ibidem, p371
34. P. Bowen, W. Jones, J. Thomas, R. Schlögl, J.C.S. Chem.Comm (1981)677
35. R. Schlögl and H.P. Boehm, Z. Naturforsch. 39 b (1984)112
36. R. Schlögl and H.P. Boehm, ibid (1984) 788
37. P. Bowen, W. Jones, J. Thomas, R. Schlögl, J.C.S. Chem. Comm (1981)679
38. R. Schlögl, H.P. Boehm, "Carbon 82", London, Ext. Abstr, (1982)73
39. H.P. Boehm, R. Schlögl, ISGIC 4, Jerusalem 1987, to be published in Synth. Met.

PART V
1. A.G. Freeman, J.C.S. Chem. Comm. (1974)74
2. P. Lagrange, A. Métrot, A. Hérold, C.R. Acad. Sci. Paris, 278 (1974)701
3. A. Hérold, G. Furdin, D. Guérard, L. Hachim, M. Lelaurain, N.E. Nadi, R. Vangelisti, Int. Symp. on GIC, Tsukuba, in Synth. Met. 12 (1985)11
4. Séminaire Franco-Japonais, Paris 1985, in Ann. Phys. Fr. 11 (1986), 3
5. E. Stumpp, H.J. Hummel, C. Ehrhardt, ISGIC 4, Jerusalem 1987, to be published in Synth. Met.
6. M. Inagaki, Z.D. Wang, Y. Okamoto, M. Okira, 1987, to be published in Synth. Met.

STATISTICAL THERMODYNAMICS OF STAGING AND INTERCALATION

S. A. Safran

Corporate Research
Exxon Research and Engineering
Annandale, NJ 08801 USA

I. INTRODUCTION

Although the original motivation for recent studies of the physical properties of intercalation compounds was technological, it was soon realized that these materials had a rich variety of novel structural properties. In particular, the in-plane structure of a single layer of intercalant have been shown to include quasi-two-dimensional solid, liquid, gas, and incommensurate solid phases, along with their associated phase transitions[1]. In addition, intercalation compounds also exhibit a modulated structure that is unique, namely staging.

The staging phenomena is characterized by a periodic sequence of intercalate layers in the host matrix. The stage number n refers to the number of host layers separating two intercalant layers. Staging is characteristic of intercalation compounds; its existence does not seem to depend on the details of the order or disorder of the in-plane atomic arrangements. Although staging was originally discussed in terms of a perfectly periodic one-dimensional structure of infinite intercalate layers, recent experiments have shown that this periodicity is not a fixed chemical property of the intercalation compound, but is rather a physical consequence of the balance between the energy and entropy of the interacting intercalate layers[2]. Thus, different stages of intercalation represent different phases of the material; phase transitions between stages can be induced by changes in temperature and or pressure[3,4]. In addition, stage ordering is not perfect; a given sample may consist of "packages" of intercalate and host that fluctuate in size[5,6]. Whether this fluctuation is a result of kinetics or a more intrinsic phenomenon is a subject of current study[7,8].

The staging phenomena has recently been reviewed by this author[9]. While the point of view is theoretical, connections and applications to experiment are emphasized. In this paper, a summary of this review is presented. In addition, the necessary statistical mechanics is introduced and some new results in the area are discussed.

II. STAGING and INTERCALATION COMPOUNDS

Intercalation of a layered host material occurs when the host contacts a reservoir of guest atoms or molecules. The guest species intercalates the host if the chemical potential difference between the reservoir (e.g. a gas of alkali atoms) and the guest atoms in the host is positive. Intercalation then proceeds and reaches equilibrium when the chemical potential difference is zero. Control of the chemical potential of the guest species is accomplished experimentally by varying the temperature of the reservoir - [e.g. the two-zone method of intercalation[10,11] where the host is maintained at one temperature and the reservoir at another] - or the electric potential in an electrochemical cell. In the former case, for a reservoir consisting of a dilute gas of intercalant, the chemical potential is proportional to $k_BT \log P$ where P is the gas pressure. In the latter case, the chemical potential is proportional to the voltage difference in the cell. In graphite intercalation compounds, the two-zone, thermal method is commonly used, although several groups have reported electrochemical data as well[12,13]. For intercalation in transition metal dichalcogenides - e.g. TiS_2, TaS_2, $NbSe_2$, - the electrochemical method predominates[14,15,16].

If there are no sharp minima in the free energy of the guest species, the chemical potential at equilibrium is a smooth function of the concentration of intercalant in the host. Structure in a plot of intercalant concentration x, vs. chemical potential μ, usually indicates the existence of ordered phases which choose a preferred stoichiometry. The presence of stage ordering is thus observed as a series of plateaus in the $(x\text{-}\mu)$ plane. A given stage exists for a certain range of chemical potential. Outside of of this range, the concentration drops sharply to that characterizing a higher or lower stage[1,10]. The finite slope of the line connecting regions of two different stages signifies that the transition between them is not sharp; there can be intermediate phases with continuously varying stoichiometries[6,17]. In some cases, these phases can be due to kinetic effects and will result in hysteresis in the $(x\text{-}\mu)$ plots.

To account for both the staging and the in-plane density in these quasi-two-dimensional materials, it is common to characterize the compounds as GH_{qn}, where G represents the guest species (atomic or molecular) and H represents the host species. The stage number is n; it represents the number of host layers which separate two neighboring intercalant layers. The in-plane stoichiometry of the guest species is characterized by the index q which is the ratio of host/guest atoms (or molecules) in a single plane. For example, in graphite, the stage 1, heavy alkali metals have an ideal stoichiometry MC_8 (M=K, Rb, Cs). For stages $n \geq 2$, the ideal stoichiometries are MC_{12n}.

While the stoichiometric notation implies the existence of a single stage compound (n) with a unique in-plane stoichiometry (q), both observation and theoretical analysis show that departures from ideal staging exist in thermodynamic equilibrium at finite temperatures[18]. In some cases, the ideal stoichiometry is perturbed due to the presence of thermally generated vacancies[2]; the intercalant layer may have a range of continuously variable in-plane stoichiometries[19].

The statistical mechanical description of this process and its implications for the integrity of stage ordering are discussed below. There it is shown that staging transitions induced by temperature occur as a result of (i) vacancies in layers that were completely filled and of (ii) extra atoms in layers that were completely empty at zero temperature. These defects disturb the zero temperature sequence of filled and empty layers and induce stage disorder as shown schematically in Figure 1a We term this type of disorder, vacancy induced stage disorder. Experimental evidence for this phenomena has been presented by Caswell and Solin[17] and by Kim et al.[19] and is also discussed below.

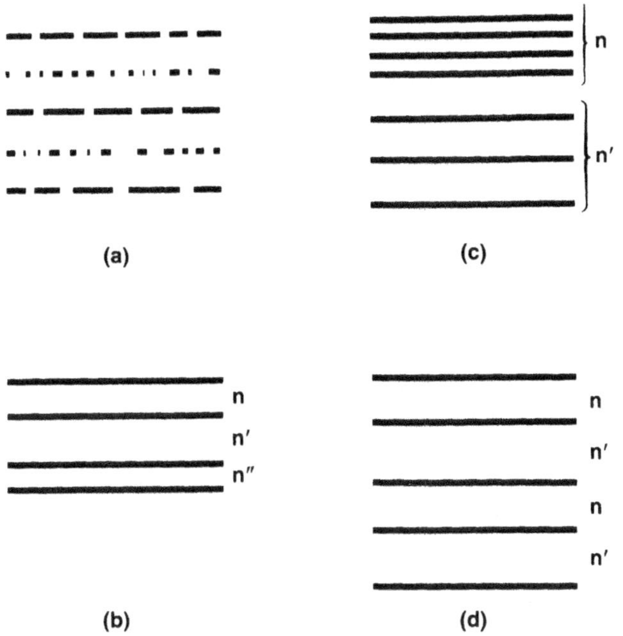

1. Types of stage ordering in intercalation compounds. The solid lines represent the intercalate layers with n and/or n' host layers between them. The dashed lines indicate layers with a lower density of intercalate atoms than the saturation density; the spacing between the dashes represents the proportion of vacancies in those layers.
(a) Vacancy induced stage disorder: Disorder is due to partial filling of layers.
(b) Random staging: Disorder is due to randomness in the interlayer spacings; all layers are at their saturation densities.
(c) A mixed phase; stages n and n' are mixed on a macroscopic length scale.
(d) A fractional stage; this is a periodic state which is a *microscopic, ordered* mixture of stages 1 and 2.

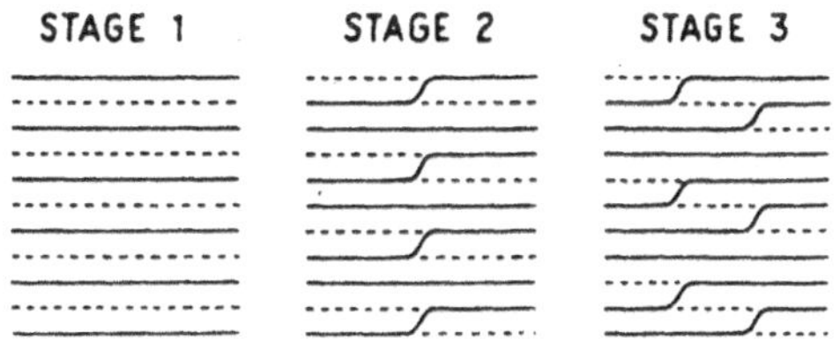

2. Staging in intercalation compounds in the domain structure suggested by Daumas and Herold. The solid lines are the graphite layers and the dashed lines are the intercalate layers.

While these considerations are relevant for infinite intercalant layers in thermodynamic equilibrium, there is considerable evidence that the intercalant layers are distributed inhomogeneously between every pair of graphite layers as shown in Figure 2. Locally, the intercalant domains or islands are arranged to form an ordered, staged structure.

This picture was first suggested by Daumas and Herold[20] as a step in understanding the kinetics of intercalation. If the intercalant layers were infinite in lateral extent, entire macroscopic planes of intercalant would have to move when a sample responds to a change in the intercalant chemical potential by a change of stage. The kinetics of such a process would seem to be prohibitively long, in comparison to the time scale of intercalation, which is typically several hours. In the domain picture shown above, only smaller scale rearrangements of the intercalant islands are needed. Other experimental evidence exists which implies that staging transitions can occur as temperature or pressure is changed with all the intercalant remaining in the host. For example, Daumas and Herold heated stage-2 KC_{24} in a carbon monoxide environment. If whole layers of intercalant had to exit the sample, the potassium vapor would have reacted with the CO, and no change of stage would be observed. The fact that they observed a change of stage from stage-2 (KC_{24}) to stage 3 (KC_{36}) was taken as evidence that the sample could restage with only internal movements of the intercalant domains. Similar conclusions were reached by Caswell and Solin[17] in their thermal experiments and by Clark[21], Solin[22], and Fischer[23] in restaging transitions that they observed as a function of applied pressure.

The existence of islands or finite domains of intercalant presents the interesting possibility of disorder in the staging sequence. For infinitely large intercalant layers, there is no staging disorder (in the absence of in-plane vacancies). This is because the entropy gain from a random reshuffling of the intercalant layers lowers the free energy by an amount proportional to $k_BT \log N_\ell$, where N_ℓ is the number of layers. The energy loss, due to the proximity of layers that would be separated in a perfectly staged structure is proportional to the number of atoms *in* the layer, N_o. In the thermodynamic limit the energy term dominates (N_o is proportional to the area of the sample, N_ℓ to its length), and perfect stage order is maintained. This argument, however, breaks down if the layers are of finite extent. Kirczenow[5] has shown how the entropy gain of the rearrangement of finite size layers in intercalation compounds can lead to stage disorder without atomic vacancies. We term this form of stage disorder - where the domain size induces the disorder - random staging. A schematic illustration is shown in Figure 1b. This form of stage disorder is distinct from a mixed phase compound which contains large domains of pure stage n and n′, as illustrated in Figure 1c. Mixed phases of pure stage materials are a natural consequence of concentration constraints of any almost any model of staging; they are absent if the intercalation compound is held at constant chemical potential (except at special values of μ). Recent experiments of Fuerst et al[24]. and of Misenheimer and Zabel[6] show evidence of random staging (apparently due to domain disorder), especially in high stage materials.

Although staging is a form of modulated structure - with the attendent phase transitions and disorderings discussed above - modulations of stage ordering itself are also possible. While the previous discussion has assumed the existence of only one periodicity - i.e. a pure stage - as the low temperature state of an intercalation compound at fixed chemical potential, more complex structures have been discussed theoretically by Safran[2,25] Millman and Kirczenow [26] and Hawrylak and Subbaswamy[27]. These structures have been observed experimentally by Fuerst and co-workers[24], and Kim et al.[28] Such arrangements, illustrated in Figure 4d, have been termed "fractional" stages since they correspond to microscopically ordered sequences of layers of intercalant with more than 1/2 of the host interlayer positions (or galleries) occupied. The stage is then generally defined as the reciprocal of the fraction of occupied galleries; a stage 3/2 would then consist of a periodic sequence of two filled layers followed by an empty layer followed by two filled layers, followed by an empty layer.

III. LATTICE GAS MODELS and STAGING

Staging phase transitions are best described by phase diagrams which show the stage (and in-plane stoichiometry) as a function of intercalant concentration, temperature, and/or pressure. To calculate these phase diagrams, the statistical thermodynamics of staging must be understood.

In mean field theory, the *free energy* of the intercalant is given by the sum of the intercalant energy and the configurational entropy of the intercalant atoms. Since the intercalant is confined to the regions between host layers, a lattice gas model can be used to describe the c-axis configurations. It is convenient, although not necessary to use a lattice gas model to describe the in-plane structures as well. This procedure is rigorously correct for intercalant layers that are commensurately ordered with respect to the host lattice. Since the focus here is on staging, the details of the in-plane structure of the intercalant are probably not crucial.

The simplest lattice-gas model to describe staging transitions was discussed by Safran[2]. This zeroth order model shows stage ordering at low temperatures, staging phase transitions (vacancy induced stage disorder) at elevated temperatures, and even fractional staging. Above a critical temperature, equilibrium states consist of stage-1 materials with continuously variable in-plane stoichiometry. This result accounts for the fact that some intercalation compounds show good stage ordering, while others may show *no* long-range order. However, the model did *not*, predict random staging (associated with domains), where the intercalant layers are at their close-packed in-plane density, but show stacking disorder (see Figure 1). This form of stage disorder requires *finite* intercalant layers, and disappears in the thermodynamic limit. In addition, the elementary model, which was based solely on *two-body* interactions between intercalant (including effective interactions mediated by the host) predicted a phase diagram which was symmetric with respect to intercalant atoms and "holes" - i.e. symmetry about 50% of the close-packed intercalant density.

The statistical mechanics of staging can be analyzed with phenomenological models. However, these models must be related to the underlying physical mechanisms so that the correct dependence of the interactions on the intercalant density, interlayer spacing etc. is included. Safran's original model assumes *infinite* intercalant layers. The free energy includes the long-ranged interlayer interaction, which was modeled as a power-law, an attractive interaction between intercalant atoms in the same plane, and a chemical potential term. The microscopic origins of the long-range interlayer interactions and in-plane attractions were discussed Ref. 9.

In this and later versions of the lattice-gas model, staging phase transitions are coupled to changes in the in-plane density of intercalant. At finite temperatures, vacancies are formed in layers that were completely saturated at T=0, while a finite density of intercalant exists in interlayer regions that were completely empty at T=0. The physical origin of this change in in-plane density is the entropy gained by disordering the intercalant layers. *In the thermodynamic limit* of intercalant layers that are infinite in their lateral extent, there is no significant entropy gained by merely disordering the stacking order of completely saturated layers.

The variable in-plane intercalant density is denoted by $\sigma_i(\mathbf{r})$, which varies from 0 to a maximum value of 1; i.e. $\sigma_i(\mathbf{r})$ is normalized to the close-packed in-plane density. The index i labels the host layer position, while **r** labels the in-plane coordinate of the intercalant atom. The fractional occupancy of the ith layer is $\sigma_i = \int dr\, \sigma_i(r) / A_o$, where A_o is the area of the layer. The mean-field free energy $F = E-TS$ where E is the internal energy and S is the entropy of mixing of the intercalant atoms-; we assume that in general, there are more possible sites than intercalant Thus,

$$F = T \sum_i [(1-\sigma_i) \log(1-\sigma_i) + \sigma_i \log \sigma_i] + \sum U(\sigma_i) + \frac{1}{2} \sum_{ij} V_{ij}\, \sigma_i\, \sigma_j - \mu \sum_i \sigma_i \qquad (1)$$

The first term represents the contribution of the entropy of mixing within a layer [it is just the logarithm of the number of ways of distributing a fraction σ_i of atoms in the possible sites of each layer]. The last term is the chemical potential of the intercalant. The in-plane effective interactions of the intercalant atoms are included in the term $U(\sigma_i)$ which in general may be a complicated function of the in-plane density. The mean-field form for the effective interlayer repulsion is quadratic in the average layer densities and is given by the term proportional to V_{ij}. As discussed in Ref. 9, the effective interlayer repulsion is approximately a power law due to the anomalous screening of the electrostatic interactions of the charged intercalate atoms - i.e $V_{ij} = V_o \,| i-j |^{-\alpha}$, where $\alpha \to 5$ as the stage $n \to \infty$.

The free energy is minimized with respect to the layer occupancies $\{\sigma_i\}$ which are then given by the self-consistent equation

$$\sigma_i = \left[\exp \left\{ -U'(\sigma_i) - \sum_j V_{ij}\, \sigma_j - \mu \right\} + 1 \right]^{-1} \qquad (2)$$

where the prime represents a derivative with respect to σ_i. The total intercalant volume fraction $x = \Sigma_i \sigma_i /N$ determines the chemical potential μ. Phase diagrams can be constructed as a function of μ and T or as a function of x and T. These two cases represent the two physical situations of (i) an intercalation compound in equilibrium with a reservoir of intercalant, where the pressure (and corresponding μ) of the reservoir is fixed, and of (ii) an isolated intercalation compound with a fixed total number of intercalant.

The phase diagram in Figure 3 is calculated for the simple case of $U = -U_o \sigma^2$ and $V_{ij} = V_o \,| i-j |^{-4}$ with $V_o = U_o/2$. The high-stage phases all terminate in a dilute, stage-1 phase ("gas") whose concentration vanishes as $T \to 0$. The transitions from stage to stage for $n>2$ are all first order; the concentration ranges over which these states are stable are within 6% of their T=0 concentrations On the other hand, the stage-2 to stage-1 transition is second order for $T>T_t$, with a large range of stoichiometries predicted for the stage-2 phase at high temperatures.

For $T<T_m$, the stage-1 phase consists of both "gas" and "liquid" phases, each having a restricted range of composition, while for $T>T_m$, the composition range is unrestricted, and no staged ($n>1$) states are possible. Finally, the existence of fractional stages, corresponding to periodic structures with multiple staging periodicities is also found.

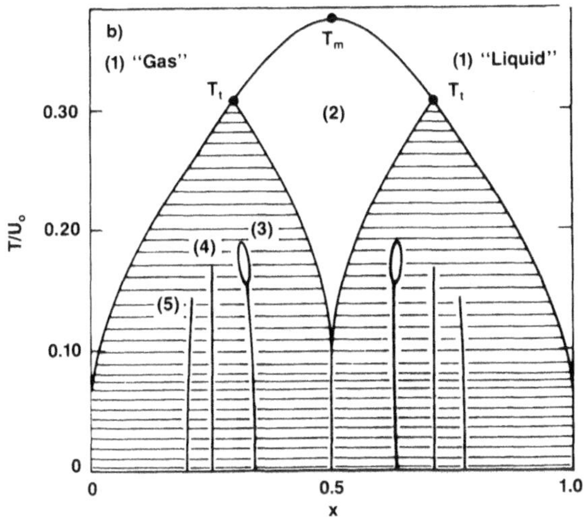

3. Phase diagram plotted as a function of the normalized temperature T/U_o and normalized concentration x. The integers in parentheses are the single-phase pure-stage states, while the cross hatched regions denote two-phase regions. The fractional stages are not shown, but are predicted by the theory. They can be found by reflecting the phase diagram about x=1/2. For clarity, only stages 1-5 are shown. [From S.A. Safran, Phys. Rev. Letts., 44,937 (1980).]

An extensive study of the temperature/concentration phase diagram has been performed by Woo et al.[19] and by Fischer et al.[3] for Li-graphite intercalation compounds. The phase diagram is shown in Figure 4. The experimental phase diagram is in agreement with the simple lattice model as far as the temperature driven destaging transitions. At high temperatures, all the phases of the material (except for x = 1/3) are unstable to a dilute stage-1 compound. Both high and low density phases of this stage-1 compound exist ("gas"/"liquid" phases discussed above). In addition, stage-2 has the largest range of stable densities at high temperatures, in agreement with theory. However, no continuous stage-1 to stage-2 transition is observed; the Li-graphite system is chemically stable only for $T<750\,^0K$.

While the experimental phase diagram for Li-graphite shows many of the features predicted by the simple lattice-gas model it shows neither fractional stages nor the symmetry about x = 1/2 predicted by the two-body (quadratic) form for the in-plane interaction, U'. There have been several attempts to modify the lattice-gas model in order to make it more realistic and physically plausible.

The major modification to the simple lattice model is the modeling of the in-plane density dependence of both the intralayer [$U(\sigma)$] and interlayer interactions V_{ij}. Deviations of these interactions from forms that are quadratic in the in-plane density yield phase diagrams that are *not* symmetric about x = 1/2. In addition, they may supress the stability of the fractional stages. The lattice-gas model was made more realistic by the modifications of Dahn[29], Millman and Kirczenow[26,30,31], and by Hawrylak and Subbaswamy[27].

DiVincenzo and Koch[32] have shown that the experimental phase diagram for Li-graphite can be understood by adding a "corrugation energy" to the simple lattice gas model of Equation (1). Although the commensurate $\sqrt{3}x\sqrt{3}$ superlattice represents the close-packed state, there also exists a commensurate 2 x 2 superlattice of the intercalant layer. There are thus two stage-2 regions in the phase diagram. The dilute (2 x 2) sttae is stable up to higher temperatues due to the larger entropy of this phase. The close packed, $\sqrt{3}$ x $\sqrt{3}$ stage-2 phase is unstable to coexistence with

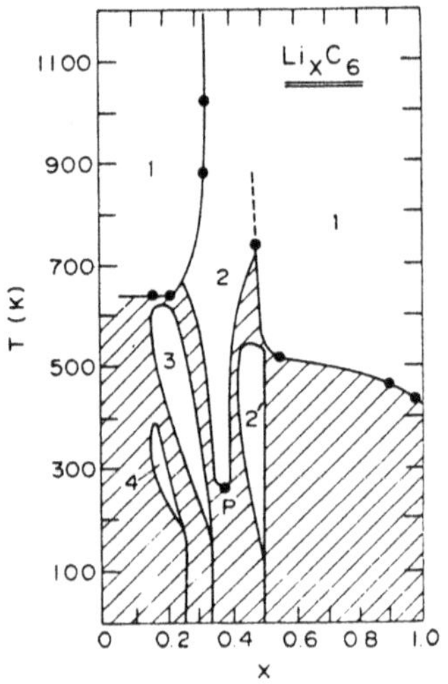

4. The experimental concentration-temperature phase diagram for Li-intercalated graphite. The concentration of intercalant normalized to its saturation value is x. Dilute stage-2 is denoted as 2, while 2' denotes a dense stage-2 phase. [From J.E. Fischer, C.D. Fuerst, K.C. Woo, Synth. Met. 7,1 (1983).]

other phases due to the small entropy of this densly packed state. The model does predict some features which are not seen experimentally. However, it does correctly explain the most striking deviations of the Li-graphite phase diagram from the predictions of the most naive lattice-gas model in terms of the in-plane structure of the Li layer.

IV. RECENT RESULTS

Pressure Induced Staging

The temperature-induced staging transitions discussed above were all coupled (at constant total intercalant density) to changes in the in-plane density. The application of external pressure has recently been shown to drive restaging transitions in an even more dramatic manner.

The effects of the application of high pressure to intercalated graphite were first demonstrated by Clarke et al.[21]. These experiments on KC_{24} showed that external pressure results in a densification of the in-plane structure of the intercalant. Since the experiments were reversible with pressure, the densification must be accompanied by an increase in the stage number. X-ray diffraction revealed that such restaging transitions did occur as a function of applied pressure. The reversibility of these transitions implies that all the transitions occurred with no loss of intercalant, providing support for the Daumas-Herold[20] picture for staging kinetics via the motion of intercalant domains.

The original experiments on KC_{24} were extended by Wada et al.[22,33,34] to Cs and Rb graphite intercalation compounds who found daging transition to stges as high as n=6. The complex behavior of K-graphite intercalation compounds at intermediate pressures has recently been studied by Kim et al.[28] and Sundqvist and Fischer[35].

The relationship between the changes in the in-plane density and re-staging transitions induced by either changes in the pressure or the temperature was elucidated by DiVincenzo et al.[36]. They found that dilute stage-2 Li-intercalated graphite showed the same phase transformation at high T and finite P as at low T and P=0.

<u>Random Staging and Intercalant Domains</u>

The lattice-gas models for staging presented in Section III, resulted in stage disorder due to the generation of vacancies and interstitials. It was assumed that the intercalate layers in the intercalation compound extend continuously over macroscopic distances. However, there is experimental evidence that in real intercalation compounds, the intercalant atoms form islands, and that every pair of host layers has an <u>equal</u> number of intercalant atoms between them[20]. Electron imaging of these islands has been reported[37,38,39,40]; a typical size is of the order of 100 Å. However, it is not certain whether these sizes are due to kinetic limitations in the intercalation process, imperfections in the host crystal, or degradation due to the electron beam.

Notwithstanding these questions about the <u>origin</u> of domains in staged intercalation compounds, several groups have studied their effect on the integrity of staging[3,5,6,18,41,42,43]. As mentioned above, stage ordering is expected to be perfect (except for vacancy induced disorder) for infinite intercalate layers. The existence[3,6,18,43] of measurable randomness in staging signifies finite domain sizes. (This kind of stage disorder refers to the existence of randomly mixed sandwiches of different stage in one compound - see Figure 1.)

Kirczenow[5,41] has recently quantified these ideas in a calculation of the degree of randomness in staging using an extension of the lattice-gas models described above. He considers a single stack of finite size domains. The entropy of disordering of the domains in different stacks and correlations between such domains are neglected; only the entropy of reshuffling the *finite* layers within a given stack is added to the free energy of Eq. (1).

Numerical solutions of the set of non-linear equations for the fractional probability of stage-i units, f_i, are found. Decreasing the range of the interlayer interaction strongly increases the staging randomness. A similar effect is obtained by decreasing the <u>strength</u> of the interlayer interaction. Thus, acceptor compounds, which generally have smaller charge transfers than donors, should show a higher degree of random staging. It is interesting to note that this model predicts a *continuous* transition between all stages; e.g. in the stage-1 to stage-2 transition, the proportion of stage-2 units grows continuously as the chemical potential μ is increased, with a concommitant decrease in the proportion of stage-1 units.

Recent x-ray measurements have begun to address the question of random staging. In the early work of Metz et al.[18] it was already noticed that stage ordering was imperfect. Again, this effect may be related to extrinsic properties of the intercalation compound (growth kinetics, defects) which lead to finite size domains. For example, Nishitani et al.[44,45], did not report any admixture of foreign stages in their study of K-graphite compounds, while other studies[6,28,46], do find evidence for admixtures of stage-(n±1) units for every n>2. Furthermore, the stage disorder seems to increase a small amount near the stage-2 to stage-3 transitions.

Kinetics of Staging and Intercalation

The existence of finite-size domains in intercalation compounds is in itself a kinetic effect as discussed in Section I. Recent theoretical studies have shown that indeed, the kinetics of intercalation do result in a domain structure - see Safran[25], Hawrylak and Subbaswamy[47], Miyazaki and Horie[48], and Kirczenow[49]. The earliest kinetic models examined temperature and chemical potential quenches and showed how the system evolves from one stage to another. A more ambitious goal is the modeling of the intercalation process itself. Kirczenow has calculated the kinetics of a full, three-dimensional model of the motion of intercalant islands as the host crystal is intercalated.

These simulations show that the Daumas-Herold domain structure eventually develops as the sample proceeds to its equilibrium stage-3 state. At intermediate times, the stage-3 ordering is not perfect and non-negligible fractions of stages 2-5 are also observed. As the stage-3 crystal is further intercalated to a final state of stage-2, considerable stage disorder is observed. The disorder is associated with the finite size of the intercalant domains[5]. In addition, kinetically induced stage disorder is observed to be more pronounced at higher stages, in agreement with the (equilibrium) model of Kirczenow[5], and with the observations of Misenheimer and Zabel[6].

Experimental studies of the kinetics of staging transitions were performed by Nishitani et al.[45], Misenheimer and Zabel[50] and by Kim and Fischer[28]. Nishitani observed that only two phases of stages n and n-1 coexist throughout the process and that the c-axis disorder is insignificant. These results are not in agreement with those of Ref.6; the reasons for discrepancy may lie in the increased resolution of these latter measurements.

REFERENCES

1. S.A. Solin, Adv. in Chem. Phys., 49, 455 (1982).
2. S.A. Safran, Phys. Rev. Letts., 44, 937 (1980).
3. J.E. Fischer, C.D. Fuerst, K.C. Woo, Synth. Metals, 7,1 (1983).
4. R. Clarke and C. Uher, Adv. Phys., 33, 469 (1984).
5. G. Kirczenow, Phys. Rev. Letts., 52, 437 (1984).
6. M.E. Misenheimer and H. Zabel, Phys. Rev. Letts., 54, 2521 (1985).
7. G. Kirczenow, Phys. Rev. Letts B, 55, 2810 (1985).
8. P. Bak and G. Forgacz, Phys. Rev. B 32, 7535 (1985).
9. S. A. Safran, in "Solid State Physics", ed. H. Ehrenreich and D. Turnbull, Academic Press, NY, in press.
10. A. Herold, Bull. Soc. Chimie Fr., 999 (1955).
11. J.G. Hooley and M. Bartlett, Carbon, 5, 417 (1967).

12. S. Aronson, F.J. Salzano, D. Bellafiore, J. Chem. Phys., 49, 434 (1968).
13. E. McRae, A. Metrot, P. Willman, A. Herold, Physica B, 99, 489 (1980).
14. J.R. Dahn and W.R. McKinnon, J. Phys. C, 17, 4231 (1984).
15. D. Kalarchichi, R.F. Frindt, Phys. Rev. B, 28, 3663 (1983).
16. K.K. Bardhan, G. Kirczenow, J.C. Irwin, J. Phys. C, 18, L131 (1985).
17. R. Clarke, N. Caswell, S.A. Solin, Phys. Rev. Letts., 42, 61 (1979).
18. W. Metz and D. Hohlwein, Carbon, 13, 87 (1975).
19. K.C. Woo, H. Mertwoy, J. E. Fischer, W.A. Kamitakahara, D.S. Robinson, Phys. Rev. B, 27, 7831 (1983).
20. N. Daumas and A. Herold, CR Acad. Sci. C, 286, 373 (1969).
21. R. Clarke, N. Wada, S.A. Solin, Phys. Rev. Letts., 44, 1616 (1980).
22. N. Wada and S.A. Solin, Physica B, 105, 268 (1981).
23. J.E. Fischer and H.J. Kim, Synth. Metals, 12, 137 (1985).
24. C.D. Fuerst, J.E. Fischer, J.D. Axe, J.B. Hastings, and D.B. McWhan, Phys. Rev. Letts., 50, 780 (1983).
25. S.A. Safran, Synth. Metals, 2, 1, (1980).
26. S.E. Millman and G. Kirczenow, Phys. Rev. B, 26, 2310 (1982).
27. P. Hawrlayk and K.R. Subbaswamy, Phys. Rev. B, 28, 4851 (1983).
28. H.J. Kim, J.E. Fischer, D.B. McWhan, J.D. Axe, Phys. Rev. B, 33, 1329 (1986).
29. J.R. Dahn, D.C. Dahn, R.R. Haering, Solid State Commun., 42, 179 (1982)
30. S.E. Millman and G. Kirczenow, Phys. Rev. B, 28, 3482 (1983).
31. S.E. Millman, G. Kirczenow, D. Solenberger, J. Phys. C, 15, L1269 (1982).
32. D.P. DiVincenzo and T.C. Koch, Phys. Rev. B, 30, 7092 (1984).
33. N. Wada, R. Clarke, S.A. Solin, Synth. Met. 2, 27 (1980).
34. N. Wada, Phys. Rev. B, 24, 1065 (1981).
35. B. Sundqvist and J.E. Fischer, Phys. Rev. B 34, 3532 (1986).
36. D.P. DiVincenzo, C.D. Fuerst, J.E. Fischer, Phys. Rev. B, 29, 1115 (1984).
37. J.M. Thomas, G.R. Millward, R.F. Schlögel, H.P. Bohm, Mat. Res. Bull., 15, 671 (1980).
38. S.F. Flandrois, J. M. Masson, J. C. Rouillon, J. Gaultier, C. Hauw, Synth. Met. 3, 1 (1981).
39. A.N. Berker, N. Kambe, G. Dresselhaus, M.S. Dresselhaus, Phys. Rev. Letts. 49, 249 (1982).
40. G. Timp, M.S. Dresselhaus, P.C. Eklund, Y. Iye, Phys. Rev. B, 26, 2373 (1982).
41. G. Kirczenow, Phys. Rev. B, 31, 5376 (1985).
42. G. Forgacs and F. Uimin, Phys. Rev. Letts., 52, 6331 (1984).
43. H.J. Kim and J.E. Fischer, Synth. Met. 12, 137 (1985); Phys. REv. B 33, 4349 (1986).
44. R. Nishitani, Y. Uno, and H. Suematsu, Phys. Rev. B, 27, 6572 (1983).
45. R. Nishitani, Y. Uno, and H. Suematsu, Synth. Met., 7,13 (1983).
46. C.D. Fuerst, T.C. Koch, J.E. Fischer, J.A. Axe, J.B. Hastings, D.B. McWhan, "Intercalated Graphite", (eds. M.S. Dresselhaus, G. Dresselhaus, J.E. Fischer, M.S. Moran), North-Holland, N.Y., 1983, p. 39.
47. P. Hawrlyk and K.R. Subasswamy, Phys. Rev. Letts., 53, 2098 (1984).
48. H. Miyazaki and C. Horie, J. Synth. Metals 12, 149 (1985); Phys. Rev. B 34, 5736 (1986).
49. G. Kirczenow, Phys. Rev. Letts., 55, 2810 (1985).
50. M.E. Misenheimer and H. Zabel, Phys. Rev. B, 27, 1443 (1983).

STRUCTURAL CHEMISTRY, PHASE EQUILIBRIA AND PHASE TRANSITIONS IN GRAPHITE INTERCALATION COMPOUNDS

John E. Fischer

Laboratory for Research on the Structure of Matter
University of Pennsylvania
Philadelphia PA 19104 U.S.A.

INTRODUCTION

This NATO Advanced Study Institute is the first time since 1978 that people working on different classes of "intercalation compounds" have been brought together. The unifying theme is the tailoring of properties by inserting guest atoms or molecules into more-or-less well-defined interstitial sites defined by a more-or-less anisotropic host lattice. The properties of interest may be primarily electronic (e.g., synthetic metals derived from doped polymers, intercalated graphite, etc.), ionic transport (solid electrolytes, superionic conductors), magnetic, etc. In all cases an important first step is to understand the rich and complex structural chemistry which results from competing microscopic interactions in low-dimensional, multicomponent systems. In particular, the most novel physical properties often depend sensitively on the effective dimensionality, a "parameter" which is not always straightforward to define.

Graphite intercalation compounds (GIC's) are one of the largest and most-studied families of guest-host materials. The variety of structures which can be synthesized results from the fact that many of the competing energies -- guest/host, guest/guest, intra- vs. interlayer, etc. -- are of the same order of magnitude. Often, a specific intercalate can be chosen to fine tune one of the interactions in order to isolate a particular thermodynamic feature. Much of the variety derives from the staging phenomenon, discussed in detail by Safran in this volume. This is loosely defined as the formation of ordered one-dimensional layer superlattices by occupying only a periodic subset of the van der Waals galleries between host layers. Another structural degree of freedom is provided by the strength of the two-dimensional corrugation potential imposed on the intercalate layers by the host layer "substrate"; the balance between interlayer guest/host and intralayer guest/guest interactions yields a two-dimensional structure which can be either commensurate or incommensurate with respect to the graphite layers.

These and other novel features have stimulated much research activity in the last ten years. Some of this is designed to study basic condensed matter phenomena -- phase transitions, 2D magnetism, superconductivity etc. - using the synthetic versatility of GIC's to produce optimum prototypes for a particular experiment. Another important subfield is the study of electronic and lattice dynamical properties and their relation to the ubiquitous charge exchange between guest and host, a necessary condition for intercalation in graphite to occur. Another major activity is the synthesis and characterization of new compounds, many of which contain more than one guest specie, as discussed by Herold in this volume. My goal in this chapter is to present the experimental information on crystal structure and phase equilibria in a manner which makes maximum contact with other families of intercalation compounds. Up-to-date general overviews are available in conference proceedings [1,2] and the text of a 1986 NATO summer school [3].

The experiments reported here were largely motivated by the pioneering theoretical work by Safran. I begin by summarizing the main features of his Ising-like model phase diagram [4], then proceed to a discussion of two of the fundamental interaction parameters, namely the elastic contribution to the in-plane guest/guest energy and the guest/host corrugation potential. The next two sections present the structural phases and phase transitions for two representative model systems, Li_xC_6 and K_xC_8, based on neutron and x-ray diffraction vs. temperature, hydrostatic pressure and intercalate concentration x. These include discussions of the one-dimensional, or stage disorder, which results from the finite magnitude and limited range of the interlayer guest-guest interaction and/or from finite size effects. A final section presents some ideas on how our knowlege of intercalated graphite may be applied to other materials.

THERMODYAMIC MODEL OF LAYER INTERCALATES

Motivated by the observation of long-period layer superlattices in GIC's (i.e., "staging"), Safran proceeded to construct a phenomenological model to explain the essential features. The problem is to minimize the free energy $F = U - TS - PV$ with respect to the configuration of n atoms occupying N sites in a layer host. For the equivalent isotropic random impurity problem with $n \ll N$, the entropy dominates at all T and P independent of the internal energy U, leading to a distribution which is uniform in all three dimensions. We will see below that some GIC's show a remnant of this "isotropic" behavior at high T. Two energies are required to explain staging: an in-plane attractive interaction U_o between pairs of atoms in the same gallery and an out-of-plane repulsive interaction V_{ij} between pairs of atoms in different galleries i and j. For $n = N$, all sites in all galleries are occupied independent of T and P; the system is a stage-1 sequence of guest and host layers. Now suppose $n = N/2$. At $T = 0$, the U_o term favors the filling of all sites in a gallery while the contribution to V_{ij} from a pair of occupied galleries leads to maximum separation between them. Summed over n the result is a periodic stage-2 sequence, two host layers followed by a maximally-dense guest layer (i.e. an empty gallery followed by a filled one). At finite T, the dimensionality of the system plays a role via the configurational entropy. The sparse filling of a gallery involves a sum over two dimensions, the third dimension being accounted for by the choice of a subset of galleries to occupy. Consequently the entropy is maximized by identical sparse filling of all galleries, since there are more ways to sparsely decorate a 2D gallery than there are ways to choose a 1D sequence of galleries to decorate. Our low-T stage-2 sequence has transformed to dilute stage-1. This result is general, independent of details of U_o and V_{ij}. The transition from dense stage-2 to dilute stage-1 is analogous to the ferromagnetic-paramagnetic transition in the Ising model of nearest-neighbor interacting spins; similar transitions are predicted to occur at all $n < N$ as shown in Fig. 1. If V_{ij} extends beyond near-neighbor filled galleries, novel "fractional stages" appear in the phase diagram; for example if V_{ij} is nonzero for $i = j + 2$, the second-neighbor coupling stabilizes a stage-3/2 sequence of the form ...HGHGHHGHGHH... where H and G denote host and guest layers. Screening beyond first neighbors limits the system to integral stages: ...HGHGHG... (stage-1), ...HHGHHG... (stage-2) etc. If V_{ij} extends to infinite neighbors and is infinite-ranged, a complete Devil's staircase of integral and fractional stages is found [5].

The temperature scale for transitions from dense high stages to dilute stage-1 is set by the magnitude of U_o. If U_o is very large they will occur at inaccessibly high T, while if U_o is small they may be obscured at low T by slow kinetics. Now consider a compound with n = N/3 at some T well above the staging transition. All galleries contain random distributions of 1/3 filled and 2/3 vacant sites, i.e. the system is dilute stage-1. If the host layers are stiff enough, the galleries will still be significantly expanded in the c-direction despite the large vacancy density per unit area. The free energy may now be reduced via the PV term by applying pressure -- the strain is primarily in the c-direction, and the volume is effectively reduced by forming dense islands and empty regions, shown schematically in Fig. 2. As discussed below, elastic strains associated with in-plane separation of filled and empty sites cause the islands and empty regions to correlate or anticorrelate along the c-axis, leading again to a staging transition at high P analogous to low T. In the simplest possible case, $U_o \neq U(T,P)$ and the low-T and high-P transitions are connected by a Clausius-Clapeyron line given by dP/dT = dS/dV where S is the entropy and dV is the volume increase per occupied site.

The Safran model in its primitive form accounts for many observations and can also be used to analyze chemical trends. The physical origins of U_o and V_{ij} are discussed in detail by Safran. We note here several important factors which were left out. First is the c-axis dilation energy; the host lattice swells to accommodate guest layers. Another is the interlayer guest/host energy which, if large enough, drives the guest layer to a commensurate structure. Yet another is the elastic distortion of host layers around dense intercalate islands (cf. Fig. 2), an idea which was introduced to explain the rapid kinetics of staging transitions (see discussions by Herold and Safran in this volume). Examples where these are important will be given later.

ELASTIC CONTRIBUTION TO U_o

Safran's model fails to explain why staging occurs in GIC's but not generally in other hosts. A simple modification solves this difficulty while keeping the simplicity [6]. We assume that U_o is mediated by the elastic properties of the host lattice and therefore differs for a given intercalate in different hosts. Different approaches may be taken. Dahn et al. [7]

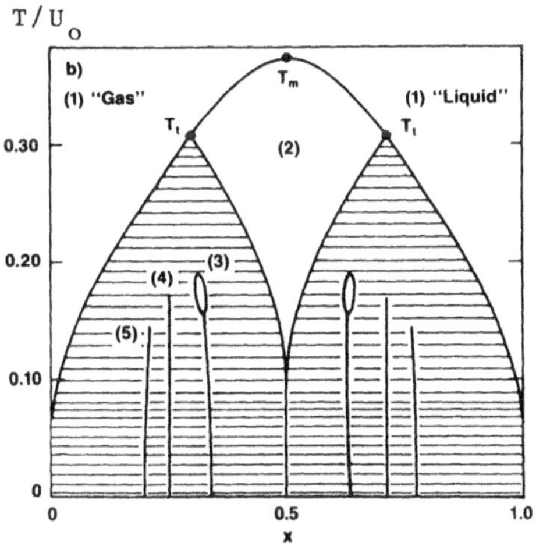

Figure 1. Staging phase diagram in the (T,x) plane with screened pair interactions, after Safran.

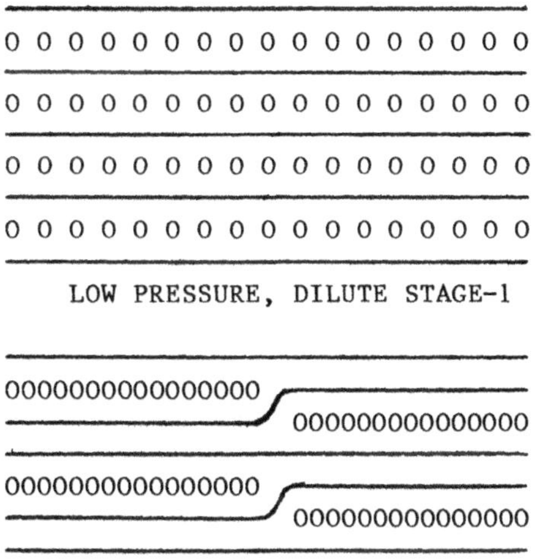

Figure 2. Pressure-induced reduction in volume, a driving force for staging transitions by in-plane condensation.

treated a system of rigid layers held together by Hooke's law springs, while Safran and Hamann envisioned microscopic dipole strain fields arising from the deformation of host layers flanking an occupied site [8]. The latter has two advantages for our purpose; it accounts in principle for contributions to both inter- and intralayer energetics, and it provides a natural connection with U_o. The relaxation of elastic distortions determines the dependence of average filled gallery thickness I_c on the fractional site occupancy, while the magnitude and range of the strain field defines the elastic contribution to U_o.

Comparison of different T-dependent staging behavior of Li intercalated into various lattices suggests that for graphite, $NbSe_2$ and TiS_2 U_o is respectively > 1500 K, of order 1000 K and < 750 K [9]. The differences may be understood in terms of different host elastic properties, by comparing the dependence of I_c on x for two sets of stage-1 compounds Li_xC_6 and Li_xTiS_2. In Figure 3, we plot I_c versus x, measured at 700K and 300K respectively to ensure that for both sets $T > T_s$ for all x where T_s is the staging transition temperature. In both cases, I_c first increases monotonically, then levels off. This implies that at low x, I_c measures an average of locally-distorted and empty regions, and thus increases as the distortions become dense and finally overlap. For Li_xC_6, saturation occurs near x = 0.7: the distortions "heal", the C layers are again flat, and I_c achieves its maximum value even though there are still 30% vacant sites. At x = 0.7, the average Li-Li separation is 5 A which we take as the decay length of C-layer distortions around an isolated Li. The more extensive in situ data for Li_xTiS_2, Fig. 3b), show even more clearly the saturation of $I_c(x)$ at lower x than in graphite; local distortions fall off more slowly with distance and thus overlap at larger separation, 8 vs. 5 angstroms. The TiS_2 "layers" are stiffer than graphite due to the nonplanar covalent bonds among constituent metal and chalcogen atoms. Neither data are consistent with the rigid layer model (dotted curves in Fig. 3 for different ratios of spring constants), supporting the local distortion viewpoint.

The strain potential goes as D/L where D is the c-axis dilation (same in both hosts) and L is the decay length. Thus, the elastic contribution to U_o goes as D/L^2; the ratio $U_o(\text{graphite})/U_o(TiS_2) \sim 2.5$, in good agreement with earlier estimates. Stiff hosts have weak, diffuse local strain fields and

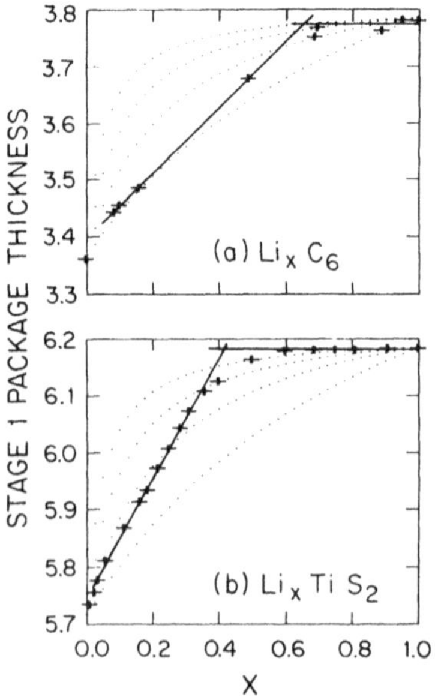

Figure 3. Comparison of experiment (+++) and theory (...) for Li in graphite and TiS_2. Rigid layer model after Ref. 7; theoretical curves for different spring constant ratios -- see Ref. 6.

weak in-plane dipole attraction between occupied sites in the same gallery. As a result, the entropy-driven transition to dilute stage-1 will occur at relatively low temperature and may not be visible at all in practice due to limited kinetics. Graphite monolayers are no doubt the most flexible of all, so competition between configurational entropy and elastic energy produces staging transitions at higher T where the kinetics are reasonable. The much stiffer host layer units in silicate clay, spinel-based superionic conductors etc. effectively rule out staging in intercalates and ion-exchanged derivatives based on these host lattices. Analogous considerations should apply to doped polymers -- in all likelihood the effective U_o will be quite large due to the highly flexible backbone, in which case T_s will exceed the thermal stability limit of the material.

CORRUGATION POTENTIAL AND COMMENSURABILITY

The magnitude of the corrugation potential plays a major role in the phase equilibria of alkali metal GIC's. The Safran model assumes commensurate local order independent of T, P and n since the number of sites N is fixed. Comparative studies of different alkali GIC's can be used to probe this interaction since the corrugation potential is species-dependent. Two approaches have been taken to compare commensurability effects of different alkali metals in graphite. DiVincenzo and Mele used a Thomas-Fermi approach to show that the commensurability energy for Li exceeds that of heavy alkalis by about a factor of 10 [10]. Similar conclusions were drawn from quasi-elastic neutron scattering studies of jump diffusion by Magerl and Zabel [11]; both results are attributed to the compact spherical charge distribution of the Li^+ core, supporting an earlier argument based on the observation that c-axis dilation relative to the ionic diameter is minimum for Li [12]. We thus have a method to isolate the effect of commensurability on phase equilibria by comparing the sturctures of Li and heavy alkali GIC's.

The results cited above do not imply that heavy alkali layers can be entirely decoupled from the graphite. Non-commensurate phases are in fact only observed over narrow ranges of T, P and n due to the balance between corrugation energy, in-plane pair energy U_o and thermal energy kT. "Floating solid" phases are rarely found [13], the more typical cases showing evidence of residual guest/host coupling. A good example is the nonstoichiometric

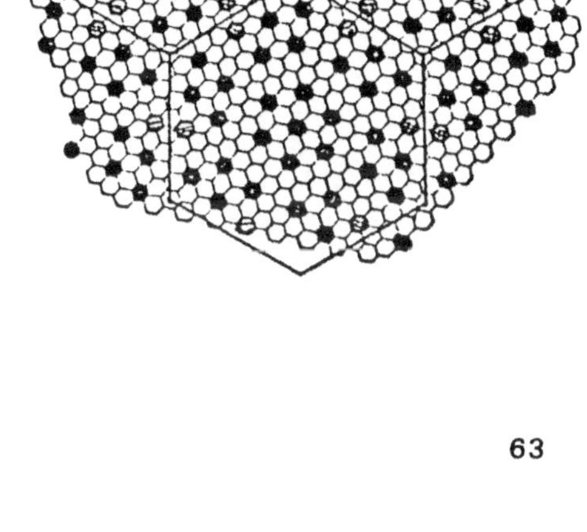

Figure 4. Discommensuration/domain lattice for the ordered phase of KC_{24}, see e.g. Ref. 14. Solid circles are filled sites within the $\sqrt{7} \times \sqrt{7}$ commensurate domains; shaded circles are K atoms belonging to the domain walls. Dashed line shows the lattice of hexagonal domains, rotated with respect to the graphite lattice.

dilute stage-2 compound "KC_{24}". The high-T in-plane structure is an oriented fluid, analogous to hexatic liquid crystals, in which the 2D Debye liquid rings show 6-fold azimuthal intensity variations indicating some short-range orientational order driven by weak host layer coupling. Long-range in-plane order sets in below 123K, and the complicated diffraction pattern [14] may be interpreted as a lattice of locally commensurate domains bounded by discommensurations (or solitons) as shown in Fig. 4. The domains are commensurate $\sqrt{7} \times \sqrt{7}$, implying a stoichiometry KC_{28} if there were no discommensurations. Thus, the average K density per unit area in the discommensurations is higher than within the domains. This structure is believed to have a large effect on lattice and single-particle dynamics, melting, elastic moduli, etc. [15].

The Li analog of "KC_{24}" is "LiC_{16}" which at STP is a dilute lattice-gas disordered stage-2 compound with roughly 2/3 occupied sites in every other gallery. According to Safran, it should phase-separate at low T or high P into a mixture of dense ordered stages 2 and 3 whose in-plane structures are locked by the $\sqrt{3} \times \sqrt{3}$ superlattice:

$$3LiC_{2x8} \rightarrow 2LiC_{3x6} + LiC_{2x6}.$$

Both predictions are borne out as shown in Figs. 5 and 6, in which we have used P- and T-dependent resistivity to locate the transitions and (00L) x-ray diffraction to identify the low-T and high-P mixed phases [16]. The kinetics of these transitions are very slow; the time constant for the stage-3 growth in response to a temperature quench from 300K to 240K is 20 hours. This is attributed to slow single-particle diffusion implied by the large activation energy, 1 eV [10,11]. The initial and final phases are both commensurate so we have all the necessary ingredients to calculate $dP/dT = dS/dV$ using the Bragg-Williams approximation for the configurational entropy in the disordered high-T/low-P phase (the same approximation is made in obtaining the (T,x) diagram, Fig. 1). We find 0.056 kbar/K which agrees very well with the experimental (P,T) slope, as shown in Fig. 7. This agreement indicates that the implied constraint $U \neq U(T,P)$ on the Clausius-Clapeyron relation is satisfied, which in turn means that the corrugation energy is large compared to kT and not greatly enhanced by a few kbar pressure.

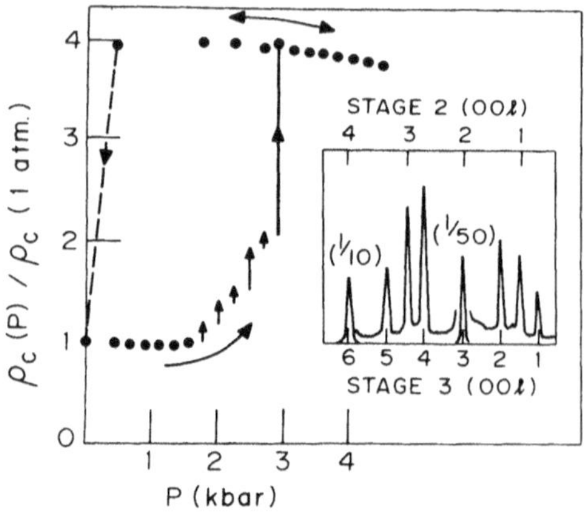

Figure 5. C-axis resistivity vs. P for LiC_{16} showing a hysteretic transition. Inset: (00L) profile of the mixed stage-2 + 3 high-P phase.

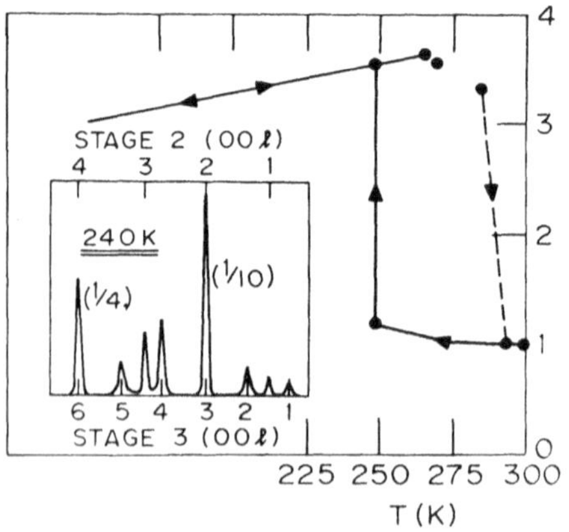

Figure 6. C-axis resistivity vs. T for LiC_{16} showing a similar hysteretic transition. Inset: (00L)'s of the stage-2 + 3 low-T phase.

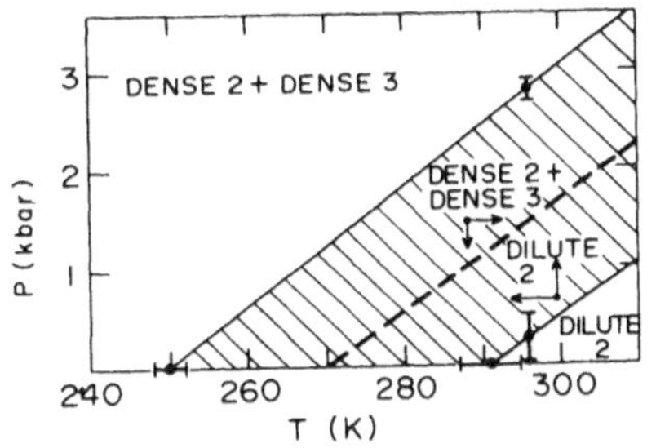

Figure 7. (P,T) phase diagram for LiC_{16}. Dots and error bars are the extremeties of the hysteresis loops in in Figs. 5 and 6. The dashed line is the equilibrium slope predicted from the Clausius-Clapeyron relation. The stable phases are dilute stage-2 (lower right) and mixed dense stage-2+ 3 (upper left); in the hatched region either is stable depending on past history.

We will see more examples later which demonstrate that Li-GIC's fit the simple picture. As noted already, this is mainly due to the fact that the Li sits over hexagon centers independent of T, P and n, but it also implies the unimportance or absence of discommensurations and other in-plane inhomogeneities. The heavy alkali analog "KC_{24}" shows more complicated behavior than "LiC_{16}"; in particular the high-P and low-T phase transitions are different. Applying pressure at 300K we find the expected staging/in-plane condensation transition [17]:

$$KC_{2x12} \dashrightarrow KC_{3x8} \quad ,$$

a transition from stage-2 to stage-3 accompanied by in-plane ordering into a 2 x 2 superlattice consistent with hard packing of K^+ spheres. However, the evolution of stage-3 with increasing P, shown in Fig. 8, is inconsistent with the simple model. The abrupt changes in relative amounts at 3.4 and 6.4 kbar [18] are quite unexpected, and the mixed-phase plateau between them implies a commensurate layer structure of intermediate density which stabilizes the stage mixture. Cooling at 1 atm., one finds in-plane ordering and "polytype" transitions [19] but not the restaging found in LiC_{16}. The (P,T) diagram, Fig. 9 [20], is much more complicated than the Li counterpart, Fig. 7. Aside from the two transitions, the high-T part actually fits the simple picture - the phase boundaries are closely approximated by straight line segments with reasonable slopes. Strong departures occur at low-T. The in-plane ordering transition at 123K is P-independent below 1 kbar implying no volume change ergo no staging transition. Above 1 kbar this "phase boundary" is T-independent implying no entropy change ergo no ordering transition.

This complicated behavior can be rationalized by invoking both the domain/discommensuration lattice and Daumas/Herold island arguments. It is certainly true that the locally-commensurate domains are smaller than the putative islands, so several discommensurations must slide past an island boundary if a low-T staging transition is to occur. The density fluctuations associated with the heavy or dense walls will be pinned at the D/H boundaries, reducing the lateral domain velocity to zero, as shown schematically in Fig. 10. In effect the complex (P,T) "phase diagram" and the lack of low-T restaging are both due to slow kinetics. The strength of the pinning depends

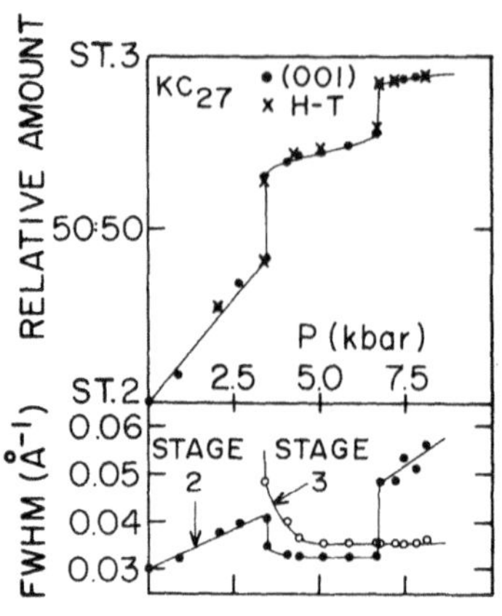

Figure 8. (top left) Evolution of stage-3 with increasing P for KC_{27}, derived from integrated (00L) intensities (dots) and from Hendricks-Teller model fits (crosses). (bottom left) Evolution of FWHM's with pressure for both phases. (right) Schematic representation of the structures at low- and high-P.

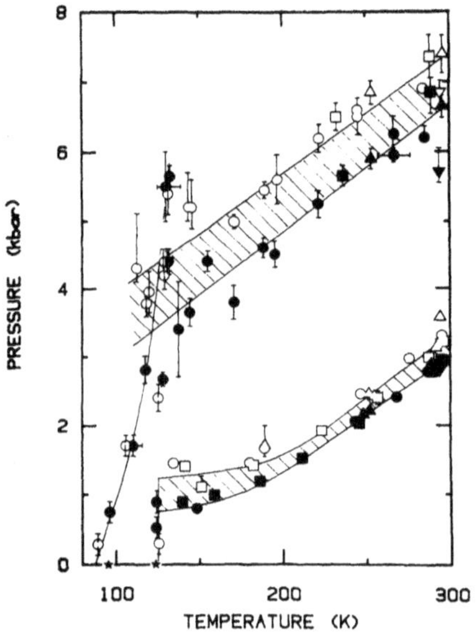

Figure 9. (P,T) phase diagram for KC_{24} based on resistivity anomalies (as in Fig. 7 for LiC_{16}). The various symbols are for different samples; open symbols denote traversing a boundary from low-P/high-T to high-P/low-T and vice versa. Intersections of the two "phase boundaries" with the right-hand P axis correspond to the two discontinuities in the stage-2/stage-3 admixture in Fig. 8, while the two stars on the T axis are the low-T structural transitions from Ref. 19.

on two factors: the amplitude of the density fluctuations and the magnitude of the D/H boundary strains. The former can be varied by exploiting the nonstoichiometry -- small variations in K/C ratio yield large variations in K density within the discommensurations -- while the latter scales with intercalate size, consistent with the absence of similar pinning in Li-GIC's. We thus prepared 3 nominally stage-2 samples with slightly different K/C ratios and looked for incipient low-T restaging [21] using Hendricks-Teller analysis of (00L) x-ray diffraction profiles [22]. Figure 11 shows 300K and 90K profiles for the densest sample. Whereas there is certainly no obvious phase transition, the fit parameters indicate the presence of 1% stage-3 "impurity" unit cells which are randomly distributed at 300K but partially phase-separated at 90K. The most dilute sample has a K/C ratio closest to the 1/28 value at which the discommensurations vanish; for this sample we find an increase from 3% to 7% stage-3 cells upon cooling from 300K to 88K. The observed trends suggest reduced pinning in the dilute sample, consistent with the process envisioned in Fig. 10.

The first observation of restaging at constant n by Clarke et al. [23] provides further support for the kinetic pinning argument. They studied a Cs-GIC sample with n/N = 0.8 and found that the low-T phase separation into mixed stage-1 + 2 only occurred if the in-plane structure was disordered in the high-T pure stage-1 phase. In terms of the kinetic pinning process described by Fig. 10, we would argue here that for Cs, the local distortion at D/H boundaries is so large that pinning occurs even without density fluctuations - the domains must melt first to facilitate uncorrelated atomic motion past the boundaries. Restaging without melting can occur in the framework of the D/H island mechanism only if the distortions are mild enough to permit solid islands to move past the boundaries. This indeed occurs for Li-GIC's over a wide range of T and n due to the smaller c-axis expansion hence weaker local strains at the D/H boundaries.

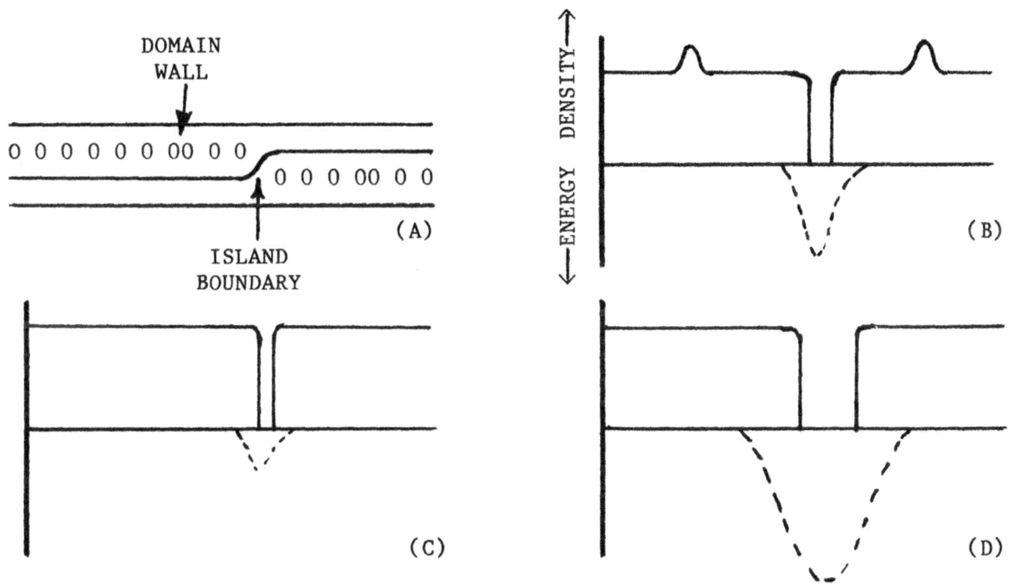

Figure 10. a) Intercalate islands consisting of locally-commensurate domains separated by dense walls; two islands must slide past each other during a staging transition and must therefore overcome the host layer strain energy at the distorted D/H island boundaries. b) Intercalate density and strain energy vs. lateral position for KC_{24}: dense walls, large D/H strain. c) Same for LiC_{16}: no walls hence no density fluctuations; weak D/H strain. d) Same for CsC_{10}: no walls but very large D/H strain.

Figure 11. (00L) x-ray profiles (dots) and HT fits (solid curves) for a "dense" stage-2 K-GIC at 300K (a) and 90K (b). Fit parameters yield constant 1% stage-3 impurity but with increased tendency to phase separation at 90K.

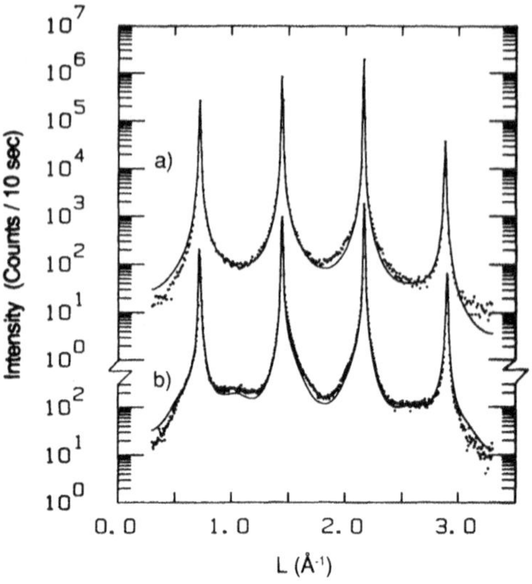

Li_xC_6: STAGING, MELTING AND THE PHASE DIAGRAM

There are many examples of staging transitions at constant n [24]. In some cases the required in-plane density change occurs by a transition to a denser commensurate in-plane structure; e.g., the pressure-induced transition in KC_8 from 2 x 2 to $\sqrt{3} \times \sqrt{3}$ [25], which drives the staging transitions via the PV term in the free energy. Li-GIC's show a new process driven by the local elastic distortions described above. Low-T mixtures of stoichiometric LiC_6 and LiC_{12} transform upon heating through the Safran restaging boundary $T_s(x)$ to nonstoichiometric Li_xC_6, an ordered pure stage-1 compound with the same in-plane $\sqrt{3} \times \sqrt{3}$ superlattice as its low-T precursors. The new phases therefore contain vacant sites throughout the otherwise long-range ordered Li layers. At higher temperatures $T_m(x)$ the long-range order disappears in a 3D melting process, the details of which vary with vacancy concentration. The Safran model neglects long-range order and thus does not account directly for the melting transitions.

We have measured staging and melting transitions on many samples in the range $0.6 < x < 1.0$. Below $T_s(x)$ we find well-ordered mixtures of LiC_6 and LiC_{12} with no stage disorder. Transitions to dilute in-plane ordered stage-1 are shown in Fig. 12, where we plot the stage-2 fraction remaining versus T derived from (00L) intensities and structure factors. For $x < 1$ the stage-2 fraction begins disappearing just above room temperature, but the transition begins at higher T if there is more stage-2 present initially. T_s increases but the transition width decreases with decreasing x; the former is expected but the latter is not. Hysteresis of 5-10K implies first-order transitions; the smearing could be due to stage disorder accompanying the transition.

Further insight into the mechanism of these transitions is obtained from the T-dependence of I_c and (001) FWHM, shown in Fig. 13 for x = 0.69. Below 350 K the stage-1 phase is stoichiometric LiC_6; I_c increases linearly with T from thermal expansion. At the 350K onset of the staging transition, stage-1 galleries collapse as Li leaves to feed the stage-2 -> stage-1 conversion. The collapse is drastic over a narrow range coinciding with 80% of the restaging. Conversely, for stage-2 $I_c(T)$ is superlinear above 350 K.

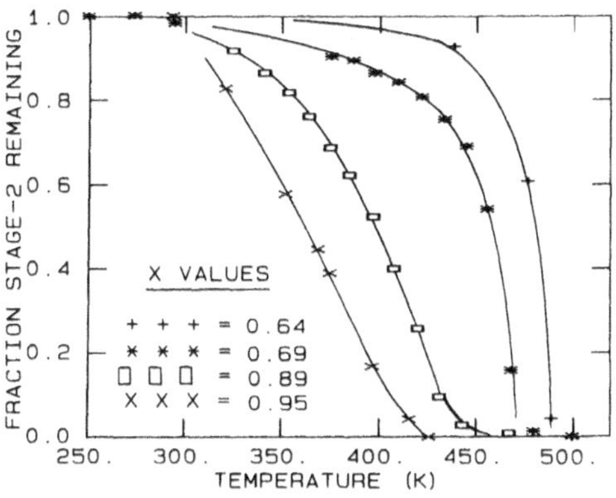

Figure 12. Staging transitions in Li_xC_6 for the indicated values of x. Solid curves are guides to the eye.

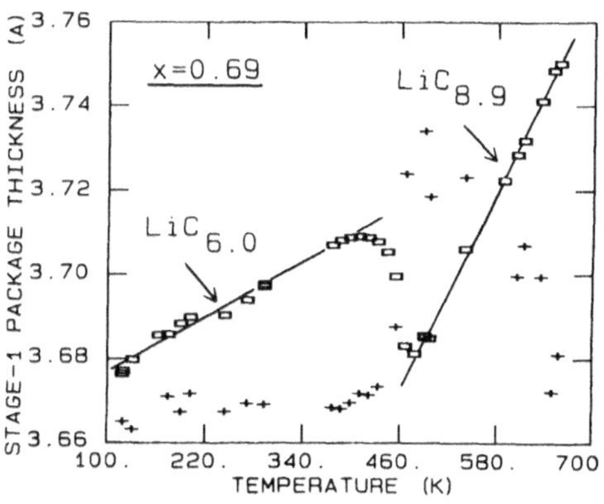

Figure 13. Stage-1 package thickness (□□□) and (001) FWHM (+++) vs. T in a sample with x = 0.69, for which T_s = 468K.

The stage-1 collapse with decreasing [Li] is due to elastic interactions as in Fig. 3, albeit on a smaller scale. The anomalous expansion of stage-2 must be a different effect since the stage-2 occupied galleries cannot get any denser. The apparent thermal expansion coefficient of stage-2 increases with T due to vacancy-enhanced anharmonicity. This is also observed in the second linear region above T_s (Fig. 13), showing a 3.6-fold greater slope for stage-1 $Li_{0.69}C_6$ relative to LiC_6, a direct consequence of the 31% vacancy concentration since both phases are ordered. Extrapolating the first linear region to 700K, the difference in I_c's becomes unresolvable (consistent with Fig. 3), while at 0K I_c = 3.66 and 3.49 A for x = 1.0 and 0.69 respectively. Consequently, $I_c(x)$ measured at 700 K must reflect both elastic interactions and anharmonicity, and we have probably overestimated the healing length L.

The (001) FWHM also shows anomalous behavior across T_s; this is shown as the (+++) in Fig. 13. The (001) is resolution-limited below T_s, but then broadens abruptly at T_s as the stage-1 phase grows. Then the FWHM gradually recovers above T_s, with significant disorder persisting well into the pure stage-1 phase. In principle, the excess width could be a Hendricks-Teller effect associated with random impurities -- completely empty galleries or stage-2 cells. One could also envision a distribution of average gallery thicknesses as the system locks into a specific in-plane density. Less likely is the decrease of the low-T c-axis correlation length by nucleation of new stage-1 cells within the pre-existing crystallites; such a process would be more likely to occur within the stage-2 phase, which in fact shows little broadening as it decays in amplitude with increasing T. The gradual out-of-plane sharpening above T_s mimics the melting transition (shown below), suggesting phase separation of filled and empty sites into dense islands and empty regions [26], in which case the excess (00L) width reflects a distribution of stage-1 package thicknesses, not a random distribution of wrong-cell packages.

Dilute Li-GIC's show sequences of restaging transitions with increasing T, exemplified by "LiC_{62}" (x $\sim$ 0.1) in Fig. 14. The ground state should be dense stage-10 LiC_{10x6} but the atomic-scale kinetics are very slow below 195K so the low-T evolution stops at stage-4. With increasing T, we cross pure stage-3 to stage-1 without observing stage-2 reflections, indicating that the

Figure 14. (OOL) neutron scans for LiC_{62} as a function of increasing T from bottom to top. The progression of pure stages is 4 -> 3 -> 1 with increasing T.

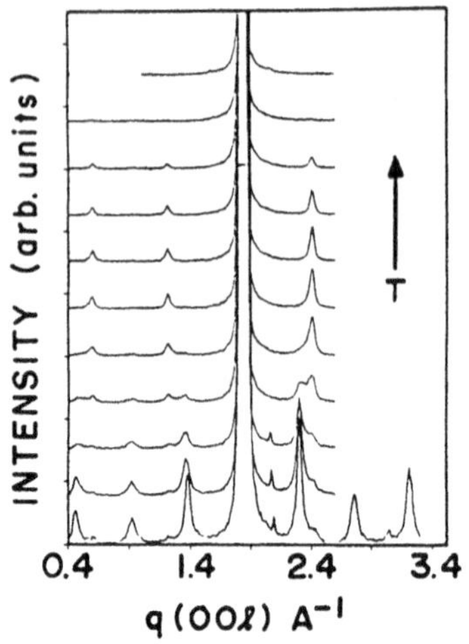

stage-2 phase field does not extend down in concentration to x = 0.1. Note the decreasing strength of superlattice reflections with increasing T. This is due to the gradual decrease in thickness difference between filled and empty galleries as the Li density per unit area decreases (cf. Fig. 3), which in turn reduces the phase shift in the structure factor for these peaks.

We return now to the concentrated stage-1 samples which exhibit novel melting behavior associated with the presence of Li vacancies in the low-T ordered phase. A random vacancy distribution would be expected to depress T_m without changing the nature of the transition. An alternative scenario consists of in-plane phase-separation into maximally-dense islands surrounded by empty canals, in which case the transition will be smeared since the Li's on the shorelines will melt first (analogous to current ideas on dislocation-mediated melting in 2D systems). The x-dependent melting was probed by measuring Li(100) integrated intensites versus T for different samples, Fig. 15. Increasing the vacancy concentration [V] (or decreasing x) decreases T_m as expected, but also smears the transition significantly. The first-order step found for x = 1.00 and 0.99 is gone for [V] > 2% but the continuous melting fits a power law exponent up to [V] = 5%. The x = 0.99 and 1.00 samples have the same T_m but the precursor is notably steeper in the latter. The exponent has been interpreted in terms of universality and tricritical behavior in the 3D-3-state Potts model [27]; since the symmetry class is unaffected by vacancies, this interpretation must be reexamined.

Another curious aspect of the Li melting is shown in Fig. 16, typical of samples with small [V]. The ordered Li remains commensurate as it melts, both lattices showing the small thermal contraction typical of graphite. At 650K, where 90% of the solid remains, both peaks shift together to higher Q -- both in-plane lattices abruptly contract by 0.2%, closely following the melting profile. We point out that the G(100) wave-vector in LiC_6 occurs at a value 1.3% smaller than in pure graphite, an effect which is attributed to the weakening of the in-plane bonds by the screening effect of the enhanced pi-electron density which results from delocalization of the Li(2s) valence electron on the graphite layers [28]. It is therefore tempting to ascribe the in-plane contraction at T_m, Fig. 16, to a small reduction in charge transfer accompanying the melting transition. An alternative explanation is that part of the well-known bond dilation in alkali GIC's is due to symmetry-driven orbital mixing, e.g. hybridization/zone folding effects, which vanish as in-plane order is lost. The abrupt contraction disappears at small x --

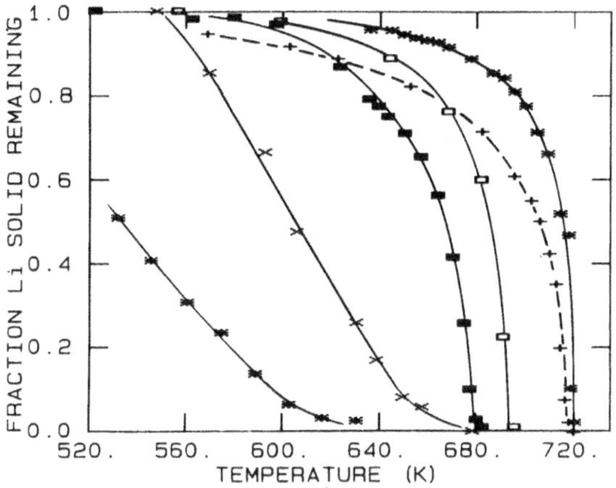

Figure 15. Melting transitions of the Li sublattice in stage-1 Li_xC_6 samples; x values from left to right are 0.64, 0.69, 0.89, 0.95, 0.99 and 1.00.

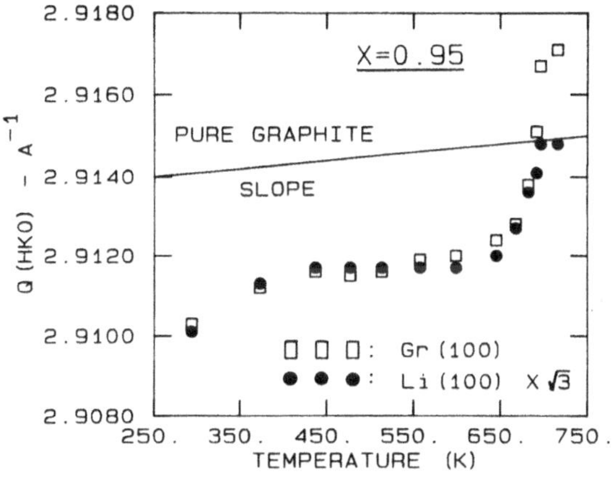

Figure 16. In-plane wave-vectors for the lowest-order reflections of the graphite and Li sublattices versus T, in a stage-1 sample with x = 0.95 for which T_m = 691K.

the graphite in-plane lattice follows the pure graphite slope through T_m but the Li expands slightly as it melts due to the large vacancy concentration; the maximum incommensurability is only 0.3%. In contrast to the (00L)'s around T_s, the in-plane FWHM's remain resolution-limited up to T_m for all x. If dense islands were present, the FWHM puts an upper bound of 350 A on the diameter. The behavior of $I_c(x)$ at fixed T (Fig. 3) suggests isolated vacancies; in-plane phase separation would lead to saturation at lower values of x. On the other hand, the excess (00L) FWHM through T_s could reflect an averaging over islands and voids, rather than Hendricks-Teller mixing of stage-1 cells with stage-2 or graphite.

Finally, in Fig. 17 we collect the $T_s(x)$ and $T_m(x)$ results into a phase diagram for x > 0.5 (see also Safran's chapter). The staging boundary agrees qualitatively with the primitive model and could be used to refine parameters in more sophisticated versions [29]. The melting boundary shows interesting structure; a simple binary mixture of filled and empty sites would exhibit a regular x-dependent liquidus, as determined by atomic-scale binding and activation energies. Here we see instead three distinct regions: at small [V] T_m drops sharply with increasing [V], consistent with single-phase melting characterized by unique energies. The plateau between 0.6 and 0.95 suggests melting of large islands from the boundaries inward, the drop in T_m at 0.6 signalling a minimum island diameter below which filled and empty sites are no longer phase-separated. The vacancy-dependent anharmonicity may also contribute to the unusual form of the melting boundary.

M_xC_8: DENSITY-DEPENDENT INTERACTIONS AND STAGE DISORDER

While the Li-GIC system is attractive for making contact with simple theories, the heavy alkali compounds exhibit a richer structural variety because the energetics are not dominated by commensurability. We give two examples which exemplify the competing interactions and then show how deviations from stoichiometry are accommodated by errors in the sequence of empty and filled galleries rather than by continuously variable in-plane density.

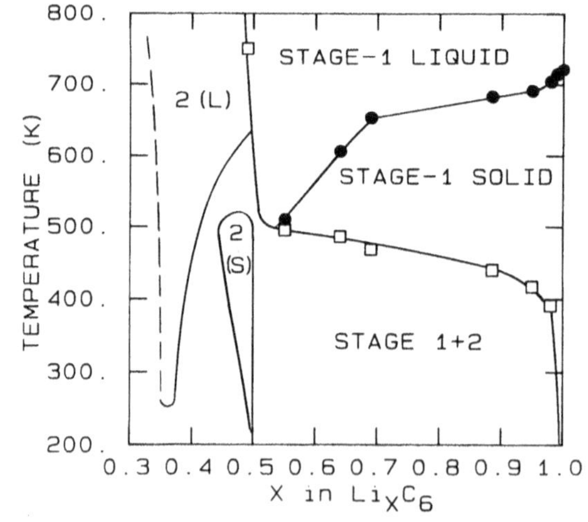

Figure 17. (T,x) phase diagram for Li_xC_6. Filled circles and open squares are from Figs. 15 and 12 respectively. The (+) at the low-T limit of stage-2 liquid is from Fig. 5. Additional data from Ref. 30 has been incorporated. All trasnsitions are first-order; the implied triple points have not been studied in detail.

TABLE I. Positions of lowest in-plane peaks in disordered and solid phases of K-GIC's. IC denotes incommensurate low-T solid.

stage	Q(liquid)	Q(solid)	T_m	structure
1	1.45-1.34	1.47	700K	2 x 2
2	1.23-1.18	1.20	123	IC
3-4	1.18-1.15	1.17-1.13	200	IC
5-7	1.13	1.11	235	$\sqrt{7} \times \sqrt{7}$

If all sites are occupied by K, Rb or Cs (n = N), we get stoichiometric stage-1 compounds KC_8, etc., with 2 x 2 in-plane superlattices and strong 3D correlations (see chapter by Herold). Unlike the Li-GIC's, there is no pure stoichiometric phase corresponding to n = N/2; the only stage-2 phase at STP has $n \approx N/3$, implying that the M ions attempt to maximize their separation in all three dimensions as the overall number density (or chemical potential) is reduced. On the other hand, at high P the heavy alkali compounds resemble Li-GIC's -- one finds stoichiometric "super-dense" stages-1 and -2, KC_{1x6} and KC_{2x6} [35] - because the commensurability energy increases with the P-induced decrease in gallery thickness. The tendency towards isotropic distributions in dilute compounds at high T continues with increasing dilution, summarized in Table I [22]. The dilute limit, stage > 4, has a commensurate $\sqrt{7} \times \sqrt{7}$ layer structure in the low-T ordered phase, while the first-order reflections of the incommensurate structures in stages 4, 3 and 2 indicate a continuously increasing density per unit area. This trend culminates in the large jump between stage-2 and stage-1, as seen in Fig. 18. Note that the same trend is followed in the disordered phases, as revealed by the stage dependence of the liquid peak positions, Table I. Viewed in this light, the incommensurate low-T phases found for stages 2-4 are consequences not merely of the weak but nonzero corrugation potential but also of the in-plane pair interactions which are controlled by n or the chemical potential.

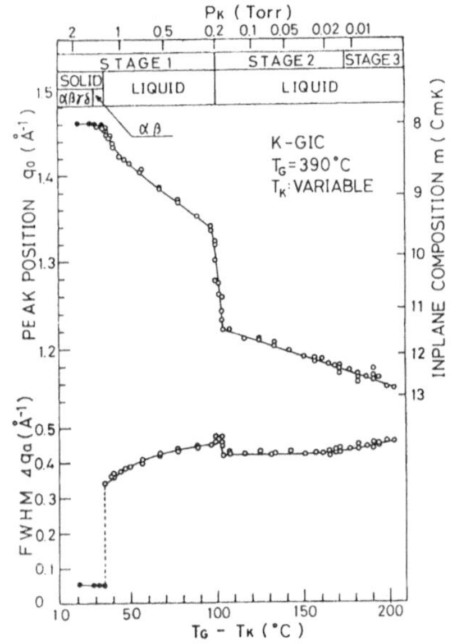

Figure 18. In-plane K sublattice peak position and width in K-GIC, versus K vapor pressure. Open and solid circles denote liquid and solid in-plane structures. Note that the liquid phases exist as pure stages for a wide range of 2D densities. From Ref. 31.

The same phenomenon is revealed in the beautiful in situ experiment of Nishitani et al. shown in Fig. 18 [31]. The sample is maintained at 390 C, in equilibrium with a variable potassium chemical potential defined by the vapor pressure P_K. The in-plane liquid peak shifts continuously to higher Q with increasing P_K; the liquid becomes denser with increasing chemical potential. The stage-3/stage-2 transition has no effect on the in-plane density, while the stage-2/stage-1 transition at higher P_K is accompanied by a stepwise increase. As n -> N the density of filled sites increases in all three directions. Commensurability is unimportant here, and the only vestige of the layer host is the jump in in-plane density as the system exhausts the available empty galleries.

Host layer corrugations favor commensurate in-plane structures at low T or high P. If V_{ij} is long-ranged, well-ordered high stages are also favored. This presents a dilemma: if n does not correspond to a stoichiometric value, how does the system adjust? In stages 2-4, variations in domain wall density compensate for diferences in metal/carbon ratio, while for higher stages the low-T in-plane density is fixed by the commensurate $\sqrt{7} \times \sqrt{7}$ 2D superlattice. If stage ordering is perfect, a set of stoichiometric "true" compounds" is implied, with phase separation if n/N does not equal a stoichiometric value. A finite-ranged V_{ij} and/or entropically-driven finite size effects [32] produce miscibility gaps in the phase diagram [33], and diffraction profiles which do not correspond to perect phases. We close this section by giving several examples.

The circles in Fig. 19 are the measured x-ray (00L) profile for nominal stage-7 K-GIC at 300K [22]. Note the semilog scale. The reflections can all be indexed on a stage-7 sequence but there are three features of the raw data which are inconsistent with a perfect 1D crystal: strong diffuse scattering between Bragg peaks, peak widths which vary nonmonotonically with index, and small but systematic deviations of the peak positions from a line defined by the average d-spacing. These are all signatures of the type of 1D disorder analyzed by Hendricks and Teller [34] and well-known in the structural chemistry of clays. The disorder results from the presence of two or more unit cells with different thicknesses and form factors, whose correlations may vary from perfectly random to phase-deparated (or alternating superlattices). The two pairs of data and solid curves in Fig. 19 represent two different approaches to modelling the disorder in this sample. The lower set is a fit

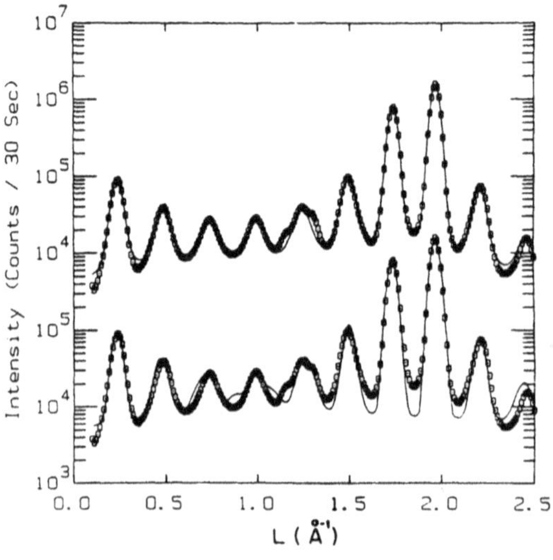

Figure 19. Data (circles) and two different HT model fits (solid curves) for the 300K (00L) x-ray diffraction profile of a high-stage K-GIC with disorder. Top curves: stage-6 + empty cells; bottom: stage-7 + 8.

to a random distribution of 75% stage-7 and 25% stage-8 cells, while the upper set is based on stage-6 plus empty galleries which provides more flexibility without increasing the number of parameters (e.g., a stage-6 cell followed by two empty galleries, viz. stage-8, is permitted). The latter gives a much better representation of the disorder; the sample contains a broad distribution of cells with $n > 6$. The fit parameters define an average stage and standard deviation, which can be tracked versus T or P to study stage disorder accompanying transitions.

Figures 20 and 21 show that this sample undergoes a complicated staging/in-plane transition at 240K. The low-T phase is characterized by a lower average stage and greater average in-plane separation as compared to the high-T phase (see Table I); the ordered $\sqrt{7} \times \sqrt{7}$ solid is less dense than the high-T liquid, compensated by a greater 1D density of filled galleries to maintain constant n. The change in average stage with T is in the opposite sense to that predicted by the Safran model, so this transition is controlled partly by commensurability and also by the entropic contribution to stage disorder [32] associated with the reduced standard deviation at low T, Fig. 20.

Similar analyses of hydrostatic pressure effects on dilute K-GIC's at 300K reveal a different process [36]. Figure 22 shows cell distributions derived from stage-6 + empty gallery fits to P-dependent (00L) neutron diffraction profiles of KC_{72}. At STP the distribution peaks sharply at stage-7 while at 9.1 kbar the distribution is broader with a long tail extending to very high cell index. Figure 23 shows the P-dependent evolution of average stage, with standard deviations given in parentheses at each data point. The abrupt jump at 7.5 kbar signals the completion of in-plane lock-in to the 2 x 2 superlattice, analogous to $\sqrt{7} \times \sqrt{7}$ at low T. The increase in average stage with P is consistent with the fact that the in-plane density increases with P. Contributions to the increasing stage disorder with P are 1) pinning of ordered islands at D/H boundaries, 2) finite range of V_{ij} and 3) finite size entropy [32]. At 8 kbar the disorder is so severe, average stage-13 with $\sigma = 5.5$, that the "superlattice" (00L)'s are essentially obliterated. Stage disorder also plays a role in the kinetics of phase transitions, as discussed in Safran's chapter.

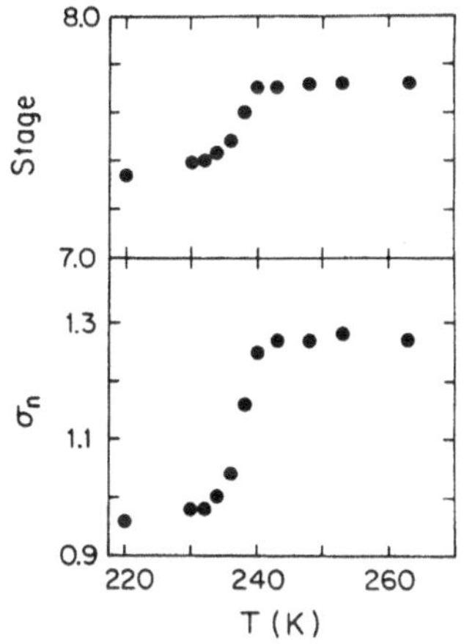

Figure 20. Mean stage (top) and standard deviation (bottom) vs. T for the sample in Fig. 19.

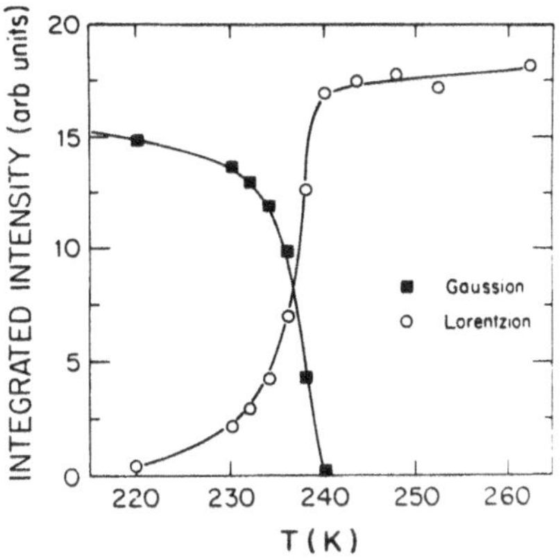

Figure 21. Integrated Gaussian (solid) and Lorentzian (fluid) fits to K(1,0) peaks vs. T for the sample in Fig. 19.

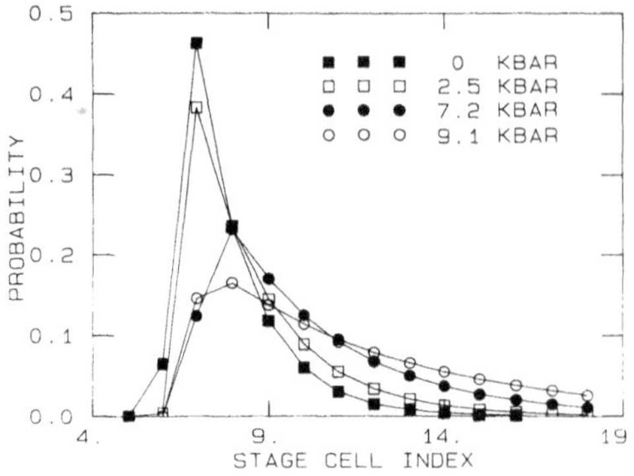

Figure 22. Cell distributions from stage-n + empty gallery HT fits to neutron (00L) profiles of a dilute K-GIC at different pressures.

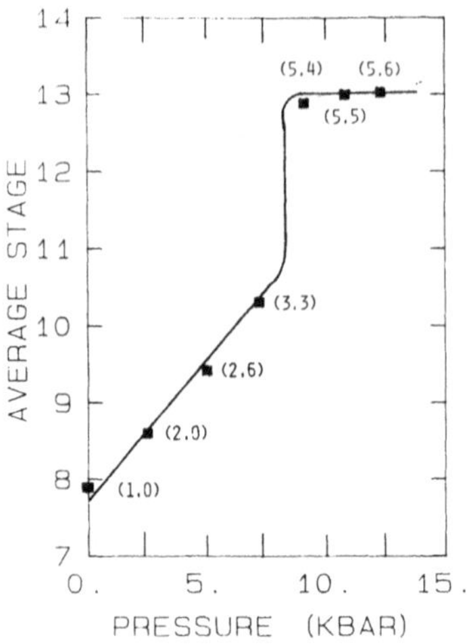

Figure 23. Average stage (■■■) and standard deviation (S.D.) versus P, deduced from Fig. 22 fit parameters.

ANALOGIES WITH DOPED POLYMERS

We have seen how a simple mean-field model with two energy parameters accounts for the principal structural features of layer intercalates, in particular the dramatic phenomena of T-, P- and n-dependent staging. We close this chapter with a speculation on what a similar model might yield for doped polymers by replacing the 1D stack of 2D galleries with a 2D array of 1D tunnels.

The dopants occupy 1D tunnels defined by the surrounding polymer backbone chains. The analog to high stages would be 2D superlattice planes normal to the polymer chains, defined by the ordered array of occupied tunnels. There will be a pairwise attractive energy between two dopants in the same tunnel, analogous to U_o, but there are only two near-neighbors rather than six in a 2D triangular lattice. The tendency to fill all sites in a tunnel at low-T (or high-P) will be relatively weaker so more tunnels will be occupied, i.e. "low stages" are favored. Conversely, the inter-tunnel repulsion analogous to V_{ij} is more important than in the layer case because there are more terms -- each tunnel has equivalent neighbors in two dimensions -- so on this basis, "high stages" will be relatively <u>more</u> stable. This all suggests an even more delicate balance of competing interactions, thus a richer phase diagram relative to layer intercalates. The dilute high-T limit will be similar in both cases, namely a nearly isotropic dopant distribution in which every tunnel contains some guest atoms. We might call this "dilute stage 1 x 1" by analogy to the Safran model. At intermediate temperatures the dimensionality of the configurational entropy becomes important. An actual Safran-style calculation would be extremely interesting.

Doped polymer systems are usually much more disordered than layer intercalates, so structural features associated with 2D staging will be harder to identify. Evidence to date suggests that all tunnels are occupied at STP, meaning that the analog to the Safran boundary is above 300K. This is consistent with our earlier argument that the elastic contribution to the pair energy, the parameter which sets the scale for staging transitions, will probably be large in linear chain polymers due to the ease of transverse deformation around an impurity. There is some evidence for 2D staging [37] and a report of a 1D incommensurate-commensurate transition [38] versus doping concentration. Thus there may be some value in further exploring the analogies between layer intercalates and doped polymers.

ACKNOWLEDGMENTS

While the speculations are entirely my responsibility, I acknowledge fruitful discussions with Paul Heiney, Eugene Mele and Samuel Safran, and the collaboration of students past and present: Carl Fuerst, Hyun-Joong Kim and Victoria Cajipe. This research is currently supported by the Department of Energy, Grant No. DEFG02-86ER45254.

REFERENCES

1. Synthetic Metals 12 (1985): Proceedings of International Symposium on Graphite Intercalation Compounds, Tsukuba Japan, May 1985.
2. Proceedings International Symposium on Graphite Intercalation Compounds, Jerusalem Israel, May 1987; to be published in Synthetic Metals.
3. Intercalation in Layered Materials, edited by M. S. Dresselhaus (Plenum NY 1986); NATO Series Vol. 148.
4. S. A. Safran, Phys. Rev. Lett. 44, 937 (1980).
5. P. Bak and R. Bruinsma, Phys. Rev. Lett. 49, 249 (1982).
6. J. E. Fischer and H. J. Kim, Phys. Rev. B35, 3295 (1987).
7. J. R. Dahn, D. C. Dahn and R. R. Haering, Solid State Commun. 42, 179 (1982).
8. S. A. Safran and D. R. Hamann, Phys. Rev. Lett. 42, 1410 (1979).
9. J. E. Fischer, C. D. Fuerst and K. C. Woo, Synth. Met. 7, 1 (1983).
10. D. P. DiVincenzo and E. J. Mele, Phys. Rev. B32, 2538 (1985).
11. H. Zabel, A. Magerl, A. J. Dianoux and J. J. Rush, Phys. Rev. Lett. 50, 2094 (1983).
12. J. E. Fischer, Mat. Sci. Eng. 31, 211 (1977).
13. N. Wada, S. Minomura and J. Pluth, p. 51 in Ref. 12.
14. M. Suzuki, Phys. Rev. B33, 1386 (1986).
15. H. Zabel, M. Suzuki, D. A. Neumann, S. E. Hardcastle, A. Magerl and W. A. Kamitakahara, p. 105 in Ref. 12.
16. D. P. DiVincenzo, C. D. Fuerst and J. E. Fischer, Phys. Rev. B29, 1115 (1984).
17. R, Clarke, N. Wada and S. A. Solin, Phys. Rev. Lett. 44, 1616 (1980).
18. H. J. Kim, J. E. Fischer, D. B. McWhan and J. D. Axe, Phys. Rev. B33, 1329 (1986).
19. J. B. Hastings, W. D. Ellenson and J. E. Fischer, Phys. Rev. Lett. 42, 1552 (1979).
20. B. Sundqvist and J. E. Fischer, Phys. Rev. B34, 3532 (1986).
21. V. B. Cajipe and J. E. Fischer, Phys. Rev. B35, 4854 (1987).
22. M. E. Huster, P. A. Heiney, V. B. Cajipe and J. E. Fischer, Phys. Rev. B35, 3311 (1987).
23. R. Clarke, N. Caswell and S. A. Solin, Phys. Rev. Lett. 42, 61 (1979).
24. J. E. Fischer and H. J. Kim, p. 137 in Ref. 1.
25. J. M. Bloch, H. Katz, D. Moses, V. B. Cajipe and J. E. Fischer, Phys. Rev. B31, 6785 (1985).
26. G. Kirczenow, p. 143 in Ref. 12.
27. D. S. Robinson and M. B. Salamon, Phys. Rev. Lett. 48, 156 (1982).
28. C. T. Chan, W. A. Kamitakahara, K. M. Ho and P. C. Eklund, Phys. Rev. Lett. 58, 1528 (1987).
29. S. E. Millman and G. Kirczenow, Phys. Rev. B26, 2310 (1982).

30. K. C. Woo, H. Mertwoy, J. E. Fischer, W. A. Kamitakahara and D. S. Robinson, Phys. Rev. B27, 7831 (1983).
31. R. Nishitani, Y. Uno, H. Suematsu, Y. Fujii and T. Matsushita, Phys. Rev. Lett. 52, 1504 (1984).
32. G. Kirczenow, Phys. Rev. Lett. 52, 437 (1984).
33. M. E. Misenheimer and H. Zabel, Phys. Rev. Lett. 54, 2521 (1985).
34. S. Hendricks and E. Teller, J. Chem. Phys. 10, 147 (1942).
35. C. D. Fuerst, J. E. Fischer, J. D. Axe, J. B. Hastings and D. B. McWhan, Phys. Rev. Lett. 50, 357 (1983).
36. J. E. Fischer and H. J. Kim, Phys. Rev. B35, 6826 (1987).
37. R. H. Baughman, L. W. Shacklette, N. S. Murthy, G. G. Miller and R. L. Elsenbaumer, Mol. Cryst. Liq. Cryst. 118, 253 (1985).
38. M. Winokur and A. J. Heeger, Phys. Rev. Lett. (in press).

MAGNETIC PHENOMENA IN LAYERED AND INTERCALATED COMPOUNDS

Denis G. Rancourt

Ottawa Carleton Institute for Physics
University of Ottawa
Ottawa, Ontario, Canada

INTRODUCTION

The intrinsic magnetism of two-dimensional (2D) lattices is of fundamental interest and has been the subject of intense theoretical studies. These studies have motivated much of the experimental work on layered and intercalated compounds. The idea of an "effective dimensionality" which can be varied from three to two as the inter-layer or inter-intercalate separation is increased has been an additional incentive. The latter notion is mainly the product of intuitive curiosity since theory has been of little help in guiding our expectation concerning such intermediate dimensionalities. That is, surprizingly little is known about local moments on quasi-2D lattices [which are essentially three-dimensional (3D) lattices with symmetries lower than cubic] compared to exactly 2D lattices. Nonetheless, one might expect to observe an h-T phase diagram such as that represented in Fig. 1 where h is the inter-layer or inter-intercalate separation and T is the absolute temperature. For example, h could be the c-axis of a tetragonal system in which the spins are on the corners of the square end faces. This hypothetical phase diagram

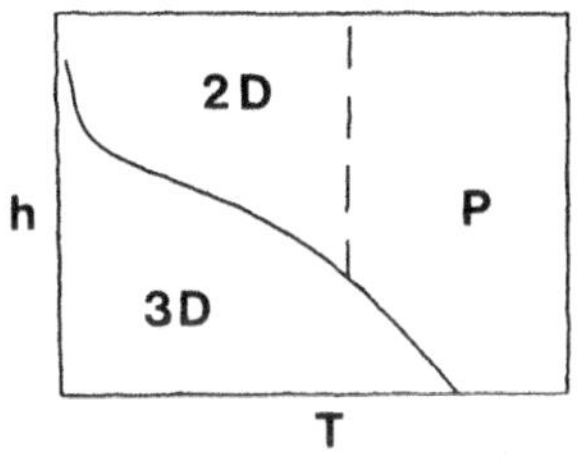

Fig. 1. Hypothetical h-T phase diagram where h is the inter-layer distance.

exhibits three regions: a high temperature paramagnetic (P) phase, a phase where some form of 2D order is stabilized and a 3D ordered phase. Particular examples of the meaning and somewhat limited usefulness of such a phase diagram are given below. It is of more consequence, however, to admit that the relation between structure and magnetism is of a far greater variety (and subtlety) than that which can be expressed by the single parameter h. This view is stressed and illustrated by many examples throughout the present paper.

It is appropriate at this point to describe qualitatively the present state of knowledge concerning magnetic model systems. For a given simple quantum mechanical model representing local moments on an infinite lattice, the relative ease with which an analytic description can be obtained depends primarily on five dimensionless quantities: i) the space dimension, D, of the lattice where D = 0 refers to the case of a single lattice point, ii) the number, N_{edf}, of "exchange degrees of freedom" where $N_{edf} = 1,2,3$ correspond, respectively, to Ising, XY and Heisenberg spins and $N_{edf} = 0$ is taken to imply the absence of exchange forces, iii) the reciprocal, N_{pg}^{-1}, of the number of elements in the point group of the lattice, iv) the number, N_{nns}, of near neighbour shells to which the magnetic interaction extends, and v) the spin, S, of the local moment. Each model system therefore corresponds to one point in the five-space of D, N_{edf}, N_{pg}^{-1}, N_{nns}, and S. The degree of completeness in our knowledge of such systems is, in a qualitative sense, greatest for those systems which are nearest to the origin. Conversely, wealth and intricacy of behaviour generally increase as one moves away from the origin. One of the furthest points for which an analytic solution is known corresponds to Ising spins on a square lattice with nearest neighbour only interactions[1]. The XY spin chain with nearest neighbour only interactions and for the case where the probing magnetic field is perpendicular to the XY plane has also been solved[2]. The infinite Heisenberg spin chain has not been solved - that is, an analytic expression for its partition function has not been obtained. Further advancement has been achieved by the use of Monte Carlo methods and by considering finite systems which can be solved exactly. 3D lattices have not been solved, although the ground states can often be predicted and the critical exponents can be obtained by renormalization. Also, various temperature expansion schemes lead to reliable results which hold over large temperature intervals. On the other hand, while it does not give exact results*, mean field theory (mft) should not be underestimated. In the Bethe-Peierls approach, finite clusters are solved exactly and are made to interact with a mean field. This method, for example, correctly predicts[3] that 2D lattices of Heisenberg spins should not possess spontaneous magnetization at $T > 0$.

Fig. 2 shows magnetic specific heats, C_m, corresponding to various model systems and calculated by various techniques, as functions of T/Θ where Θ is the mft value of the Curie-Weiss temperature. All curves are for nearest neighbour only interactions and $S = \frac{1}{2}$. Fig. 2(a) is for Ising spins whereas 2(b) is for Heisenberg exchange. Curves labelled 1 correspond to one-dimensional (1D) antiferromagnetic (AF) lattices whereas all other curves pertain to ferromagnetic (F) interactions. The curves (not shown) for F interactions and 1D lattices diverge to infinity as $T \to 0$. Curves labelled 2 correspond to square lattices and those labelled 3 are for simple cubic lattices. The curves labelled "mft" show the mean field theory result

*The common statement that "mean field theory is exact in four dimensions" refers to correct critical exponents being obtained in Landau theory when D = 4. This reflects the fact that critical fluctuations decrease in importance (and can more easily be ignored) as the dimensionality is increased. Landau theory includes a magnetization gradient term in the local free energy whereas mft does not.

which is the same for both types of exchange considered and all lattices (whether 1D, 2D, or 3D). R is the universal gas constant.

In conclusion, the space of simple model magnetic systems is gradually being "filled" and the observed behaviours of an increasing number of real systems can be compared to calculated behaviours such as those shown in Fig. 2. It should be emphasized however that many characteristics of real systems are not taken into account by the simple model systems described above. For example, the actual functional dependence, $J(r_{ij}) \equiv J_{ij}$, of the magnetic exchange strength versus inter-spin separation might be such that N_{nns} will be large and J_{ij} will alternate in sign for successive shells thereby leading to complicated frustration effects[5]. Spin-orbit coupling leads to crystal field effects which in turn give rise to effective exchange anisotropies which, for two particular cases, are modelled by Ising and XY exchange couplings. In a correct treatment, the exchange proper is found to be isotropic (i.e. of the Heisenberg type) and the crystal field terms are explicitly included in the spin Hamiltonian. Furthermore, in many magnetic materials, the spin is not a whole or half integer (due to itinerate electron effects) and S itself has a temperature (and structure and local environment) dependence. Many materials contain several kinds of magnetic ions which are on distinct sublattices, or are distributed at random, or participate in compositional modulations. Real crystals contain defects and are of finite size - in antiferromagnets, in particular, vacancies and surface effects cause local residual magnetizations. Many intrinsically interesting magnetic materials are not crystalline, even to a first approximation, for example: amorphous alloys, very small particles, magnetic liquids, colloidal suspensions, various fractal structures and aggregates, etc. **Dipole-dipole** forces are always present and complicate things immensely. Resulting domain structures cannot usually be predicted a priori. These forces are known to be of significant relevance in understanding the so-called metamagnets (which are layered compounds with F in-layer exchange couplings) and are possibly the only magnetic forces[6] between the intercalate layers of many graphite intercalation compounds (GIC's). Finally, in real systems, the spin degrees of freedom are coupled to other electronic and to the lattice degrees of freedom. Spin-lattice coupling can lead to the well known Spin-Peierls dimerization in infinite spin chains[7]. Such coupling has recently been shown to play a governing role in the spin dynamics of Ising chains[8] and is believed to constitute a mechanism which explains the existence of magnetic short range order (MSRO) in a broad temperature interval above the magnetic ordering temperatures of many materials[9].

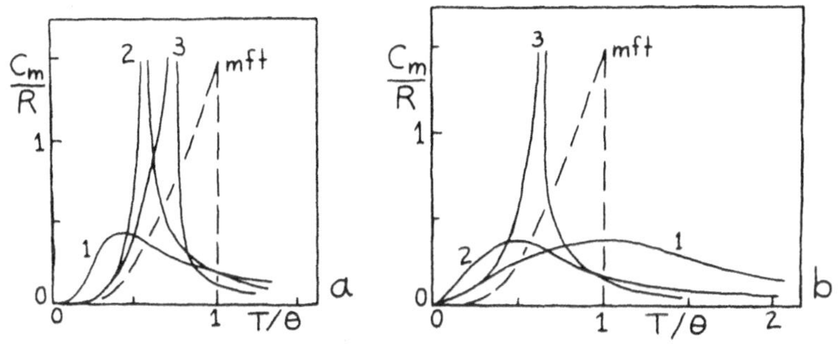

Fig. 2. Theoretical magnetic specific heats for various simple magnetic model systems - after de Jongh and Miedema[4].

Let us now briefly review the most common types of measurements in solid state magnetism. As can be surmised from Fig. 2, much information concerning the existence and position of a critical temperature, the extent of spin-spin correlations, the nature of the ordered state, etc., is contained in the magnetic contribution to the specific heat. Essentially the same information is contained in the initial dc susceptibility, χ_{dc}. For example, in antiferromagnets the relation

$$C_m(T) \simeq f(T) \cdot \partial[T \cdot \chi_{dc}^{\parallel}(T)]/\partial T, \tag{1}$$

where f(T) is a slowly varying function of T and $\chi_{dc}^{\parallel}$ is the parallel initial susceptibility, often holds[10]. Another accessible bulk magnetic property is the equilibrium magnetization, M(H,T), induced in the sample by a constant field H and at a temperature T. If very slow spin relaxation mechanisms are present, attaining equilibrium experimentally will be difficult. In such circumstances, the time dependence, M(t), of the magnetization can be measured, leading to information about the relaxation mechanisms and the magnetic path (or bond) morphology. If, in addition, the trajectory to equilibrium is plagued by metastable states, these can be explored by trying various other trajectories. Such time, memory and metastability effects appear pervasively in the measured magnetizations of transition metal chloride-GIC's although they have only recently begun to be explored systematically[11]. Other bulk-type measurements which give information on the magnetism are, for example, magneto-optics and magneto-resistivity. Magnetic ac susceptibility (χ_{ac}) measurements yield complete information concerning both spin response and intrinsic spin dynamics over a vast frequency range. The χ_{ac} technique should be considered a strectroscopic technique [along with neutrons, nuclear magnetic resonance (NMR), electron spin resonance (ESR), and Mössbauer effect spectroscopy (MES)] and its relation to MES will be discussed separately below. NMR and MES are based on hyperfine interactions between probe nuclei and their electronic environments. They are true microscopic techniques in that contributions from all the chemical/crystallographic environments are seen. Neutron diffraction (ND) gives magnetic correlation lengths and ordered state spin structures. Inelastic neutron scattering (INS) probes high frequency ($\sim 10^{12}$Hz) "spin wave" excitations. The ability of MES to bridge the frequency gap between INS and χ_{ac} measurements (especially in ordered magnetic materials) is now being recognized[12]. We stress that, in all the measurement types mentioned above, validity in interpreting the data is highly dependent on the choice of microscopic models for both the structure and the magnetism. Modelling of the structure must allow for defects, non-homogeneities, finite crystallographic correlation lengths, etc. - particularly in intercalation compounds. The collective magnetic behaviour is, in turn, most sensitive to this structure and advancement in understanding the magnetism will usually entail considerable refinement of the structure.

SOME STRUCTURAL GENERALITIES

Superlattices, Layered Materials, and Intercalation Compounds

In this section we define and compare three classes of planar systems: superlattices, layered materials, and intercalation compounds. Superlattices are built layer by layer using epitaxial depositing techniques such as molecular beam epitaxy. They are structurally metastable in that they do not represent equilibrium structural phases of alloys with the same overall compositions. Those which are interesting magnetically have so far been constituted of transition metals where the individual layers are between ~3 Å and ~100 Å in thickness. Most are 3D since either the magnetic interlayer interactions are strong or the layers are much thicker than the lattice spacings - in which case the "superlattice" is essentially a two phase solid mixture. When metals with different intrinsic types of magnetism are combined, complicated magnetic phase diagrams and novel collective behaviours are predicted[13]. Superlattices (more than layered materials and intercalation compounds) belong to a large class of synthetic and natural 3D materials known as modulated structures[14,15].

Layered materials are substances in true equilibrium structural states. Examples are $FeC\ell_2$, $NiC\ell_2$, and $CoC\ell_2$ which are metamagnets[16] and K_2MF_4 (M = Ni, Mn, Cu) and $BaM_2(XO_4)_2$ (M = Co, Ni; X = P, As) which are considered to be quasi-ideal 2D magnets[4,17]. In these materials, the crystallography is such that (i) intra-layer inter-spin distances are shorter than inter-layer distances and/or (ii) intra-layer magnetic interactions are much stronger than inter-layer magnetic forces, and/or (iii) magnetic frustration effectively reduces the inter-layer magnetic interactions. Whereas superlattices and intercalation compounds contain specific types of defects associated with their synthesis (e.g., commensurability faults, inter-boundary diffusion, step faults, stage mixing, filling non-uniformity, chemical in-plane charge transfer non-homogeneity, etc.) layered compounds contain only the more usual crystalline defects (e.g., dislocations, vacancies, etc.).

Finally, intercalation compounds distinguish themselves in that they are made from two parts: a host which itself is a layered compound prior to intercalation and an intercalate which is induced to chemically insert itself between the layers of the host. The resulting intercalate layers are often found, on a local scale, to be crystallographically essentially equivalent to layers in a corresponding pristine material. They can also form "amorphous" or "liquid" layers and novel intercalate structures can be stabilized by the host-intercalate interaction. It is difficult to discuss the structural stability of intercalation compounds or even to ascertain theoretically whether maximum equilibrium filling is achieved. This arises from the fact that intercalation products are usually reaction path and host sample (shape, size, defect density, etc.) dependent. That is, resulting structures (stage and intercalate domain distributions, etc.) depend more on the intercalation dynamics than on free energy considerations because of the non-permeability of the host layers.

Daumas-Hérold Domains versus Elementary Islands

It now seems clear[6,18-21] that intercalation of transition metal di-chlorides into graphite involves the formation of intercalate islands, that there is roughly one excess chlorine atom per metal atom on the perifery of these islands and that this extra chlorine is predominantly responsible for the charge transfer from the host graphite. The measured chlorine excess is corroborated by the existence of a threshold pressure of chlorine gas for intercalation to occur and by a decrease in chlorine gas partial pressure during intercalation and is consistent with the charge transfer deduced from the C-C bond length[20,21]. Most measurements lead to island diameters between ~100 Å and ~200 Å which is much smaller than the expected Daumas-Hérold domains, based on direct observations in $SbC\ell_5$-GIC's[22,23]. We propose that many intercalate islands (typically $\sim 10^2$ or more) aggregate to form large intercalate domains. This is analogous to a mechanism based on "elementary islands" and proposed by Kirczenow[24] except that we expect the island shapes and sizes to be predominantly determined by the chlorine excess[25] rather than by host lattice elastic energy. Also, we expect that in the mono-atomic donor-type intercalates, the intercalation might correspond more to dendritic "flow" rather than diffusion of islands. Given (i) that Kirczenow's model does not lead to high stages ($n > 2$) or to high stage purities as are commonly observed in donor-type GIC's and (ii) the apparent impossibility of obtaining pure stages and the relative difficulty in obtaining high stages in MCl_2-GIC's, we feel that Kirczenow's model is particularly applicable to MCl_2-GIC's but not to donor-type compounds as claimed. Based on a conservation of the chlorine excess and the possibility of inter-island bridging, we have calculated[25] an optimum intra-intercalate-domain filling coefficient of $\alpha_c = 2/3$; the measured value is ~0.7[20,21].

Intercalate Structure and Charge Transfer: Why $FeCl_2$ has not been Intercalated into Graphite

Although intercalation of transition metal di-chlorides into graphite is understood in terms of the chlorine excess discussed in the previous section, the same is not true for $FeCl_3$-GIC. While a chlorine excess is measured, it is typically ten times smaller per metal atom than in the di-chlorides[26,27]. In further contrast to the di-chlorides, the in-plane filling coefficients are very high; typically between 0.8 and 0.92[28]. Also, $FeCl_3$ intercalates at a lower temperature than the di-chlorides and $FeCl_3$-graphite was the first stable GIC to be synthesized[29]. These facts suggest that the intercalation and charge transfer mechanisms in $FeCl_3$-GIC are different from those which operate in MCl_2-GIC's with M = Ni, Co, Mn, Cu, and others. Indeed, there is conclusive evidence[26,30-32] that, although the in-plane stoichiometry is very nearly $FeCl_3$ in $FeCl_3$-GIC, about 20% (±2%) of the iron is in a 2+ rather than 3+ ionization state. The corresponding charge transfer of ~0.2 electrons per iron atom seems to be independent of stage, filling coefficient and biintercalation partner[32] which suggests that it is **primarily a consequence of the** (particular) **intercalation mechanism.** In contrast, any such "disproportionation" occuring in MCl_2-GIC's will likely correspond to metal atoms having **higher** degrees of ionization than that inferred from the pristine stoichiometry, due to their proximity to local chlorine excesses.

We propose that, in $FeCl_3$-GIC, ~20% of the iron atoms occupy sites whose local structures are such that the 2+ state is stabilized and that these sites arise from intercalate layer defects which are generated during intercalation. We further propose that the preferred defect for creating such sites might be the vacancy/interstitial pair represented in Fig. 3(a) where the interstitial atom sees a local structure similar to what an iron atom sees in $FeCl_2$. This defect is plausible given the crystallographic

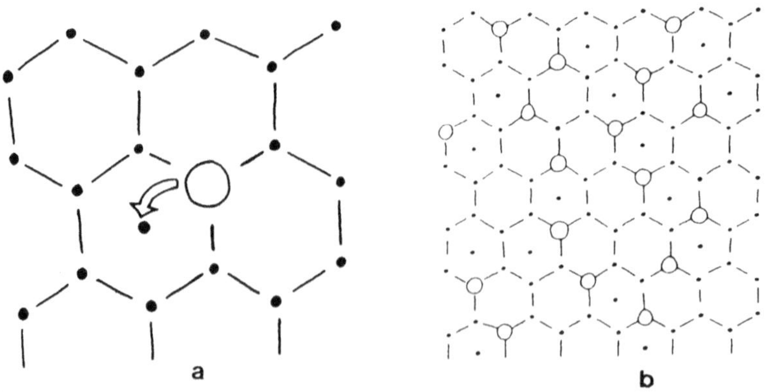

Fig. 3. (a) Proposed defect which acts as a vehicle for charge transfer in intercalated $FeCl_3$ layers. (b) Example of an iron ion distribution resulting from an ~ 20% population of vacancy/interstitial pairs such as the one shown in "a".

relation between $FeCl_2$ and $FeCl_3$ layers: a pristine $FeCl_3$ layer is essentially an ordered vacancy structure of the pristine $FeCl_2$ layer. We take these intrinsic vacancies to be less ordered in an intercalated layer of $FeCl_3$ such that, with an ~20% concentration of defects, the in-layer iron distribution might be as represented in Fig. 3(b).

In conclusion, we propose that the 2+/3+ ionic instability of iron with respect to local structure gives rise to a low activation energy defect-driven intercalation which constitutes an effective charge transfer mechanism in $FeCl_3$-GIC. This mechanism cannot occur with $FeCl_2$ which would depend on a chlorine excess for its charge transfer. The latter being energetically more difficult (as indicated by the relatively high intercalation temperatures for MCl_2-GIC's), $FeCl_3$-GIC is always formed instead of $FeCl_2$-GIC under the reaction conditions which have been tried so far. Note that the existence of structurally prefered Fe^{2+} sites does not preclude the Fe^{2+}/Fe^{3+} electron hopping which is observed in the Mössbauer spectra of $FeCl_3$-GIC[30]. Different electron dwell times for sites with different local structures should however be considered, although this is not expected to significantly affect the analysis at the highest temperatures

RELATION BETWEEN STRUCTURE AND MAGNETISM

This is the main section of the present paper, in which we present theoretical and experimental examples ranging from the possibility of observing a Kosterlitz-Thouless phase to the connection between reentrant spin glass behaviour and the magnetism of $CoCl_2$-GIC. It is hoped that the non-specialist reader will thereby acquire some feeling for both the main features and some of the intricacies which are involved.

Long Range Magnetic Order and Lattice Dimensionality

Excellent reviews of the most common (Ising[33,34], XY[35] and Heisenberg[36]) simple quantum models exist. A thorough review of 1D model systems has been given by Steiner et al.[37] and many specialized monographs, such as ref. 7, also exist. Within the context of such models, one can ask: When will some form of long range order be stabilized at a temperature $T > 0$? Historically, the answer seemed to be that only 3D lattices could support magnetic long range order at non-zero temperatures. Much interest has recently been generated by the suggestion, from Kosterlitz and Thouless[38] and based on a classical analogue of the XY model, that a novel ordered state will occur in 2D at $T > 0$.

Block first argued[39] from spin wave theory that ferromagnetism could not exist at $T > 0$ in 1D and 2D magnets. More recently, Mermin and Wagner[40] rigorously proved the absence of ferromagnetism or antiferromagnetism in Heisenberg 1D and 2D models. Various authors have argued[41], however, that a small amount of exchange anisotropy, as is inevitable in real system, will permit the usual type of long range order to be established in 2D. That the presence of a phase transition in the Heisenberg 2D model is clearly indicated, from high-temperature expansions[42], suggests that novel forms of magnetic long range order, corresponding to zero magnetization, might occur in this and the XY model at $T > 0$. Kosterlitz and Thouless[38] have suggested that such an ordered state will correspond to bound spin vortex/antivortex pairs in the 2D plane-rotator model (i.e. classical XY model) and that a similar ordered state will not exist in a 2D classical Heisenberg model. These conclusions are supported by Monte Carlo calculations[43-47] on both quantum and classical XY models. Of relevance to real systems, however, in-plane symmetry-breaking crystalline fields have been shown[48] to significantly affect the critical behaviour of plane-rotator models.

We end this subsection by noting that, in most real situations, the above discussion in terms of overly simple classical or quantum models[8] will probably not be of primary relevance. For example, we have argued[8] that long range order, corresponding to regular arrays of solitons in 1D Ising chains, will occur due to strong inter-soliton repulsion arising from spin-lattice coupling.

Frustration as a Means of Reducing Dimensionality: $CuC\ell_2$ and $CuC\ell_2$-GIC.

Frustration can occur with particular spin arrangements and refers to a competition between different exchange bonds. The simplest example is that of a trimer of Ising spins on an equilateral triangle with antiferromagnetic interactions. Even at the lowest temperatures, this system must contain at least one kink (or soliton) which can be in one of three energetically equivalent positions. The ground state, therefore, is three-fold degenerate. The two spins which bracket the kink are said to be "frustrated". The ground state of a larger system containing N frustrated Ising spins will be $\sim 2^N$-fold degenerate; corresponding to a topological spin glass*. Another classic example is that of an fcc lattice with near neighbour only antiferromagnetic bonds. It is well known that this system cannot order in the usual two-sublattice antiferromagnetism and that any ordering is suppressed far below $T \sim J_{ij}/k_B$, where k_B is Boltzman's constant. $K_2PbCu(NO_2)_6$, for example, is such a system[4] where spin-1/2 Cu^{2+} ions lie on an fcc lattice. Whereas the known exchange strength corresponds to a mft Curie-Weiss temperature of ~10 K, a transition, which is presumably due to next near neighbour interactions, is not seen until T ~ 0.5 K.

In this section, we explore the possibility that frustration can lead to an effective dimensionality which is lower than the dimensionality of the original lattice. Such a situation has been suggested[50], on the bases of Monte Carlo results, for a particular 3D lattice which shows 1D behaviour at low temperatures. We have argued[19], as follows, that the 1D behaviour which is observed in the layered compound $CuC\ell_2$ and in $CuC\ell_2$-GIC also arises from frustration.

The layers in $MC\ell_2$ (M = transition metal) are essentially identical crystallographically to those which intercalate into graphite to form $MC\ell_2$-GIC's. The in-plane crystallographic structures are always based on a six-edge-sharing side by side stacking of more or less distorted chlorine octahedra, such that the resulting metal ion lattice is closely triangular. This is shown in Fig. 4 where the edges of top triangular faces of octahedra are drawn with full lines, the edges of bottom triangular faces of octahedra are drawn with dashed lines and the shared edges are drawn with dotted lines. The metal ions are at the centers of the octahedra and therefore find themselves at a height of half the distance between the top and bottom octahedral faces. The latter top to bottom distance is typically 5.8 Å and is also the inter-metal-layer distance (h) in the pristine materials. In $CuC\ell_2$, the octahedra are distorted such that the metal-layer lattice is constructed with isosceles rather than equilateral triangles. The resulting in-plane structure corresponds to parallel chains of Cu^{2+} ions and an intra-chain near neighbour distance d_1 = 3.30 Å. The next near neighbour distance, d_2 = 3.80 Å, is to intra-plane ions in adjacent chains. The third near neighbour distance is d_3 = 5.98 Å and is measured between adjacent metal layers. In $CuC\ell_2$-GIC's, d_1 and d_2 are the same as in $CuC\ell_2$, to within ~1%[51], whereas distances corresponding to d_3 are greatly increased. The

*We distinguish between "topological spin glasses" which arise from frustration on regular lattices and "cluster spin glasses" which arise either from mictomagnetism (including any form of chemical clustering) or from percolation cluster considerations[49].

metal ions are linked by superexchange bonds via chlorine ions. Admitting that the resulting first and second near neighbour exchange parameters (J_1 and J_2) are comparable in size ($|J_1| \gtrsim |J_2|$) and antiferromagnetic, the crucial point is that consecutive chains within a plane are translated by $d_1/2$ with respect to each other and are consequently frustrated. The result is a very effective magnetic isolation of intra-plane chains.

The powdered initial susceptibility of $CuC\ell_2$ is modelled quite well by a spin-1/2 Heisenberg chain with an antiferromagnetic coupling of $J_1 = 54.6$ K, down to the vicinity of the 3D ordering temperature ($T_{3D} = 23.9$ K). The specific heat also shows characteristic 1D behaviour[52]. Given Eq. 1, therefore, $\chi_{dc}(T)$ for $CuC\ell_2$ closely resembles the $C_m(T)$ corresponding to curve 1 in Fig. 2(b) when $T > 23.9$ K or $T/\Theta \gtrsim 0.04$. In particular, a large broad maximum in χ_{dc} occurs at $T_{max}/J_1 \approx 1.282$. We hasten to add, however, that in both $CuC\ell_2$ and $CuC\ell_2$-GIC's χ_{dc} shows significant anisotropy for fields applied in different crystalline directions[19]. The latter is inconsistent with purely Heisenberg 1D behaviour and reflects the difference between a truly 1D Heisenberg lattice and an effectively 1D lattice which arises from competition between **Heisenberg** bonds on a 2D "isosceles" lattice.

In $CuC\ell_2$-GIC's, while the main contribution to the initial susceptibility is the 1D behaviour seen in pristine $CuC\ell_2$, additional features arise from the intercalation. These are (i) the supression of 3D ordering due to a loss of inter-layer interactions, (ii) the disappearance of a low temperature spin flop transition which occurs at $H_{SF} = 41$ kG in $CuC\ell_2$ and (iii) the presence of finite size effects which give dominant contributions to χ_{dc} at low temperatures. The latter point is discussed more fully below in the subsection on "bimodal in-plane structure". Here, it suffices to say that the contributions from finite size effects are unambiguously recognized and can therefore be removed from the measured χ_{dc} in order to reveal a bulk-like contribution $\chi_{b1}(T)$. χ_{b1} is found to be almost exactly the same as the measured χ_{dc} for pristine $CuC\ell_2$ at $T \gtrsim T_{3D}(CuC\ell_2)$. This shows that the "1D behaviour" in $CuC\ell_2$ is indeed entirely independant of inter-layer interactions and that the in-plane (intra-intercalate) exchange interactions in $CuC\ell_2$-GIC are the same as those in pristine $CuC\ell_2$, to within ≈0.1%.

We end by pointing out that if T_{max} is associated with a "2D transition", then T_{3D} and T_{max} for both $CuC\ell_2$ and $CuC\ell_2$-GIC's can be used to

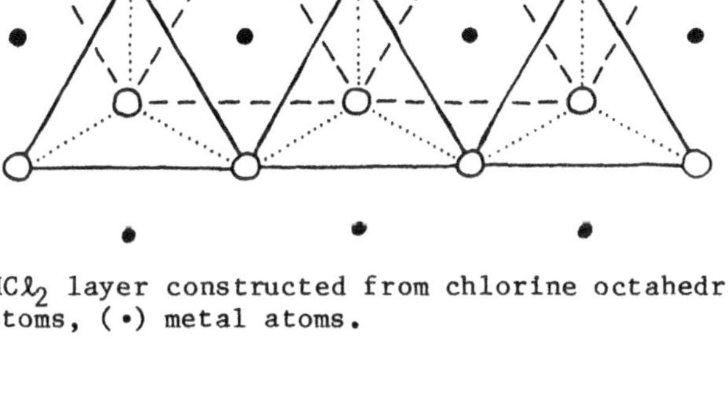

Fig. 4. $MC\ell_2$ layer constructed from chlorine octahedra: (o) $C\ell$ atoms, (•) metal atoms.

construct a phase diagram analoguous to Fig. 1 where T_{3D} is indeed decreased by intercalation and the dashed line is indeed vertical - having no h-dependence.

Spin Dynamics from AC-Susceptibility and Mössbauer Effect Spectroscopy: Magnetic Solitons in One-Dimension

Given the pervasiveness of soliton (kink or domain wall) excitations in magnetic systems and the possibility of having 1D behaviour in layered and intercalated compounds, it is of relevance to describe how, in the simplest possible case of Ising chains, the solitons give rise to the spin dynamics observed by MES and in χ_{ac} measurements.

A kink is a point, line, surface, or region which separates two equivalent ground states of a given system. Common examples are domain walls in ferromagnets and grain boundaries in crystals. A soliton, in the most restrictive definition, is a moving kink whose velocity is a constant of the motion. We also admit solitons whose motions are diffusive and/or random walk-like and also often refer to kinks as solitons. A magnetic soliton on an Ising ferromagnetic chain is represented in Fig. 5(a). Flipping one spin on either side of the soliton will not change the energy and will result in the soliton moving by one lattice spacing. A soliton on an Ising antiferromagnetic chain is represented in Fig. 5(b) and two analogues of the latter are shown in Figs. 5(c) and 5(d).

Thiel et al.[53] have pointed out that the linebroadening which occurs in the Mössbauer spectra of many quasi-1D magnetic materials must be due to solitons. Instead of the reciprocal, $1/\tau_{ac}$, of the spin autocorrolation time going as the soliton density ($n_S \sim \exp(-E_S/k_BT)$ where $E_S \simeq 2J_{ij}$ is the soliton energy) as would be expected from a free soliton gas, however, it was found that $1/\tau_{ac} \sim \exp(-2E_S/k_BT)$. It was suggested[54] that this could be understood if a strong enough inter-soliton repulsion existed, however no known mechanism led to such forces. We showed[8] that these forces must arise from the lattice non-rigidity and that the relevant spin fluctuation mechanism arises from soliton/antisoliton pair creation followed by "rapid running apart". Based on this physical picture, it is possible to predict how the same soliton dynamics will also determine the measured ac-susceptibility.

In an Ising ferromagnetic chain, the rate of change, dm/dt, of the magnetization has two contributions: a relaxation term, $-A_pm/\tau_p$, which arises from the pair creation and a force term, $A_FH_o\cos(\omega t)$, which arises from the sinusoidally varying applied field of angular frequency ω and magnitude H_o. The equation of motion, therefore, is:

$$dm/dt = -A_pm/\tau_p + A_FH_o\cos(\omega t) \tag{2}$$

which has a steady state solution given by:

$$m(t) = \chi' H_o \cos(\omega t + \phi) \tag{3}$$

where:

$$\chi' = A_F \tau_p / A_p[1 + (\omega\tau_p/A_p)^2],$$

$$\tan(\phi) = \chi''/\chi',$$

and

$$\chi'' = A_F \omega\tau_p^2 / [A_p^2 + \omega^2\tau_p^2] \tag{4}$$

A_p is the number of spins (typically $10 - 10^3$) which are over turned during the rapid running apart of thermally created soliton/antisoliton pairs and τ_p must be interpreted as the mean time between pair creation events on a given site. The force term arises as follows:

$$\partial m/\partial t\,]_F = g\mu_B\, n_S(T)\ \nu_o\ [\exp(-E_b^+/k_BT) - \exp(-E_b^-/k_BT)] \qquad (5)$$

where, E_b^+ and E_b^- are barriers for soliton motion to, respectively, lower and higher energies - given the instantaneous value of the driving field. $n_S(T)$, as before, is the soliton density and ν_o is the soliton vibration (or trial) frequency. We define $E_b^o = (E_b^+ + E_b^-)/2$ and note that for fields applied along the Ising direction: $E_b^- - E_b^+ = \delta = 2g\mu_BH_o\cos(\omega t)$. The condition $\delta/2k_BT << 1$ is satisfied at all attainable temperatures such that A_F is given by:

$$A_F = 2\ n_S(T)\ (g^2\mu_B^2\ /\ k_BT)\ \nu_o\ \exp(-E_b^o/k_BT) \qquad (6)$$

Given that $1/\tau_p \simeq f_p\exp(-2E_S/k_BT)$ and that $n_S(T)$ is known, Eq. (4) represents complete expressions for the real and imaginary susceptibilities as functions of T and ω. An example of these, at constant ω, is shown in Fig. 6. Our approach has assumed that f_p, $\nu_o >> \omega/2\pi$ and is therefore valid for moderate and low driving frequencies.

χ' (Eq. 4 and Fig. 6) is quite different from the initial susceptibility, χ_{dc}, of an Ising ferromagnetic chain with a constant probing field applied in the Ising direction, in that it does not diverge as $T \to 0$. Instead, it exhibits a maximum (not to be confused with the maxima in χ_{dc} occurring in antiferromagnetic chains) followed by an inflection point as the temperature is lowered. The inflection point in χ' coincides with a maximum in χ''. This maximum occurs at $\omega_{max} = A_p/\tau_p$ such that the temperature dependence of τ_p can be measured by mapping either the temperatures of the maxima for different probing field frequencies or the frequencies of the maxima for different temperatures. One should observe the frequencies of the maxima to go as:

$$\omega_{max} = A_pf_p\ \exp(-2E_S/k_BT) \qquad (7)$$

which is also the predicted temperature dependence of the inverse spin autocorrelation time ($1/\tau_{ac}$) resulting from the same soliton/antisoliton pair creation mechanism and measured by MES. This means that when both $\ln(\omega_{max})$ (where a ω_{max}, measured by χ_{ac}, can be between 0 Hz and $\sim10^6$ Hz) and $\ln(1/\tau_{ac})$ (where a $1/\tau_{ac}$, measured by Fe-57 MES, can be between $\sim10^6$ Hz

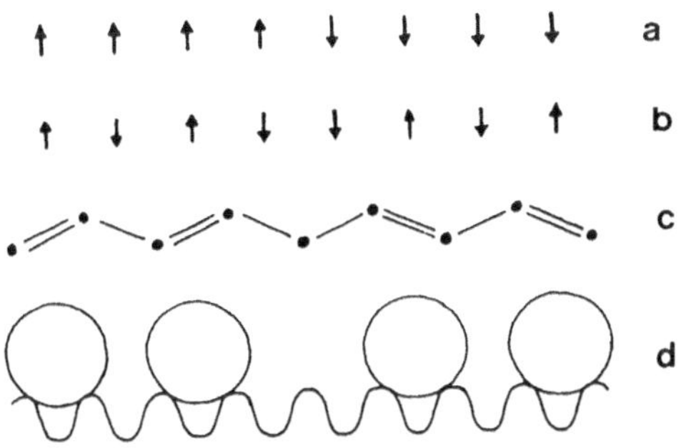

Fig. 5. 1D kinks, solitons, or domain walls: a) on an Ising ferromagnetic chain, b) on an Ising antiferromagnetic chain, c) on a double-bond/single-bond polymer chain, and d) in a system with large adsorbed atoms on a small wavelength substrate potential.

and $\sim 10^{10}$ Hz) are plotted as functions of $1/k_BT$, they will fall on the same straight line with slope $-2E_S$ and intercept $\ell n(A_p f_p)$! This is an elegant example of how χ_{ac} and MES, which operate in different frequency windows, can complement each other. In this case, showing a unified behaviour arising from a unique soliton pair creation and repulsion mechanism. Exactly such a behaviour has been observed in $FeC\ell_2 \cdot (NC_5H_5)_2$ by ElMassalami whose work[55] was the author's first contact with this problem.

Various additional comments should be made. One is that the pair creation energy is actually greater than twice the soliton energy because of the interaction energy which is initially greatest when the soliton and antisoliton are created with only one lattice spacing of separation. This is seen in the data of de Groot et al.[54] where the slope of $\ell n(1/\tau_{ac})$ versus $1/k_BT$ is always ~10% greater than $4J_{ij}$ as obtained from susceptibility measurements. Also, given defects, the chains will be of finite length such that single solitons can be created on the free ends. This leads to a second maximum[55] in χ'' whose position in frequency goes as $\sim\exp(-E_S/k_BT)$. In addition, one should keep in mind that J_{ij} itself will have a temperature dependence from normal lattice expansion and that A_p depends on $n_S(T)$. Finally, since the solitons are imagined to order into a regular array with spacing $d \approx a_o/n_S(T)$ where a_o is the inter-spin distance, they will have soliton-lattice vibrations about their mean positions. This should lead to small contributions to the Mössbauer spectra, corresponding to a population of spins with comparatively short autocorrelation times. Such contributions have recently been recognized as such by Smit et al.[56], in the Mössbauer spectra of many quasi-1D materials.

Structure, Exchange Mechanism and Moment Formation

The situation in pure transition metal chlorides, fluorides and oxides, is relatively simple and is characterized by well defined ionic states and local moment magnitudes of the metal atoms and by the usual superexchange bonds via the anions. Intercalation, however, leads to various complications since (i) the charge transfer may depend on a coexistence of more than one ionic type (as is discussed above for the case of $FeC\ell_3$-GIC), and (ii) defects are generated which alter the local charge densities, chemical environments, and symmetries. In addition, there may be electron hopping to and from defect sites, corresponding to localized valence fluctuations and, as always, the amount of magnetocrystalline coupling, the

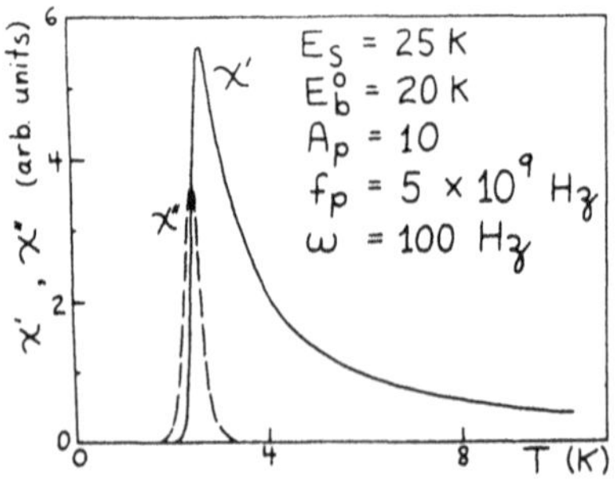

Fig. 6. Real and imaginary parts (χ' and χ'') of the calculated ac-susceptibility for an Ising ferromagnetic chain with probing field applied along the Ising direction. Resulting from a mechanism based on soliton dynamics, as described in the text.

moment size, and the exchange coupling strength, depend on the ionic state.

Insight into these problems can be gained by comparing with an intrinsically itinerate electron system such as a pure metal. The dominant exchange mechanism in a metal is generally thought to be an RKKY-like indirect exchange via conduction electrons. The moment formation mechanism is not completely understood and is referred to as the "Kondo lattice problem". Given this limited theoretical understanding, metallic iron nonetheless offers a beautiful example of the inter-relations between the mechanism for moment formation, the exchange mechanism, and the magnetocrystalline coupling. α-Iron (bcc structure) is strongly ferromagnetic with a Curie temperature of 1044 K and has a magnetic saturation moment of 2.2 μ_B whereas γ-iron (fcc structure) is strongly antiferromagnetic with Curie-Weiss temperature of $\Theta = -1340$ K and has a moment of 2.8 μ_B! The antiferromagnetism of γ-iron leads to the coexistence of ferromagnetism and antiferromagnetism in fcc Fe-Ni alloys[57] and such coexistences are in turn responsible for reentrant spin glass behaviour which is discussed below in connection with $MC\ell_2$-GIC's.

Most intercalates, until now, have been relatively well behaved, however, the above suggests that some intercalates (particularly metals; e.g., Europium[58]) might have completely different magnetisms than their pristine counterparts - with different size moments and even exchange couplings of opposite signs.

Islands, Superparamagnetism, Inter-Island Bridging and Superferromagnetism

In a ferromagnet with Heisenberg exchange, the ground state is infinitely degenerate since the magnetization can point in any direction. In a real ferromagnet, the magnetocrystalline coupling defines a finite number (typically 2 to 12, depending on the local symmetry) of easy directions. The magnetization is then effectively frozen into one of these directions since jumping between the easy directions is energetically inaccessible in most circumstances. If the volume, V, of the ferromagnet is made small, however, the total magnetocrystalline energy (which is proportional to V) will eventually become comparable to the thermal energy, k_BT. Thermal hopping of the ferromagnet's entire magnetic moment will then occur and the resulting infinite time average of this moment will be exactly zero. This phenomenon is well known as superparamagnetism[59,60] (SP) and occurs at room temperature when the F (or AF or ferrimagnetic) particle diameter is typically between ~10 Å and ~1000 Å. In an ensemble of such particles, there may be inter-particle (i.e. inter-moment or inter-supermoment) interactions which cause the individual supermoment time averages to be non-zero. We have referred to the latter situation as superferromagnetism[61] (SF) and this term is now widely used as such. The best example of SF which has been studied in detail is that of Fe-Ni Invar in which 7.2 Å diameter Ni-rich clusters form supermoments at ~1000 K: these are superparamagnetic down to an ordering temperature of ~460 K, below which their behaviour is superferromagnetic[57]. In all such systems, the supermoment (i.e., intra-particle magnetic order) forms at a temperature which is the order of the near neighbour exchange strength and its dynamics is then determined by the relative importance of inter-particle interactions, magnetocrystalline coupling, shape anisotropy terms, magnetodamping mechanisms, possibility of "buckling" and "curling", etc.

We have argued[6] that, in those $MC\ell_2$-GIC's which have F in-plane interactions (M = Co,Ni), large supermoments form at higher temperatures than the temperatures at which ordering phenomena are observed. This follows from (i) the existence of ~150 Å diameter islands in these compounds and (ii) the fact that the in-plane interactions are the same as in the corresponding pristine materials (to within ~1-2%), as can be surmised from the Curie-Weiss behaviours of the measured susceptibilities. We showed that, assuming such supermoments, the complicated two-step ordering behaviour[11,62-69] of $MC\ell_2$-GIC's (M = Co, Ni) could be explained by admitting only four physical parameters: a supermoment size, μ_S, an inter-supermoment inter-plane AF coupling, J_{AF}, and two inter-supermoment intra-plane F coupling constants, J_{F1} and J_{F2}, which apply to two equal populations of intercalate layers. We then proposed[25] that the two F coupling constants (denoted E_F and $E_F - E_o$ in ref. 6) arise from two distinct in-plane morphologies: one of mostly isolated islands and another of linked (or connected or bridged) islands. An intercalation mechanism in which a "bridging transition" occurs may be the cause of this bimodal intercalate morphology[25]. An entire section (below) of the present article is devoted to this and other evidence for such a bimodal morphology in $MC\ell_2$-GIC's. We end this section with a calculation in which the F coupling strength, for a given in-plane morphology, is obtained.

The fact that, in $MC\ell_2$-GIC's (M = Co, Ni), the spins can be taken to be constrained to the plane at low temperatures, simplifies the calculation somewhat. A supermoment is taken to interact with z neighbours via bridges of length ℓ and width w. If it makes an angle ϕ with the in-plane magnetization, m, then its average energy of interaction can be shown to be given by:

$$E \simeq - z\, m\, (\ell w/a_o^2)\, J_{ij}\, \cos(a_o\phi/\ell) \qquad (8)$$

where a_o is the atomic lattice parameter and J_{ij} is the near neighbour inter-spin intra-plane exchange constant. In deriving the above expression for E, it was assumed (given the comparatively sluggish dynamics of supermoments) that the bridging spins couple neighbouring supermoments by Néel wall configurations in which neighbouring intra-bridge spins are rotated by only $(a_o/\ell)\phi$ with respect to each other. It then follows, since the supermoments can be treated classically, that m is given by:

$$m = \int_o^{2\pi} \cos\phi\, \exp(-E/k_BT)\, d\phi \;/\; \int_o^{2\pi} \exp(-E/k_BT)\, d\phi \qquad (9)$$

which exhibits an ordering temperature, T_F, given by:

$$T_F \sim z\, (w/\ell)\, (J_{ij}/k_B) \qquad (10)$$

It follows from (10) that the mft F coupling constant J_F (E_F in ref. 6) is approximately given by:

$$J_F \sim 2z\, (w/\ell)\, J_{ij} \qquad (11)$$

Eq. (11) shows the essential features of the relation between J_F and the in-plane morphology which is characterized by a mean number (= z/2) of bridges per island, an effective bridge width, w, and an effective bridge length, ℓ. This shows clearly that different morphologies lead to different F in-plane coupling constants. In particularly, z is expected to be especially sensitive to the filling coefficient, α, in the vicinity of optimum filling[6,25].

Spin Structures of Pristine versus Intercalated Materials: $FeC\ell_3$ versus $FeC\ell_3$-GIC.

Among exotic spin structures, helical spin structures are actually quite common in the rare earths[70]. Such spin structures can be understood if the nature, strength, range and orientations of the exchange bonds are known, and also require dipole-dipole forces and magnetocrystalline anisotropy to be considered. The angle of rotation usually has a temperature dependence and first order transitions can occur - on either side of which completely different spin structures are stabilized. Given the delicate balance of forces which conspire to give these spin structures, intercalation can be expected to lead to substancially perturbed (or dramatically different) spin structures - even if one assumes neither the exchange mechanism nor the moment size or ionic state to be affected. A case in point is that of $FeCl_3$-GIC and pristine $FeCl_3$ (see ref. 32 for an overview).

Pristine $FeCl_3$ has a particularly complicated spiral structure, a magnetic ordering temperature of 8.76 K, exhibits a transition to two-sublattice antiferromagnetism at a relatively weak applied field of ~1.5 T (at T = 4.2 K), has a spin flop at ~4.0 T, and displays strong MSRO up to T ≃ 40 K. In comparison, stage-1 $FeCl_3$-GIC orders magnetically at ~4 K, shows MSRO up to T ≃ 20 K and has an in-plane two-sublattice AF ground state in which more than ~60% of the iron has spins which are closely confined to the plane. This shows that the tendency within an $FeCl_3$ layer is towards a simple in-plane AF order and that the complicated spin structure in pristine $FeCl_3$ therefore arises from inter-layer interactions. Given the relative translation of successive planes and the AF inter-layer superexchange bonds in pristine $FeCl_3$, frustration prevents the prefered intra-layer spin order from being established. The resulting compromise is a complicated spin arrangement which is particularly sensitive to thermal lattice expansion, forced lattice contractions, density of defects or impurities, etc.

Bimagnetism in $FeCl_3$-C-$NiCl_2$-C- Biintercalation Compound

We have introduced[32] the term "bimagnetism" to describe the situation where two magnetic sublattices, within a given compound, behave independently to the extent that they order magnetically at two different temperatures. This occurs despite the physical proximity and interwoven organization of the two sublattices. The phenomenon is, not surprisingly, quite rare but has been shown to occur[71-80] in $M(HCOO)_2 \cdot 2H_2O$ with M = Fe, Co, and Mn. Rancourt et al. have shown[32] that bimagnetism also occurs in the biintercalation compound $FeCl_3$-$NiCl_2$-GIC. Graphite biintercalation compounds contain a regular sequence of two kinds of intercalated layers[81,82]. In the case of $FeCl_3$-$NiCl_2$-graphite, true biintercalation is possible because $FeCl_3$ intercalates at a much lower temperature than $NiCl_2$ such that stage-2 $NiCl_2$-GIC can be synthesized first in order to serve as a chemically stable host for the $FeCl_3$ intercalation. In the resulting compound, the $NiCl_2$ layers order magnetically at T = 20.5 K whereas the $FeCl_3$ layers remain paramagnetic down to $T < 4.2$ K. This is seen unambiguously by combining SQUID magnetometry with Fe-57 MES which sees only the iron moments. The $FeCl_3$ layers behave essentially as they do in $FeCl_3$-GIC of all stages at $T > 4.2$ K. The $NiCl_2$ layers, on the other hand, order at 1.5 K lower than in stage-2 $NiCl_2$-GIC. This is believed to be due to the increased inter-$NiCl_2$-layer distance rather than to the presence of intervening $FeCl_3$ layers. The only magnetic interaction between $NiCl_2$-layers, in the bi- and singly intercalated compounds, is argued to be dipole-dipole forces. These forces are not felt by the $FeCl_3$ layers in $FeCl_3$-$NiCl_2$-GIC for two reasons: i) the intra-$FeCl_3$-layer exchange interactions are AF such that, even in the presence of MSRO, the local intra-$FeCl_3$-layer magnetization density is expected to be low everywhere and at all temperatures of interest, and ii) the inter-$NiCl_2$-layer dipole-dipole forces are such that the F $NiCl_2$ layers stack antiferromagnetically resulting in net near-zero dipole-dipole forces at the $FeCl_3$ layers. By analogy, $FeCl_3$-C-$CoCl_2$-C- biintercalation compound should also be bimagnetic. This is suggested by χ_{ac} measurements which show the $CoCl_2$ layers to behave essentially as they do in $CoCl_2$-GIC[62].

Evidence from Magnetism for a Bimodal In-Plane Structure in MCl_2-GIC's

The double peak structures of the low temperature ac-susceptibilities (and the dc-magnetization bumps which occur at the same temperatures), in those $MC\ell_2$-GIC's which contain F layers[11,32,62-69], can be understood in terms of two intra-plane F coupling constants[6]. Given the likely structural origin of these coupling constants (as discussed above), these behaviours constitute strong evidence for a bimodal in-plane morphology in $MC\ell_2$-GIC's. We have proposed[25] that any given $MC\ell_2$-GIC which is filled to nominal saturation will contain two types of planes which are roughly equal in number but which differ in their degrees of inter-island connectedness. The distribution along the c-axis of the two types of planes is expected to be mostly random yet may be more regular. Bunching of similar planes probably does not occur since the AF reflections seen by neutron diffraction[83-85] seem not to grow appreciably until the magnetically weaker planes have also "ordered" ferromagnetically.

It has recently occured to us that the magnetic behaviour of $CuC\ell_2$-GIC[19,86] offers additional strong evidence for a bimodal in-plane morphology in $MC\ell_2$-GIC's. The low temperature ($T \lesssim 15$ K) and low field ($H \lesssim 1$ kG) dc-susceptibility is very closely given by:

$$\chi_{dc} = \chi^o + A_f/T \tag{12}$$

where χ^o and A_f are sample dependent constants. χ^o is a bulk-like contribution which should be roughly proportional to the amount of $CuC\ell_2$ per gram of GIC. On the other hand, the A_f/T term arises from finite size effects and, in an isolated chain model (see above section on frustration), is approximately given by:

$$A_f \simeq g^2 \mu_B^2 N_A n_{ch} / 3 k_B M_{mol} \tag{13}$$

where N_A is Avogadro's number, M_{mol} is the molecular mass for a molecule with one Cu atom (the stoichiometry is typically $C_{10}CuC\ell_{2.2}$ for stage-2), and n_{ch} is the number of effectively finite (or "short") odd chains per metal ion. A_f, therefore, should be proportional to χ^o since its sample dependence arises from n_{ch} which, for a given morphology, is proportional to the amount of $CuC\ell_2$. Rancourt et al.[19] have indeed found that A_f is often proportional to χ^o with $A_f/\chi^o \simeq 10.1$ K. Surprisingly, however, another group of $CuC\ell_2$-GIC's gives $A_f/\chi^o \simeq 10.1\text{ K} + (4.0 \times 10^{-5}\text{ emuK/g})/\chi^o$. This is shown in Fig. 7 where A_f versus χ^o is plotted for nine stage-2 samples made

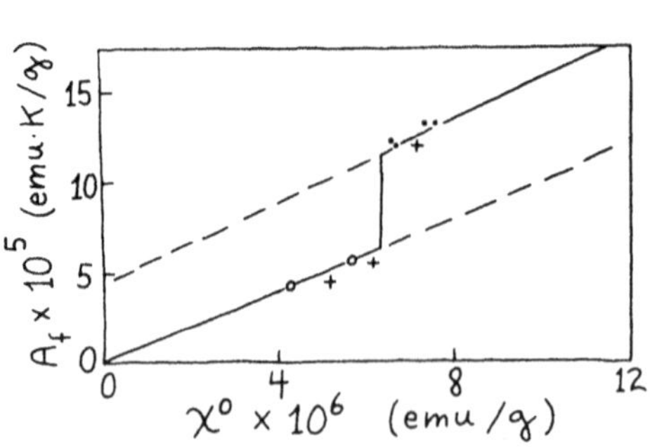

Fig. 7. A plot of A_f versus χ^o for different samples of stage-2 $CuC\ell_2$-GIC based on different host graphite materials. A dc field is applied in the plane and the corresponding susceptibility at $T < 15$ K is found to go as $\chi_{dc} = \chi^o + A_f/T$. Based on the data of Rancourt et al.[19].

from three different host graphite materials: (o) graphite fibers, (•) natural graphite crystals, and (+) highly oriented pyrolitic graphite (HOPG). All of the samples are distributed into two groups:

$$\text{group A with } A_f \simeq (10.1\ \text{K})\chi^o \tag{14}$$

and

$$\text{group B with } A_f \simeq (10.1\ \text{K})\chi^o + (4.0 \times 10^{-5}\ \text{emuK/g}) \tag{15}$$

The mean measured stoichiometry for group A is $C_{12}CuCl_{2.2}$ whereas for group B it is $C_{10}CuCl_{2.2}$. Group A samples have $CuCl_2$ fillings corresponding to χ^o between ~4 × 10^{-6} emu/g and ~6.4 × 10^{-6} emu/g whereas the group B samples are spread over χ^o values from ~6.4 × 10^{-6} emu/g to ~8 × 10^{-6} emu/g. Another important feature which distinguishes the group A and B samples is the absence and presence, respectively, of a low temperature anomaly at T ≃ 9 K. This is shown in Fig. 8 where "corrected" susceptibilities for samples from groups A and B are given as functions of T. By "corrected" we mean that A_f/T has been removed from the measured data. The low temperature anomaly occuring in group B samples is also understood in terms of finite size effects[19] and should not be attributed to a phase transition. The large broad maximum occuring at T ≃ 72 K in all samples is, of course, the maximum discussed above which also occurs in pristine $CuCl_2$ and which is attributed to effectively infinite 1D AF S=1/2 chains isolated by frustration. The question now is: Why are there two distinct sample behaviours and, in particular, what is the significance of $\chi^o \simeq 6.4 \times 10^{-6}$ emu/g at which a step of $\Delta A_f \simeq 4.0 \times 10^{-5}$ emuK/g occurs in A_f? We propose that group A samples have completely "de-bridged layers" (i.e. layers containing only isolated/disconnected elementary islands) whereas group B samples have a maximum number (≃ 50%) of "bridged layers" (i.e. layers containing a bridged network of connected elementary islands). The intra-bridge short odd chains are in excess of the intra-island short odd chains and cause the observed ΔA_f step. $\chi^o \simeq 6.4 \times 10^{-6}$ emu/g is then understood to correspond to a critical filling at which bridging must occur[25]. The low temperature anomaly at T ≃ 9 K is caused by the short odd intra-bridge chains - longer chains do not give rise to such an anomaly[19]. In conclusion, our proposition fits all of the available data: group A samples have less filling, lower χ^o values, no low temperature anomaly at T ≃ 9 K and lower A_f values than group B samples which exhibit a large low temperature anomaly. In addition, we note that the sample classification is highly dependent on the host graphite used: all of the fiber based material was group A, all of the natural graphite crystal based compounds were group

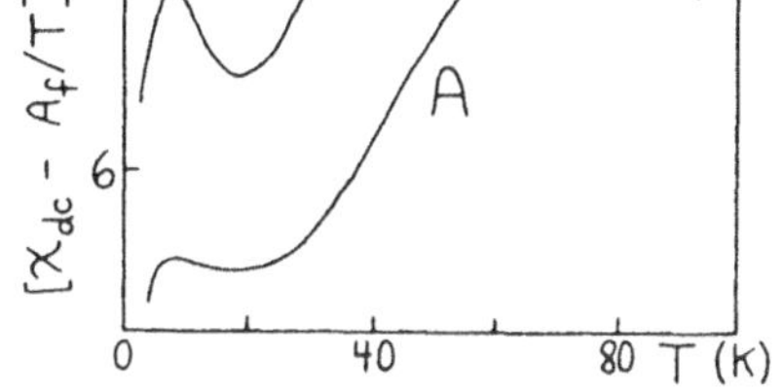

Fig. 8. "Corrected" susceptibilities (= $\chi_{dc} - A_f/T$) for a group A sample (here, a graphite fiber based $CuCl_2$-GIC sample) and a group B sample (here, a natural graphite crystal based $CuCl_2$-GIC sample). Based on the data of Rancourt et al.[19].

B and the smaller HOPG based samples were group A whereas the largest HOPG based compound was group B[19]. This suggests that by choosing an appropriate host graphite or by doing a sample to sample statistical study of a given intercalation, it may be possible to observe F layer $MC\ell_2$-GIC (M = Co, Ni) samples which are "A-type" and contain only de-bridged layers. Such samples would exhibit only one peak in χ_{ac} - at the temperature which has been termed $T_{c\ell}$ ($\simeq$ 18.0 K in $NiC\ell_2$-GIC and $\simeq$ 8.0 K in $CoC\ell_2$-GIC's) - and would not exhibit the characteristic "hump" in the low field dc magnetization. Rancourt et al. have, in fact, attributed an abnormally low ordering temperature, in a powdered sample of $FeC\ell_3$-$NiC\ell_2$-GIC, to de-bridged $NiC\ell_2$ layers[32].

Other experiments which are suggested to test our model of a bimodal morphology are as follows. (i) De-bridging studies, were partial and controlled de-intercalation is intentionally induced, may reveal F layer $MC\ell_2$-GIC's which undergo discontinuous drops in ordering temperatures from an upper value (T_{cu}) to $T_{c\ell}$. (ii) In-plane conductivity measurements might reveal a bimodal distribution (sample to sample) of resistivities. It seems that the ratio of GIC sample resistance to pre-intercalation host sample resistance is a useful parameter for such studies[87]. And (iii) Co-57 MES source experiments with $CoC\ell_2$-GIC should reveal two different predominant subspectra which would arise from the two types of layers. This would be seen at temperatures where the super-moment dynamics predominantly determine the lineshape since each type of layer would have its own exchange barrier for super-moment fluctuations[12,57].

Structure as the Origin of "Heavy Fermion" Behaviour

The so-called "heavy fermion" or "heavy electron" systems are compounds with rare earths (mainly Ce, Yb, U, Np, and Pu), such as $CeCu_2Si_2$, $NpBe_{13}$ and $CeA\ell_3$, which display enormous specific heat γ values typically ranging from $\gamma \simeq 10^2$ to $\gamma \simeq 10^3$ mJ/mol K^2. So far, roughly a dozen such compounds have been identified and recent reviews have been given[88,89]. If one assumes the measured γ values to be purely electronic in nature (the phonon contribution to the specific heat being taken into account by a BT^3 term), then one concludes from the free electron formulae ($\gamma_{fe} = m^*k_F k_B^2/3\pi^2\hbar^2$) that the effective electron masses are of order $m^*/m \simeq 10^3$ - which is indeed "heavy". We wish to point out that the layered compound $CuC\ell_2$ behaves in an exactly analogous way to those "heavy fermion" systems which have magnetically ordered ground states: it has a very large slope of its lattice corrected specific heat ($\gamma \simeq 200$ mJ/mol K^2) and a magnetic ordering λ-point anomaly which i) occurs at a "surprisingly" low temperature compared to the measured Curie-Weiss constant and ii) corresponds to a total entropy which is lower than that expected, based on the effective moment value deduced from the high temperature susceptibility. Yet the specific heat of $CuC\ell_2$ (ref. 52 and as discussed above) is well understood in terms of ordinary electronic, magnetic and lattice contributions. We propose that the large γ values in heavy fermion systems can be understood in a similar fashion and that, therefore, they are only effective γ values which arise from the magnetism of local moments rather than from heavy electrons. Many of the heavy fermion systems clearly have structures with 1D or 2D arrays of rare earth ions. In these cases and given AF exchange interactions, the magnetic contributions to the specific heats should display large Schottky anomalies (as shown in Fig. 2) which, at low temperatures, simulate large γ values. Such Schottky anomalies also occur in frustrated local moment lattice systems. In all cases, any 3D magnetic ordering is strongly suppressed below the near neighbour exchange strength and much of the local moment entropy occurs at the Schottky barrier temperatures rather than at lower temperatures.

As a further and quantitative illustration of how low-dimensionality can lead to large effective γ values, consider a Heisenberg chain of spins with near neighbour only AF interactions. (Given the known structure of, for example, U_2Zn_{17}, this is a relevant starting point.) The initial susceptibility is known exactly[90] for the case $S \to \infty$ and can, for our purpose, be used to calculate C_m via Eq. 1. The resulting effective γ value is surprisingly independent of T, at temperatures below the Schottky-like maximum, and is given by:

$$\gamma_{eff} \; (= \partial C_m/\partial T) \;\; \simeq \;\; N_A \, k_B^2 \, / \, |J| \qquad (16)$$

which, for $|J| \simeq (10\ K)k_B$, corresponds to $\gamma_{eff} \simeq 10^3$ mJ/mol K^2. The question then arises: "Given the common occurrence of low-D materials in nature, why have "heavy electrons" not been discovered earlier?" We must also keep in mind that complicated helical spin structures are common in rare earth compounds such that the local moment magnitudes cannot reliably be obtained by fitting magnetic neutron reflections with simple AF lattice models. There is a danger that such suspect procedures be used to obtain anomalously small local moments and thereby "confirm" the existence of a Kondo instability.

As a final concrete example, we believe that the detailed behaviour[91,92] of URu_2Si_2 (above its superconducting transition at $T \simeq 0.8$ K) can be understood in terms of ordinary magnetism with preserved local moments on U ions. The natural starting point is the realization that URu_2Si_2 is a layered compound with AF in-plane interactions. The necessity to include a large magnetocrystalline anisotropy term in the spin Hamiltonian is seen in its (not unusually) large susceptibility anisotropy[91,92].

Reentrant Spin-Glass Behaviour and $MC\ell_2$-GIC's with M = Co,Ni: Spins on a Sponge

This section is intended mainly to draw the reader's attention to the fact that the highly unusual and intricate ac-susceptibility behaviours of those $MC\ell_2$-GIC's which contain F layers[11,62-66,68] are not unique. Remarkably similar $\chi_{ac}(T,H)$ curves are measured in reentrant spin-glass ferromagnets near their tri-critical concentrations (e.g., refs. 93-95). These $\chi_{ac}(T)$ curves and their applied field dependences are qualitatively understood[94,96] within the context of the Sherrington-Kirkpatrick model[97]. The latter model is based on an entirely different physical picture than our own model[6] which explains the $MC\ell_2$-GIC differential susceptibilities. This serves as a reminder that even a model which can reproduce measured quantities is not necessarily correct and that similarity in behaviour does not imply common causes. It seems clear to us[6,12,57], however, that the cooperative magnetic phenomena seen in both reentrant spin-glass ferromagnets and $MC\ell_2$-GIC's occur between already magnetically short range ordered clusters which arise from elementary intercalate islands in the latter and from chemical clusters in the former. We note, also, that in metallic spin-glass materials, magnetic clustering is expected even in the absence of chemical inhomogeneities[49]. We hope, therefore, that considerations of superpara/ferromagnetism will eventually acquire their rightful place in discussions concerning these materials. This leads to the problems of calculating the temperature dependence of the supermoment magnitude (or degree of MSRO) and of determining how the short range and longer range magnetic orders are coupled. These problems are related to current investigations of spins on a fractal[98,99] and, more directly, to what might be called "spins on a sponge". The question of the influence of the morphology (which, for example, will be partly characterized by the mean pore size relative to the inter-spin separation) is then of direct relevance to many of the magnetic systems mentioned in this article. Many other magnetic systems are sponge-like - in that they contain holes (pores) or magnetically weak regions - yet are not fractals. Some GIC's may best be described as ideal simple model systems for studying such structures.

CONCLUSION

By the use of examples, we have illustrated many aspects of the magnetism of layered and intercalated compounds. The dependence of the observed magnetism on structure is of primary importance in these materials. We have not defined "structure" only in the restrictive crystallographic sense but take it to be a complete characterization of the physical substance, which includes a quantitative description of defects, faults, impurities, domain structures, exchange path geometries, etc. Such features often predominantly determine the magnetism and always at least perturb the ideal behaviour - especially in GIC's which contain novel intrinsic "defects". We have shown the magnetism of certain layered and intercalated compounds to be more closely related to topics of current interest such as spin-glass magnetism and "heavy fermion" behaviour rather than to considerations of ideal 2D magnetism. We have described situations (i.e. structures) which give rise to: i) the possibility of reducing dimensionality by means of magnetic frustration, ii) solitons being the preferred low temperature excitations and predominantly determining the spin dynamics, iii) superpara- and superferromagnetism, iv) two-step ordering processes characterized by successive peaks in ac-susceptibility, v) true bimagnetism, and vi) effective heavy fermion behaviour with a specific heat constant $\gamma \sim 10^3$ mJ/mol K^2 of purely local moment origin. We have also briefly reviewed both the field of simple magnetic models and the usefulness of some of the most common experimental probes in magnetism. Finally, we have suggested various specific experiments and have expressed the view that hysteresis and time-effects have not sufficiently been explored.

The field of magnetism in planar systems is an exciting source of new phenomena and will continue to arouse the fundamental research interests of many - as though only the tip of the iceberg has so far been seen... It may also attract the research interests of the more practically minded since the very new high-T_c superconductors[100] are also planar magnetic systems (of copper ion spins). Will some form of "itinerate-magnetoelastic coupling" in planar geometry prove to be a viable Cooper-pairing mechanism?

ACKNOWLEDGEMENTS

I am very greatful to my many collaborators and colleagues: J.M. Daniels, B. Hun, S.R. Julian, C. Meschi, H.H.A. Smit, R.C. Thiel and especially S. Flandrois. Financial support from the Natural Sciences and Engineering Research Council of Canada is greatfully acknowledged.

REFERENCES

1. L. Onsager, Crystal statistics I. A two-dimensional model with an order-disorder transition, Phys. Rev. 65:117 (1944).
2. S. Katsura, Statistical mechanics of the anisotropic linear Heisenberg model, Phys. Rev. 127:1508 (1962). Note error in Fig. 3 of this reference: curve labelled "i" for $|J|\chi_{\perp}/Nm^2$ should intersect the axis of ordinates at $2/\pi \simeq 0.6366$ rather than drop off to zero as $T \to 0$.
3. P.R. Weiss, The application of the Bethe-Peierls method to ferromagnetism, Phys. Rev. 74:1493 (1948).
4. L.J. de Jongh and A.R. Miedema, Experiments on simple magnetic model systems, Adv. Phys. 23:1 (1974).
5. D. Mattis and W.E. Donath, Role of Fermi surface and crystal structure

in theory of magnetic metals, Phys. Rev. 128:1618 (1962).
6. D.G. Rancourt, New theory for magnetic graphite intercalation compounds: superferromagnetism in two dimensions, J. Magn. Magn. Mat. 51:133 (1985).
7. J.W. Bray, L.V. Interrante, I.S. Jacobs and J.C. Bonner, in: Extended Linear Chain Compounds, J.C. Millder, Ed., Plenum, New York (1982).
8. D.G. Rancourt, Low temperature behaviour of Ising magnetic chains: decorated solitons, locally enhanced exchange and diffusive propagation, Sol. State Comm. 58:433 (1986).
9. D.G. Rancourt, Magnetoelastic mechanism for magnetic short range order above T_c, unpublished.
10. M.E. Fisher, Relation between the specific heat and susceptibility of an antiferromagnet, Phil. Mag. 7:1731 (1962).
11. M. Matsura, Hierarchy nature of magnetic ordering in $MC\ell_2$-GIC's, Ann. Phys. 11:C2-117 (1986).
12. D.G. Rancourt, Pervasiveness of cluster excitations as seen in the Mössbauer spectra of magnetic materials, Proceedings of the International Conference on the Applications of the Mössbauer Effect (ICAME-87), Melbourne, Australia, 1987, to be published in Hyp. Int.
13. e.g., L.L. Hinchey and D.L. Mills, Magnetic properties of superlattices formed from ferromagnetic and antiferromagnetic materials, Phys. Rev. B 33:3329 (1986).
14. Modulated Structures - 1979, AIP conference proceeding number 53, J.M. Cowley, J.B. Cohen, M.B. Salamon and B.J. Wuensch, eds., American Institute of Physics, New York (1979).
15. Synthetic Modulated Structures, L.L. Chong and B.C. Giessen, eds., Academic Press, New York (1985).
16. I.S. Jacobs and P.E. Lawrence, Metamagnetic phase transitions and hysteresis in $FeC\ell_2$, Phys. Rev. 164:866 (1967). Includes a review of metamagnetism.
17. L.P. Regnault and J. Rossat-Mignod, Phase transitions in quasi-2D planar magnets, in: Magnetic Properties of Layered Transition Metal Compounds, L.J. Jongh and R.D. Willet, eds., unpublished.
18. S. Flandrois, A.W. Hewat, C. Hauw and R.H. Bragg, Neutron diffraction and small-angle scattering studies of graphite-nickel chloride intercalation compounds, Synth. Met. 7:305 (1983).
19. D.G. Rancourt, C. Meschi and S. Flandrois, S = 1/2 antiferromagnetic finite chains effectively isolated by frustration: $CuC\ell_2$-GIC, Phys. Rev. B 33:347 (1986).
20. S. Flandrois, J.-M. Masson, J.-C. Rouillon, J. Gauthier and C. Hauw. Intercalation compounds of graphite with nickel chloride: synthesis, structure, and mechanism of intercalation, Synth. Met. 3:1 (1981).
21. F. Baron, S. Flandrois, C. Hauw and J. Gauthier, Charge transfer and islands in metal halide GIC's: new evidence from X-ray diffraction of intercalated $MnC\ell_2$, Sol. State Comm. 42:759 (1982).
22. D.M. Hwang, X.W. Qian and S.A. Solin, Physics and chemistry of spatially resolved microdomains in $SbC\ell_5$-GIC, Phys. Rev. Lett. 53:1473 (1984).
23. R. Levi-Setti, G. Crow, Y.L. Wang, N.W. Parker, R. Mittleman and D.M. Hwang, High-resolution scanning-ion-microprobe study of graphite and its intercalation compounds, Phys. Rev. Lett. 54:2615 (1985).
24. G. Kirczenow, Kinetics of stage ordering and stage transitions, Phys. Rev. Lett. 55:2810 (1985).
25. D.G. Rancourt, B. Hun and S. Flandrois, Magnetic properties of intercalated transition metla chlorides which have ferromagnetic in-plane coupling, Ann. Phys. 11:C2-107 (1986).
26. M. Prietsch, C. Wortmann, C. Kaindl and R. Schlögl, Mössbauer study of stage-2 $FeC\ell_3$-graphite, Phys. Rev. B 33:7451 (1986).
27. R. Schlögl, P. Bowen, G.R. Millward, W. Jones and H.P. Boehm, Microstructure and stability of iron chloride graphite, J. Chem. Soc. Faraday Trans. 1 79:1793 (1983).

28. J.M. Cowley and J.A. Ibers, Structure of $FeC\ell_3$-GIC, Acta. Crystallogr. 9:421 (1956).
29. H. Thiele, Die vorgänge bei dem aufblähen von graphit, Z. Anorg. Allg. Chem. 207:340 (1932).
30. D. Hohlwein, P.W. Readman, A. Chamberod and J.M.D. Coey, Magnetic properties of two-dimensional iron chloride GIC, Phys. Stat. Sol. (b) 64:305 (1974).
31. S.E. Millman and G. Kirczenow, Mössbauer analysis of the acceptor site for the donated electrons in $FeC\ell_3$ GIC, Phys. Rev. B 28:5019 (1983).
32. D.G. Rancourt, B. Hun and S. Flandrois, New biintercalation compound, $FeC\ell_3$-$NiC\ell_2$-graphite, studies by Mössbauer effect spectroscopy and SQUID magnetization measurements: an ideally decoupled bimagnetic system, submitted for publication, and extended abstract at 18th Biennial Conference on Carbon, Worcester, Massachusetts, July 19-24, 1987.
33. S.G. Brush, History of the Lenz-Ising model Rev. Mod. Phys. 39:883 (1967).
34. C. Domb, Ising model, in: Phase Transitions and Critical Phenomena, vol. 3, C. Domb and M.S. Green, eds., Academic Press, London (1974).
35. D.D. Betts, X-Y model, in: Phase Transitions and Critical Phenomena, vol. 3, C. Domb and M.S. Green, eds., Academic Press, London (1974).
36. G.S. Rushbrooke, G.A. Baker, Jr. and P.J. Wood, Heisenberg model, in: Phase Transitions and Critical Phenomena, vol. 3, C. Domb and M.S. Green, eds., Academic Press, London (1974).
37. M. Steiner, J. Villain and C.G. Windsor, Theoretical and experimental studies on one-dimensional magnetic systems, Adv. Phys. 25:87 (1976).
38. J.M. Kosterlitz and D.J. Thouless, Ordering metastability and phase transitions in two-dimensional systems, J. Phys. C 6:1181 (1973).
39. F. Block, Zur theorie des ferromagnetismus, Zeits f. Physik 61:206 (1930).
40. N.D. Mermin and H. Wagner, Absence of ferromagnetism or antiferromagnetism in one- or two-dimensional isotropic Heisenberg models, Phys. Rev. Lett. 17:1133 (1966).
41. e.g., D.J. Breed, Experimental investigation of two two-dimensional antiferromagnets with small anisotropy, Physica 37:35 (1967).
42. H.E. Stanley and T.A. Kaplan, Possibility of a phase transition for the two-dimensional Heisenberg model, Phys. Rev. Lett. 17:913 (1966).
43. M. Suzuki, S. Miyashita, A. Kuroda and C. Kawwabata, Monte Carlo simulations of the two-dimensional quantal and classical spin systems - A new type of phase transition with vortices, Phys. Lett. 60A:478 (1977).
44. C. Kawabata and K. Binder, Evidence for vortex formation in Monte Carlo studies of the two-dimensional XY-model, Sol. State Comm. 22:705 (1977).
45. J. Tobochnik and G.V. Chester, Monte Carlo study of the planar spin model, Phys. Rev. B 20:3761 (1979).
46. D.H. Lee, J.D. Joannopoulos, J.W. Negele and D.P. Landau, Discrete-symmetry breaking and novel critical phenomena in an antiferromagnetic planar (XY) model in two dimensions, Phys. Rev. Lett. 52:433 (1984).
47. H. De Raedt, B. De Raedt and A. Lagendijk, Thermodynamics of the two-dimensional spin-1/2 XY model, Z. Phys. B 57:209 (1984).
48. J.V. José, L.P. Kadanoff, S. Kirkpatrick and D.R. Nelson, Renormalization, vortices, and symmetry-breaking perturbations in the two-dimensional planar model, Phys. Rev. B 16:1217 (1977).
49. D.A. Smith, The zero-frequency susceptibility of spin glasses and mictomagnets, J. Phys. F 5:2148 (1975).

mictomagnets, J. Phys. F 5:2148 (1975).
50. O. Nagai, Y. Yamada and H.T. Diep, Linear-chain-like excitations in a three-dimensional Ising lattice with frustration: Monte Carlo simulations, Phys. Rev. B 32:480 (1985).
51. C. Hauw, J. Gauthier, S. Flandrois, O. Gonzalez, D. Dorignac and R. Jagut, Étude par diffraction des rayons-X et mocroscopie électronique à haute résolution de quelques composés d'insertion graphite-chlorure métalique, Synth. Met. 7:313 (1983).
52. J.W. Stout and R.C. Chisholm, Heat capacity and entropy of $CuC\ell_2$ and $CrC\ell_2$ from 11° to 300° K: Magnetic ordering in linear chain crystals, J. Chem. Phys. 36:979 (1962).
53. R.C. Thiel, H. de Graaf and L.J. de Jongh, Soliton contribution to the Mössbauer linewidth in quasi-one-dimensional antiferromagnets with Ising anisotropy: theory versus experiment, Phys. Rev. Lett. 47:1415 (1981).
54. H.J.M. de Groot, L.J. de Jongh, M. ElMassalami, H.H.A. Smit and R.C. Thiel, Mössbauer relaxation studies of non-linear dynamical excitations in low-dimensional magnets, Hyp. Int. 27:93 (1986).
55. M. ElMassalami, Relaxation Studies on Non-Linear Excitations in Pure and Impure Ising-Type Magnetic Systems, thesis, Kamerlingh Onnes Laboratorium, Leiden, (1987).
56. H.H.A. Smit, H.J.M. de Groot, M. ElMassalami, R.C. Thiel and L.J. de Jongh, Fe-57 Mössbauer effect study of non-linear excitations in anisotropic quasi-1D magnetic systems, (unpublished), and personal communications.
57. D.G. Rancourt, H.H.A. Smit and R.C. Thiel, Metastable compositionally and magnetically modulated state of Fe-Ni Invar and the associated super-moment dynamics from Mössbauer spectroscopy, J. Magn. Magn. Mat. 66:121 (1987).
58. H. Suematsu, K. Ohmatsu, T. Sakakibara, M. Date and M. Suzuki, Magnetic properties of europium-graphite intercalation compound C_6Eu, Synth. Met. 8:23 (1983).
59. L. Néel, Théorie du trainage des ferromagnêtiques en grains fins avec application aux terres cuites, Ann. Geophys. 5:99 (1949).
60. W.F. Brown, Jr., Thermal fluctuations of a single domain particle, Phys. Rev. 130:1677 (1963).
61. D.G. Rancourt and J.M. Daniels, Influence of unequal magnetization direction probabilities on the Mössbauer spectra of superparamagnetic particles, Phys. Rev. B 29:2410 (1984).
62. M. Suzuki, I. Oguro and Y. Jinzaki, Magnetic phase transition of stage-1 ternary $CoC\ell_2$-$FeC\ell_3$-graphite intercalation compound, J. Phys. C 17:L575 (1984).
63. M. Elahy and G. Dresselhaus, $CoC\ell_2$-intercalated graphite: A quasi-two-dimensional magnetic system, Phys. Rev. B 30:7225 (1984).
64. G. Chouteau, R. Yazami and Ph. Touzain, Synthesis and new magnetic properties of stage-1 $CoC\ell_2$-GIC, presented and published as an extended abstract, Carbone-86, Baden Baden.
65. Y. Murakami, M. Matsuura, M. Suzuki and H, Ikeda, Ordering process in $CoC\ell_2$-GIC, J. Magn. Magn. Mat. 31-34:1171 (1983).
66. S.T. Chen, K.Y. Szeto, M. Elahy and G. Dreselhaus, Magnetic phase transitions in $CoC\ell_2$-GIC, J. Chim. Phys. +Phys. Chim. Biol. 81:863 (1984).
67. H. Suematsu, R. Nishitani, R. Yoshizaki, M. Suzuki and H. Ikeda, Magnetization measurements of the second stage $NiC\ell_2$-GIC, J. Phys. Soc. Jpn. 52:3874 (1983).
68. M. Suzuki, H. Ikeda, Y. Murakami, M. Matsuura, H. Suematsu, R. Nishitani and R. Yoshizaki, Magnetic phase transitions in $NiC\ell_2$-GIC, J. Magn. Magn. Mat. 31-34:1173 (1983).
69. Yu. S. Karimov and Yu. N. Novikov, Phase transitions in a two-dimensional ferromagnet with "easy-plane" anisotropy, JETP Lett. 19:159 (1974).

70. Y.-A. Rocher, L'Étude de la structure électronique des métaux des terres rares, Adv. Phys. 11:233 (1962).
71. A.F. Cohen, S.A. Friedberg and G.R. Wagner, The low temperature magnetic susceptibility of $Mn(HCO_2)_2 \cdot 2H_2O$, Phys. Lett. 11:198 (1964).
72. H. Abe and M. Matsuura, Proton nuclear magnetic resonance study in manganese formate dihydrate, J. Phys. Soc. Jpn. 19:1867 (1964).
73. H. Abe and K. Torii, Susceptibility of manganese formate dihydrate, J. Phys. Soc. Jpn. 20:183 (1965).
74. R.D. Pierce and S.A. Friedberg, Heat capacity of $Mn(HCOO)_2 \cdot 2H_2O$ between 1.4 and 20 K, Phys. Rev. 165:680 (1968).
75. P. Burlet, P. Burlet, E.F. Bertaut, G. Roult, A. de Combarieu and J.J. Pillon, Étude de l'antiferromagnétisme à très basse température dans le formiate de manganese dihydrate, Sol. State Comm. 7:1403 (1969).
76. K. Takeda, T. Haseda and M. Matsuura, A study of magnetic phase transition in a two-dimensional lattice by heat capacity measurements, Physica 52:225 (1971).
77. M. Matsuura, H.W.J. Blöte and W.J. Huiskamp, Heat capacity and magnetic behaviour of cobalt and manganese formate dihydrate, Physica 50:444 (1970).
78. Y. Ajiro, N. Terata, M. Matsuura and T. Haseda, NMR study of magnetic ordering in two-dimensional antiferromagnet, $Mn(HCOO)_2 \cdot 2H_2O$, J. Phys. Soc. Jpn. 28:1587 (1970).
79. K. Takeda and S. Matsukawa, Magnetic susceptibility of antiferromagnetic $Co(HCOO)_2 \cdot 2H_2O$, J. Phys. Soc. Jpn. 30:887 (1971).
80. P. Burlet, P. Burlet and E.F. Bertaut, Étude par diffraction neutronique du formiate de fer dihydrate, Sol. State Comm. 9:1633 (1971).
81. A. Hérold, G. Furdin, D. Guérard, L. Hachim, N.E. Nadi and R. Vangelisti, Some aspects of graphite intercalation compounds, Ann. Phys. 11:C2-3 (1986).
82. A. Hérold, Synthesis of GIC, invited contribution in the present monograph.
83. M. Suzuki, H. Ikeda and Y. Endoh, Magnetic neutron scattering from second-stage $CoC\ell_2$-GIC, Synth. Met. 8:43 (1983).
84. H. Ikeda, Y. Endoh and S. Mitsuda, Two-dimensional spin correlations in high stage $CoC\ell_2$-GIC, J. Phys. Soc. Jpn. 54:3232 (1985).
85. D.G. Wiesler, M. Suzuki, H. Zabel, S.M. Shapiro and R.M. Nicklow, Physica 136B:22 (1986); M. Suzuki, D.G. Wiesler, P.C. Chow and H. Zabel, Magnetic phase transitions in intercalated graphite, J. Magn. Magn. Mat. 54-57:1275 (1986); D.G. Wiesler, M. Suzuki and H. Zabel, Ordering in quasi-2D planar ferromagnets: a neutron scattering study of graphite intercalation compounds, Phys. Rev. B (1987).
86. H. Nishihara, I. Oguro, M. Suzuki, K. Koga and H. Yasuoka, High-temperature magnetic susceptibility of $CuC\ell_2$-GIC, Synth. Met. 12:473 (1985).
87. H. Oshima, J.A. Woollam, A. Yavrouian and M.B. Dowell, Electrical and mechanical properties of copper chloride-intercalated pitch-based carbon fibers, Synth. Met. 5:113 (1983).
88. G.R. Stewart, Heavy-fermion systems, Rev. Mod. Phys. 56:755 (1984).
89. P.A. Lee, T.M. Rice, J.W. Serene, L.J. Sham and J.W. Wilkins, Theories of heavy-electron systems, Com. Mod. Phys. B 12:99 (1986).
90. M.E. Fisher, Magnetism in one-dimensional systems - the Heisenberg model for infinite spin, Am. J. Phys. 32:343 (1964).
91. T.T.M. Palstra, A.A. Menovsky, J. van den Berg, A.J. Dirkmaat, P.H. Kes, G.J. Nieuwenhuys and J.A. Mydosh, Superconducting and magnetic transitions in the heavy-fermion system URu_2Si_2, Phys. Rev. Lett. 55:2727 (1985).
92. T.T.M. Palstra, Magnetism, supercondictivity and their interplay. A study of three novel intermetallic compounds: $La(Fe,A\ell)_{13}$ UNiSn

URu_2Si_2, thesis, Kamerlingh Onnes Laboratorium, Leiden (1986).
93. M.B. Salamon, K.V. Rao and Y. Yeshurun, Spin-glass/ferromagnetic transitions and multicritical points in amorphous transition metal alloys, J. Appl. Phys. 52:1687 (1981).
94. S.M. Bhagat, J.A. Geohegan, M.L. Spano and H.S. Chen, Ferromagnetic vs spin glass behaviour in amorphous Fe-Ni based alloys, J. Appl. Phys. 52:1741 (1981).
95. A.Z. Men'shikov, G.A. Takzey and A. Ye. Teplykh, Spin-glass in γ-$(Ni_{80-c}Cr_{20})Fe_c$ alloys, Phys. Met. Metall. 54:41 (1982).
96. G.J. Nieuwenhuys, H. Stoker, B.H. Verbeek and J.A. Mydosh, Differential susceptibility of spin glass-ferromagnetic systems, Sol. State Comm. 27:197 (1978).
97. D. Sherrington and S. Kirkpatrick, Solvable model of a spin-glass, Phys. Rev. Lett. 35:1792 (1975).
98. L.M. Sander, Fractal growth, Sci. Am. Jan:94 (1987).
99. M. Continentino and A.P. Malozemoff, Dynamical susceptibility of spin glasses in the fractal cluster model, Phys. Rev. B 34:471 (1986).
100. A. Khurana, Superconductivity seen above the boiling point of nitrogen, Phys. Today April:17 (1987).

APPLICATIONS OF GRAPHITE INTERCALATION COMPOUNDS AND RELATED PROCESSES FOR SYNTHESIS

Michio Inagaki

Materials Science, Toyohashi University of Technology
Tempaku-cho, Toyohashi 440, Japan

INTRODUCTION

A balanced development of synthesis and application technologies of materials is necessary for modern industries. In former days, new materials either synthesized by chance or found in nature were subjected to search their applications. Nowadays, however, the applications of materials spread so widely that the application technology asks new materials which need for development of new science and technology.

The history of graphite intercalation compounds (GICs) goes back to the middle of 19th century. Excellent works have been done and attracted the interests of a number of scientists since. The year 1976 is an epoch in the history of GICs. Vogel(1976; 1977) announced that the GICs of antimony and arsenic pentafluoride had higher electrical conductivity than the metallic copper, which promoted the interests of people and accelerated the development of application technology on GICs. Actually, however, graphite fluoride has been used as cathode material of commercial lithium battery since 1973(Watanabe & Fukuda, 1970). And Ubbelohde (1972) already reported that some GICs had rather high electrical conductivities and called them "synthetic metals".

In the first part of the present lecture, I will review the possibilities of applications of GICs in different fields and trends to approach the practical applications. In the latter part, I will discuss on the requirements on GICs and their synthesis for application development and then explain our recent experimental results on synthesis of metal chloride-GICs by using molten salts of chlorides which is believed to give us the possibility to develope the practical process for synthesis corresponding to the industrial applications of these GICs.

The possibilities of application of GICs seem to be roughly divided into six fields (Inagaki, 1982), as summarised in Table 1 with representative intercalants in each field.

Table 1. Possibilities of Applications of GICs

1. Highly-conductive materials		
	HNO_3, ICl	Ubbelohde(1972)
	SbF_5, AsF_5	Vogel(1976,1977)
	AsF_5	Murakami(1983)
	$CuCl_2$	Oshima et al.(1982)
	$CuCl_2$, $FeCl_3$	Chieu et al.(1982)
	F_2-MF_n	Nakajima et al.(1982)
	K-Bi	Lagrange et al.(1985)
2. Electrode materials in batteries		
Primary battery	$(CF)_n$	Watanabe & Fukuda(1970)
	metal chlorides	Armond & Touzain(1977)
	$(C_2F)_n$	Kita et al.(1979)
	F_2-MF_n	Nakajima et al.(1982)
	graphite oxide	Yazami & Touzain(1985)
Secondary battery	$NiCl_2$, $MnCl_2$	Flandrois et al.(1980)
	80%graphite+20%PP	Beck & Krohn(1982)
	$NiCl_2$-$FeCl_3$	Inagaki et al.(1987)
Thermo cell	Br_2-KBr aq.	Lalancette & Roussel(1976)
	Br_2-KBr aq.	Endo et al.(1980)
	NO_x-HNO_3	Inagaki et al.(1983)
3. Catalyzers for organic synthesis		
	Polymerization	Li, K, K-Hg
	Ammonium synthesis	K-$FeCl_3$
	Isomerization	SbF_5
	Bromination	Br_2
	Esterification	H_2SO_4, HNO_3
4. Materials for hydrogen-storage and isotope-separation		
Storage	$KH_{2/3}C_8$	Saeher & Herold(1965)
	$K(H_2)_{2.1}C_{24}$	Tamaru et al.(1971)
	$K(H_2)_{1.9}C_{24}$	Lagrange et al.(1975)
Separation	K-H_2	Terai & Takahashi(1981)
5. Exfoliation of graphite		
	H_2SO_4, HNO_3	
	K, Na-THF	Inagaki et al.(1983)
6. Others		
Lubricant	$(CF)_n$	Fussaro & Sliney(1969)
Electrochromic	Li-DMSO	Pfluger et al.(1979)

Highly-conductive materials

The synthesis of highly-conductive GICs are carried out in four different families of compounds, which are summarized in Table 2 with the reported maximum conductivities and some comments on stability in air.

Table 2. Highly-Conductive GICs

Pentafluorides	SbF_5, AsF_5
higher than metallic copper,	10^4 S/m
unstable in air	
Chlorides	$CuCl_2$, $FeCl_3$
$CuCl_2$	$7.8 \cdot 10^2$ S/m (pitch-based fibers)
$FeCl_3$	$1.4 \cdot 10^3$ S/m (vapor-grown fibers)
stable in air,	
comparable with Cu in figure of merit	
Fluorides	F_2+CuF_2, MgF_2, LiF_2
$1 \cdot 10^3$ S/m,	relatively stable in air
Donner ternary systems	K + Bi
$2 \cdot 10^3$ S/m,	relatively stable in air

The field of highly-conductive materials was accelerated by the work on arsenic pentafluoride-GICs and its high conductivity was reproduced in 1983, seven years later from the first report, by the Japanese group (Murakami, 1983). The enhancement of conductivity strongly depends on the crystallinity of the host graphite. In Fig. 1, the conductivity after intercalation of AsF_5 is plotted against the conducivity of the host carbon fibers(Shioya et al., 1986). In order to have high conductivity, comparable to copper, very high crystallinity of the host, which can be achieved by using the vapor-grown carbon fibers and very high temperature heat treatment, is essential. These GICs are unstable in air so that to avoid their decomposition the use of sheaths of either copper or teflon have been proposed (Vogel, 1976; Kalnin & Goldberg, 1983). The compounds from the vapor-grown graphite fibers were reported to keep high conductivity of about 1×10^3 S/m during two years after exposing them to air(Shioya et al., 1986). It seems to be due to the formation of protective graphite layers at the surface of the fiber by the preferential decomposition, as reported on the nitric acid-GICs from the same graphite fibers(Endo et al., 1981).

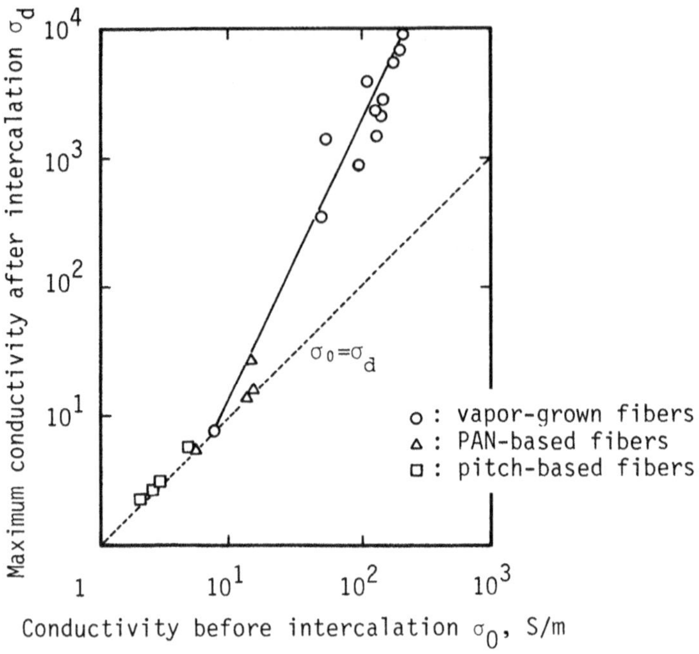

Fig. 1 Relation in electrical conductivity before and after intercalation of AsF_5 on various carbon fibers heat-treated at high temperatures

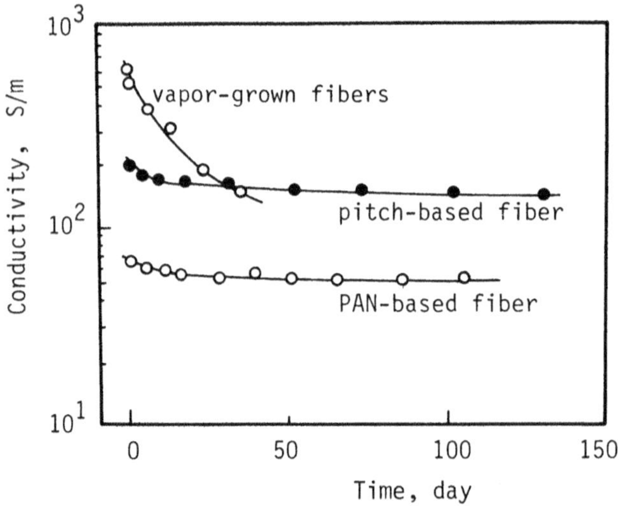

Fig. 2 Changes in electrical conductivity of the fluorine-GICs with time in air.

The GICs prepared in the presence of CuF_2 from different carbon fibers heat-treated above 2800°C.

Oshima et al.(1982) reported that the $CuCl_2$-GICs made from the graphitized mesophase-pitch-based carbon fibers had relatively high conductivity($7.8x10^2$ S/m) and proposed to discuss by using figure of merit (conductivity/density). Chieu et al.(1982) reported the conductivity of about $1.4x10^3$ S/m on $FeCl_3$-GICs from the graphitized vapor-grown fibers. These compounds were reported to be stable in air.

Another approach to highly-conductive GICs has been shown by the group of Watanabe(Nakajima et al., 1982; 1983). In the presence of some metal fluorides, graphite reacts with fluorine gas at relatively low temperatures, much lower than without metal fluorides, and the products give electrical conductivity as high as $2x10^3$ S/m. The detailed structure of these GICs is still under investigation, but they are claimed to be acceptor-type intercalation compounds, only a part of fluorine intercalated making a covalent bond with carbon atoms in the graphite layer planes. Nakajima (1984; 1985) reported the compounds thus synthesized from vapor-grown carbon fibers showed high conductivity but low stability in air, and the compounds from PAN-based carbon fibers had relatively low conductivity but high stability in air, as shown in Fig. 2.

Although the GICs which showed high conductivity had been limited in the acceptor-type compounds, Lagrange et al.(1985) reported that the donner-type compounds of alkali metal-bismuth ternary GICs had also as high conductivity as $2x10^3$ S/m. They announced that the GICs obtained were relatively stable in air.

Electrode materials in batteries

Graphite materials have been used in different batteries, but they have only a role to conduct the current, for example, graphite powder being mixed with nickel hydroxide powder in the cathode of Ni-Cd secondary battery. In the GICs, the active materials are intercalated as intercalants in between graphite layers, in other words, layer by layer mixing. Therefore, high activity and low inner resistance of batteries can be expected on the electrodes of GICs. Representative GICs which have been used for electrode materials in different batteries are summarized in Table 1.

The great success has been achieved by the use of graphite fluoride $(CF)_n$ as cathode material in primary lithium battery. This battery is commercially distributed in the world, and has advantages of light weight, high voltage, high energy density, long preservative, low self-discharging, etc. Kita et al.(1979) synthesized a covalent compound $(C_2F)_n$, corresponding to the second-stage, and found that its battery had higher open circuit voltage(OCV) than that of the first-stage $(CF)_n$. Also they made a battery from the fluorine-GICs prepared in the presence of metal fluoride and got a good performance, high capacity and flat discharge curve(Nakajima et al., 1982).

A lithium battery with the cathode of graphite oxide was constructed(Yazami et al., 1983), which had OCV of 2.2 V and high theoretical energy density as 1200 Wh/kg. The same group(Armond & Touzain, 1977) examined different GICs as the cathode materials of lithium battery, but none of them gave a better performence than those with the covalent compounds, graphite fluoride and graphite oxide.

The secondary batteries with the active material of nickel hydroxide are widely used. The group of Flandrois (1981) synthesized the $NiCl_2$-GICs under Cl_2 gas pressure of more than 600 Torr and constructed the battery with those GICs as cathode material. The conversion of nickel chloride to its hydroxide and the improvement of performance of the battery were found with the repetition of charge-discharge cycles. They used also $MnCl_2$-GICs as the cathode material (Flandrois & Baron, 1983). Flandrois & Herran (1986) discussed on the GICs as electrode materials. Recently Inagaki et al.(1987) synthesized air-stable $NiCl_2$-$FeCl_3$-GICs by

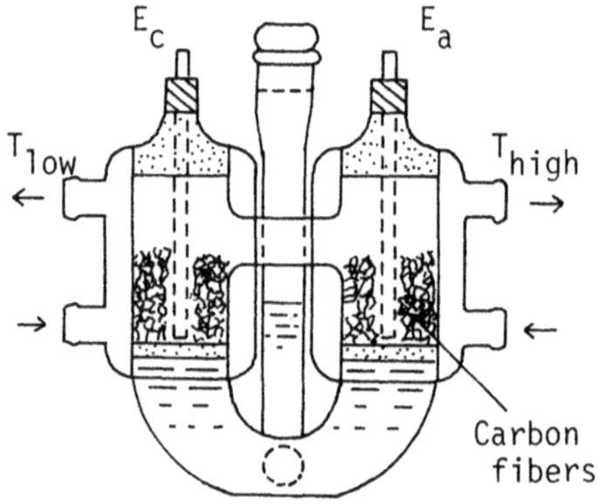

Fig. 3 Set-up of the thermo cell

using molten salt method, as will be explained in the next section, and used them as a cathode of battery (Inagaki et al., 1987). Beck & Krohn (1982; 1983) used a composite of 80 wt% graphite powder and 20 wt% polypropyrene(they called it CPP) as a cathode in sulfuric and fluoric acids, and more than 2500 cycles of charge-discharge were possible.

Lalancette and Roussel(1976) proposed a new type of cell which is a kind of concentration cell, and converts small temeprature difference between two electrodes to electrical power. So, we call this kind of cell as a thermo cell. A fundamental set-up of the cell is shown in Fig. 3, where one electrode is kept at a low temperature near room temperature and the other at a high temperature, and these two electrodes are connected by two bridges; the lower one being filled with liquid electrolyte and the upper one with gas.

Lalancette & Roussel(1976) used natural graphite powder in two electrodes and 10% KBr aqueous solution saturated with bromine as electrolyte, and obtained OCV of 69 mV from the temperature difference between 8 and 100°C. Although they announced the OCV value was close to the theoretical value, Endo et al.(1980; 1983) obtained much higher OCV, 200 mV, from the temperature difference of 80°C by using well-graphitized vapor-grown fibers as electrode materials and by improving the geometry of the cell. Short circuit current (SCC) they obtained was 10 mA/cm^2.

Inagaki et al.(1983; Maeda et al., 1983; 1987) used a cloth of PAN-based carbon fibers as electrodes and 30 % nitric acid as electrolyte, and obtained the OCV of 150 mV and SCC of 1 mA. In their cell, the 1300° C-treated PAN-based carbon fibers gave the highest output, but disruption of the fiber was observed after 2 months. By using the 2000 °C-treated fibers long life of the cell was realized and the cell was working even after two years, though the output power of the cell lowers a little bit. In this cell, the active matter, e.g. bromine, circulates between the high temperature and the low temperature electrodes, as gas species through the gas-phase bridge and as ions through the liquid-phase bridge. Therefore, the biggest advantage of this thermo cell is maintenance-free, except keeping the temperature difference between two electrodes.

Catalyzers for organic synthesis

There have been done many works to use GICs in the synthesis of various organic compounds and many excellent review papers on this subject (Boersma, 1974; Kagan, 1976; Setton, 1982). The roles of these GICs in organic synthesis process may be divided into two categories, one being catalyzers for synthesis reaction and the other one of reactants which are kept in between graphite layers as intercalants. In the former, also two cases can be considered; 1) GICs act as real catalyzers to accelerate the reaction without any change in themselves and 2) the intercalants in GICs do not get into the final products of the reaction but GICs themselves are decomposed to help the formation of intermediate compounds.

In Table 3, the reported results on the polymerization of various hydrocarbons by using the alkali metal-GICs as catalyzers are summarized. Not only intercalant but also stage of GICs affect the results; from butadiene 1-4 polymers being obtained in as high yield as 90 % by using KC_{12}. The solvent used also modifies the results; from methylmethacrylate no polymers being obtained in toluene, but polymers in dimethoxyethane with high yield close to 80 %. Alkali metal-GICs in binary and ternary systems were used for the polymerization of caprolactam, which had strong effect on the molecular weight distribution in the polymers obtained. The first-stage SbF_5-GIC was found to catalyze the isomerization of hydrocarbons.

Keeping the reactants in the interlayer space of graphite as intercalants, that is, the latter case in the above categories, is often effective to make the volatile, hygroscopic and unstable reactants easy to handle, and also to improve the selectivity and conditions for the reaction. The bromine- and SbF_5-GICs were excellent reactants for bromination and fluorination, respectively, of hydrocarbons. $SbCl_5$-GICs had the ability to exchange bromine in hydrocarbon molecules to chlorine with high selectivity and high yield. Sulfuric and nitric acid-GICs promoted the estirification of various hydrocarbons.

Table 3. Polymerization by using Alkali Metal-GICs

Monomer	GIC	Yield (%)	Polymers	Conditions
ethylene	KC_8	--	trans monoolefine	200°C, 68 bar 21 hr, iso-octene
butadiene	KC_8	40-80	51%:1-2 49%:trans 1-4	30°C, 15 hr cyclohexane
	KC_{12}	51.3	90%:1-2, 10%:1-2	15°C, 100 hr cyclohexane
	KC_{12}	21.5	88%:1-4, 12%:1-2	15°C, 100 hr toluene
	KC_{24}	--	66%:1-2, 35%:trans1-4	75°C, 12 hr cyclohexane
isoprane	KC_8	90	65%:3-4, 5%:1-2 35%:trans1-4	25°C, 12 hr n-heptane
	LiC_{12}	80	44%:cis1-4, 21%:3-4 35%:trans1-4	15°C, 168 hr cyclohexane
	KC_{37}	95	43%:1-4, 57%:3-4	15°C, 76 hr cyclohexane
methyl-metha-crylate	LiC_{12}	80	4%:iso, 63.6%:syndio 32.4%:atactic	-63°C, 48 hr DME

<u>Materials for hydrogen-storage and isotope-separation</u>

Potassium-GICs react with hydrogen gas to form two kinds of ternary compounds, represented by $KH_{2/3}C_8$ and $K(H_2)_2C_{24}$. The former $KH_{2/3}C_8$ was synthesized from the 1st stage binary compound KC_8 at room temperature(Saeher & Herold, 1965). The latter compound was synthesized from the 2nd stage binary compound KC_{24} near liquid nitrogen temperature by Tamaru et al.(Watanabe et al., 1971) and was reported to have the saturated composition of $K(H_2)_{2.1}C_{24}$. Lagrange & Herold(1975) studied these compounds in detail and showed the saturated composition of $K(H_2)_{1.9}C_{24}$.

Their characteristics are summarized in Table 4.

The compound $K(H_2)_2C_{24}$ keeps the second-stage structure, in which hydrogen molecules are absorbed in the space among potassium atoms intercalated in between graphite layers. The absorption of hydrogen was affected by the structure of host graphite; a low-temperature treated coke could absorp more hydrogen in certain composition of K/C than the natural graphite, as shown in Fig. 4. This compound has some advantages for the storage of hydrogen gas; (A1) no appreciable change in size during absorption-desorption cycle, (A2) perfectly reversible and fast reaction with hydrogen gas, (A3) desorption by simple heating or evacuation, and (A4) light weight. But also they have some disadvantages; (D1) absorption and storage near liquid nitrogen temperature and (D2) a little low capacity of hydrogen (13.7 ℓ per 100 g of the compound, but 15.5 ℓ per 100 g of $LaNi_5$).

Table 4. Potassium-Hydogen-GICs

	High-temperature form	Low-tepmerature form
Composition	$KC_8H_{2/3}$	$KC_{24}H_4$
Starting GIC	KC_8	KC_{24}
Reaction temp.	room temp.	liquid nitrogen temp.
Structure	2nd stage, K-H-K triple layer	2nd stage, H_2 molecules between K
Formation energy	15 - 17 kcal/mol	2.2 - 2.9 kcal/mol
Isotope effect	concentration of H	concentration of D
Exchange	formation of HD	no formation of HD

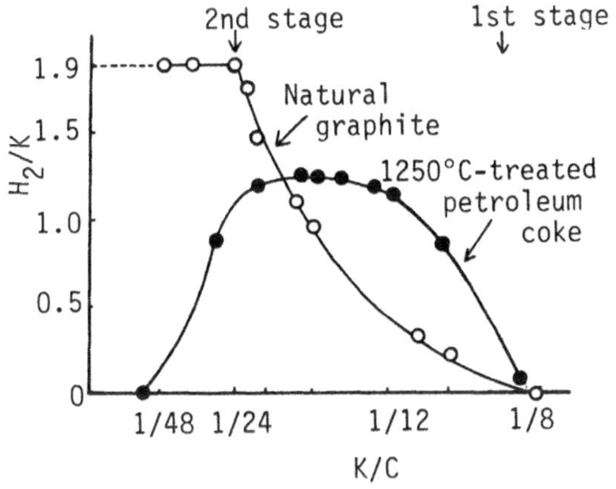

Fig. 4 Saturated amounts of hydrogen in K-GICs

The binary potassium-GICs showed high values of separation coefficient of hydrogen-isotopes(Terai & Takahashi, 1981; 1983; 1984). The separation coefficients on H_2-D_2 and H_2-HT mixed gases are summarized in Table 5. It is worth while to mention that GICs have much higher coefficient than zeorite(MS-5A). The exfoliation of the starting graphite was found to improve the coefficient(Akuzawa et al., 1984).

Takahashi(1982) used the K-GICs as a column of gaschromatography for the separation between hydrogen and deuterium. Using 5 cm length of the GIC-column, better separation between H_2 and D_2 was obtained than the Al_2O_3-column which had been commonly used.

Table 5. Isotope-Separation Coefficient and Abosorbed Volume of Hydrogen by KC_x

Sample	$a(H_2/D_2)$	$a(H_2/HT)$	v (mlH_2,NTP)
KC_{12}(HTT 1500)	8.7	5.0	92
KC_{24}(HTT 1500)	5.7	3.7	111
KC_{12}(HTT 2300)	7.1	4.5	72
KC_{24}(HTT 2300)	5.7	3.7	120
KC_{22}(Grafoil)	5.9	---	119
molecular sieve(MS-5A)	2.6-2.8	1.7	102

Exfoliation of graphite

The residue compounds of sulfuric and nitric acid-GICs contain large amount of intercalants remained in between graphite layers. By quick heating of these residue compounds, so-called exfoliated graphite have been produced in industrial scale. A typical appearance of exfoliated graphite is shown in Fig. 5. Flexible graphite sheets prepared from exfoliated graphite by roll-forming get recently increasing demand to use as gaskets, packing and heat insulators because of its high thermal resistivity, high thermal condutivity along the sheets, stability to compression, relaxation of stress, self-lubricity and high anisotropy in thermal conductivity. Some problems, however, have been pointed out; need concentrated acids, formation of poisonous gases during exfoliation and erosion of metals by minute amount of residual acids in graphite sheets.

Production of exfoliated graphite through the GICs containing organic molecules were proposed, which might solve the problems mentioned above(Inagaki et al.,1983; 1985). By starting from cobalt-tetrahydrofuran-GICs(Inagaki et al., 1984) the graphite sheets were obtained, in which minute metallic cobalt particles were dispersed and consequently stuck to a magnet(Inagaki et al., 1987).

Fig. 5 SEM micrographs of exfoliated graphite

Other applications

Graphite fluoride has a potentiality as solid lubricant which is announced to be used under severe conditions, high temperature, high speed, corrosive atmosphere (Fussaro & Sliney, 1969). It has also large contact angle to any kinds of liquid, which may promise as a coating for the protection from the liquids (Watanabe et al., 1982).

The color change from black to blue by the formation of the ternary GICs of alkali metal-dimethylsulfoxide was tried to use the electrochromic device(Pfluger et al., 1979). Writing and erasing times were reported to be the order of 0.2 seconds.

REQUIREMENTS FOR APPLICATION DEVELOPMENT OF GICs

From the view point of the development of practical applications, the GICs have to have specialized characteristics, such as high-conductivity, electrochemical activity, etc. In addition, the process for their synthesis have to satisfy the following conditions;

1. enough stability in air,
2. simple process for their synthesis,
3. mild conditions for their synthesis,
4. capability to synthesize in large quantity.

$CuCl_2$-KCl system (Inagaki & Wang, 1987)

The $CuCl_2$-GICs have been reported to have relatively high electrical conductivity and to be rather stable in air. Different processes have been used to prepare these GICs: the reaction of graphite with $CuCl_2$ vapour , the reaction with metallic copper in a flow of chlorine gas, and the reaction between graphite and $CuCl_2$ in a CCl_4 solution.

The $CuCl_2$-KCl system is selected to prepare $CuCl_2$-GICs, one of the components $CuCl_2$ having been known to produce GICs, as mentioned above, but the other KCl not. This molten system has two eutectic points, as its phase diagram is shown in Fig. 6. One of eutectic compositions, 55/45 in $CuCl_2$/KCl ratio, was used, because changes in molten state with proceeding of the interalation reaction could be avoided. The procedure and equipment for the synthesis is very simple, as shown in Fig. 7, which is common in the following synthesis in the molten salts of metal chlorides. A largely excess amount of the chloride mixtures was used: graphite/ chlorides ratios of either 0.67/1.00 or 1.33/1.00. The unreacted chlorides were repeatedly washed out by water until no trace of chloride was detected in the filtrate.

The compounds obtained in this molten salt system was $CuCl_2$-GICs, in which chlorine and copper were found to be distributed homogeneously in a flake of the compound, while no potassium was detected under X-ray microprobe analyzer. The identity period I_c determined from X-ray powder patterns was 1.286 nm for the 2nd stage compound, which coincided with the value reported. The weight gain after the intercalation gave a molar ratio of graphite/cupric chloride of 14.0 and 22.0 on the 2nd and 3rd stage compounds,respectively. By taking account of the coexistence of a small amopunt of graphite, the formula of the GICs can be written as $C_{6s}CuCl_2$ (where s is the stage number), as reported before (Oshima et al., 1983).

The 2nd stage $CuCl_2$-GICs were synthesized at a temperature just above the eutectic point, 380°C, in two hours, as shown in Fig. 8. By the prolonged heating, the diffraction lines at low angles become sharp, suggesting the improvement of structural order in the compounds. The stage of the compound can be controlled by changing the ratio of cupric chloride to graphite and also the temperature of synthesis; by using a graphite/$CuCl_2$ ratio of 1.22/1.00 the 2nd stage and by 2.44/1.00 the 3rd stage, and below 400°C the 2nd stage and at 450°C the 4th stage were obtained.

The products obtained by the present method always contain graphite in addition to the $CuCl_2$-GIC, even before washing with water. By the process of washing, the amount of the coexisting graphite increased, very remarkably in the graphite with small particle size (46 µm) but not appreciably in that with large particle size (400 µm size powder and kish graphite). The compounds wahsed out from molten salts were very stable.

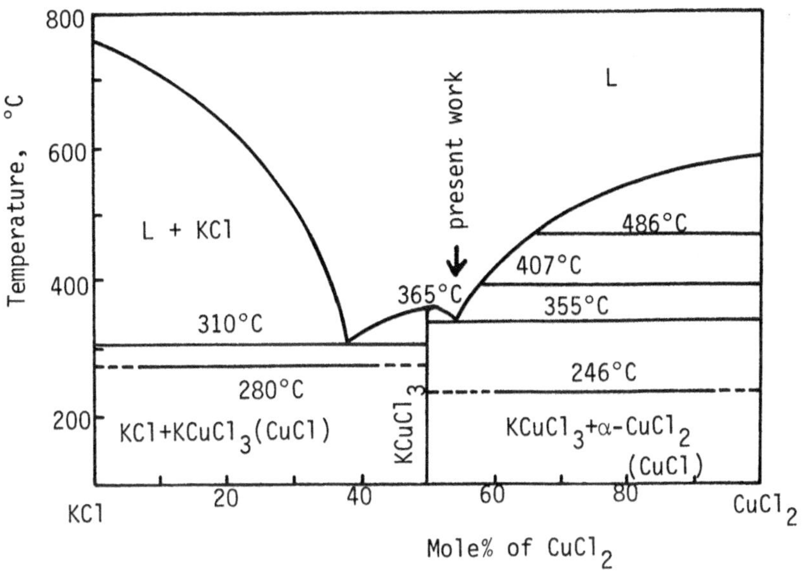

Fig. 6 Phase-diagram of $CuCl_2$-KCl molten salt system

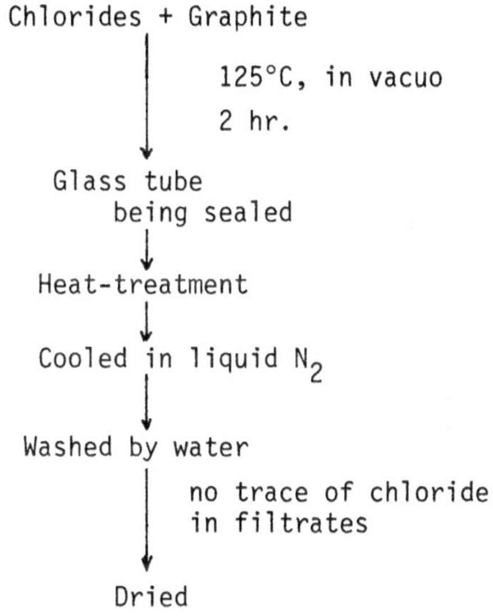

Fig. 7 General Procedure for the synthesis of GICs by molten salt method

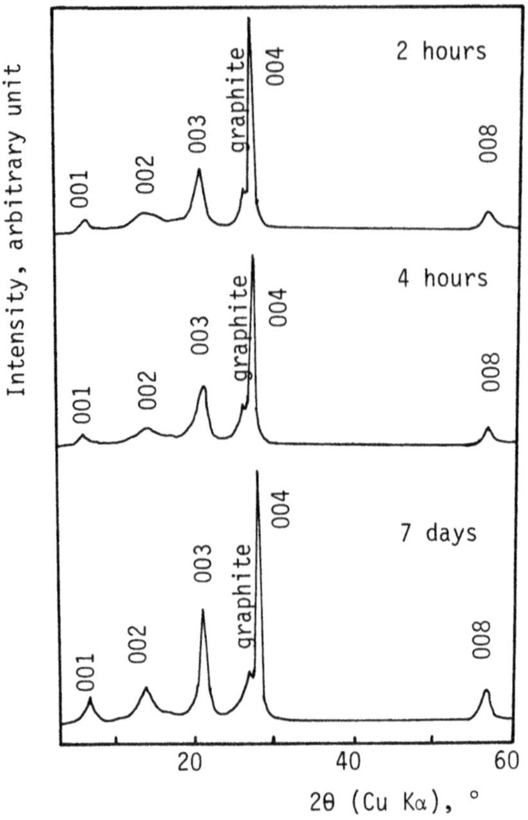

Fig. 8 X-ray powder patterns of the $CuCl_2$-GICs prepared from the 400 μm graphite powder at 380°C in different reaction times

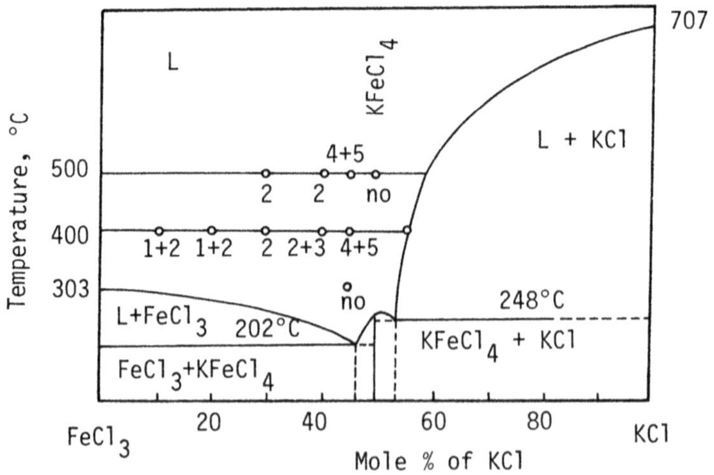

Fig. 9 Phase diagram of the $FeCl_3$-KCl molten salt system and experimental points with the resultant stage number of GICs

No change in X-ray powder patterns, no increase in relative intensity of graphite, no trace of higher stage structure and also no broadening of diffraction lines, were detected on the samples kept for more than one year in open air and even on the samples boiled in water for one hour. Therefore, the coexsting graphite is supposed to be formed at the periphery of the flaky particles and to be one of the factors for the high stability of these washed samples.

Other system such as $CuCl_2$-NaCl was also used for preparing the same GICs, in which a little higher temperature was needed because of higher eutectic point than in $CuCl_2$-KCl system.

$FeCl_3$-KCl system (Wang & Inagaki, 1987)

The $FeCl_3$-GICs has been extensively studied in many respects; their synthesis, structure, and properties. They are reported to have relatively high electrical conductivity, but usually not very stable in air.

The $FeCl_3$-KCl molten salt system was used to synthesized the $FeCl_3$-GICs, which were also stable in air as $CuCl_2$-GICs did. The stage structure of the compounds obtained in the molten salts with different $FeCl_3$/KCl ratio is summarized in Fig. 9, showing the experimental points in the phase diagram of this system and the stage numbers of the GICs obtained.

In this system, no intercalation reaction was detected at the chloride composition of 55/45 in $FeCl_3$/KCl and at a temperature of 300°C, even though $FeCl_3$-GICs were easily obtained at the same temperature by the reaction with $FeCl_3$ vapour. Even at higher temperatures as 400 - 500° C, only high stage compounds, probably 4 - 5th stage, were obtained at this composition. This result seems to be related to the formation of the compound $KFeCl_4$ in solid in this system. At the composition of $KFeCl_4$, i.e. 50/50 in $FeCl_3$/KCl, no GIC was obtained even at 500°C.

Apart from the eutectic composition and with the increase in $FeCl_3$ content, low stage compounds were obtained at 400°C. In the molten salt with the $FeCl_3$/KCl ratio of 90/10 the 1st stage with small amount of the 2nd stage, in 70/30 the single phase of 2nd stage, and in 60/40 the mixture of 2nd and 3rd stage compounds were obtained.

$FeCl_3$-$NiCl_2$ system (Inagaki et al., 1987)

The $NiCl_2$-GICs have attracted great interest because of its stability in air and also in different kinds of solvents, and it have been used as a cathode in a secondary battery, as stated in previous section. However, their synthesis required a pressure of chlorine gas above 600 Torr. On the other hand, the $FeCl_3$-GICs were prepared rather easily, because of low melting point as 303°C and giving off the chlorine gas by itself.

By using mixtures of $NiCl_2$ and $FeCl_3$ we have synthesized GICs containing both chlorides. The ratio of graphite/chlorides was changed in the range from 3/1 to 6/1, and the ratio of $NiCl_2/FeCl_3$ from 1/9 to 6/4. The reaction was carried out at 400 and 420°C for three days. On this system, phase diagram has never reported, and so the melting of chloride mixtures were confirmed by visual observation.

Starting with the mixtures of $NiCl_2/FeCl_3$ ratios from 1/9 to 3/7, the 1st stage compounds were obtained, of which identity period along the c-axis, I_c, was measured as 0.946 nm from X-ray powder pattern (Fig. 10a). From the X-ray microprobe analysis, both elements Ni and Fe were detected in the same flakes of the compounds obtained and were found to be distributed homogeneously in the compounds. Therefore, the compounds thus prepared were considered to be intercalation compounds of the mixtures of $NiCl_2$ and $FeCl_3$ with graphite, i.e., the $NiCl_2$ and $FeCl_3$ were mixed randomly on a microscopic level in the intercalant layers. By using $NiCl_2/FeCl_3$ ratio of 5/5, the mixture of the 1st and 2nd stage compounds and by 6/4 the 2nd stage compound were obtained, as shown in Fig. 10b and c. The compounds obtained as a single stage show sharp 00ℓ peaks, whose widths are comparable with instrumental broadening.

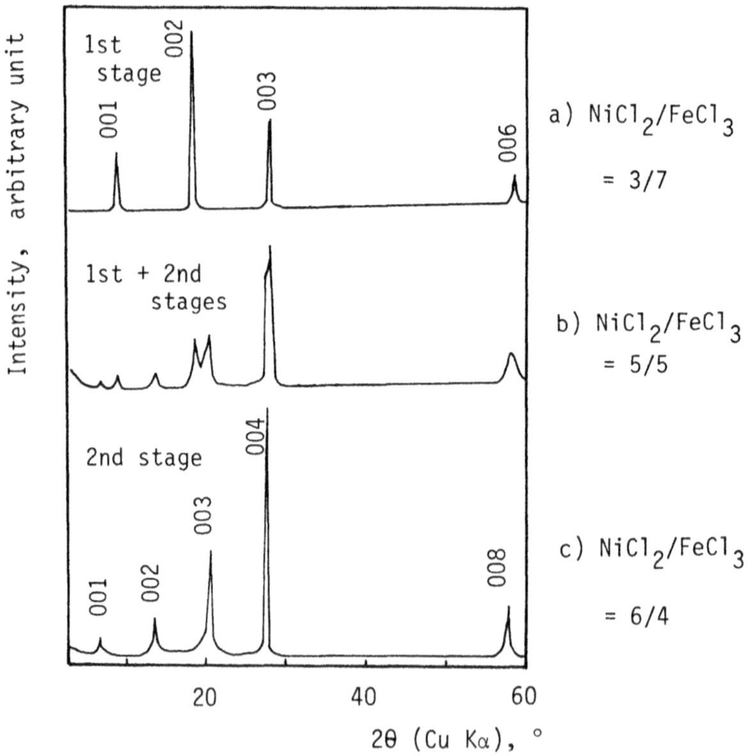

Fig. 10 X-ray powder patterns of the GICs prepared by using chloride mixtures with different $NiCl_2/FeCl_3$ ratios at 420°C in 3 days

The relative intensity of the Kα line of Ni to that of Fe on the energy dispersion spectra which were measured on a flake of the compounds prepared with different starting $NiCl_2/FeCl_3$ ratios showed that by changing the starting ratio of $NiCl_2/FeCl_3$, the composition of the intercalated chlorides also changed, as shown in Fig. 11.

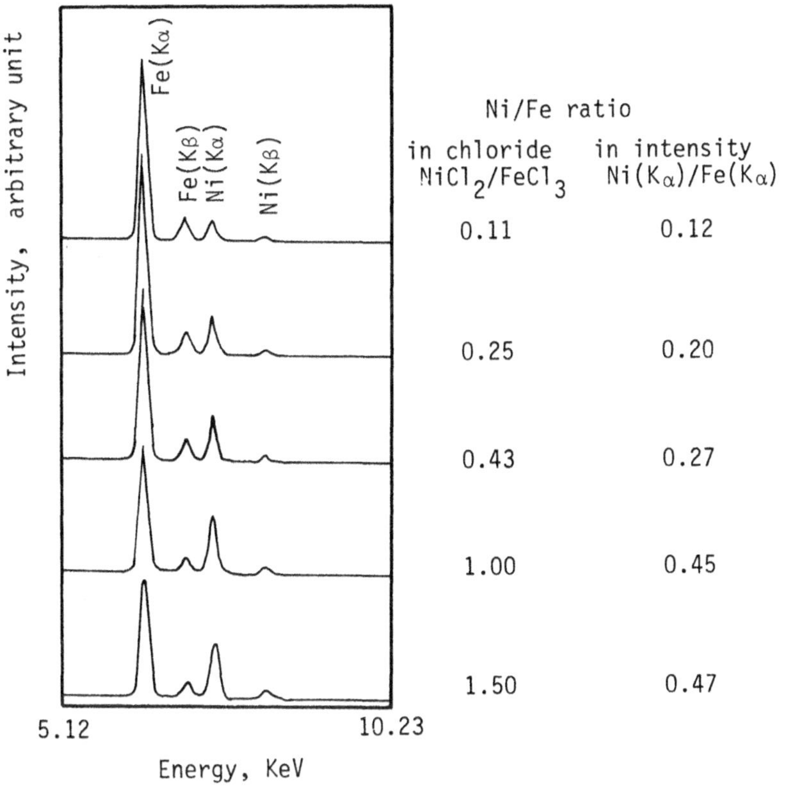

Fig. 11 Energy dispersion spectra on the GICs prepared from different $NiCl_2/FeCl_3$ ratios in molten salts

On the compounds synthesized by using $NiCl_2$-$FeCl_3$ molten systems, no coexistence of graphite was detected in X-ray powder patterns, unlike the case of $CuCl_2$-GICs. These compounds were very stable in air, for example, the compound obtained with a starting $NiCl_2/FeCl_3$ ratio of 3/7 showing no change in the powder pattern even after kept for one month in water. High stability of these compounds seems to be related to the presence of $NiCl_2$.

Roles of chloride components for synthesis of GICs

By using molten salts of metal chlorides, either binary or ternary GICs of metal chlorides were synthesized at relatively low temperatures in relatively short time. From our studies, the roles of the metal chlorides on the intercalation reaction in molten salt method are classified as follows;

1. acting as intercalants which give specialized characteristics to the resultant GICs,
2. serving as a resource providing chlorine gas to the system,
3. lowering the melting point of the system,
4. giving the stability to the resultant GICs.

In the system of $CuCl_2$-KCl, KCl component in molten salt plays a role to lower the melting point of the system and consequently make the intercalation reaction easier at low temperatures. $CuCl_2$ component intercalates into interlayer space of graphite, and gives off chlorine gas by its disproportionation. On the $CuCl_2$-GICs obtained through the molten salt method, small amount of graphite always coexisted. Relative amount of coexisting graphite increased with the decrease in average particle size of the starting graphite powder used. It must be pointed out, however, that all of the compounds, even with small particle size, prepared by the present procedure are very stable in air and even in boiling water. Therefore, the coexisting graphite is supposed to be formed at the periphery of the flaky particles of graphite and to be an important factor for the high stability of these compounds obtained.

In the system of $NiCl_2$-$FeCl_3$, $FeCl_3$ component plays various roles, intercalating to give high conductivity to the compounds obtained, giving off chlorine gas by disproportionation of itself, and lowering the melting point. On the other hand, $NiCl_2$ component intercalates into graphite to give electrochemical activity as the cathode material of secondary battery and also high stability of the compounds. The ternary compounds obtained from small flakes of kish graphite with roughly 20 mm^2 area and 0.3 mm thick by using the molten salt with the ratio $NiCl_2/FeCl_3$ of 4/6 gave the electrical conductivity of around 10^3 S/m at room temperature. The compounds synthesized from natural graphite powder with 400 µm by using the motlen salt with the same composition showed good performance as the cathode in alkali secondary battery (Inagaki et al., 1987).

Advantages of molten salt method for synthesis of GICs

The present method using the molten salts of metal chlorides has following advantages for the synthesis of various GICs;

1. being possible to synthesize in large quantity,
2. obtaining homogeneous compounds even in large quantity,
3. simple process at low temperatures in short time by using simple equipment,
4. being air stable in most cases,

The air stability of the compounds obtained might be closely related to the process of washing-out of excess chlorides after intercalation reactions. By wahing, the intercalants at the edge of particles seem to be abruptly expelled and the peripheral part of the particles returned to graphite which sealed the intercalants inside of the particles, being called "quenching". Similar quenching effect was observed on the Co-THF-graphite ternary compounds after their synthesis in THF solution to preserve the magnetic property (Inagaki et al., 1984).

So far, we have been used only metal chlorides but will be able to expand to other halides, nitrates, etc.

SYNTHESIS OF DONNER-TYPE GICs BY USING ALLOYS

Different alloy systems have been used to synthesize the binary and ternary GICs. By using K-Cs and Na-Ba alloys, ternary GICs with different ratios of metals were synthesized (Furdin, et al., 1974; Billaud & Herold, 1975), but binary Li-GICs by using Li-Na alloy (Billaud et al., 1979). The reaction of graphite with amalgams of alkali metals and alloys of alkali metals with thallium and bismuth gave ternary compounds (El-Makrini, et al., 1980; 1980), in which the layer of mercury, thallium and bismuth was sandwiched by alkali metal layers in the interlayer space of graphite. The structure of these GICs were discussed in the relation with that of potassium-hydrogen-GICs.

The ternary GICs of alkali metals and bismuth, particularly K-Bi-GICs, are interesting from the view point of applications because they have rather high electrical conductivity and is relatively stable in air(Lagrange, et al., 1985; Bendriss-Rerhrhaye, 1986), as shown in the previous section. Different stages of the compounds, $KBi_{0.6}C_{4s}$, were prepared by the direct reaction of HOPG with the molten K-Bi alloys with different compositions at a temperature just above their melting points (400-700°C). The unreacted excess alloys were scraped off from the flakes of GICs formed. The stage of the compounds was controlled by changing the composition of alloys, but the ratio Bi/K in the metals intercalated was a constant at 0.6. The compounds were stable in air during one month.

REFERENCES

Akuzawa, N., Amemiya, T., Terai, T, and Takahashi, Y., 1984, Annual meeting of Carbon Society of Japan (1984) Extended Abstracts p.44
Armand, M., and Touzain, Ph., 1977, Mat. Sci. Eng. 31:319
Beck, F., and Krohn, H., 1982, DECHEMA-Monographien 92:57
Beck, F., and Krohn, H., 1983, Synth. Met. 7:193
Bendriss-Rerhrhaye, A., 1986, Thesis, Univ. Nancy I.
Billaud, D., and Herold, A., 1975, Compt Rend. C 281:305.
Billaud, D., McRae, E., and Herold, A., 1979, Mat. Res. Bull. 14:857.
Boersma, M. A. M., 1974, Cat. Rev. Sci. Eng. 10:243
Chieu, T. C., Dresselhous, M. S., and Endo, M., 1982, Phys. Rev. B 26:5867
El-Makrini, M., Lagrange, P., Guerard, D., and Herold, A., 1980, Carbon 18:211.
El-Makrini, M., Lagrange, P., and Herold, A., 1980, Carbon 18:374.
Endo, M., Koyama, T., and Inagaki, M., 1980, Oyo Buturi 49: 563
Endo, M., Koyama, T., and Inagaki, M., 1981, Synth. Met. 3: 177
Endo, M., Yamagishi, Y., and Inagaki, M., 1983, Synth. Met. 7:203
Flandrois, S., Masson, J.-M., and Rouillon, J.-C., 1981, Synth. Met. 3:195
Flandrois, S., and Baron, F., 1983, Mat. Res. Soc. Symp. Proc. 20:237.
Flandrois, S., and Herran, J., 1986, Synth. Met. 14:103.
Furdin, G., Carton, B., Billaud, D., Zeller, C., and Herold, A., 1974, Compt Rend. C 278:1025.
Fussaro, R. L., and Sliney, H. E., 1969, NASA TND-5097
Inagaki, M., Matsumoto, A., Ssakai, M., and Maeda, Y., 1983, Nihon Kagaku Kaishi 1983:309.
Inagaki, M., 1982, Hyoumen 20:130.
Inagaki, M., Muramatsu, K., Maeda, Y., and Maekawa, K., 1983, Synth. Met. 8:335.
Inagaki, M., Shiwachi, Y., and Maeda, Y., 1984, J. Chim. Phys. 84:847.
Inagaki, M., Muramatsu, K., Maekawa, K., and Tanabe, Y., 1985, Tanso 1985:160.
Inagaki, M., and Wang, Z. D., 1987, Synth. Met. 20:1.
Inagaki, M., Wang, Z. D., Okamoto, Y., and Ohira, M., 1987, Synth. Met. 20:9.
Inagaki, M., Iwashita, N., Wang, Z. D., and Maeda, Y., 1987, J. Eelectrochem. Soc. (to be published).
Inagaki, M., Mine, H., and Sakai, M., 1987, Zairyou (in press).
Kagan, H. B., 1976, Chemtech. 1976:510.
Kalnin, I. L., Goldberg, H. A., Synth. Met. 8:277.
Kita, Y., Watanabe, N., and Fujii, Y., 1979, J. Am. Chem. Soc. 101:3832.
Lagrange, P., and Herold, A., 1975, Compt Rend. C 281:381.
Lagrange, P., Bendriss-Rerhrhaye, A., Mareche, J. F., and McRae, E., 1985, Synth. Met. 12:201.
Lalansette, J.-M., and Roussel, R., 1976, Can. J. Chem. 54: 3541.

Maeda, Y., Matsumoto, A., Itoh, E., and Inagaki, M., 1983, Synth. Met. 7: 211.
Maeda, Y., Itoh, E., and Inagaki, M., 1987, Synth. Met. 20: 73.
Murakami, S., 1983, Symposium on High Performance Polymeric Materials.
Nakajima, T., 1982, Eelectrochemica Acta 27: 1535.
Nakajima, T., Kawaguchi, M., and Watanabe, N., 1982, Carbon 20: 287.
Nakajima, T., Kawaguchi, M., and Watanabe, N., 1983, Synth. Met. 7: 117.
Nakajima, T., 1984, "Sciences and New Applications of Carbon Fibers" Toyohashi Univ. Tech. p. 101.
Nakajima, T., Watanabe, N., Kameda, I., and Endo, M., 1985, Carbon, 24: 343
Oshima, H., Woollam, J. A., and Yavrouian, A., 1982, J. Appl. Phys. 53: 9220.
Oshima, H., Woollam, J. A., and Khan, A. A., 1983, Synth. Met. 5: 113.
Pfluger, P., Kuenzi, H. U., and Guentherodt, H.-J., 1979, Appl. Phys. Lett. 35: 771.
Setton, R., Beguin, F., and Piroele, S., 1982, Synth. Met. 4: 299.
Shioya, J., Matsubara, H., and Murakami, S., 1986, Synth. Met. 14: 113.
Takahashi, Y., 1982, Postsymposium on Graphite Intercalation Compounds, Toyohashi Univ. Tech.
Terai, T., and Takahashi, Y., 1981, J. Nucl. Sci. Tech. 18: 643.
Terai, T., and Takahashi, Y., 1983, Synth. Met. 7: 49.
Terai, T., and Takahashi, Y., 1984, Carbon 22: 91.
Ubbelohde, A. R., 1972, Chem. Ind. 5: 588.
Vogel, F. L., 1976, Bull. Am. Phys. Soc. 21: 262.
Vogel, F. L., 1977, J. Mat. Sci. 12: 982.
Wang, Z. D., and Inagaki, M., 1987, Synth. Met. (to be published).
Watanabe, N., and Fukuda, M., 1970, U.S. Patent, 3,536,532; 3,700,502
Watanabe, K., Soma, M., Onishi, T., and Tamaru, K., 1971, Nature 233: 160.
Watanabe, N., Nakajima, T., and Ohsawa, N., 1982, Bull. Chem. Soc. Jpn, 55: 2029
Yazami, R., and Touzain, Ph., 1985, Synth. Met. 12: 499.

REACTIVITY OF LOW DIMENSIONAL TRANSITION METAL DERIVATIVES TOWARDS CATIONIC INTERCALATION CHEMISTRY

Jean Rouxel

Laboratoire de Chimie des Solides, U.A. CNRS n°279
Université de Nantes, 2, rue de la Houssinière
44072 Nantes Cédex 03, France

Cationic intercalation compounds are usually formed through a reversible redox process involving a guest species and a host.[1a,b,c,d] These topochemical ion-electron transfer processes[2] were initially developped in the case of two dimensional host structures, which actually justified utterly the name of intercalation used for these reactions. Later on, and due to the variety of fundamental topics involved and to the numerous possible applications, the intercalation concept was extended to reactions concerning three dimensional channel structures. However, in that case, there will be strongly limiting geometrical constraints in connection with a rigid framework. The ideal host structure would be, on the contrary, a deformable one, able to adapt itself to each type of ions, which was precisely the case of low dimensional solids. Indeed, two-dimensional and one-dimensional structures remain basically the ideal host structures when one wants to better understand the intercalation reaction itself. The present paper will consequently deal with such materials like transition metal di- or trichalcogenides, halides and oxihalides. A particular attention will be paid to the electronic exchange between guest and host, this aspect being sometimes somewhat neglected as compared to purely geometrical topics. The reverse of the intercalation process will be also considered in this contribution. It leads to a modification of already known host structures, or to new ones, which are generally metastable and very reactive.

1. SOME GENERAL CONSIDERATIONS ABOUT CHALCOGENIDES HOST STRUCTURES AND THE REDOX PROCESS

Let us consider a low dimensional solid. It can be regarded as built up from a stacking of slabs or a juxtaposition of fibers. The van der Waals gap that separates slabs or fibers is bonded on each side by atomic layers of the same nature (S^{2-} anions in the case of chalcogenides). For a given structural type to be stable the repulsion between these similar anionic layer must not be two strong, i.e. the system must not be too ionic. If one looks at an ionicity-structure diagram such as a Mooser-Pearson diagram,[3] two-dimensional solids are to be found in the lowest ionicities regions. In the case of oxides the repulsion is quite strong and very destabilizing. Most of the MO_2 oxides have the rutile structure and not the layer structure of the parent dichalcogenides. Layered oxides can be found only in the case of the highest oxidation states of the metal, which means a strong polarization of the electronic density of the anions towards the inside of the slabs (this is the MoO_3 case for example). Otherwise hydrogen bonds should develop through the van der Waals gap, or extra cations, between the slabs, have to increase the stability of the structure by minimizing the slab to slab repulsion through their separation and by introducing attractive coulombic forces. The existance of Na_xMO_2 phases[4] isotypic with Na_xTiS_2 phases is a consequence of this. Such phases form also the basis of low-temperature deintercalation processes leading to layered oxides that would not be obtained through usual solid state techniques. What has also to be remembered here, is that the structure will be more two dimensional or one dimensional if an increase of the ionic character introduces increased repulsive interactions between slabs or chains. Of course this will induce some instability for the structure and the compound may become unstable. But as far as intercalation chemistry is concerned, the van der Waals gap which has to accommodate the guest species, is largely determined by this slab to slab repulsion, i.e. by the ionicity of the bonds in the slabs. It may be possible that at the very beginning, the intercalation process is also favored by the electronic transfer which determines an opening of the slab to slab separation prior to any ion intercalation.

Now let us have a first look at the redox process. Figure 1 shows a general band scheme for transition metal dichalcogenides (T.M.D.C.). Between a valence band essentially sp anionic in character and antibonding levels formed from the corresponding cationic levels, the d orbitals of the metal, split by the crystal field play an essential role concerning the chemical reactivity as well as the physical properties.This leads to introduce the two classes of TMDC. In both cases the slabs are made of three atomic layers with two atomic layers framing the cationic one, but the coordination of the metal in the slabs can be either octahedral like in TiS_2 or trigonal prismatic like in 2H NbS_2. Representing the anionic layers by A, B, C, and the cationic planes by a, b, c, the CdI_2 structure of TiS_2 will be described by the AbC, AbC

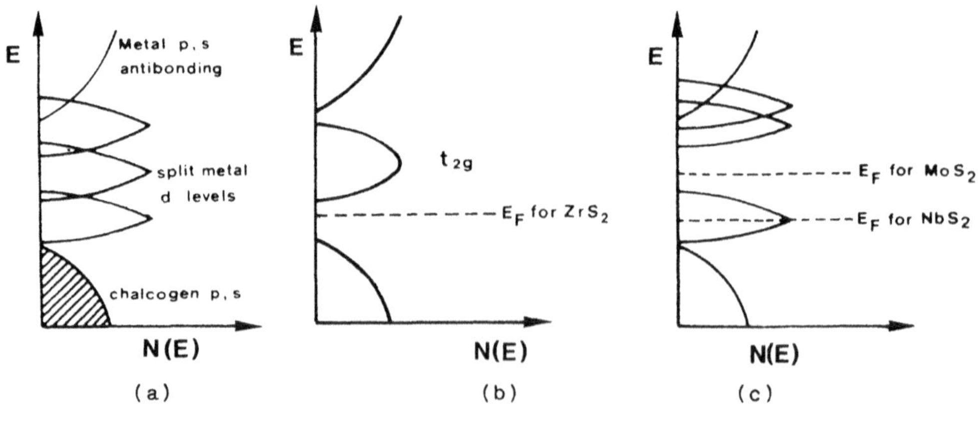

Fig.1. Band structure models for dichalcogenides : (a) general scheme, (b) T.M.D.C. with octahedral slabs, (c) T.M.D.C. with trigonal prismatic slabs.

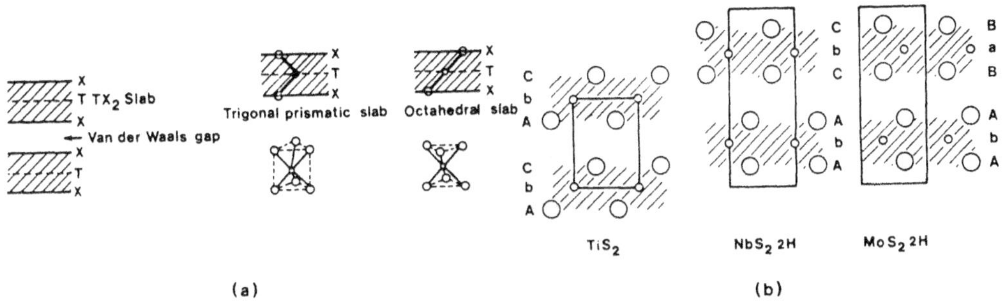

Fig. 2. (a) Basic features for a representation of lamellar dichalcogenides. (b) TiS_2, 2H NbS_2 and MoS_2 structural types.

sequence with a $(AC)_n$ close packing of the anions (Figure 2). The 2H NbS_2 structure will appear as AbACbC, the anionic stacking consisting of blocks of two AA or CC superposing anions layers. In both cases (TiS_2 and NbS_2) octahedral sites exist in the van der Waals gap between C and A (TiS_2) or A and C (NbS_2). MoS_2 shows a first difference in respect of stacking of trigonal prismatic slabs, the atomic sequence being AbABaB with B sulfur layer above the b cationic layer and a similar arrangement for a and A layers. More generally gliding motions of the slabs one over the other explain the large number of polytypes that can be observed and are referred to in the Ramsdell notation.[5] 2H means two slabs with hexagonal symmetry, 3R is for three slabs with rhombohedral symmetry, etc.

In the case of IVA elements we have octahedral slabs and the lowest d levels built a broad t_{2g} band able to accommodate six electrons. Then, the VA elements with a +4 oxidation state, will bring one d electron. A lowering of the symmetry of the coordination polyhedron with subsequent splitting of the bands is energetically favorable. We move from octahedral symmetry to trigonal prismatic symmetry and the lowest d levels are now involved in a narrow a'_1 (dz^2) band. This band is half filled for niobium in NbS_2 which is metallic. It is completely filled for its neighbor, molybdenum in the semi-conducting MoS_2. Going further to the right the next column would present once more "octahedral" dichalcogenides with a half filled t_{2g} band (no stabilization in trigonal prismatic symmetry). But this is not observed because there is an additional phenomenon which is very important and concerns the decrease in expansion of d orbitals and their progressive lowering. At a given moment they will enter the sp anionic valence band. If a filled d level is in this situation there is no drastic change in the properties of the material. But if an empty d level enters the sp anionic band it will fill up at the expanse of this band at the higher part of which holes appear (Figure 3). The result is a reduction of the cation along with the formation of $(S_2)^{2-}$ pairs as illustrated by pyrite and marcasite (6).

The question there arises to know where are the redox centers and we shall have to consider this redox competition between cations and anions in connection with intercalation chemistry. Niobium triselenide provides a very fine example of the redox competition, with a segregation in the same structure of chains illustrating various possibilities. Its structure (Figure 4) is built up from trigonal prismatic ($NbSe_3$) chains[7] running along the b axis of a monoclinic unit cell and formed from a stacking of $[NbSe_6]$ prisms through the sharing of Se_3 triangular basis.

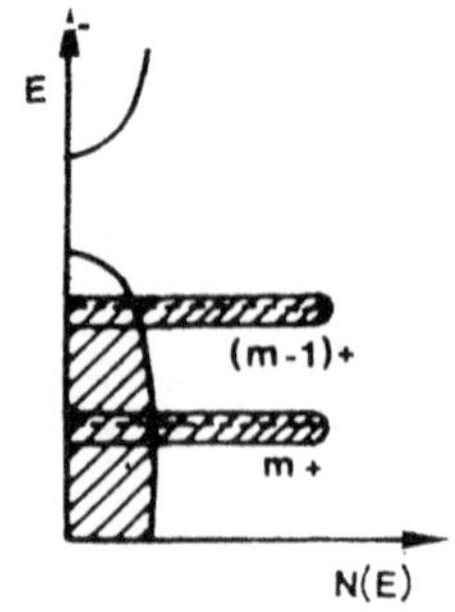

Fig. 3. d levels ans sp anionic band at the end of a period.

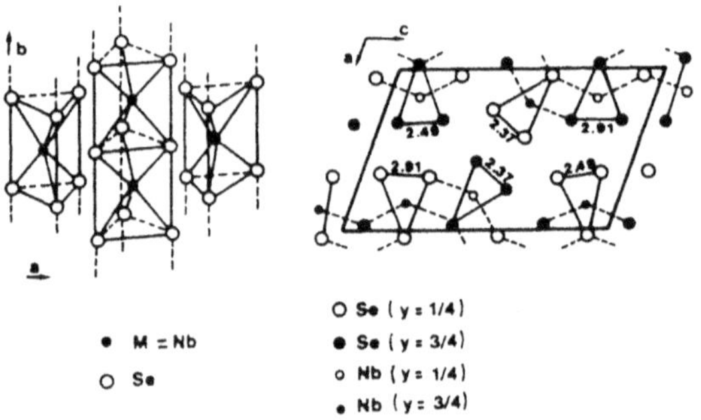

Fig. 4. The three types of chains and their relative arrangement in $NbSe_3$.

Three types of chains, each being repeated twice, are found in the unit cell according to the length of a Se–Se bond (respectively 2.37, 2.48 and 2.91 Å). Neighboring chains are displaced by $\vec{b}/2$. Figure 5 illustrates the electronic structure of a $[NbSe_6]$ prism in D_{3h} (regular prism with an equilateral basis) and C_{2v} (irregular prism) geometries.[8] With the D_{3h} symmetry we have the classical splitting of the d orbitals with a'_1 (dz^2) and then e' (dxy, dx^2-y^2) below e" (dxy, dyz). The highest orbital set, e", derived from the six selenium atoms is above a block of filled Se levels. The electronic situation of $[NbSe_6]^{6-}$ corresponds there to Nb^{5+} with all d levels being empty, and a hole in the Se e" levels. The distortion to C_{2v} symmetry illustrates the formation of a Se–Se bond, with a stabilized Se–Se bonding level and a destabilized antibonding Se–Se level. Now one electron occupies the a'_1 level of a cation reduced at the 4+ oxidation state. The 2.37 Å chain is in the latter situation. It is a conducting chain with a d^1 electronic configuration. The 2.91 Å chain is an insulating chain with Nb^{5+}, d^0 configuration. The 2.48 Å chain represents an intermediate situation. Of course the extension of the calculation to the overall band structure of $NbSe_3$ will change this first approach, particularly because $NbSe_3$ with strong lateral bonds is not purely 1D. But this scheme shows nicely the complexity of things.

To stay with the host structures themselves let us say that the band structure allows also to derive some considerations about the stability and stoichiometry of the host structures. This is clearly the case if one considers the MPS_3 family of layered materials.[9,10] These compounds

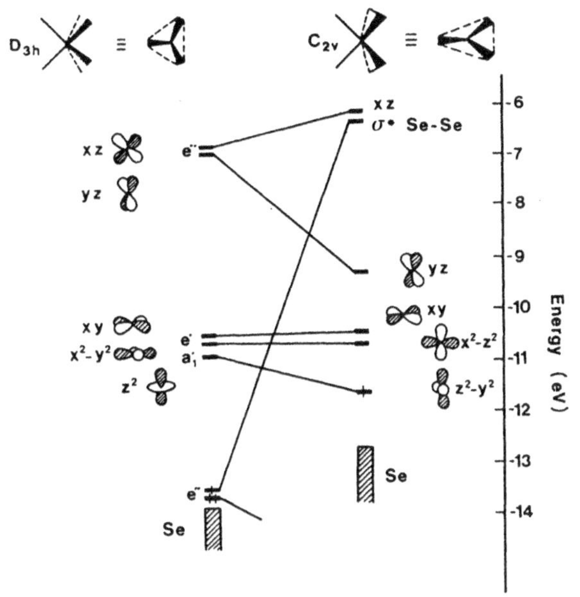

Fig. 5. Electronic structures of a $[NbSe_6]$ prism in D_{3h} and C_{2v} geometries.

are built up from "octahedral" slabs like TiS_2 except that the octahedral holes in the [SMS] sandwiches are shared by M^{2+} cations and (P_2) pairs in the proportions of 2/3 and 1/3. $NiPS_3$ at the end of the first transition period is rather stable. But at the beginning one will get for example $V_{0.78}PS_3$, and not VPS_3 as published earlier. Due to the stability of V^{3+} in respect to the top of the valence band the stoichiometry is affected by the presence of V^{3+} ions. At the beginning of a period one can also find extra cations in between the slabs ($Ti_{1+x}S_2$, $Nb_{1+x}S_2$) with a correlative reduction of the high oxidation states that are theoretically possible for the cation involved in the slabs.

An other very important point concerning the host structures is the crystal field stabilization. It explains the non reproducibiity of certain discharge curves when cycling. Indeed, if we conseder once more the MPS_3 family which has the advantage of being spread out over a complete transition metal period of the classification, it appears that some MPS_3 phases allow to substitute the M^{2+} cations either by alkali-metal or organometallic cations. This can be done at moderate temperature (20–100°C) in aqueous or alcoholic solvents, with or without the use of complexing agents.[11] Basically the reaction can be written :

$$M^{II}PS_3 + 2x\ M^{I}X \longrightarrow M^{II}_{1-x}\ M^{I}_{2x}\ PS_3 + x\ M^{II}\ 2X^-$$

It appears clearly that substitution is easily achieved for d^5 or d^{10} cations such as Mn^{2+}, Cd^{2+}, Zn^{2+} without crystal field stabilization in octahedral symmetry. It remains possible with the help of a complexing reagent in the case of $FePS_3$ and even $CoPS_3$. But it is impossible in $NiPS_3$. The thermal agitation factors of the cations already gave an indication of this behavior.[10] The cations with no crystal field stabilization have clearly the higher thermal agitation. These results explain why $NiPS_3$, with no deviation from stoichiometry due to the non stability of Ni^{3+} in front of sulfur, a high crystal field stabilization preventing from any exchange, represents the most reliable MPS_3 host structure and consequently the best cathodic material with a very good behavior in cycling.

2. ELECTRONIC TRANSFER AND REACTIVITY

In the search of host structures to do intercalation chemistry two conditions have to be fullfilled (i) to get a skeleton presenting cages, tunnels or large van der Waals gaps between fibers or layers, in which guest ions are easily moving (ii) to have redox centers able to exchange electrons with the intercalated species within that skeleton. The latter condition plays an essential role. As we shall see it may prevent any intercalation process to occur in spite of the presence of available sites with suitable size and geometry.

In the case of alkali-metals intercalates electrons are donated to the host and three different situations can be recognized corresponding to three different steps in the delocalization of the transferred electrons. Indeed the accepting level can be a discrete atomic one, it can be a molecular level of a discrete polyatomic entity existing in the structure, and finally it can be situated in a conduction band. We shall illustrate that with a few examples.

2.1. Localized redox centers

The first case mentioned just above corresponds to a reduction of a given ion in the host network, for instance Fe^{3+} to Fe^{2+} in FeOCl intercalates.[12,13] At low temperatures the transferred electron is effectively localized on Fe^{2+} centers and the number of such centers, that can be found by Mössbauer studies, equals the number of Li^+ ion in Li_xFeOCl. At increasing temperatures a classical hopping mechanism, associated to a small polarons exchange between Fe^{3+} and Fe^{2+}, appears. Figure 6 represents the FeOCl structural type with slabs built up from double metal-oxygen layers bonded on each side by chlorine layers.

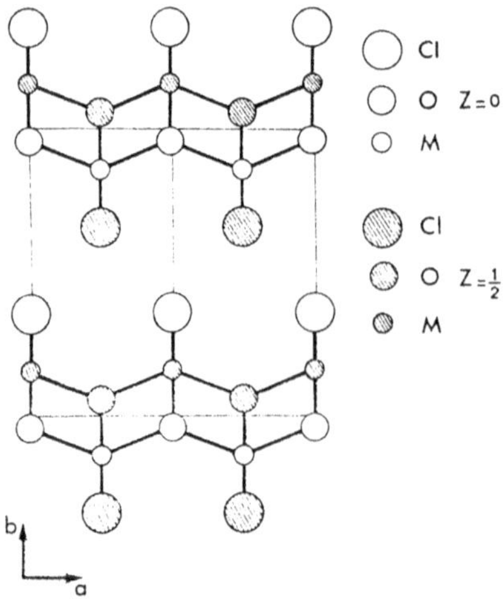

Fig. 6. The FeOCl structural type.

Intercalation in such frameworks depends on factors which are the rigidity of the lattice and the stability of the +3 oxidation state of the metal.[12] The rigidity of the MOX lattice of course means the more or less easy separation of the slabs. Indications concerning this behavior are to be found in thermal dilatation coefficients of the crystalline parameters. The unit cell parameters of FeOCl, VOCl, VOBr,[14] CrOCl and CrOBr have been refined and studied according to temperature[15] with the help of a very precise camera. Table I gives the corresponding values which allow to calculate thermal dilatation coefficients. Substitution of bromine by chlorine leads to a Δb difference of 0.49 Å in the direction perpendicular to the slabs both in VOCl-VOBr and CrOCl-CrOBr systems. A very interesting point is the value observed for the α_b thermal coefficient in the direction perpendicular to the slabs. This value is much higher in the case of FeOCl (α_b = 21.7) than in the case of VOCl (α_b = 15.7) or CrOCl (α_b = 15). In addition the α_a and α_c coefficients (associated to the slabs) express a higher rigidity of the MOX slabs in the case of CrOCl. This last point is probably related to a most symmetrical electronic population around Cr^{3+} ions. In fact, using identical positions in the Pmnm space group, it is possible to calculate more homogeneous distances in the $[MO_4Cl_2]$ octahedron in the case of chromium than vanadium for example (Table II). The higher value of α_b in the case of FeOCl agrees with the fact that molecular intercalation, which needs a large Δb separation, is easier in FeOCl than in any other oxyhalide.

Table I. Crystalline parameters and thermal coefficients in some oxihalides. Mean thermal dilatation coefficients between 78 and 300 K ($\times 10^6$ K^{-1}).

	a (Å)	b (Å)	c (Å)	γ (°)	α_a(1)	α_b(1)	α_c(1)	α_v(1)
FeOCl								
300 K	3.7729(5)	7.9104(3)	3.3026(4)	90	-	-	-	-
78 K	3.7667(5)	7.8733(3)	3.2961(4)	90	7.6(6)	21.7(6)	9.0(7)	33(2)
VOCl								
300 K	3.7734(5)	7.926(1)	3.2992(5)	90	-	-	-	-
78 K	3.7660(5)	7.900(1)	3.2890(5)	90	8.1(5)	15.7(5)	13.5(5)	37(1.5)
VOBr								
300 K	3.778(1)	8.411(1)	3.405(1)	90	-	-	-	-
78 K	3.772(1)	8.377(1)	3.401(1)	90	6.7(5)	18.6(5)	5.0(5)	30(1.5)
CrOCl								
300 K	3.855(1)	7.737(1)	3.166(1)	90	-	-	-	-
78 K	3.852(1)	7.714(1)	3.163(1)	90	3(1)	15(1)	5(1)	23(1)
CrOBr								
300 K	3.861(1)	8.191(4)	3.230(1)	89.94	-	-	-	-
78 K	3.858(1)	8.163(4)	3.224(1)	89.67	3(1)	20(1)	8(1)	31(3)
VOCl + p-methyl-pyridine								
300 K	3.792(5)	27.646(11)	3.225(5)	90	8(1)	39(2)	15(1)	62(4)
78 K	3.785(5)	27.409(11)	3.236(5)	90	-	-	-	-

Table II. Distances in the MO_4Cl_2 octahedron (Angströms).

$V-O_1$	1.97	$Cr-O_1$	2.00
$V-O_2$	2.03	$Cr-O_2$	2.01
V-Cl	2.35	Cr-Cl	2.28

Table III Redox potentials (volts) of some M^{3+}/M^{2+} couples.

Fe^{3+}/Fe^{2+}	0.77
V^{3+}/V^{2+}	- 0.25
Cr^{3+}/Cr^{2+}	- 0.41

As far as the reduction of the M^{3+} cation is concerned, FeOCl is once more the most favorable case (Table III). This observation will also have drastic effects on the type of molecules that could be intercalated in the various hosts. Intercalation in FeOCl may occur for molecules presenting a pk_a factor at least equal to 2.8 whereas the corresponding values are 5.2[a] and 10.5 respectively for VOCl and CrOCl.[16] Similar considerations have been introduced for transition metal dichalcogenides and MPS_3 phases.[17] Various intercalates of FeOCl and other oxyhalides have been prepared, including amines,[18] metallocenes,[19] a crownether[13b] and TTF[20] types or 1D organic phases.

Lithium intercalates were prepared using the butyllithium technique or electrochemically.[21] Figure 7 shows various voltammetric sweeps corresponding to electrochemical intercalation processes. In the case of FeOCl two well-defined waves corresponding to the intercalation and deintercalation processes are appearent. The value of the potential at the cathodic current peak gives E_0 (FeOCl) = -270 mV vs.Ag/Ag^+ and indicates a reduction of Fe^{3+} species. The reoxidation wave at -80 mV corresponds to a one cycle charge recuperation in excess of 80 % which is in agreement with the good reversibility observed chemically. The two chromium oxyhalides are characterized by the stability of the Cr^{3+} ions. The chloride gives E_0 (CrOCl) = 1340 mV and the bromide E_0 (CrOBr) = 940 mV. As expected, the Cl^- ions stabilize the higher valency of chromium more effectively than the Br^- ions. The system shows a bad reversibility because no reoxidation peak is manifested. Similar conclusions can be drawn for VOCl.

An other very interesting examples concerns the layered semiconductor $\alpha RuCl_3$. It has a vacant layered dichalcogenide structural type (except that the stacking sequence of anions is ABC and not AB like in TiS_2), with an octahedral slab which can be represented by $[Cl\ \square_{1/3}Ru_{2/3}Cl]$. $\alpha RuCl_3$ undergo intercalation redox reactions.[22]

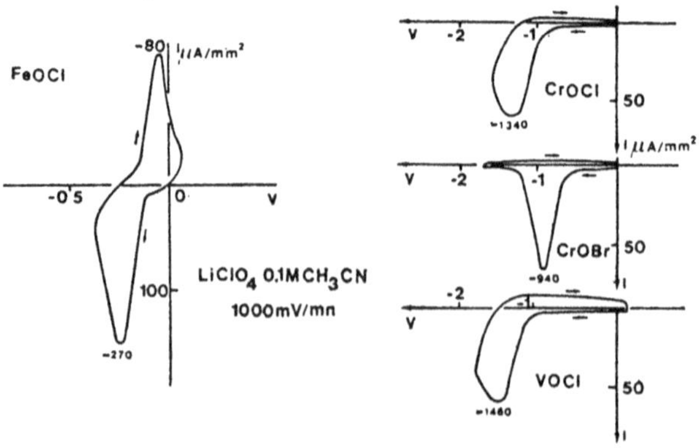

Fig. 7. Cyclic voltammetric sweeps for oxyhalides.

Formally the number of octahedral vacant sites is equal to one per slab formula representation as above, i.e. 1.5 per ruthenium atom. But the maximum charge transfer observed is $[RuCl_3]^{-1}$. Reduction below Ru^{II} is not possible. It is a first example of an electronically limited intercalation reaction. The reduction of α $RuCl_3$ by lithium in $LiClO_4$/propylène carbonate leads to a very large change in interlayer spacing in relation with the uptake of Li^+ solvated ions. $RuCl_3$ presents a very rich intercalation chemistry of solvated ions that will be developped in another chapter of this volume (see R. Schlöllhorn)[10]. Unsolvated phases can be obtained with the less electropositive d cations Cu^+ and Ag^+. $Cu_{0.50}RuCl_3$ is a black mixed valence compound whereas Cu_1RuCl_3 is a yellow opaque phase rather unstable in air and easily oxidized to black $Cu_{0.50}RuCl_3$ under formation of copper oxide Cu_2O. Obviously the electronic delocalization in $Cu_{0.50}RuCl_3$ stabilizes the situation.

2.2. Partial delocalization

Chevrel phases[23] built up from Mo_6 octahedral clusters enclosed in pseudo cubes of eight chalcogen atoms illustrate this situation with transferred electrons that are delocalized among a finite number of directly bonded metal atoms. Intercalation can be performed in a Mo_6X_8 host obtained from deintercalation of a true $M_xMo_6X_8$ Chevrel phase (X = S, Se) or from a direct preparation from the elements of the thermodynamically stable binary phase Mo_6Se_8. However in the latter case products with low reactivity are obtained, very likely as a consequence of defects blacking the empty lattice channels. In the $M_xMo_6X_8$, M cations occupy some of the numerous empty sites between Mo_6X_8 pseudo cubic asemblies. $Cu_2Mo_6X_8$ samples[24,25] are usually the starting materials in the deintercalation process.

For small ions the structure[26] provides formally 12 potential lattice sites in the unit cell, which lead in fact to a maximum of six cations expected per Mo_6X_8 unit when taking into account coulombic repulsions between monovalent cations. However the molecular levels of the Mo_6 cluster have 20 electrons. Four additional electrons can be taken up to reach the maximum allowed population of 24 electrons. This gives an electronic limit associated to a maximal intercalation of 4 monovalent cations or 2 bivalent cations and so on.

Du to its importance towards batteries and due to the fact that it has the smallest size, thus minimizing the steric constraints, intercalation of Li^+ ions has been thoroughly studied.[27-29] A critical and complete study is given by R. Schöllhorn.[30] Indeed the end point of the reaction is attained after a transfer of 4 electrons per Mo_6X_8 group. The whole intercalation domain is marked by the presence of $Li_1Mo_6S_8$ and $Li_3Mo_6S_8$ phases. There is no continuous intercalation $Li_xMo_6X_8$, as proposed by some authors[31] on the basis of in-situ X rays studies. These transfers of an even number of electrons should find a significance in the population of the Mo_6 cluster levels. Between $Li_3Mo_6S_8$ and $Li_4Mo_6S_8$ things are a bit more complicated with a semi continuous process interrupted maybe by a two phase region ($3.6 \leqslant x \leqslant 3.8$). The band structure of the host may also shed some light on the pseudo plateau observed in the discharge curves (see section 2.3). A maximum transfer of 4 electrons is also observed for Mo_6Se_8.

In addition to this situation of an electronically limited intercalation process the study of $Li_xMo_6S_8$ phases brought interesting problems about the existence of Li_3^{3+x} clusters in $Li_3Mo_6S_8$. The large Knight shift observed (δ = -100 ppm at 135 K), the most likely geometrical arrangement of Li atoms are in favor of such lithium aggregates. The terminal situation[32] $Li_4Mo_6S_8$ corresponds to the expected electronic situation $(Li^+)_4Mo_6S_8^{4-}$.

2.3. Delocalization of the transferred electron

Let us come back to the band structures of transition metal dichalcogenides. It is much easier to intercalate in octahedral chalcogenides with the broad t_{2g} band (TiS_2, ZrS_2) than in trigonal prismatic chalcogenides (NbS_2, TaS_2) with a half filled narrow dz^2 band. Intercalation compounds formed with MoS_2 appear to be most unstable : this can be related to the fact that the dz^2 band is already filled and the electron has to be accommodated on a higher level. The case of TaS_2 is of particular interest. It shows an octahedral form as well as a trigonal prismatic one. In both cases similar octahedral sites exist in the van der Waals gap. Consequently the geometrical conditions for intercalation are similar and differences are to be attributed to the band structure. Indeed it appears that there is no problem to intercalate up to x = 1 in the broad band 1T form (all available sites occupied which means a geometrical limitation), but it is rather difficult to reach this limit in the narrow band trigonal prismatic structure of 2H TaS_2.[33] Electrochemically it seams possible to force intercalation up to x = 1[2,34] but longly annealed samples reach an equilibrium at x = 0.72 which agrees well with previous solid state chemistry work.[35]

Indeed it is worthwile to notice that an intercalation process with a complete delocalization of the electron will be influenced not only by the number of empty levels but also by the structure of the band itself, i.e. by the density of states. A pseudo-plateau in the potential variation of a discharge curve may be related to the electronic filling of a high density of states zone. This plateau would not mean a two-phase region but illustrates the fact that the difference in energy of electrons in the alkali metal, i.e. before intercalation, and in the host, i.e. after intercalation, remains quite constant. Such an idea[36] can be readily illustrated in the case of $NbSe_3$ or $FeNb_3Se_{10}$.[37] It is also proposed in the case of A_xMO_2 phases.[38]

The $FeNb_3Se_{10}$ structural type (figure 8) associates two different types of chains running two by two parallel to the b axis.[39] One is

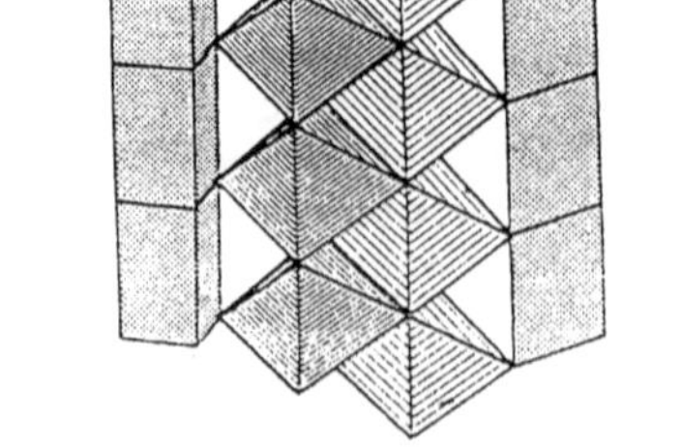

Fig. 8. The $FeNb_3Se_{10}$ structural type.

exactly the trigonal prismatic chain of $NbSe_3$ with the shortest Se-Se bong (2.37 Å), the second shows an octahedral distribution of selenium atoms around both niobium and iron, which allows some non stoichiometry of the $Fe_{1+x}Nb_{3-x}se_{10}$ type (x < 0.33). On contrary to $NbSe_3$ (see part 4.5) the selenium pairs do not participate in the redox process. The oxidation formalism :

$$[Nb^{4+}_{1-x}\ Nb^{5+}_{x}(Se_2)^{2-}Se^{2-}_{T.P.Chains}]_2\ [Fe^{3+}_{1+x}Nb^{5+}_{1-x}(Se^{2-})_4\ \text{oct. chains}],$$

shows that a reduction of Nb^{5+} to Nb^{4+} and Fe^{3+} to Fe^{2+} will lead to an electronic transfer of 2.66 e^- for x = 0.33. This fits well with the limit indicated by the discharge curve (Figure 9a).

In the band structure (Figure 9b) calculated by Whangbo et al[40] four bands labelled (a), (b), (c), (d), appear in the vicinity of the Fermi-level. These bands have indeed a metal atoms d orbitals character. Band

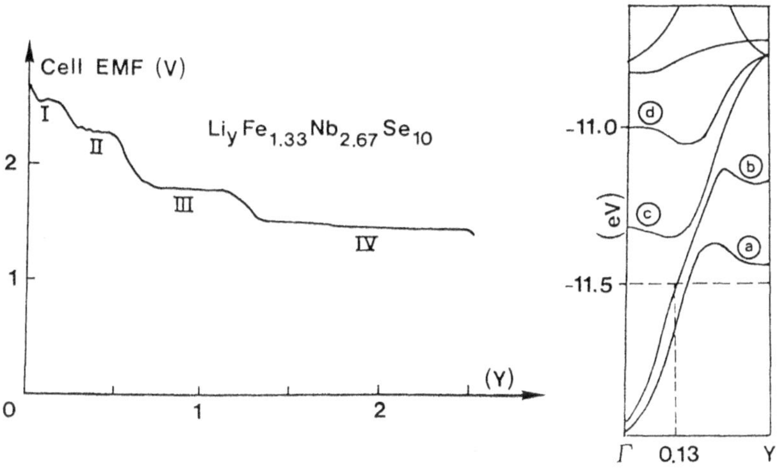

(a) (b)

Fig. 9. (a) discharge curve, (b) band structure of $FeNb_3Se_{10}$.

(a) (and also band (b)) is at first dispersive (along the b* direction). This band, first concerned by the electronic transfer should then induce a negative slope in the discharge curve (part AB of this curve). Then, band (a) becomes flat. It should correspond at that point a great density of states and thus the corresponding filling is done at nearly the same potential (pseudo plateau I). A second slope has to be observed between the top of (a) and the bottom of (c). Reaching the bottom of (c) the potential does not vary (plateau II). From the bottom of band (c) to the top of (b) both bands exhibit the greatest dispersion. Again at the top of (b) one observes a flateness of that band which correspond to plateau III. Between (b) and (d) the potential is again varying and finally the last plateau (IV) corresponds to the filling of band (d).

3. GEOMETRICAL FACTORS

Intercalation usually results in geometrical effects directly manifested through lattice parameter variations, essentially a parameter expansion in the direction perpendicular to the slabs of the host. This expansion is a global effect which may hide the fact that intercalated A^+ ions determine various structures,[16] and may be ordered or not.[41] It also includes an electronic component.

In layered transition metal dichalcogenides A^+ ions can occupy either octahedral or trigonal prismatic sites between the slabs. The first ones are already present in the host structure and do not need any rearrangement. Trigonal prismatic sites imply a shifting of one slab out of two in order to get on each side of the van der Waals gap sulfur layers that are directly superimposables. It implies some cost in energy that has to be compensated as we shall see in next section. The occupancy of octahedral or trigonal prismatic sites stems from three factors : the size of the alkali metal, the amount of intercalation, the ionicity of the host structure.[42] An octahedron can accommodate higher charges on the anions, it represents the best equilibrium between such charges. Thus the general ionization scheme xA^+, MS_2^{x-} explains why for a given alkali metal, the octahedral form may appear for the higher values of x (higher charge transfer), whereas the trigonal prismatic forms are to be observed for the lower values of x. With a bigger alkali metal the sulfur layers are more distant and the trigonal prismatic structure is favored. With a more ionic host structure the formal charge on the anions is increased, which gives an initial advantage to octahedral arrangements. For example, obtained from the more ionic host ZrS_2, $KZrS_2$ is octahedral, but $KTiS_2$ is trigonal prismatic. All these remarks can be rationalized and illustrated in a general ionicity-structure diagram.[43] By plotting the $r_{A^+}/r_{S^{2-}}$ ratio versus a function related to the amount of intercalation x and the fractional ionicities (f_i and f'_i) of the A-S and M-S bonds, it is possible to draw an unambiguous limit between octahedral and trigonal prismatic regions (Figure 10).

Let us consider now the Δc parameter expansion in the direction perpendicular to the slabs of the host structure. It is often considered as the proof that an intercalation process has really taken place. But one should notice that (i) there may be no Δc at all in a given intercalation process and (ii) this parameter expansion has not a purely geometrical origin but comes also from electronic effects. The geometrical contribution depends on the size of the A^+ ion : for a similar van der Waals gap, the same amount of intercalation and the same structural type, it is directly proportional to the radius of the A^+ ion (Figure 11a). It depends also on the initial size of the van der Waals gap which may be represented by the sulfur-sulfur distance on each side (Figure 11b). One can extrapolate a critical value of about 3.20 Å for which no Δc would be observed. This is the case of $NiPS_3$ and, indeed, there is no Δc expansion when Li^+ ions are intercalated in $NiPS_3$. Now, the electronic transfer from guest to host contributes to Δc through two opposite effects, i.e. an increasing repulsion between slabs that are now negatively charged and an attraction between these slabs and the A^+ positive layers. The balance is certainly in favor of the latter as shown by the existence of Na_xMO_2 phases with a layer structure stabilized by intercalation.[4,44] But, if the A^+ ion is small, it may not separate sufficiently the slabs to minimize the repulsions and the slab to slab repulsion will be the essential contribution to Δc. Such a situation has been proposed for $Li-TiS_2$.[45] More generally, it can provide a representation of the very first step of the intercalation process with an opening of the interslab space from electronic origin, before the ions

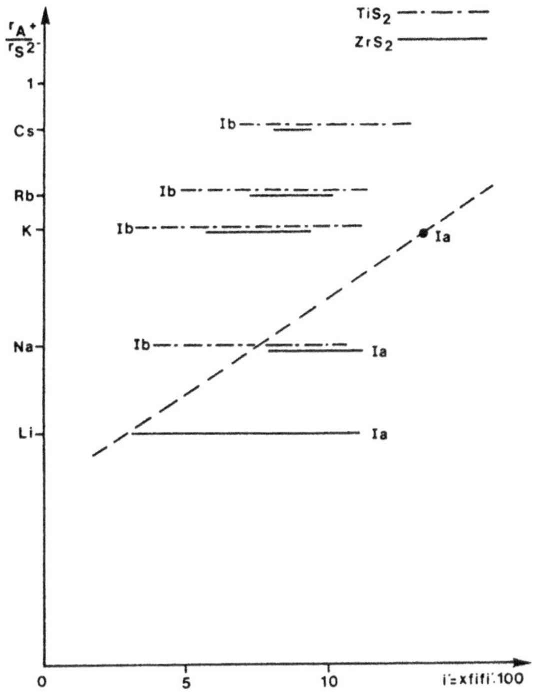

Fig. 10. Ionicity-structure diagram for A_xTX_2 intercalates in CdI_2 like hosts structures.

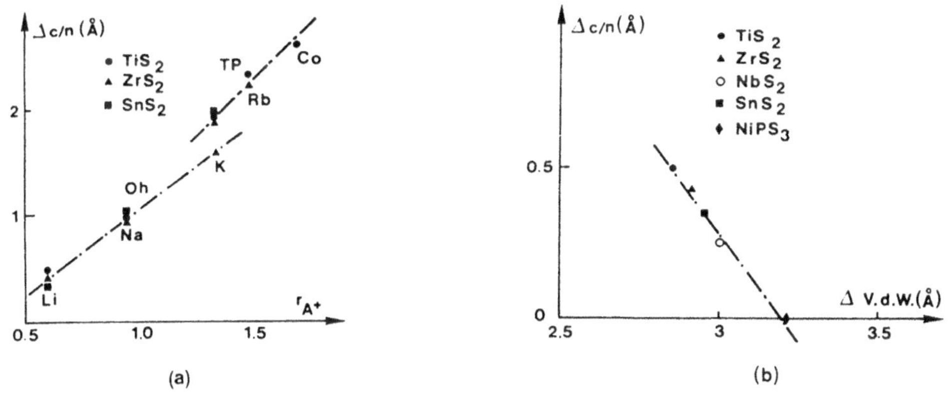

Fig. 11. (a) c parameter expansion for one slab (Δc/n) as a function of the alkali metal size (r_{A}+) and its site symmetry (trigonal prismatic or octahedral) ; (b) (Δc/n expansion for lithium in various hosts as a function of the initial size of the van der Waals gap.

enter the van der Waals gap. It is also highly favorable to obtain a dispersion of the slabs, like a dispersion of polymers, in a solvent. This was nicely illustrated for the first time in the case of $Li_2Mo_6Se_6$. When exposed to highly polar solvents such as dimethyl sulfoxide or N-methylformamide, $Li_2Mo_6Se_6$ and $Na_xMo_6Se_6$ can be dissolved.[46] The absence of swelling for the $M_2Mo_6X_6$ compounds with M = K, Rb, Cs, is clearly the result of a competition between a cationic solvation energy and a lattice energy which, as we mentioned above, will increase with the size of the cationic specy.

Finally, concerning this overview of reactivity and geometrical factors, let us recall that low dimensional chalcogenides are oftentime non-stoichiometric due to the presence of extra metal atoms in the van der Waals gap. These metal atoms link the slabs together and impede the diffusion of intercalation species ($Ti_{1+x}S_2$ for example). Most of the advantages of layered structures associated with the elasticity and local deformation of the slabs are lost.

4. STRONGLY COUPLED ELECTRONIC-GEOMETRICAL EFFECTS

4.1. Δc and long range information

We have just mentioned above that Δc cannot be reduced to a purely geometrical effect. It appears as the result of a competition between geometrical and electronic factors. In fact Δc is practically deduced from X rays measurements that are average long range observations, which implies a good transfer of information over the whole structure. Precisely an electronic delocalization acts as a good information transmitter. When the electron transferred from guest to host is trapped in a particular site of a non metallic host structure the situation is rather different. There, the host may be able to accept little amounts of a small cation without any important parameter expansion. A local distortion is progressively smoothened and non observed through long range observations. But if the system becomes metallic, immediately the usual expansion is observed. The Li-$ZrSe_2$ system[47] provides a very nice example of that situation (see paper by P. Deniard in this volume).

4.2. Jahn-Teller systems

If Jahn-Teller distortions are involved inducing strains in the host lattice, largely biphased systems $A_\varepsilon MS_2$-A_1MS_2 will result.[48,49] This is for example the case of the $LiVSe_2$ system. The creation or the removal of the distortion reflects in an important cost in elastic energy which has to be overcome by a sufficient gain in electronic energy. One can also suppose that the slabs do not allow the existence of distorted and undistorted sites simultaneously (no dynamic Jahn-Teller effect).

4.3. Site occupancy modification in the host

When the electronic transfer to the host results in a reduction of Fe^{3+} to Fe^{2+} it can determine a change in site symmetry (Fe^{3+} in tetrahedral environment, Fe^{2+} in octahedral sites). This is well illustrated by the case of $Li_{1+x}FeCl_4$ with a progressive reduction of Fe^{3+} to Fe^{2+} when x increases from 0 to 1.

4.4. The origin of the Na_xTiS_2 phase transition

In the sodium rich region of this system one finds a phase with sodium in trigonal prismatic sites (0.40 < x < 0.64) and a phase with sodium in octahedral sites between TiS_2 slabs ($0.75 < x \leqslant 1$). A critical discussion of this phase transition stems from a combined study by X rays and ^{23}Na NMR,[51] and band structure calculations.[52] When going from the T.P. phase to the octahedral one, the cell volume remains nearly constant, but this is due to two opposite behaviors, i.e. a 3.9 % decrease of the c parameter and a 2.2 % increase of the a parameter. Such a transition costs elastic energy because of the strong anisotropy of the structure with a much larger stiffness in the slabs planes (a parameter) than along the c direction. To overcome this cost in elastic energy the transformation must lead to a lowering of the electronic energy. It appears that the sodium knight shift undergoes a 48 ppm increase when passing from the T.P. to the octahedral phase. In addition K_{axial} is negligible in the T.P. phase but not in the octahedral one. This implies an increase of the density of states during the transition and a anisotropic charge distribution around sodium in the octahedral phase. These statements have been confirmed by band structure calculations performed as a function of a parameter θ (Figure 12a) which characterizes the TiS_2 slab deformation at the transition. For the T.P. phase the energy is the lowest for special point M (1/2, 0)a* as in TiS_2, but for the octahedral phase this is observed for Γ point (0, 0)a*. As a function of θ the e_M and e_Γ levels vary and e_Γ becomes more stable for θ > 56.4° (Figure 12b). This critical value corresponds exactly to the structural discontinuity associated with the transition. The crossing of levels is directly responsible for the transition. Band structure calculations confirm also the increase in density of states that was suggested by K_{iso} measurements. Furthermore the analysis of the contribution of the d orbitals to the e_M and e_Γ levels shows a contribution of all d orbitals to the e_M level whereas e_Γ has no dz^2 contribution, which explains the variation of K_{axial} at the transition.

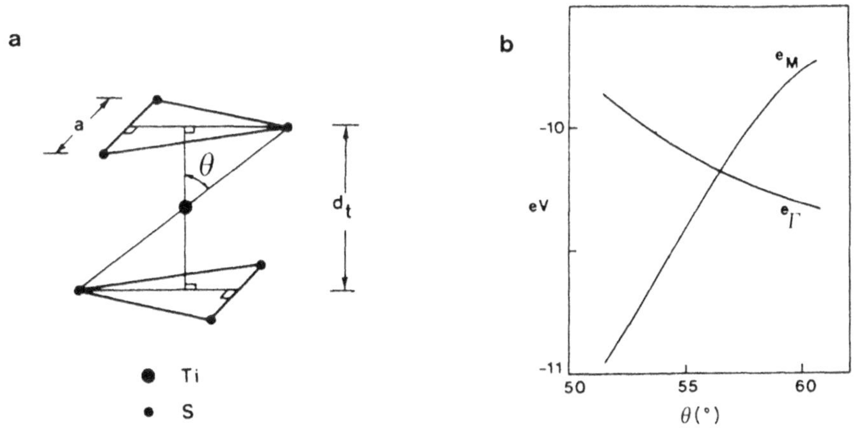

Fig.12. (a) The angle θ is a measure of a TiS_2 slab deformation during intercalation, (b) e_M and e_Γ levels as a function of θ .

4.5 Intercalation in $NbSe_3$

A critical discussion of the intercalation process in $NbSe_3$ includes both geometrical constraints and electronic considerations.

Three lithium are accommodated in the structure. Murphy and Trumbore[53] observed that at the beginning the host structure is preserved ($x \lesssim 1$). Then the structure seems to change drastically with a disappearance of many X-ray lines and the broadening of the remaining ones for $x \approx 2$. Finally, the third Li atom would cause the break up of the structure, leading to a different spectra. On the contrary, Chianelli[54] was able to index Li_3NbSe_3 powder pattern with a monoclinic unit cell $\sim$30 % larger than that of $NbSe_3$, and proposed a localization of lithium atoms all around the $NbSe_3$ chains.

Better than purely one-dimensional and due to rather strong lateral bonds, $NbSe_3$ has to be considered as a two dimensional compound with a high 1D anisotropy witing its slabs[55]. One has to consider genuine slabs made of coupled fibers (figure 13). Table IV gives the size of the empty octahedral and tetrahedral sites that can be found in the structure.

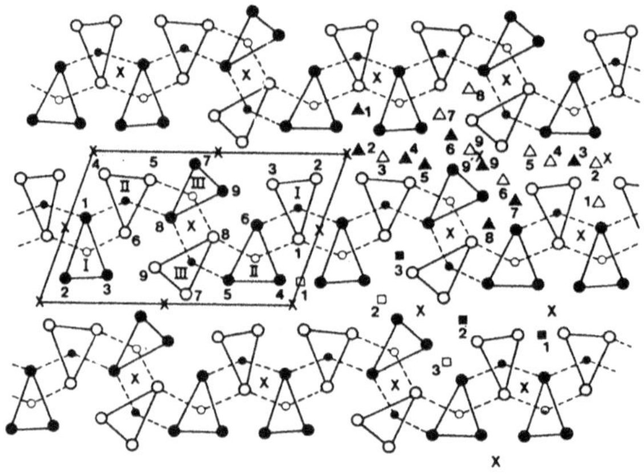

Fig. 13. Octahedral and tetrahedral sites available in the $NbSe_3$ structure.

Table IV The average distances from the center of the Os or Td sites to the Se atoms.

Os1–Se	2.43	Ts4–Se	2.29
Os2–Se	2.51	Ts5–Se	2.18
Os3–Se	2.37	Ts6–Se	2.17
Ts1–Se	2.04	Ts7–Se	2.06
Ts2–Se	2.20	Ts8–Se	2.05
Ts3–Se	2.26	Ts9–Se	2.13

The $LiNbSe_3$ composition will be reached with a total occupancy of the octahedral sites OS (1,2,3). These sites are the usual sites for a lithium intercalation in chalcogenides. Here, their sizes fit very well with a Li^+ accommodation which would occur without any large structural change. The agreement with the number of octahedral sites and the x value up to which the structure is perfectly preserved is a strong argument in favour of an occupancy of these sites. Then Li^+ ions will have to occupy tetrahedral sites. Most of these sites are unstable towards a lithium occupancy because they are too close to each other and also to octahedral sites ; only TS_4 sites can be considered. In addition a distorted octahedral site $OS^{9,9'}$ can also be occupied along with TS_4. Further intercalation will force the structure to distort. The upper limit allowing intercalation without disruption of the host corresponds to 9Li per unit cell ($Li_{1.5}NbSe_3$). These assumptions are in good agreement with the structural observations made by Murphy[53].

Now, the electronic transfer can be discussed on the basis of both the discharge curve and band structure calculations.[56] If we take into consideration a reduction of both the cations and the Se-Se pairs as suggested by the results of Chianelli, a first estimation of the global possible transfer stems from simple structural considerations. The insulating chain (Se-Se = 2.909 A) corresponds to $Nb^{5+}(d^0)$. A reduction to Nb^{4+} allows to accommodate one electron, i.e., 0.33 e^- for the whole structure. The two others chains show C.D.W. phenomenon. Each of them presents 0.5 e^- per niobium in agreement with a q vector of the C.D.W. distortion close to 0.25 b*. Reduction of niobium to "4+" oxydation state leads to a transfer of 0.5 e^- on each chain, i.e., one electron for two chains (among the three), and then 0.33 e^- relatively to the $NbSe_3$ formulation. Concerning the $(Se_2)^{2-}$ pairs, a pure Se-Se pair can accommodate two electrons. This will lead to a filling of the Se-Se antibonding states. The lengthening of the Se-Se distance which results is an indirect measure of the population of the antibonding states. Two selenium pairs are close to the classical value of 2.34 Å for the Se-Se distance in a true Se_2 pair. But the third one (2.909 Å) is much longer (though still shorter than the value of 3.30 A in $NbSe_2$ which corresponds to well separated anions). It will accommodate less than 2 e^-. For the three chains and the three pairs we shall not reach the maximum of six electrons (less than 2 e^- per chain in average). Finally the maximum allowed electronic transfer will be slightly below 2.66, which fits rather well with the 2.50 Li. Of course this can be taken only as a rough estimation. One could imagine a further reduction of the cation. But the fact that a rather good agreement with experiments is obtained is indicative of the validity of this approach.

Now a more precise analysis comes from the band structure calculations. Indeed, the discharge curve is directly related to the position of the Fermi-level and its shape will depend of the density of states that are being progressively filled during the electronic transfer associated with the intercalation process. Referring to the band structure calculations reported by Whangbo et al[57] (see figure 14b), the band filling is starting from the c,d,e,f bands. It is noticed that the bands d and e have mainly type III chain character (Se-Se = 2.37 Å) while the bands c and f have type I chain character (2.48 Å). However, the type II chain character at the Fermi-level is not vanishing, but small. This indicates that the band filling is concomitant and as these bands (c,d,e,f,g) show a large dispersion the filling is illustrated by an important variation of potential (part 1 of figure 14a). Arriving just below the h band, the smooth curvature of this band implies the filling at nearly the same potential (part 2 of figure 14a). After that the six bands are spread again indicating a continuously variation of potential (part 3). For x = 0.66, the Fermi-level is likely located in the middle of the h band, i.e., in the energy zone where bands start to flatten. Thus, the plateau (part 4) observed up to the breaking of potential (limit of intercalation) could be due to the filling of top of the c,d,e,f,g and h bands and also of bands of Se character.

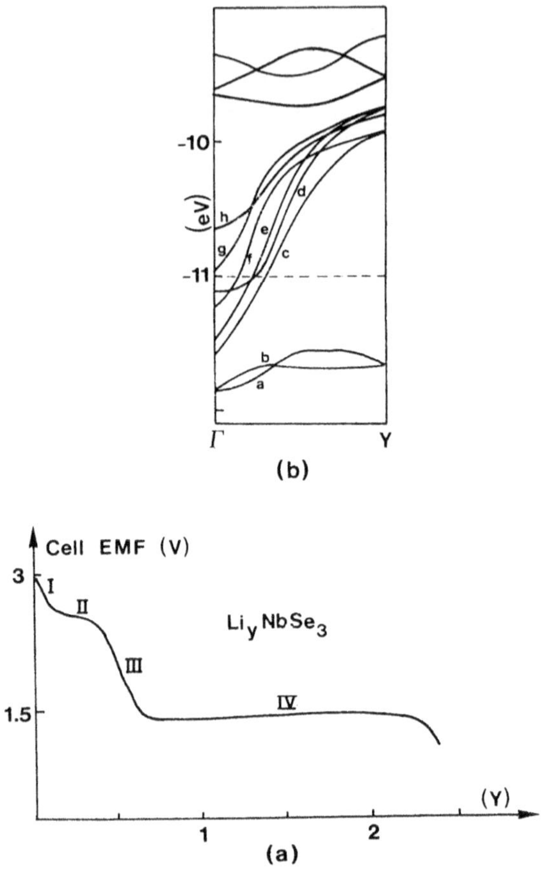

Fig. 14. (a) discharge curve and (b) band structure of $NbSe_3$.

5. DEINTERCALATION CHEMISTRY

5.1. Back to the intercalation reaction

Finally intercalation chemistry appears as a competition between a cost in strain energy, needed for example to open the van der waals gaps of a two dimensional structure, and a gain of energy which is a gain in electronic energy. Of course other factors are to be considered such as the configurational entropy of the intercalated species, but those are by far the most important contributions. There, is the driving force for intercalation chemistry.

A desintercalation process may occur if an oxidizing reagent presents accepting levels lower than the last occupied levels of the host. From the relative position of a given host structure respective to various reagents in a redox scale such as the one proposed by D. Murphy[58] it is thus possible to intercalate or deintercalate in the considered phase. This appears to be a powerful tool in chemical synthesis. Starting from a $A_xM_yS_z$ ternary sulfide obtained at high temperature, it is possible to remove the alkali metal and to obtain a new MySz binary phase, either previously unknown or presenting a new structural type. Highly reactive and generally metastable solids result from these methods. One of the first example was VS_2 [59]. There, the redox process concerns very simply the vanadium atoms. Let us consider more in details two other examples which show more complex situations.

5.2. A new FeS_2 from Li_2FeS_2 deintercalation : a mixed-redox-centers system

Li_2FeS_2 that is prepared from Li_2S and FeS at 800°C has a hexagonal unit cell based on a hexagonal close packing of S^{2-} anions with tatrahedral Fe^{2+} ions and Li^+ ions scattered on tetrahedral and octahedral sites [60]. A 2a 2b superstructure illustrates an ordering between occupied an empty sites [61].

Lithium removal from Li_2FeS_2 to $LiFeS_2$ results in the oxidation of Fe^{2+} into Fe^{3+}, the remnant lithium cation of Li_1FeS_2 being in octahedral site. The subcell changes very little, but an accident is observed for x = 1,5, on the quasi equilibrium discharge curve of the Li_xFeS_2/Li system. A narrow biphased domain could be supposed to exist there. But the Mössbauer studies show clearly that some iron ions shift at that moment from tetrahedral to octahedral symmetry and this may explain the discontinuity of the discharge curve as well [62]. From Li_1FeS_2 to FeS_2, a potential plateau reveals a biphased domain contributed by the end compounds, and the infra-red spectra [60] indicate that this further removal of lithium results in the oxidation of sulfur with formation of $(S_2)^{2-}$ pairs. The final product can be written as $Fe^{3+}S^{-2}(S_2^{-2})_{½}$ with the following site repartition :
$Fe^{3}_{0.5(Oh)}Fe^{3+}_{0.5(Td)}S^{-2}(S_2^{-2})_{½}$. Electrochemically it is possible to reintroduce two lithium per FeS_2 mole, but the system cycles well for 1,5 lithium.

This system shows how deintercalation chemistry can be complicated. The redox process once more is of first importance. The FeS_2 one gets is structurally similar to the model which has been proposed (but not directly established for IrS_2). It is a phase in which the charge balance is between the classical 2D ($Ti^{IV}S_2$) phases types and the 3D ($Fe^{II}S_2$) pyrites or marcasite ones. Once more it appears as the result of a competition between the relative stability of the metal d orbitals and the top of the sp anionic band.

5.3. Another very interesting system is the one based on the use of spinel structures. Spinel structures present vacant tetrahedral and octahedral sites and therefore satisfy to geometrical conditions for the uptake and transport of additional cations. The electronic redox condition will be of great importance. In the case of oxides, cationic levels are simply involved [63]. But when chalcogenides are concerned the redox competition between cations and anions may induce structural changes and at least large differences from one system to another. For example $CuCr_2Se_4$ can only be reduced to $Cu_2Cr_2Se_4$ while $CuTi_2S_4$ can only be oxidized to Ti_2S_4 [64]. This could be reasonably well explained in terms of stability of Ti^{4+} in presence of sulfur and of instability of Cr^{4+} in presence of selenium. The oxidation process of $CuTi_2S_4$ leading to a new form of TiS_2 [64-65] is well understood according to this simple model. The case of $CuCr_2Se_4$ is more complicated. Considering the initial oxidation scheme to be $Cu^+(Cr^{3+})_2(Se^{2-})_3Se^{\bullet -}$ rather than $Cu^+C^{3+}Cr^{4+}(Se^{2-}{}_4)$, the reduction process will not concern chromium but the hole on the anion.

References

1a. Intercalation in layered Materials, M.S. Dresselhaus Ed. Nato Asi series B148, Plenum Press, New York 1987.

b. J. Rouxel in : "Intercalated Layered Materials", F. Levy Ed., D. Reidel, Dordrecht (1979) p.201.

c. Intercalation Chemistry, M.S. Wittingham, A. Jacobson, Ed. Academic Press, New York (1982).

d. R. Schöllhorn in "Inclusion Compounds" Eds J.L. Atwood, J.E.D. Davies and D.D. mac Nicol vol. 1, Academic Press, New York (1984).

2. R. Schöllhorn, Angew Chem. Int. Ed. Engl., 19:983 (1980).

3. E. Mosser and W.B. Pearson, Acta Cryst., 12:1015 (1959).

4. C. Delmas, J.J. Braconnier, A. Maazaz and P. Hagenmuller, Rev. Chim. Min., 19:343 (1982).

5. L.S. Ramsdell, Am. Mineralogist, 32:64 (1947).

6. F. Jellinek in "Inorganic Sulfur Chemistry" G. Nickless Ed. Elsevier Amsterdam (1968).

7. A. Meerschaut and J. Rouxel, J. Less Common Metals, 39:498 (1975).

8. R. Hoffmann, S. Shaik, J.C. Scott, M.H. Whangbo, M.J. Poshee, J. Solid State Chem., 34:263 (1980).

9. W. Klingen, R. Ott and H. Hahn, Z. Anorg. Allg. Chem., 396:271 (1973).

10. R. Brec, Sol. State Ionics, 22:3 (1986).

11. R. Clément, J. Chem. Soc., Chem. Comm., 647 (1980).

12. J. Rouxel and P. Palvadeau, Rev. Chim. Min., 19:317 (1982) and G.A. Fatseas, P. Palvadeau and J.P. Venien, J. Solid State Chem., 51:17 (1984) ; L. Coic, P. Palvadeau, F. Menil, L. Fournes, Mat. Res. Bull., 16:1055 (1981)

13a. R.H. Herber and Y. Maeda, Inorg. Chem., 19:3411 (1980)

b. R.H. Herber and R.A. Cassel, Inorg. Chem., 21:3713 (1982).

14. P. Palvadeau, J.P. Venien and G. Calvarin, C.R., series II, 292:1259 (1981).

15. J.F. Berar, G. Calvarin and D. Weigel, J. Appl. Cryst., 13:201 (1980).

16. F. Kanamaru, M. Shimada and M. Koizumi, Chem. Lett., 373 (1974).

17. J.P. Audière, R. Clément, Y. matthey and G. Mazières, Physica B, 99:133 (1980).

18. F. Kanamaru, M. Shimada, M. Koizumi, M. Takano and T. Takada, J. Solid State Chem., 7:297 (1973).

19. T.R. Halbert, D.C. Johnston, L.E.M. Mc Candlish, A. Thompson, J.C. Scanlon, J.A. Dumesic, Physica B+C, 99(1-4):128 (1980).

20. B.A. Averill, S.M. Kanzlarich, B.K. Tes and J. Faber, Mol. Cryst. Liq. Cryst., 120:259 (1985).

21. A. Armand, L. Coïc, P. Palvadeau and J. Rouxel, J. Power Sources, 3:137 (1978).

22. R. Steffen and R. Schöllhorn, Solid State Ionics, 22:31 (1986).

23. R. Chevrel, M. Sergent, J. Solid State Chem., 3:8807 (1971).

24. R. Schöllhorn, Angew. Chem. Int. Ed. Engl., 19:983 (1980).

25. B.C.H. Steele, G.J. Dudley, K.Y. Cheung, Proceeding 2nd Int. Meeting Sol. Electrolytes St Andrews (1978) aabstract 7-3.

26. K. Yvon, "Current Topics in Material Science" Kaldis Ed., North Holland (1979) vol. II.

27. R. Schöllhorn, M. Kumpers, J.O. Besenhard, Mat. Res. Bull., 12:781 (1977).

28. J.M. Tarascon, F. Di Salvo, D.W. Murphy, G.W. Hull, E.A. Rietman, J.V. Waszczak, J. Solid State Chem., 34:204 (1984).

29. W.R. Mc Kinnon, J.R. Dhan, Phys. Rev. B, 31:3084 (1985).

30. E. Gocke, R. Schöllhorn, G. Aselmann and W. Müller-Warmuth, Inorg. Chem, in press.

31. S.T. Coleman, W.R. Mc Kinnon, J.R. Dahn, Phys. rev. B, 29:4147 (1984).
32. R. Schöllhorn, Personal communication and ref 30.
33. J. Rouxel, P. Molinié, L.H. Top, J. Power Sources, 9:345 (1985).
34. S. Wittingham, Personal communication.
35. W.P.F.A. Omloo and F. Jellinid, J. Less Common Metals, 20:121 (1970).
36. J. Rouxel and R. Brec, Annual Review Mat. Science, 16:137 (1986).
37. A. Ben Salem, A. Meerschaut, M.H. Whangbo, J. Rouxel, to be published.
38. J. Molenda, Solid State Ionics, 21:263 (1986).
39. A. Meerschaut, P. Gressier, L. Guémas and J. Rouxel, Mat. Res. Bull. 16:1035 (1981) ; R.J. Cava, V.L. Himes, A.D. Mighell and R.S. Roth, Phys. Rev. B, 24-6:3634 (1981).
40. M.H. Whangbo, R.J. Cava, F.J. Di Salvo and R.M. FLeming, Sol. State Comm., 43:277 (1982).
41. T.J. Hibma, J. Solid State Chem., 34:94 (1980).
42. A. Le Blanc, M. Danot, L. Trichet and J. Rouxel, Mat. Res. Bull., 9:191 (1974).
43. J. Rouxel, J. Solid State Chem., 17:223 (1976).
44. C. Fouassier, C. Delmas and P. Hagenmuller, Mat. Sci. Eng., 31:2 (1977).
45. A.H. Thompson, C.R. Symon, Solid State Ionics, 3-4:175 (1981).
46. J.M. Tarascon, F.J. Di Salvo, C.H. Chen, J.P. Caroll, M. Walsh, L. Rupp, J. Solid State Chem., 58:290 (1985).
47. P. Deniard (in this volume).
48. D.W. Murphy, F.J. Di Salvo, G.W. Hull and J.V. Waszczak, Inorg. Chem., 15:17 (1976).
49. M.S. Whittingham and F. Gamble, Mat. Res. Bull., 10:363 (1975).
51. P. Molinié, L. Trichet, J. Rouxel, C. Berthier, Y. Chabre and P. Segransan, J. Phys. Chem. Solids, 45-1:105 (1984).
52. M.H. Whangbo, L. Trichet and J. Rouxel, Inorg. Chem., 24:1824 (1985).
53. D.W. Murphy and F.A. Trumbore, J. Crystal Growth 39:135 (1977).
54. R.R. Chianelli and M.B. Dines, Inorg. Chem., 14(10):2417 (1975).
55. J.L. Hodeau, M. Marezio, C. Roucau, R. Ayroles, A. Meerschaut, J. Rouxel and P. Monceau, J. Phys. C, 11:4117 (1978).
56. A. Ben Salem, A. Meerschaut, J. Rouxel, M. Evain and M.H. Whangbo, submitted Sol. State Ionics.
57. M.H. Whangbo and P. Gressier, Inorg. Chem., 23:1305 (1984).
58. D.W. Murphy and P.A. Christian, Science, 205:651 (1979).
59. D.W. Murphy, J.N. Carides, F.J. Di Salvo, C. Cros, J.V. Waszczak, Mat. Res. Bull., 12:825 (1977).
60. P. Gard, C. Sourisseau, G. Ouvrard and R. Brec, Solid State Ionics, 20:231 (1986).
61. A. Le Méhauté, R. Brec, A. Dugast and J. Rouxel, Solid State Ionics, 314:185 (1981).
62. L. Blandeau, G. Ouvrard, Y. Calage, R. Brec and J. Rouxel, to be published
63. J.B. Goodenough, M.M. Thackeray, W.I.F. David, P.G. Bruce, Rev. Chim. Min., 21:435 (1984).
64. R. Schöllhorn and A. Payer, Angew. Chem. Int. Ed. English, 22:905 (1986).
65. S. Sinha, D.W. Murphy, Solid State Ionics, 20:81 (1986).

LAYERED HOST LATTICES: STRUCTURE, BONDING, PHYSICAL PROPERTIES AND CHEMICAL REACTIVITY

Robert Schöllhorn

Technische Universität Berlin
Institut für Anorganische und Analytische Chemie
D-1000 Berlin 12, Germany

INTRODUCTION

The primary base for the description of a solid is its chemical composition in terms of the constituent elements and their relative proportions. For a more specific characterization it is necessary to establish the geometrical structure, to describe the quality of interaction between the constituent atoms (chemical bonding) and to determine the physical properties of the compound under consideration. Presently we are at a level of knowledge that gives access to a basic understanding of the mutual relationship between these static properties of solids for many systems. One additional quality of a system that is of utmost interest in chemistry represents a dynamic aspect: the reactivity of a given solid under defined conditions. In many respects, however, our knowledge on the correlation of structure, bonding and physical properties with the chemical reactivity is still rather unsatisfying. This is the reason, why certain simple systems which allow within certain limits the prediction of their reactivity on grounds of their basic properties have attracted particular interest. Among these are the so-called intercalation systems which are characterized by their topotactic reactivity, i.e. the structural integrity of the reacting solid is maintained in the course of the chemical process. The reversible uptake of mobile guest species can be observed with host lattices of different structural dimensionality: framework (3D), layer (2D) and chain type (1D) lattices. Layered host lattices have always found, however, preferred attention due to their high structural flexibility, the ability to adapt to the geometry of the inserted guest species and the particular

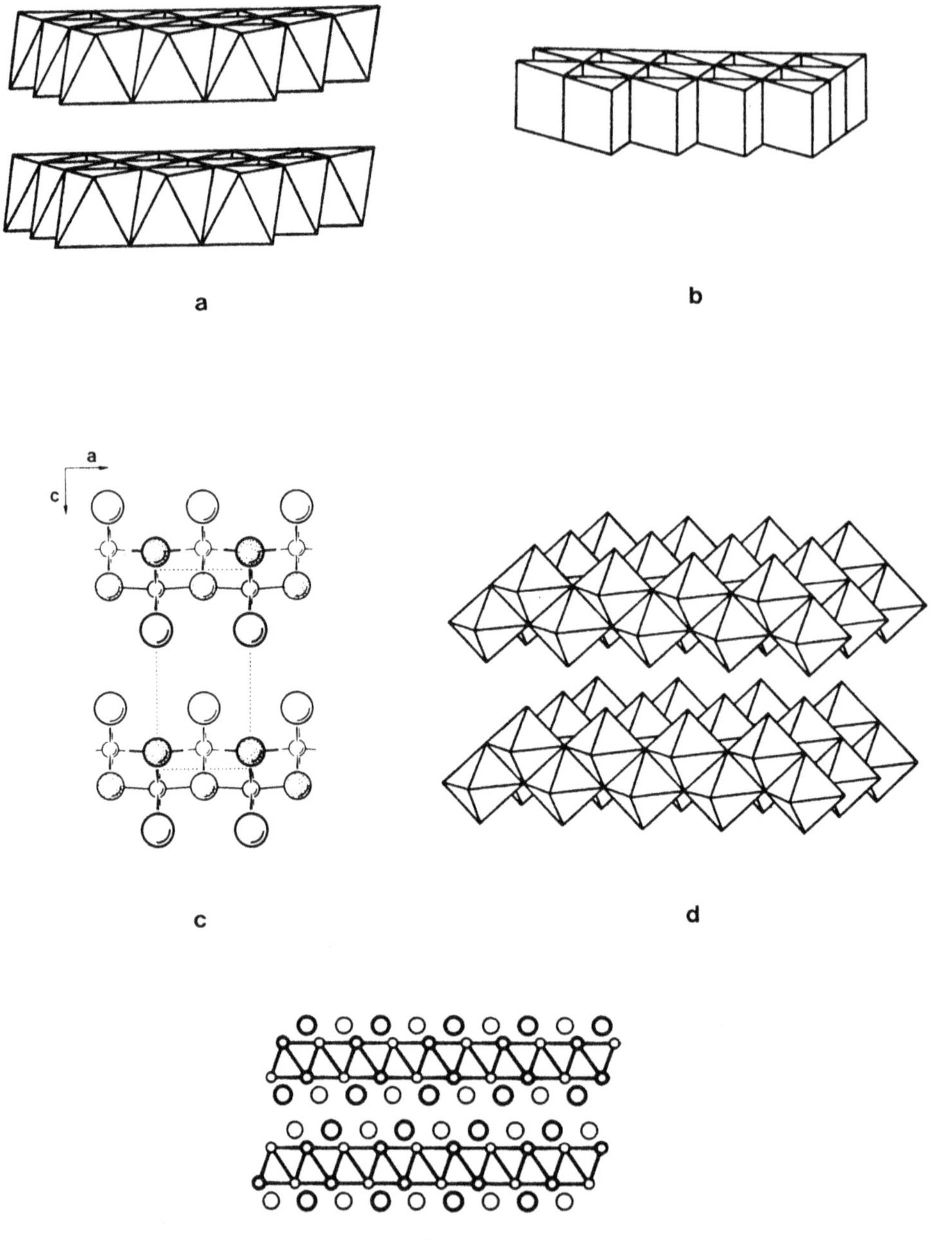

Fig. 1. Structure schemes of some host lattices with layer structure. (a) CdI_2 type, edge sharing MO_6 octahedra; (b) 2H-MoS_2 type, edge sharing MO_6 trigonal prisms; (c) FeOCl, large circles: dotted = O, open = Cl; (d) MoO_3, edge and vertex sharing MO_6 octahedra; (e) ZrBr: layers of edge sharing M_6 clusters with the anions (large circles) above and below.

types of structural transformations observed as a consequence of the insertion or exchange of guest species[1-4]. A particularly strong correlation between physical properties and chemical reactivity is observed in those cases, where the intercalation process is coupled with redox reactions. Here the interplay between the geometrical structure and the electronic properties of the host lattice in terms of (i) electronic transport and (ii) band structure plays an essential role in determining the chemical reactivity, the appearance of intermediate phases and the intercalation limit (maximum amount of electrons or guest species, respectively, that can be inserted).

The structure schemes of a series of layered host lattices (transition metal compounds) able to undergo intercalation via redox reactions are illustrated in Fig. 1. It is amazing that in spite of the differences in composition and details of the sheet unit structure the basic chemical reactivity of these phases turns out to be closely related. The different kinds of guest species that can be intercalated via redox or exchange reactions can be divided into three categories: (i) simple metal cations, (ii) solvated metal cations and (iii) inorganic or organic molecular cations. The intercalation of strongly electropositive ions like Li^+, Na^+ has been the subject of a large number of investigations which is in part due to the potential application of these phases, e.g. in galvanic cells for energy storage. We shall discuss in the following a series of characteristic aspects associated with the correlation between structure,bonding and properties of layered host lattices and the chemical reactivity of these solids at ambient temperature. The emphasis will be put on layered chalcogenides with electronic conductivity able to undergo redox reactions by electron/ion transfer.

REACTION MECHANISM AND SPECIFIC STRUCTURAL CHANGES

The most important structural transitions encountered in intercalation processes of layered systems are (i) changes in interlayer spacing d, (ii) stacking changes,i.e. gliding of neighbouring layers relative to each other, (iii) the appearance of "staging" phenomena, i.e. the selective filling up of the van der Waals gaps between the host layers in a regular sequence and (iv) order/disorder transitions in the guest sublattice (Fig. 2). These effects have been well investigated e.g. on graphite intercalation compounds[5] and on alkali metal intercalation compounds of transition metal dichalcogenides MX_2 [6]. Besides the alkali metals there is a group of mono-

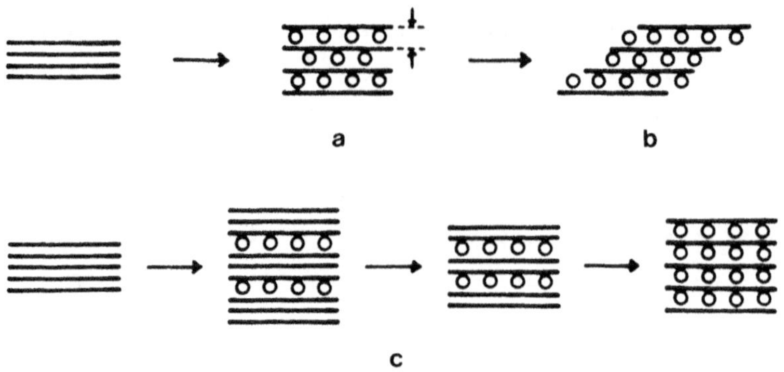

Fig. 2. Principal geometrical transitions of layered host lattice matrices upon intercalation of guest species (indicated as small circles).(a) change in interlayer spacing d; (b) change in stacking mode of the layers; (c) intermediate phases at low guest concentrations (sequence: host lattice, 3rd stage phase, 2nd stage phase, 1st stage phase)[29].

valent transition metal ions which can be intercalated under similar conditions as the electropositive metals. Recent more detailed investigations on the reaction of layered dichalcogenides with these ions have shown that similar effects with respect to structural transitions can be observed. This is to some extent surprising, since Cu^+, Ag^+ and Tl^+ are d^{10} ions which are known to establish partiallycovalent bonding to chalcogenide host lattices[7], while alkali ions can be described in a first approximation as hard sphere ions with electrostatic interaction with the host lattice. ESCA measurements have clearly established that Cu is always formally monovalent in chalcogenide lattices[7].

The phase diagram for the electrochemical intercalation of Cu^+ in TiS_2 from $CH_3CN/CuCl$ electrolytes is shown in Fig. 3. A nonstoichiometric phase range Cu_xTiS_2 I with $0 \leq x \leq 0.5$ appears in the first part of the reaction with a rather small increase in interlayer spacing due to the fact that Cu^+ is located on large octahedral sites. A second one phase range Cu_xTiS_2 II with $0.7 \leq x \leq 0.95$ follows at high Cu concentration with significantly higher interlayer spacing. Cu^+ would prefer to reside on tetrahedral sites; however, at low

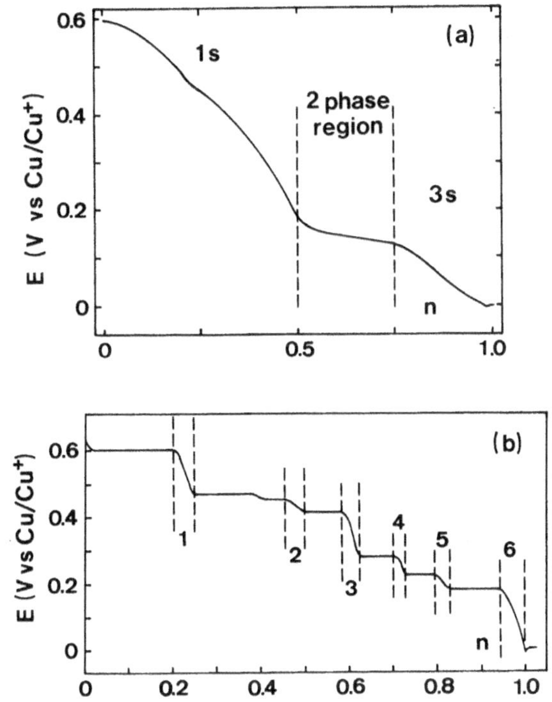

Fig. 3. Intercalation of copper in layered chalcogenides MS_2. (a) potential/charge transfer curve for the cathodic reduction of TiS_2 in Cu^+/CH_3CN; n = charge transfer (e^-/TiS_2); 1s, 3s = number of TiS_2 sheets per unit cell in the one phase regions; (b) cathodic reduction of VS_2 under the same conditions, numbers indicate the different phases appearing in the course of the reaction[8,13].

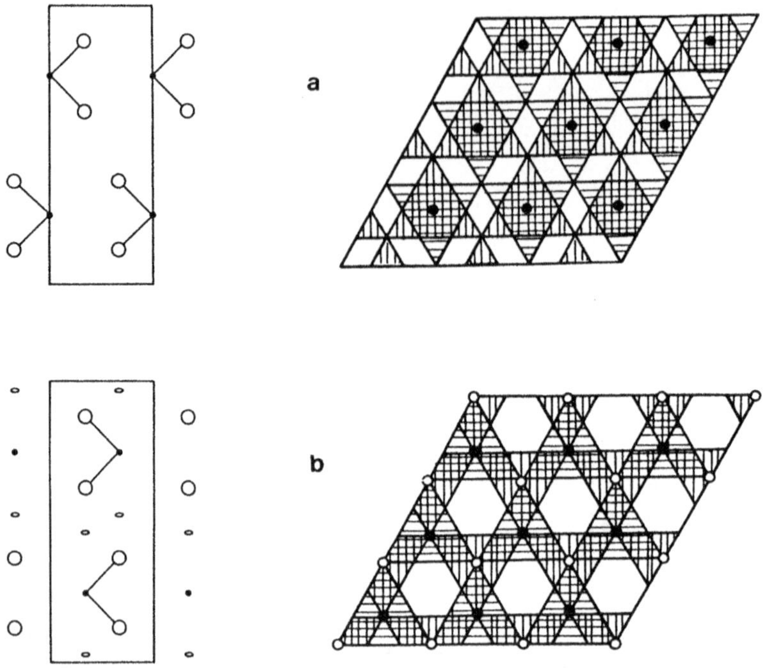

Fig. 4. Projection of the 2H-NbS_2 structure (a) and the structure of $Cu_{0.5}NbS_2$ with 2H-MoS_2 stacking (b); left hand $11\bar{2}0$ section; right hand projection along 001; ● Nb at z = 1/4; ○ Nb at z = 3/4. Both structures have $P6_3/mmc$ symmetry; in $Cu_{0.5}NbS_2$ Cu atoms occupy statistically 1/4 of the interlayer tetrahedral sites at 300 K. The structural transition can be described formally as a shift of adjacent layers in $11\bar{2}0$ direction of $a \cdot \sqrt{3}/3$.[11]

concentrations of Cu on tetrahedral sites the loss in energy by higher d values favours the octahedral geometry as a metastable site (c.f. below). At a critical threshold value of Cu concentration I transforms into II in a two phase region:

$$Cu_{0.5}TiS_2 + 0.2\ Cu^+ + 0.2\ e^- \rightleftharpoons Cu_{0.7}TiS_2 \qquad (1)$$

1s stacking	3s stacking
octahedral	tetrahedral
metastable	stable

This transition is connected with a change in layer stacking from 1s to 3s (3s ≙ three TiS_2 sheets per unit cell)[8,9]. The latter form is thermodynamically stable and structurally identical with $Cu_{0.7}TiS_2$

prepared at high temperatures[9,10]. The transition described by the equation above is reversible via reoxidation of $Cu_{0.7}TiS_2$ or of phases with higher Cu content.

A related change in stacking is observed on Cu intercalation in $2H\text{-}NbS_2$. In this case a transition

$$2H\text{-}NbS_2 \text{ type} \rightleftharpoons 2H\text{-}MoS_2 \text{ type}$$

is found[9,11](Fig. 4). The copper ions reside on tetrahedral sites in $Cu_{0.5}NbS_2$ with statistical distribution; at low temperatures a disorder/order transition is observed in the guest ion sublattice[12]. A stacking transition in the inverse sense appears, if the thermal ternary phase $Cu_{0.5}NbS_2$ with $2H\text{-}MoS_2$ structure is oxidized anodically to $2H\text{-}NbS_2$. There are two possible mechanisms by which this transition may take place: (i) gliding motion of entire layers with respect to each other and (ii) local cooperative site change of niobium ions to neighbouring trigonal prisms within the NbS_2 sheets. One may easily estimate that both processes require a considerable activation energy. While the first type of mechanism should be strongly dependent upon crystallite size (diameter of basal planes), the second mechanism should be independent of dimensional parameters. Experimentally one can prove in fact that the transition is inhibited, if single crystals of $2H\text{-}NbS_2$ (ca. 3-5 mm in diameter) are used as the starting material for intercalation which favours the layer gliding model[11]. It is interesting, however, to note that in spite of the estimated high activation energy the migration of V^{3+} ions from intralayer sites in $LiVO_2$ to interlayer sites was observed recently on oxidation of $LiVO_2$ to $Li_{0.3}VO_2$ at ambient temperature[15].

A case of particular interest is the intercalation of Cu in vanadium disulfides[13]. VS_2 is a metastable host lattice that can be prepared by the deintercalation of the ternary phase $LiVS_2$[14]. The phase diagram (Fig. 3b) is unusual, since it displays six different compounds; there is no other intercalation system known so far that contains a comparable number of intermediate phases. The phase ranges are $0.21 < x < 0.25$ (1), $0.47 < x < 0.50$ (2), $0.59 < x < 0.63$ (3), $0.71 < x < 0.73$ (4), $0.80 < x < 0.83$ (5), $0.95 < x < 1.0$ (6). In the hexagonal phase (1) the increase in interlayer spacing Δd is rather small (32 pm); Cu^+ ions are on octahedral sites and the compound is metastable. The transition (1) $\rightarrow$ (2) is correlated with a symmetry change hexagonal $\rightarrow$ monoclinic and a change for Cu coordination octahedral $\rightarrow$ tetrahedral. The following phases with higher Cu content remain monoclinic and are thermodynamically stable. The reason for the appearance of a large series of intermediate phases is most likely due to the difference in ionicity of the Cu-S and Cu-V bonding and Jahn/Teller effects leading to small distortions of the lattice with low energy difference. Similar effects have been described in the Na_xVS_2 and Na_xVSe_2 systems[16,17].

The dependence of the mechanism of intercalation upon the polarity of the solvent medium is demonstrated by the intercalation of Cu in the defect CdI_2 type lattice of the halide α-$RuCl_3$. Reaction in aprotic solvents such as CH_3CN yields two line phases $Cu_{0.5}RuCl_3$ and Cu_1RuCl_3. In aqueous Cu^{2+} electrolyte in the first intercalation step hydrated Cu^{2+} ions are intercalated; after a critical Cu^{2+} concentration in the lattice a transition to the desolvated phase $Cu^+[RuCl_3]^-$ with monovalent Cu^+ ions takes place[18].

Structurally much more complex systems are found, if the intercalation is connected with the appearance of staging phenomena (Fig. 2) in the initial states of the reaction. The idealized staging model describes the observation of intermediate structures with a regular sequence of occupied and non occupied interlayer regions and has been extensively studied for graphite as the host lattice[5]. It has also been observed for intercalation compounds of transition metal dichalcogenides with alkali metals[19,20] and hydrated ions as the guest species[21]. For the larger alkali metals stage 4, 3 and 2 have been observed depending upon the preparation procedures. For intercalation systems with Ag^+, e.g. Ag_xTiS_2, Ag_xTaS_2 second stage phases have been reported which undergo a transition to the first stage at higher Ag concentration[22-24](Fig. 5).

For hydrated phases $A^{n+}_{x/n}(H_2O)_y[MS_2]^{x-}$ (A^{n+} = alkali, alkaline earth or transition metal ion ; M = Ti, Nb, Ta) a sequence 3rd stage $\longrightarrow$ 2nd stage $\longrightarrow$ 1st stage is observed regularly[21]. Staging is not found, however, for VS_2, 1T-TaS_2 (isostructural with TiS_2) or halide host lattices, e.g. α-$RuCl_3$[18], YOCl[25], FeOCl[26]. If even closely related host lattices like TiS_2, VS_2, 1T-TaS_2, 2H-TaS_2 display strong differences in reactivity with respect to staging, the question arises as to whether the origin of this phenomenon is due to structural or electronic reasons. A most simple model explains staging in terms of a gain in lattice energy at low guest concentration, if the guest species are concentrated in a limited number of van der Waals gaps rather than being homogeneously distributed over all interlayer sites. In the latter case large regions of void space

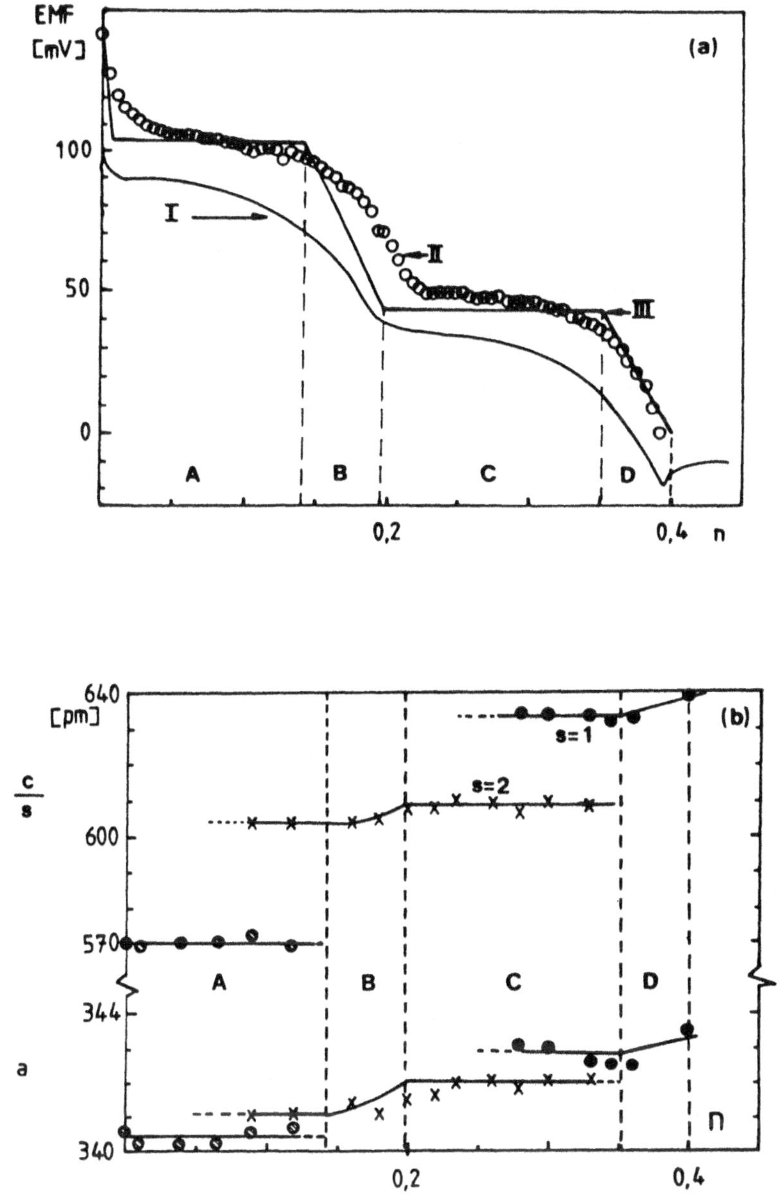

Fig. 5. Formation of Ag_xTiS_2 by cathodic reduction of TiS_2 in aprotolytic Ag^+ electrolyte; (a) change in potential E (vs Ag/Ag^+) of the working electrode with the charge transfer n (e^-/TiS_2); (b) variation of the hexagonal parameters a, c with n; s = stage number.

would be created due to the increase in interlayer spacing upon insertion of guest species. Electronic aspects are involved in so far as those host lattices which do not form higher stage intermediates show no formation of intermediates at low charge transfer (i.e. low guest concentration), e.g. $1T\text{-}TiS_2$, VS_2. A more realistic model proposed by Hérold[5] explains staging in terms of a domain scheme with all van der Waals gaps being involved but local concentration of the guest species across the basal plane in certain regions. Additional kinetic aspects to be considered are the increased activation energy for stage formation with increasing stiffness (elasticity) of the layers or increasing layer thickness (e.g. MoO_3). The size of the guest species may also affect the appearance of higher stage intermediates. Li^+ and Cu^+ ions with rather low effective ionic radii do not induce stage formation even at low concentration due to the small increase in interlayer separation upon intercalation.

The influence of kinetics upon staging is clearly demonstrated by the fact that the halfwidth of the Bragg lines is significantly increasing upon formation of these phases; after the transition to the first stage phase the line width decreases again.

Severe influence of kinetics is also observed with respect to reactions under non equilibrium conditions. If the intercalation of hydrated Na^+ ions into $2H\text{-}NbS_2$ is carried out by slow electrochemical reaction (galvanostatic mode or potentiostatic screening), a sequence 3rd stage $\rightarrow$ 2nd stage $\rightarrow$ 1st stage is found[21]. If the reaction is performed, however, with polycrystalline material by electrochemical reduction at high current densities or by chemical reduction with reducing species of low potential such as $S_2O_4^{2-}$, a different mechanism is observed[27]. No higher stage phases can be detected under these conditions; the reaction takes place as a simple two phase process

$$2H\text{-}NbS_2 \longrightarrow Na^+_{0.5}(H_2O)_x[NbS_2]^{0.5-}$$

under direct formation of the terminal phase (Fig. 6b). The influence of particle size upon the reaction mechanism is demonstrated by performing the same process with large single crystals (ca. 8 mm in diameter) which leads again to a different mechanism (Fig. 6a). At the initial stage four different phases are present simultaneously: (1) host lattice, (2) the hydrated Na^+ phase with bilayers of water, (3) the hydrated phase with monolayers of water and (4) 3rd stage and 2nd stage compound. The X-ray data reveal that (4) is present only in low concentration at the phase boundary to $2H\text{-}NbS_2$ and disappears in the course of the reduction process almost simultaneously with the host lattice. After quantitative reduction (disappearance of $2H\text{-}NbS_2$, Fig. 6a) only phases (2) and (3) are present. In the further reaction a slow transition of the monolayer phase (3) to the bilayer phase (2) takes place by diffusion of water into the crystal with quantitative formation of (2) which represents the terminal product. The process is thus characterized by a "synchronous multistage mechanism" that

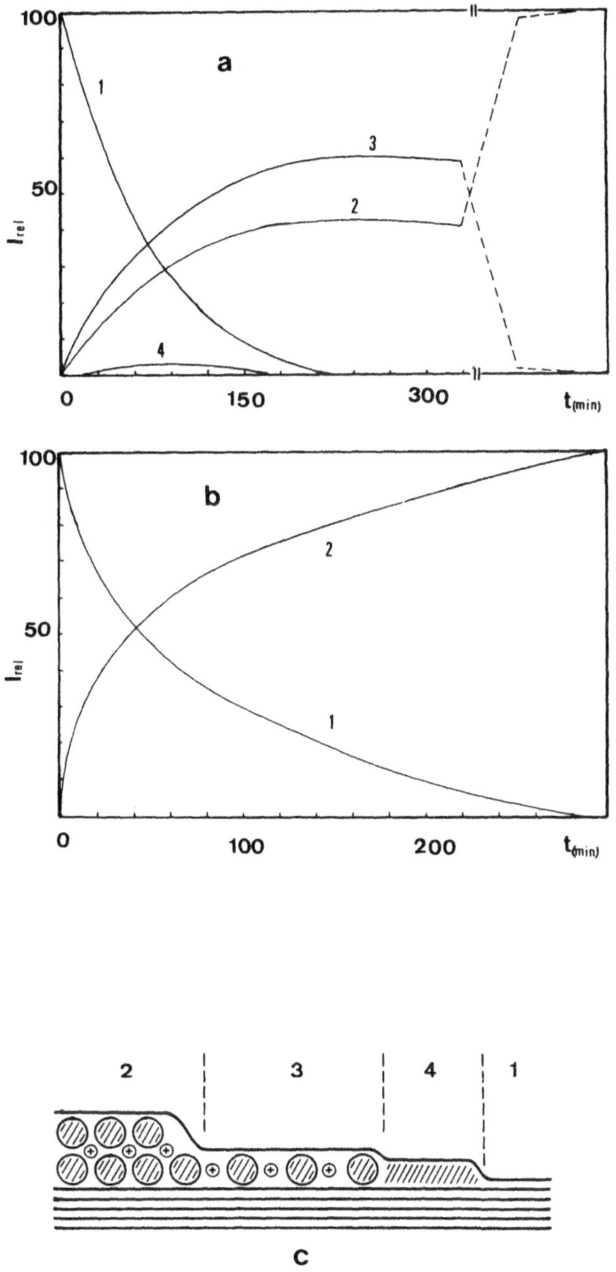

Fig. 6. Mechanism of the chemical reduction of 2H-NbS_2 to $Na^{+}_{0.5}(H_2O)_n$ $[NbS_2]^{0.5-}$ with aqueous $Na_2S_2O_4$ under nonequilibrium conditions. (a) Reduction of large single crystals; the change of the relative intensity of characteristic 001 reflections with time is indicated; (1) host lattice, (2) bilayer hydrate, (3) monolayer hydrate, (4) 3rd stage + 2nd stage phase; the section with broken lines is on a different time scale (ca. 60 h in total). (b) Reduction of polycrystalline material (Ø 0.1 mm); (1) host lattice, (2) bilayer hydrate. (c) Scheme of synchronous multistage mechanism, numbers refer to Fig. 6a[11].

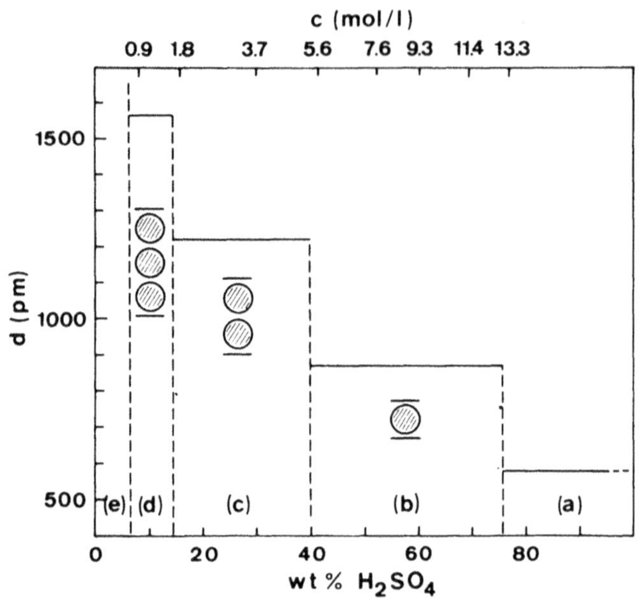

Fig. 7. Formation of hydronium phases $(H_3O^+)_x(H_2O)_y[RuCl_3]^{x-}$ by cathodic reduction of α-$RuCl_3$ in aqueous H_2SO_4: dependance of product structure and hydration upon the electrolyte concentration. (a) hydrogen bronze H_xRuCl_3, (b) monolayer hydrate, (c) bilayer hydrate, (d) trilayer hydrate, (e) colloid region; d = interlayer spacing[18].

demonstrates the complexity of reactions which can appear in layered intercalation systems.

Hydrated intercalation phases $A_x^+(H_2O)_y[MX_2]^{x-}$ are also appropriate examples for demonstrating the influence of the guest ion type upon the product structure[28]. For monovalent and bivalent metal ions the cation/solvent structure is directly related to the charge/radius ratio e^2/r (solvation energy) of the cation; at a critical value of e^2/r a transition monolayer hydrate/bilayer hydrate is found. Similarly the hydration state can be affected by the concentration of the guest cation in the electrolyte, e.g. in the case of H_3O^+ compounds which may form mono-, bi- and triple layers of water with decreasing concentration of H_3O^+ in the equilibrium electrolyte. The same sequence was found for the layered halide α-$RuCl_3$ as the host lattice (Fig. 7)[18]. At very low concentrations of H_3O^+ the layers separate from each other and colloidal phases can be obtained via one-dimensionally disordered intermediate states.

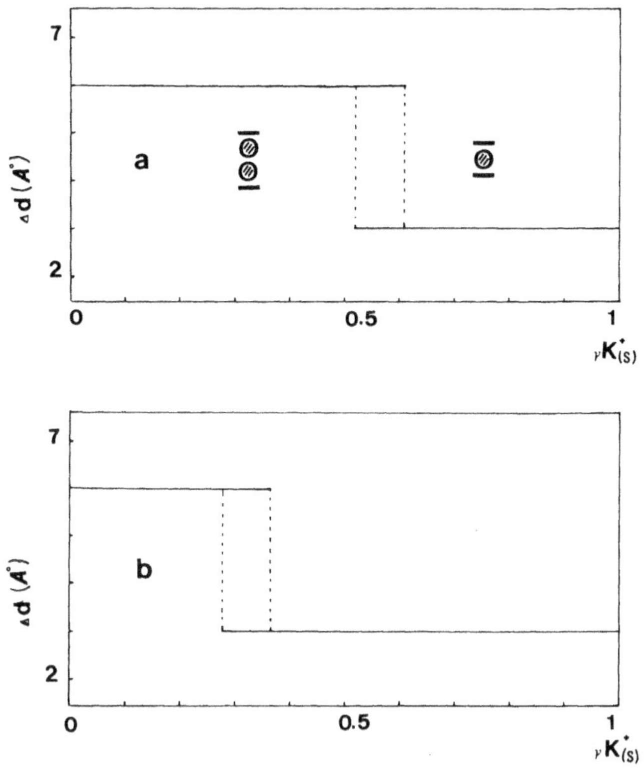

Fig. 8. Cooperative cation exchange and exchange hysteresis for the system $A^+_{0.5}(H_2O)_y[NbS_2]^{0.5}$ with $A^+ = Na^+, K^+$: change in interlayer space height Δd with the molar ratio γ_{K^+}. (a) exchange starting from pure Na^+ phase, (b) exchange starting from K^+ phase[11,29].

Ion exchange reactions which proceed with a stacking change in the layer sequence can lead to cooperative structural transformations[28]. Examples are ion exchange equilibria Na^+/K^+ in hydrated phases $A^+_x(H_2O)_y[MX_2]^{x-}$ [29] or $A^+_x(H_2O)_y[RuCl_3]^{x-}$ [18]. Fig. 8 illustrates the results for the $A^+_x(H_2O)_y[NbS_2]^{x-}$ phases. It can be seen that the transition from the Na^+ phase (bilayer hydrate) with an interlayer space height Δd of 6 Å to the K^+ phase ($\Delta d = 3$ Å, monolayer hydrate) does not proceed via a simple two phase region extended over the entire accessible molar ratio range Na/K. X-ray diffraction and analytical data reveal that the transformation proceeds in a rather narrow region at a critical Na/K threshold concentration. This cooperative behaviour must be attributed to the fact that the process is correlated not only with a change in interlayer space height Δd but also with a stacking transition from a 2s layer sequence to a 6s sequence:

$$Na^+ \text{ phase (bilayer hydrate)} \rightleftharpoons K^+ \text{ phase (monolayer hydrate)}$$

$$\Delta d = 6\ \text{Å} \qquad\qquad \Delta d = 3\ \text{Å}$$

$$2s \qquad\qquad 6s$$

Recent studies[27] have shown that this cooperative ion exchange system displays a clear hysteresis. If the exchange is started with the pure Na^+ compound (Fig. 8a), the position of the transition region is different as compared to the inverse case, where the exchange is started with the pure K^+ compound (Fig. 8b) which means that the activation energy for this transformation is dependent upon the direction of the reaction pathway. Similar cooperative transitions with hysteresis have been described for anion exchange reactions of layered hydroxides of the hydrotalcite type with positive layer excess charge[36].

Hysteresis phenomena are characteristic for solid state reactions proceeding at low temperatures, where the thermal energy of the system becomes a limiting factor. It is surprising, however, at a first glance that even the intercalation of hydrogen can be related to strong hysteresis. This has been observed e.g. for the cathodic reduction of molybdenum trioxide MoO_3 in H^+ electrolytes which results in the formation of a sequence of five different hydrogen bronzes H_xMoO_3. From these three only can be identified upon reduction of MoO_3, while the two others appear as intermediate phases only on the way back i.e. reoxidation of the terminal phase $H_{2.0}MoO_3$[29].

The temperature dependent dynamics of guest species are also frequently displaying hysteresis effects. Recent investigations on the chalcogenide hydrogen bronze $H_{0.3}TaS_2$ have shown that hydrogen is located on tetrahedral sites in the van der Waals gap. At low temperatures a structural transition appears which can be interpreted as the formation of a $3a_o \times 3a_o$ superstructure with formation of metal clusters in the TaS_2 layer units[30] (Fig. 9). Simultaneously the motional narrowing of the rigid lattice spectrum (^1H NMR) is observed for the protons. The transformation temperature is found, however, to be dependent upon the direction, i.e. whether the spectra are measured during a heating or a cooling period. The appearance of a hysteresis proves in a direct way that the electronic features of the host lattice (charge density wave distortion) are strongly related with the temperature dependent dynamics of the guest species. A similar hysteresis is observed in the order/disorder type transition of $Cu_{1/2}NbS_2$ [31-33]. At low temperatures the electrical resistivity shows a hysteresis of the transformation of 10 K in width (Fig. 10).

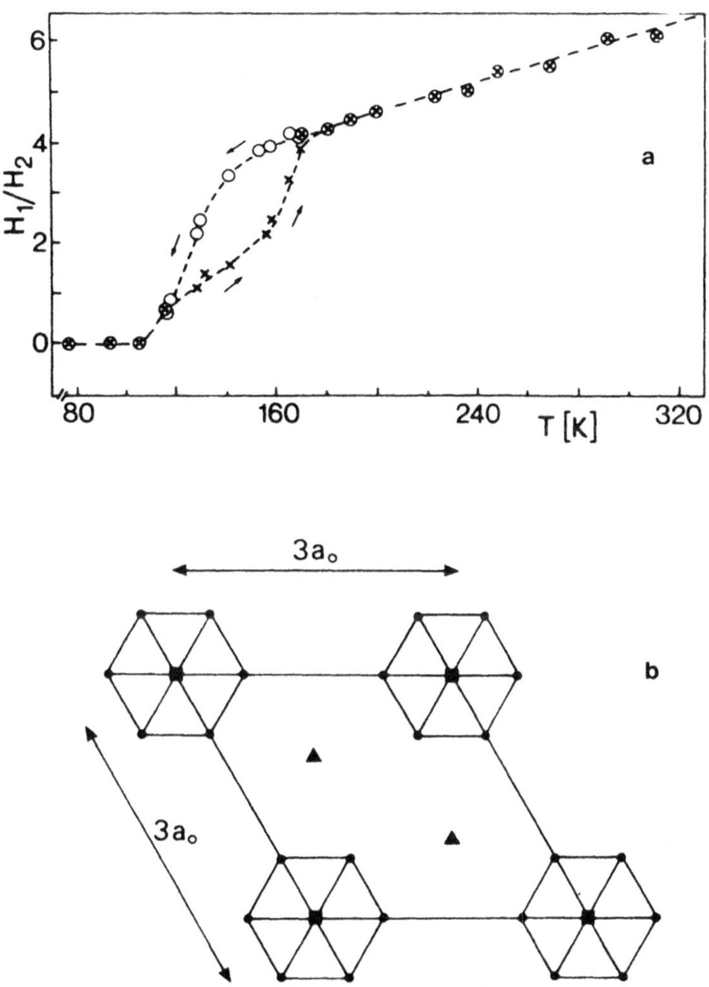

Fig. 9. Hysteresis of proton mobility in 1T-$H_{1/3}TaS_2$. (a) relative intensity of the two line signal ratio versus temperature, (b) scheme of $3a_0 \times 3a_0$ superstructure for Ta atoms in the TaS_2 layers.[30]

IRREVERSIBLE REACTIONS

All intercalation reactions are associated basically with the irreversible formation of a finite concentration of defects in the host lattice that will increase with cycling processes intercalation $\rightleftharpoons$ deintercalation. Frequently the defect concentration may be rather low, while in some cases totally irreversible reactions can be observed. For layered systems stacking disorder is a characteristic defect, in extreme cases total one-dimensional disorder can be found

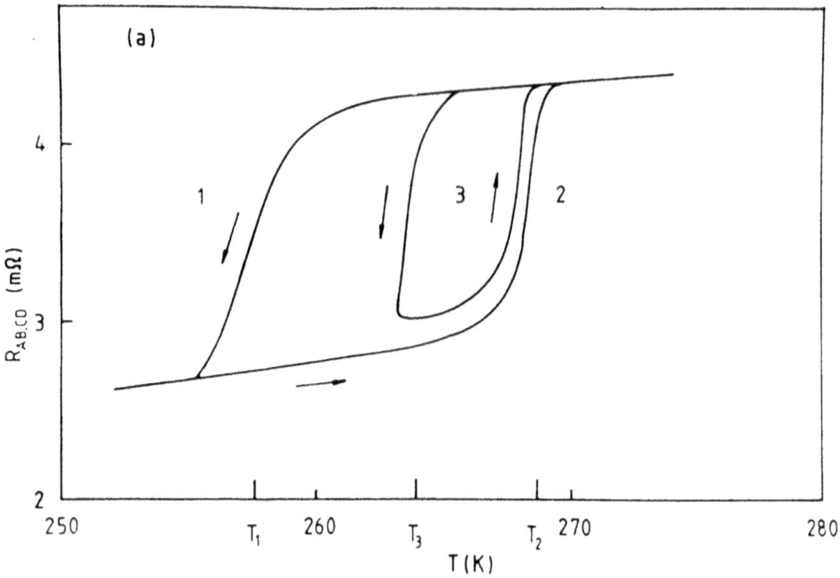

Fig. 10. Hysteresis of the electrical resistance of $Cu_{1/2}NbS_2$ in the temperature range of the order/disorder phase transition in the Cu sublattice[12].

or even desintegration to colloid systems. The intercalation mechanism requires a limited reaction zone at the phase boundary with elastically deformed layer regions leading to X-ray line broadening in intermediate reaction states. This effect is, however, essentially reversible; the final products exhibit usually a line width comparable with that of the original host lattice.

Local defects by chemical reaction of host species with the lattice is particularly strong in chalcogenide systems with guest ions containing polar solvent shells (e.g. H_2O, NH_3), where irreversible nucleophilic displacement of the layer chalcogenide ions takes place in a slow process at 300 K. Spontaneous irreversible desolvation of the guest ions has been found upon the electrochemical intercalation of Ni^{2+}/H_2O into 2H-NbS_2. The hydrated phase $Ni^{2+}_{0.25}(H_2O)_y[NbS_2]^{0.5-}$ transforms into $Ni_{0.25}NbS_2$ upon further reduction at low electrode potential[34]. Irreversible processes can also be found in ion exchange reactions of layered hydrated alkali phases of dichalcogenides with chalcophilic cations of low solvation energy. Exchange of e.g. $K^+_x(H_2O)_y[NbS_2]^{x-}$ with Cu^+ or Ag^+ results in the formation of the corresponding unhydrated Cu and Ag phases which are resistent to further exchange with ions that are more electropositive due to the strong covalent interaction of Cu^+ and Ag^+ with the lattice chalcogen anions. An unusual phenomenon that has not yet been explained is the observation that cobaltocenium ions in $(Cp_2Co^+)_{0.25}[MX_2]^{0.25-}$ cannot be removed quantitatively from the chalcogenide lattice by ion exchange at room temperature, although they can easily be intercalated and remain mobile at room temperature. The intercalation of neutral Lewis base molecules with high ionization energies is also related to an irreversible reaction of a part of the intercalated guest species (c.f. below).

At higher temperatures intercalation compounds - which usually are metastable systems - transform irreversibly into new phases. A typical example is the thermal transition of certain layered copper intercalation compounds Cu_xMX_2 which undergo a change from layered phases to framework phases and a simultaneous change in Cu coordination geometry[9,29,35]:

$$Cu_{0.5}TiS_2 \xrightarrow{330^\circ C} CuTi_2S_4 \quad \text{cubic spinel}$$

$$Cu_{0.5}VS_2 \xrightarrow{460^\circ C} CuV_2S_4 \quad \text{cubic spinel}$$

$$Cu_{0.5}VSe_2 \xrightarrow{420^\circ C} CuV_2Se_4 \quad \text{tetragonal spinel}$$

layered structure (Cu oct.) ⟶ framework structure (Cu tetr.)

The copper vanadium selenide system is of particular interest as the reaction leads to a new spinel which again is metastable and decomposes at higher temperatures.

REACTION MECHANISM AND CHARGE TRANSFER

Besides structure and bonding the electronic properties play a major role among the factors which determine the reactivity and the reaction mechanism of host lattices which undergo intercalation reactions via electron/ion transfer[3,37]. We shall discuss here two specific points:

(i) the electronic transport as a threshold property in intercalation/deintercalation processes (kinetic aspect)

(ii) the band structure properties as a threshold for the maximum degree of intercalation (thermodynamic aspect)

Adequate electronic transport properties are obviously a basic prerequisite for intercalation reactions via redox processes[3,4]. They may become limiting factors in terms of kinetics even under thermodynamically favourable conditions. A characteristic example is the formation of hydrogen bronzes H_xMoO_3. The host lattice molybdenum trioxide is a wide gap semiconductor. The electrochemical reduction process takes place, however, without problems at the triphase boundary host lattice/product/electrolyte, since the primary product $H_{0.3}MoO_3$ is a metal in terms of electronic transport. The following intermediate phases and the terminal product similarly exhibit high conductivity. On reoxidation, however, the original state i.e. the "empty" host lattice MoO_3 cannot be reached, since the last two phase region MoO_3/$H_{0.3}MoO_3$ produces a thin layer of insulating MoO_3 at the interface to the current lead that is blocking further electron transport[38]. The same is true for other cases (e.g. WO_3 monoclinic, WO_3 hexagonal), where an insulator/metal transition appears at the initial stage of the reduction.

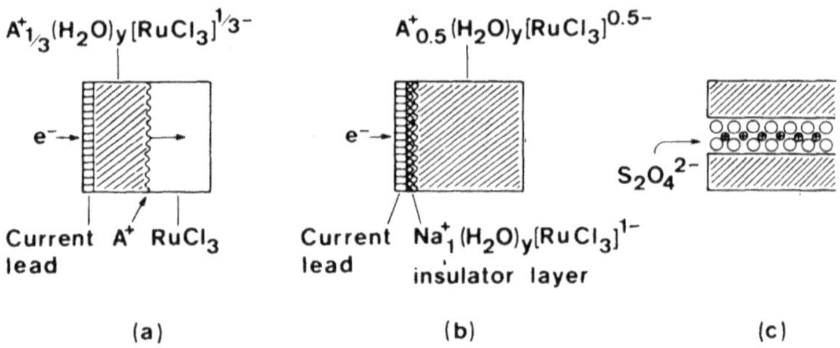

Fig. 11. Influence of the electronic transport properties on the intercalation reaction of α-$RuCl_3$. (a) electrochemical reduction, electronic transport via $[RuCl_3]^{0.3-}$ layers (scheme of working electrode), (b) inhibition of further electrochemical reaction by blocked electron transport at the phase boundary $[RuCl_3]^{0.5-}/[RuCl_3]^{1-}$, (c) chemical reduction with aqueous $Na_2S_2O_4$: electron transport via diffusion of the reducing species in the interlayer region[18].

The inverse case is observed for layered hexagonal α-$RuCl_3$. The halide is a semiconductor and can easily be converted by cathodic reduction in aqueous electrolytes to hydrated black hexagonal phases e.g. $Na^+_{0.5}(H_2O)_y[RuCl_3]^{0.5-}$. Further cathodic reduction to yellow monoclinic $Na^+_1(H_2O)_y[RuCl_3]^{1-}$ with wide gap semiconductor properties is blocked by the lack of electron transport. Chemical reduction with sodium dithionite $Na_2S_2O_4$ as the reducing agent to the yellow phase with integral Ru valency is possible, however (Fig. 11). This is due to the diffusion of the $S_2O_4^{2-}$ or $SO_2^{\bullet -}$ anions in the interlayer space. This reduction mechanism represents a case of electron transport in the lattice via diffusion of electron donor species which effect local reduction at the ruthenium sites in the $RuCl_3$ layers[18]. A similar situation is given for layered vanadium phosphate $VOPO_4(H_2O)_2$ which can only be reduced by chemical means due to the low electron transport rate[39].

A basic problem that concerns thermodynamic aspects is the prediction of the maximum possible uptake of electrons by a given host lattice, i.e. the stoichiometry of the terminal phase[3,40]. The simple model treats the host lattice as an electron sink; if a rigid band model applies and the band structure is known, the maximum potential electron transfer per formula unit of the host compound can be predicted. This model obviously does not take into account a series of factors that may limit the electron uptake, e.g. kinetic effects, steric effects and Coulomb repulsion between the intercalated guest cations.

In solvated systems the maximum electron transfer is frequently limited by the potential of solvent decomposition. For the intercalation of metal ions the maximum charge transfer is limited in electrochemical terms by the equivalence of the redox potentials of the host lattice (Z) and of the corresponding metal/ion system:

$$E\ (Z^{o}/Z^{x-})\ \hat{=}\ E\ (M^{o}/M^{+})$$

This is demonstrated e.g. by TiS_2 metal intercalation systems: sodium can be intercalated up to a limit of $x = 1$ in Na_xTiS_2, the limit for Cu is $x \sim 0.95$, while the terminal phase in the case of the more electropositive Ag^+ ions corresponds to $x = 0.4$[24]. A similar situation is observed for α-$RuCl_3$ as the host lattice: the upper limit for Cu^+ concentration in Cu_xRuCl_3 is $x = 1$, which corresponds to a maximum electron transfer, since all Ru^{3+} ions are now reduced to Ru^{2+}. The silver phase shows, however, a maximum stoichiometry limited to $x = 0.4$[18].

Recent studies have confirmed that the simple model which describes the redox intercalation process by quantitative transfer of the electrons to the host lattice according to

$$xA^{+} + xe^{-} + Z \rightleftharpoons A_x^{+}[Z]^{x-}$$

is an idealized description which has to be replaced by more refined models in particular for host lattices with low electron affinity. In the latter case partial covalent host/guest or guest/guest bonding with localized electrons (partial charge transfer) may play a role. In some cases more electrons than estimated by the band structure or by chemical bonding arguments may be accepted by the system

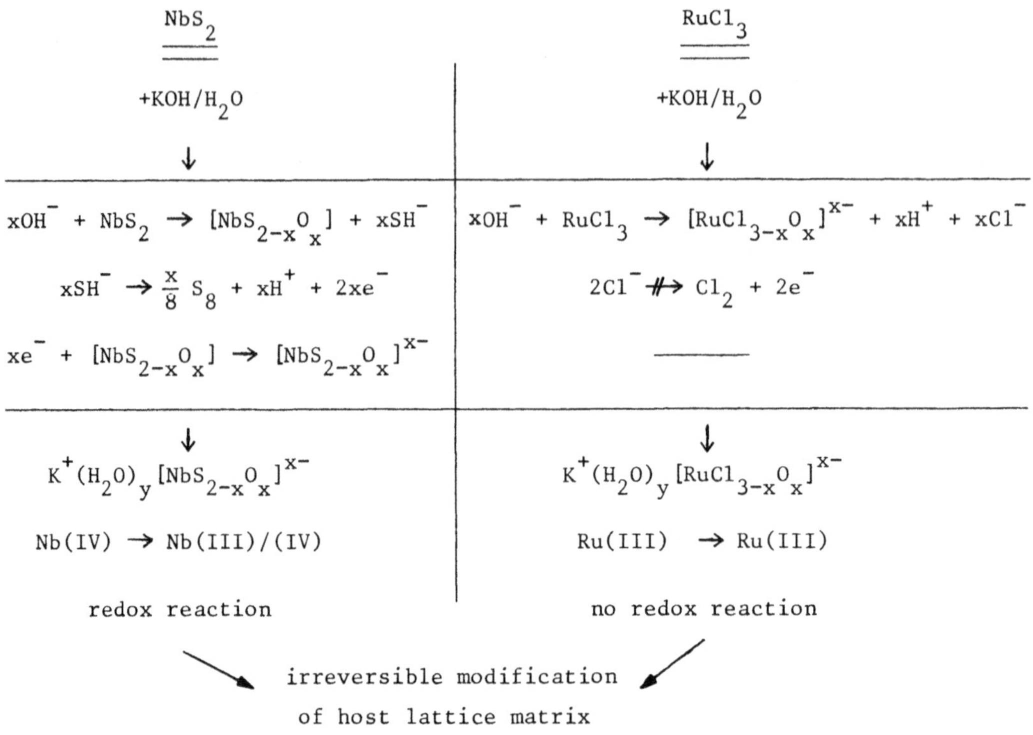

Fig. 12. Scheme of intercalation via solvolysis reactions: left side NbS_2 (hydrolysis/redox mechanism); right side α-$RuCl_3$ (hydrolysis/protolysis mechanism).[18]

[3,40]. Examples are $Ni_2H_2Mo_6S_8$[41], $Li_3Mo_6S_8$[42] and Li_2VSe_2[43], where guest metal-metal and guest metal-hydrogen bonding have been discussed; for TlV_5S_8 the redox reaction appears to proceed exclusively via guest ion redox processes[44].

The charge transfer in the case of the intercalation of neutral Lewis base guest species with high ionization energy E_I has been shown earlier to be associated with a partially irreversible redox reaction[3,4,40]. Recent investigations[45] on the system TiS_2/NH_3 have led to a confirmation of these results. Studies on the $RuCl_3/NH_3$, $RuCl_3$/pyridine[46] and FeOCl/pyridine[47] system have verified the validity of the general redox intercalation model proposed earlier[3].

A further specific problem concerning electron transfer in intercalation reactions of host lattices with electronic conductivity pertains to the observation of the formation of hydrated phases upon the reaction of layered hosts with aqueous solutions of alkali hydroxides or molecular Lewis bases with appropriate pK_a. It has become clear now that two different mechanisms may be responsible depending upon the host lattice: (i) the hydrolysis/redox mechanism with partial solvolysis (nucleophilic anion replacement), if species are formed in this process which exhibit reducing properties (the host layers become modified with respect to composition) and (ii) the hydrolysis/protolysis mechanism without redox reaction provided the resulting M-O-H groups in the host layers have strongly acidic character (Fig. 12)[18].

CONCLUSIONS

Intercalation chemistry represents a particularly instructive example for the strong correlation between the chemical reactivity and the factors characterizing a solid in static terms, i.e. composition, structure, chemical bonding and physical properties. Since these reactions proceed at low temperature, kinetic effects play a dominant role and allow thus the synthesis of a large variety of new metastable solids not accessible by thermal preparation; this aspect is obviously a most attractive line for future research. The reaction of layered solids with neutral guest species e.g. pyridine[48] demonstrates that reorganization of molecules can proceed at low temperatures in the constrained environment of the interlayer space which may induce highly specific processes. The potential of application of redox type intercalation compounds as heterogeneous catalysts has, however, not really been evaluated so far.

REFERENCES

1. F. Lévy, ed., "Intercalated Layered Materials", D. Reidel, Dordrecht (1979).
2. M.S. Whittingham and A.J. Jacobson, eds., "Intercalation Chemistry", Academic Press, New York (1982).
3. R. Schöllhorn, in "Inclusion Compounds", Vol. I, J.L. Atwood, J.E.D. Davies and D.D. Mac Nicol, eds., Academic Press, New York (1984), p. 249.
4. R. Schöllhorn, Angew. Chem. 92, 1015 (1980); Angew. Chem. Int. Ed. Engl. 19, 983 (1980).
5. A. Hérold, in "Intercalated Layered Materials", F. Lévy, ed., D. Reidel, Dordrecht (1979), p. 321.
6. J. Rouxel, in "Intercalated Layered Materials", F. Lévy, ed., D. Reidel, Dordrecht (1979), p. 201.
7. J.C.W. Folmer and F. Jellinek, J. Less-Common Metals 76, 153 (1980).

8. R. Schöllhorn, in "Physics of Intercalation Compounds", L. Pietronero and E. Tosatti, eds., Springer Verlag, Berlin/New York (1981), p. 33.
9. M.M. Schulte, Dissertation, University of Münster (1987).
10. N. Le Nagard, O. Gorochov and G. Collin, Mat. Res. Bull. 10, 1287 (1975).
11. W. Paulus, M. Schulte and R. Schöllhorn, to be published.
12. B.W. Pfalzgraf, H. Spreckels, W. Paulus and R. Schöllhorn, J. Phys. F: Met. Phys. 17, 857 (1987).
13. P. Rathner, Dissertation, University of Münster (1986).
14. D.W. Murphy, J.N. Carides, F.J. Di Salvo, C. Cros and J.V. Waszak, Mat. Res. Bull. 12, 825 (1977).
15. M.M. Thackeray, L.A. de Picciotto, W.I.F. David, P.G. Bruce and J.B. Goodenough, J. Solid State Chem. 67, 285 (1987).
16. C.F. van Bruggen, C. Haas and G.A. Wiegers, J. Solid State Chem. 27, 9 (1979).
17. C.F. van Bruggen, J.R. Bloembergen, A.J.A. Bos-Alberinck and G.A. Wiegers, J. Less-Common Metals 60, 259 (1978).
18. R. Steffen and R. Schöllhorn, Solid State Ionics 22, 31 (1986).
19. J. Rouxel, M. Danot and J. Bichon, Bull. Soc. Chim. Fr. 3930 (1971).
20. T. Hibma, J. Solid State Chem. 34, 97 (1980); Physica 99 B, 136 (1980).
21. C. Riekel, H.G. Reznik and R. Schöllhorn, J. Solid State Chem. 34, 253 (1980).
22. G. Scholz and R. Frindt, Mat. Res. Bull. 15, 1703 (1980).
23. A.G. Gerards, H. Roede, R.J. Haange, B.A. Bouekamp and G.A. Wiegers, Synth. Met. 10, 51 (1984).
24. J. Schramke and R. Schöllhorn, Solid State Ionics 23, 197 (1987).
25. J.E. Ford and J.D. Corbett, Inorg. Chem. 24, 4120 (1985).
26. H. Meyer, A. Weiss and J.O. Besenhard, Mat. Res. Bull. 13, 913 (1978).
27. G. Haidari, W. Paulus and R. Schöllhorn, to be published.
28. R. Schöllhorn, in "Intercalation Chemistry", M.S. Whittingham and A.J. Jacobson, eds., Academic Press, New York (1982), p. 315.
29. R. Schöllhorn, in "Chemical Reactions in Organic and Inorganic Constrained Systems", R. Setton, ed., D. Reidel, Dordrecht/Boston (1986), p.323.
30. C. Ritter and R. Schöllhorn, Solid State Comm. 61, 117 (1987).

31. R. De Ridder, G. van Tendeloo, J. van Landuyt, D. van Dyck and S. Amelinckx, Phys. Stat. Sol. a, 37, 591 (1976).
32. W. Thulke, R. Frahm, R. Haensel and P. Rabe, Phys. Stat. Sol. a, 75, 501 (1983).
33. B.W. Pfalzgraf, H. Spreckels, W. Paulus and R. Schöllhorn, to be published.
34. P. Buller, Dissertation, University of Münster (1985).
35. P. Rathner, A. Payer and R. Schöllhorn, in preparation.
36. R. Schöllhorn and B. Otto, J.C.S. Chem. Comm. 1987, 1222
37. J. Rouxel, J. de Phys. 83, 841 (1986).
38. R. Schöllhorn, Pure and Appl. Chem. 56, 1739 (1984).
39. J.W. Johnson, and A.J. Jacobson, Angew. Chem. 95, 422 (1983); Angew. Chem. Int. Ed. Engl. 22, 412 (1983).
40. R. Schöllhorn, Comments in Inorg. Chem. 2, 271 (1983).
41. W. Schramm, E. Gocke and R. Schöllhorn, Mat. Res. Bull. 21, 929 (1986).
42. G. Aselmann, W. Müller-Warmuth, E. Gocke and R. Schöllhorn, Z. Phys. Chem. N.F. 151, 103 (1987); E. Gocke, R. Schöllhorn, G. Aselmann and W. Müller-Warmuth, Inorg. Chem. 26, 1805 (1987).
43. Y. Watanabe, G.A. Wiegers and C.F. van Bruggen, Synth. Met. 10, 1 (1984).
44. W. Schramm, R. Schöllhorn, H. Eckert and W. Müller-Warmuth, Mat. Res. Bull. 18, 1283 (1983); H. Eckert, W. Müller-Warmuth, W. Schramm and R. Schöllhorn, Solid State Ionics 13, 1 (1984).
45. L. Bernard, M. McKelvy, and W. Glaunsinger, Solid State Ionics 15, 301 (1985); H.J.M. Bouwmeester and G.A. Wiegers, to be published.
46. N. Janzen, Dissertation, University of Münster (1987)
47. H. Eckert, R.H. Herber, J. Chem. Phys. 80, 4526 (1984).
48. R. Schöllhorn, H.D. Zagefka, T. Butz and A. Lerf, Mat. Res. Bull. 14, 369 (1979).

PREPARATION METHODS FOR ALKALI METAL INTERCALATION COMPOUNDS OF OXIDES AND CHALCOGENIDES

D. W. Murphy, S. A. Sunshine and S. M. Zahurak

AT&T Bell Laboratories
600 Mountain Avenue
Murray Hill, New Jersey 07974

INTRODUCTION

There have been numerous advances in the understanding of intercalation systems in the last few years. These advances have been largely driven by recognition of the utility of these materials in applications such as batteries, sensors, etc. which have brought a critical mass of scientists together to work on the subject.[(1)] One of the aspects which has been fruitful is the area of synthesis. Improved reaction methods have allowed the preparation of many old intercalation compounds in high purity and has led to the elucidation of many new systems. Traditional thinking of intercalation systems as layered systems has been broadened to include three dimensional and even amorphous hosts. Techniques learned in intercalation chemistry have been used to prepare systems which are of interest for all the same reasons that any solid state system is of interest including structure, bonding, physical properties and their interrelation. There has been, in effect, increased recognition that the traditional rules of chemistry such as preferred oxidation states and coordination geometries for particular ions still generally apply, but in many cases the limits may be extended in intercalation systems. The purpose of this article is to outline some of the synthetic techniques which have advanced the field of intercalation in oxides and chalcogenides. We will concentrate mainly on compounds of alkali metals since these presented formidable challenges and because they have been of considerable practical interest as battery electrode materials.

Two principles are key to a discussion of synthetic reactions. First, as in most solid state systems, separation techniques are lacking so it is extremely important that reactions are free of inseparable impurity phases. Second, the reactions involve oxidation-reduction chemistry. This second point is most clearly seen by expressing an overall reaction in terms of two half-cell reactions as illustrated in Eqn. 1 for the reaction of Li and TiS_2.

$$Li \rightarrow Li^{+} + e \qquad (1a)$$

$$Li^+ + e + Ti^{4+}S_2 \rightarrow Li^+Ti^{3+}S_2 \quad (1b)$$

$$Li + Ti^{4+}S_2 \rightarrow Li^+Ti^{3+}S_2 \quad (1)$$

These half-cell reactions form the basis of intercalation battery systems and are discussed in depth elsewhere in this volume. Electrochemistry has been used as a synthetic technique. It offers advantages over many other methods in control of the redox potential, stoichiometry, and rate of the reactions. In addition, it provides a convenient geometry for many in situ experiments. Elegant examples of the power of the in situ study of reactions include mapping out phase boundaries in complicated systems,(2) and suggestions of ion ordering within a single phase.(3) Electrochemical reactions are not ideal for preparation of bulk samples for ex situ studies. This is largely due to the inability to prepare a perfect electrode in which both electrical and ionic contact are simultaneously provided. In general, there will be electrically isolated regions. It is often necessary to add an electronic conductor, which is difficult or impossible to later remove, to provide sufficient electron transport within an electrode. Electrolyte solvents may also cointercalate resulting in a large volume increase and exfoliation of the electrode.

CHEMICAL REAGENTS FOR INTERCALATION

Chemical reactions are most suitable for preparing large amounts of material and for broad based studies of the materials. The simplest concept is the direct reaction (Eqn. 2) of the intercalant (M′) with the host MX_n

$$M' + MX_n \rightarrow M'MX_n \quad (2)$$

Such reactions are often successful for less reactive metals such as post transition elements and have been widely used in these cases. In the cases where direct reaction is successful it is generally the method of choice. For reactive metals such as the alkalies it is generally not very useful. The pitfalls include reaction of M′ with reaction vessels, thermal instability of the product at reaction temperatures, and over reaction to give non-intercalation processes. Reaction with containers can usually be circumvented, but it is often easier to use an alternate reaction. The other problems can only be avoided by alternate reactions. Alternative methods, which tend to be much milder, have been termed "chemie douce" or "soft chemistry".

An intercalant that illustrates the need and benefits of "chemie douce" is Li. The possibility of using Li intercalation as an electrode reaction in batteries generated considerable interest in lithium intercalation chemistry. Metals such as Fe and Ni can be used as reaction vessels in direct reactions using Li metal, but there have been many intercalation compounds studied that are not stable at the temperatures required for direct reaction (500-600 °C). The first reagent used extensively for Li intercalation was Li dissolved in liquid NH_3.(4) While offering many advantages over direct reaction, it is often accompanied by cointercalation of ammonia. Heating to remove ammonia may result in ammonoloysis of the host yielding products of less than the highest quality. The first reagent for Li intercalation to combine the attributes of simplicity of set up and operation, general applicability, and high purity products was n–BuLi.(5,6) This reagent has become the most common chemical method for the preparation of Li intercalation compounds. The reaction proceeds as shown in Eqn 3.

$$xn\text{-}BuLi + MX_n \rightarrow Li_xMX_n + [xBu\text{-}] \rightarrow Li_xMX_n + \frac{x}{2}C_8H_{18} \qquad (3)$$

The reaction is typically run as a heterogeneous reaction with the n-BuLi dissolved in hexane in which the MX_n is insoluble. Following the reaction, which can be complete in minutes at room temperature or require several days at reflux depending on the host, the product is simply isolated by filtration. The only precautions necessary are to protect n-BuLi and the product from exposure to air, but this is necessary in any method since, in general, most $LiMX_n$ are reactive towards atmospheric moisture.

Several other reagents are available and offer advantages in particular cases. A major consideration in choice of reagent is the ease of reduction of the host. Figure 1 shows the relative reduction potentials of some intercalation hosts and some of the lithium intercalation reagents. The importance of using a reagent just on the reducing side of the host is illustrated by a comparison of the reaction products of V_2O_5 with LiI and n-BuLi. The reaction with LiI[7] (Eqn 4)

$$V_2O_5 + LiI \rightarrow LiV_2O_5 + \frac{1}{2}I_2 \qquad (4)$$

yields a well characterized intercalation compound even with a large excess of LiI, whereas the more reactive n-BuLi in excess reacts further to give products which are not intercalation compounds.

The discussion above centers on Li, but many of the reagents are applicable to other cations as well. The major exception is n-BuLi. This widely used reagent does not have an analog with other cations. A class of reagents that appears to be both relatively convenient and general for a variety of cations is borohydrides, ABH_4.[10] Organic as well as inorganic cations have been intercalated by this method in TaS_2, FeOCl, and MoO_3.

DESINTERCALATION

The term intercalation has as part of its definition reversibility of the chemical reaction. The reverse reaction is generally called "desintercalation". Most alkali metal intercalation reactions are too exergonic to be thermally reversible. However, such reactions can often be "reversed" either by another chemical reaction or electrochemically. An example is the charging of a positive intercalation electrode which corresponds to desintercalation. Many of the reagents in Fig. 1 can be used as chemical oxidants to accomplish desintercalation. The general principle is illustrated for the I_2/I^- redox couple in Eqn 5.

$$LiI + MX_n = LiMX_n + \frac{1}{2}I_2 \qquad (5)$$

This reaction may proceed spontaneously in either direction depending on the reduction potential of MX_n. The reaction was illustrated previously going to the right with V_2O_5, but with chalcogenides the reaction generally proceeds to the left and is one of the most widely used desintercalation reactions.[8] The reaction with ReO_3 (Eqn 6) can be driven in either direction with an excess of reagent.[9]

$$0.25LiI + ReO_3 \rightarrow Li_{.25}ReO_3 + 0.125I_2 \qquad (6)$$

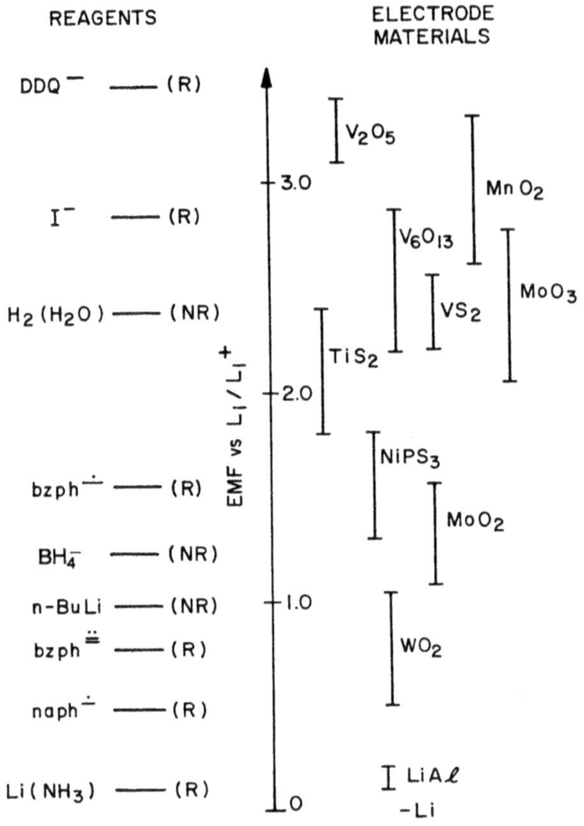

Fig. 1. A comparison of the redox potentials of a variety of intercalation reagents and hosts. The scale is an estimate based on observations from Li intercalation. (R) denotes reagents for which the redox couple is reversible and (NR) denotes irreversibility.

Water itself is listed as a reagent and leads to desintercalation in highly reduced intercalation compounds with formation of H_2 and LiOH. With higher oxidative stability only partial desintercalation or hydration of the intercalated ions may occur.

Desintercalation is a key concept, not only in completing the reversibility concept, but also as a synthetic route to new host materials. Some key examples of host materials synthesized by the desintercalation route include VS_2[8], CoO_2[15], Mo_6S_8[16,17] and c-TiS_2[18,19]. In each of these cases, the compound is stable and well characterized at room temperature, but can not be synthesized directly because of the greater stability of some alternate phase. There is, however, a ternary phase of each of these hosts that can be synthesized directly and the host obtained by a subsequent desintercalation.

The stable ternary phase $LiVS_2$ has the same structure as $LiTiS_2$ and following desintercalation with I_2 the VS_2 formed has the same 1T structure as TiS_2. For VS_2 the stability is not unexpected because the structure type is well known. However, the layered $LiCoO_2$ on desintercalation yields a layered CoO_2 that is unprecedented among oxides. Another unprecedented structure type occurs on desintercalation of the spinel $CuTi_2S_4$. Desintercalation of Cu affords a cubic(c) TiS_2. The resulting c-TiS_2 undergoes reversible lithium intercalation comparable to that of 1T-TiS_2. This last example and that of Mo_6S_8 illustrate the concept that desintercalation of a ternary phase may lead to a host suitable for intercalation with another ternary element.

We have synthesized c-ZrS_2 by a combination of methods. Desintercalation of Cu from $CuZr_2S_4$ resulted in only partial removal of Cu. However, it had been noted that the layered compounds Li_xZrS_2 ($0.3<x<0.5$) when heated above 250 °C converted to a spinel structure.[20] In our hands, the conversion occurred but the temperature was substantially higher (600 °C). This c-Li_xZrS_2 could then be intercalated by n-BuLi and desintercalated by I_2 to afford products with $0 < x < 1$.

INTERCALATION IN SINGLE CRYSTALS

Desired intercalation stoichiometries can generally be obtained in polycrystalline powders through stoichiometric addition of appropriate reagents followed by an equilibration step. However, these same reagents have been less successful for making single crystal intercalation compounds. The special problems encountered with single crystals relate to cointercalation of solvents, relatively slow diffusion of the alkali in the host, changes in volume which tend to break the crystals, and most importantly, lack of control over stoichiometry. Control over stoichiometry is lacking because it is virtually impossible to measure out the appropriate amount of reagent for a crystal that may be only a few milligrams. There have been successful intercalation reactions on small perfect crystals of TiS_2 with n-BuLi[11], and crystals of $A_{.5}MS_2$(A=Li, Na, K, Rb, Cs; M=Ti, Nb) have been grown directly from alkali halide melts.[12] However, relatively few physical properties have been measured on single crystals. These measurements are crucial to a more complete understanding of these materials. A method for obtaining crystals of controlled stoichiometry in AMS_2 (A=Li, Na, K, Rb; M=Ti, Ta) has been reported[13] that utilizes the small, but finite vapor pressure of the alkali metal in equilibrium with the intercalation compound at 300-700 °C. Typically an excess of a polycrystalline powder that has the composition desired for the crystal is sealed in a fused silica ampoule along with a crystal. The tube is heated for a few days to equilibrate the alkali metal between powder and crystal via the vapor.

Although lithium has a substantial vapor pressure at moderate temperature (10^{-2} torr at 650 °C), it is somewhat surprising that the intercalation compounds Li_yMX_2 have a significant vapor pressure since they are generally formulated as $Li_y^{+}M^{y-}X_2$. Estimates of the degree of charge transfer in $LiMX_2$ from ^7Li nmr data indicate ~90% charge transfer from Li to M.[14] To a first approximation, the Li vapor pressure should be reduced from pure Li by the percent of charge transferred. This would still leave a substantial vapor pressure of 10^{-3} torr at 650 °C. In general, we expect that the Li vapor pressure will depend on the activity of Li in the intercalation compound. The Li activity, and thus the vapor pressure, is related to the EMF of $Li/Li^{+}/Li_yMX_2$ cells, with lower cell voltages giving higher activities and vapor pressures. Intercalation systems exist with a wide range of Li activities and, thus, Li vapor pressures.

CONCLUSION

In this article we have attempted to show that there are a wide variety of techniques for accomplishing intercalation and desintercalation. Judicious choice of technique is necessary to accomplish the desired goal. Examples have been shown that illustrate the principles for selective intercalation, intercalation of crystals. and desintercalation.

REFERENCES

[1] See for example, "Solid State Batteries," Marinas Nighoff Publishes, Dordrecht, C. A. C. Sequeira and A. Hooper, ed., 1985 and "Solid State Ionics - 85," North-Holland, Amsterdam, J. B. Boyce, L. C. De Jonghe, and R. A. Huggins, ed., 1985 and M. S. Whittingham, Prog. Solid State Chem., *12*, 1 (1978).

[2] Dahn, J. R. and Mckinnon, W. R., Phys. Rev., *B32*, 3003 (1985).

[3] Thompson, A. H., Phys. Rev. Lett., *40*, 1511 (1978).

[4] See, for example, M. Danot, A. Leblanc, and J. Rouxel, Bull. Soc. Chim. Fr., 2670 (1969).

[5] Dines, M. B., Mater. Res. Bull., *10*, 287 (1975).

[6] Murphy, D. W., DiSalvo, F. J., Hull, G. W., Waszczak, J. V., Inorg. Chem., *15*, 17 (1976).

[7] Murphy, D. W., Christian, P. A., DiSalvo, F. J., and Waszczak, J. V., Inorg. Chem., *18*, 2800 (1979).

[8] Murphy, D. W., Carides, J. N., DiSalvo, F. J., Cross, C., and Waszczak, J. V., Mater. Res. Bull., *12*, 825 (1977).

[9] Cava, R. J., Santoro, A., Murphy, D. W., Zahurak, S. M., and Roth, R. S., J. Solid State Chem., *50*, 121 (1983).

[10] Kanatzidis, M. G., and Marks, T. J., Inorg. Chem., *26*, 783 (1987).

[11] Chianelli, R. R., J. Cryst. Growth, *34*, 239 (1976).

[12] Schollhorn, R., and Lerf, A., J. Less Common Met., *42*, 89 (1975).

[13] Zahurak, S. M., and Murphy, D. W., J. Solid State Chem., in press.

[14] Silbernagel, B. G., Solid State Comm., *17*, 361 (1975).

[15] Mizushima, K., Jones, P. C., Wiseman, P. J., and Goodenough, J. B.,

Mater. Res. Bull., *15* (783) 1980.
[16] Schollhorn, R., Kuempers, M., and Besenhard, J. O., Mater. Res. Bull., *12*, 781 (1977).
[17] Tarascon, J. M., Waszczak, J. V., Hull, G. W., DiSalvo, F. J., and Blitzer, L., Solid State Comm., *47*, 973 (1983).
[18] Schollhorn, R., and Payer, A., Angew. Chem. Int. Ed. Eng., *24*, 67 (1985).
[19] Sinha, S., and Murphy, D. W., Solid State Ionics, *20*, 81 (1986).
[20] Deniard, P., Chevalier, P., Trichet, L., and Rouxel, J., Synth. Metals, *5*, 14 (1983).

ELECTRONIC STRUCTURE OF TRANSITION-METAL CHALCOGENIDES AND THEIR INTERCALATION COMPOUNDS

W. R. McKinnon

Solid State Chemistry, Division of Chemistry
National Research Council of Canada, Ottawa, Canada K1A 0R9

INTRODUCTION

Intercalation compounds are solids where host atoms provide a structure into which guest atoms can be inserted. Generally, charge is transferred between the outer orbitals of the guest atoms and the electron bands of the host during intercalation. In the transition-metal chalcogenides, the guest gives up charge to become a positive ion, and the electrons donated by the guests enter the host band structure. Thus the electronic structure of the host should, together with the sites available for the guest ions, determine whether a given guest can be inserted, and to what extent.

In this article, I review the electronic structure of transition-metal chalcogenides and their intercalation compounds. First I relate the local arrangement of the atoms to the details of the electronic structure. The simplest model to explain the electronic properties of intercalation compounds is the rigid-band model, which assumes the band structure of the host is unchanged by intercalation. I discuss the evidence for charge transfer and rigid-band models. There is a misconception that rigid-band models imply that the energy of electron states is fixed during intercalation; rather, they imply only that the shape of the bands is fixed. The bands actually move down in energy during intercalation, as discussed by Friedel (1954) over 30 years ago. This motion is not observable in experiments that measure energy differences (spectroscopic techniques like photoemission) or that measure the density of states at the Fermi surface (specific heat, magnetic susceptibility). The motion can be inferred from the behavior of the chemical potential of the guest, as measured in electrochemical cells, but only after the contributions from the ions have been accounted for. I discuss to what extent the chemical potential of the guest, as measured in electrochemical cells, reflects the electronic structure.

GENERAL FEATURES OF THE STRUCTURE

Most of the intercalation compounds of transition-metal chalcogenides are formed from hosts with transition metals from groups 4 (Ti, Zr, Hf), 5 (V, Nb, Ta), or 6 (V, Mo, W) of the periodic table. Intercalation compounds of the transition-metal chalcogenides have several common features. The transition metal, M, and the guest atom, A, sit in cages of polarizable chalcogens, X. The polarizability of the chalcogens is important in determining the structure (Haas 1985). The Coulomb interaction of point ions favors a close-packed, symmetric structure, but the energy of polarization of the chalcogens is lowered in structures of lower symmetry like layered compounds. Thus layered structures are more common in compounds of the larger, more polarizable chalcogens (X = S, Se, Te) than in oxides.

The coordination of M by X, and in some cases the second-neighbor coordination of M by M, determine the electronic structure, as discussed below. This coordination, then, is the most important aspect of the structure for this discussion. In the layered transition-metal dichalcogenides MX_2, two types of M-X coordination are important: octahedral and trigonal prismatic (Fig.1). In both coordinations, 6 X surround each M, forming either an octahedron or a trigonal prism. Layers of octahedrally-coordinated MX_2 usually stack one above the other to form the 1T structure; layers of trigonal prismatic MX_2 are generally staggered to form a two-layer unit cell, as in the 2H-NbS_2 or 2H-MoS_2 structures (Wilson and Yoffe 1969). Other more complicated stackings also occur.

Molybdenum chalcogenides adopt another structure that forms many intercalation compounds. In compounds of the form $A_xMo_6X_8$, called Chevrel compounds, cubes of Mo_6X_8 with X on the corners and Mo on the centers of the faces (Fig.2) are arranged on a nearly cubic lattice (Chevrel and Sergent 1982, Yvon 1979). The cubes rotate to bring Mo atoms closer to X atoms on adjacent cubes and bond the structure. Again the A atoms sit in cages of X atoms.

Transition-metal oxides adopt other structures, such as rutile and perovskite. In many of these the transition-metals sit in octahedra or distorted octahedra of oxygen.

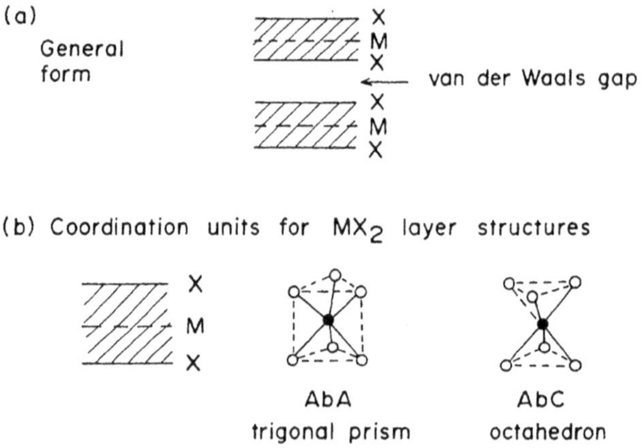

Fig.1. Coordination of transition-metal atoms M by chalcogen atoms X in transition-metal dichalcogenides MX_2.

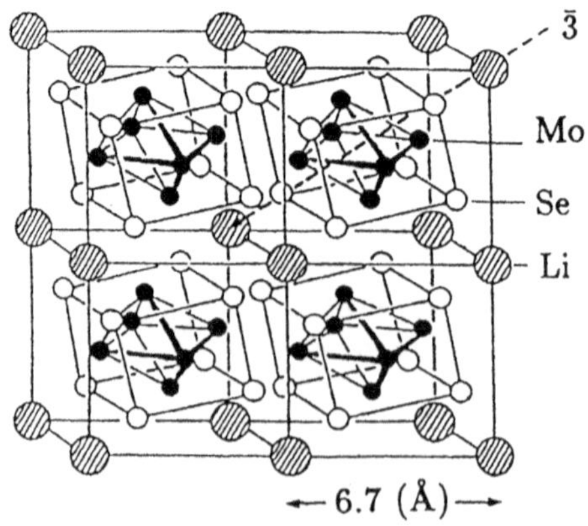

Fig.2. Structure of the Chevrel compound $LiMo_6Se_8$.

GENERAL FEATURES OF THE ELECTRONIC STRUCTURE

In spite of the impressive successes of band structure, which exploit the long-range periodicity of solids, most of the details of the electronic structure of solids is determined by the local arrangement of atoms (Heine 1980). The effects of this coordination are best explained with molecular-orbital theory (Huisman et al 1971). The important atomic orbitals are the outermost p orbitals from the chalcogens X, and the s, p, and d orbitals from the transition metal M. Fig.3 shows schematically how the atomic orbitals combine. The s and p orbitals of M overlap with the p orbitals of X to form bonding and antibonding sp orbitals. Since the X orbitals have a lower energy, the bonding combinations are formed largely from X-p, and the antibonding ones from M-s and M-p. Thus we may loosely refer to the bonding orbitals as X-p orbitals, and the antibonding ones as M-s or M-sp.

The gap between the bonding and antibonding sp orbitals is several eV, and it contains the d-orbitals from M. These d-orbitals hybridize with the sp-bonding orbitals, forming spd bonding and antibonding orbitals. Again we may loosely identify the bonding orbitals are X-p and the antibonding ones as M-d.

The degree of hybridization of a particular d-orbital depends on whether that orbital points toward the chalcogens. If it does, it will be pushed to higher energies than those that don't. (This happens because the d-orbitals are actually antibonding orbitals; their bonding counterparts in the group of orbitals loosely called X-p will be pushed to lower energy.) Thus the details of the splittings of the d-orbitals depends on the coordination of M by X. Fig.4 shows those splittings in octahedra and trigonal prisms (Huisman et al 1971). In an octahedron, the d-orbitals separate into three lower and two upper orbitals. (If the octahedron expands or contracts along the vertical direction, the three lower orbitals also separate, but not enough to concern us here.)

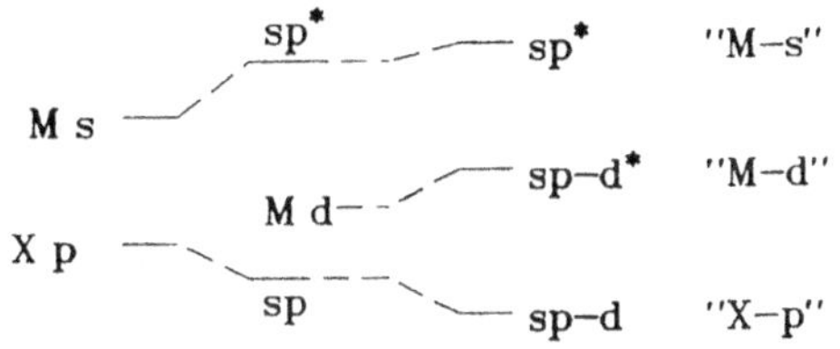

Fig.3. How the atomic orbitals in compounds of transition metals (M) and chalcogenides (X) combine to form bonding and antibonding orbitals.

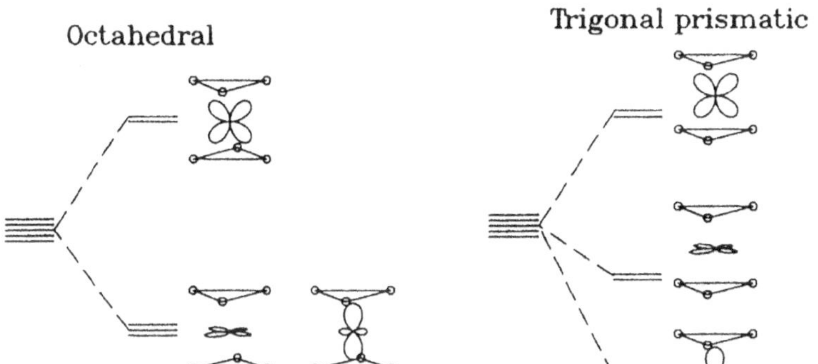

Fig.4. The splitting of the d-orbitals in trigonal prisms and octahedra of chalcogens.

In a trigonal prism, the d-orbitals split into a lowest orbital (largely of d_z^2 symmetry), two intermediate orbitals, and two upper orbitals.

The loose interpretation of the orbitals as being pure X or M gives the expected oxidation states of the components. The X-p orbitals are completely full, so X is in an oxidation state of -2. Thus in MX_2, M is in an oxidation state of +4. Atoms from group 4 have a valence state d^0, those in group 5 d^1, and those in group 6 d^2, corresponding to 0, 1, or 2 electrons in the d-orbitals in Fig.4.

Calculations of the band structures of the compounds support these simple molecular-orbital arguments. (For a recent review of band structures of layered compounds see Doni and Girlanda 1986.) When the clusters are brought together, the orbitals in Fig.4 broaden into bands. The d-bands broaden both because of direct Md-Md overlap and indirect Xp-Md-Xp overlap. The schematic band structure of the MX_2 compounds with octahedral and trigonal prismatic coordination are shown in Fig.5. Each orbital in Fig.4 can hold two electrons per MX_2 formula unit, corresponding to spin-up and spin-down states. Thus in MX_2 compounds from group 4,

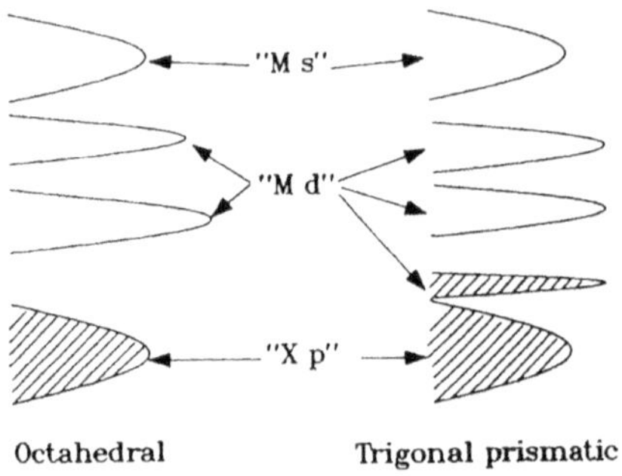

Fig.5. Schematic band structures for all the transition-metal dichalcogenides MX_2.

the p-bands are full and the d-bands are empty, so these compounds are semiconductors or semimetals. Group 6 MX_2 compounds with trigonal-prismatic coordination are semiconductors, with the Fermi level in the gap above the lowest d-band, whereas in those with octahedal coordination are metallic. Group 5 compounds are metals whether they form octahedral or trigonal prismatic coordination. Photoemission experiments also support this simple picture (see, for example, Coehoorn et al 1987 and references there).

More support for this electronic structure comes from the stability of the trigonal-prismatic structure. Often the energy of a compound can be lowered if the structure can be altered to produce a gap in the density of states at or near the Fermi energy. Thus the trigonal-prismatic coordination is most stable in the group 6 MX_2 compounds, when the lowest d-band is full and the next one empty (Haas 1975, Kertesz and Hoffman 1984). Both trigonal prismatic and octahedral coordination are seen in group 5 and group 6 MX_2 compounds, but only octahedral coordination is found in unintercalated group 4 compounds. In $MoTe_2$, the octahedral coordination is stable at high temperatures, but the structure changes to trigonal prismatic (and the material goes from a metal to a semiconductor) as the temperature is lowered (Vellinga et al 1970).

The d-bands can also be split if the M atoms move to form stronger d-d overlaps. In the rutile oxide VO_2, where the V is octahedrally coordinated, the V atoms move together to form pairs below 340K. This pairing pulls one orbital per double formula unit, or one electron state per formula unit, out of the lower d-band in Fig.5 (Goodenough 1971). A similar formation of bonds and energy gaps occurs in many of the layered MX_2 compounds, accompanying the formation of so-called charge-density waves (Haas 1985).

In the Chevrel compounds Mo_6X_8, clusters of M atoms in each unit cell form an octahedron, so the d-orbitals along the edges of the octahedron form bonding combinations and lie below the other d orbitals (which tend to point more toward the X atoms and so are pushed to higher energies as anti-bonding orbitals (Nohl et al 1982). An octahedron has 12 edges, so there are 12 bonding d-orbitals, which can hold 24 electrons per Mo_6X_8 formula unit. Since the X-p bands are full, (so the X atoms are again formally -2), there are 20 electrons left to fill those lower d-orbitals. Thus the Chevrel compounds Mo_6X_8 are metallic, and the compound $Ru_2Mo_4Se_8$ is semiconducting, because each of the two Ru adds two extra electrons, filling the lower d-band (Perrin et al 1980).

Band theory considers electrons moving in the average potential from all the electrons. If there are strong electron-electron repulsions, the theory breaks down. The standard example of this breakdown is in a chain of s-orbitals, with one electron per orbital. In band theory, the s-orbitals would form a half-full band and so the chain would be metallic. But if it costs too high an energy to add a second electron to any one atom, the chain will be insulating; the band splits into two subbands, corresponding to one or two electrons per atom. The energy cost to remove an electron from one atom and double it up with an electron on a second is known as the Hubbard energy, U. This energy is generally much lower for atoms in solids than for isolated atoms, because in solids electron states spread out over neighboring atoms. It can, however, be appreciable in small transition metal atoms, in particular those to the right in the first row in the periodic table. When U is large compared to the width of the d-bands, we must speak of localized d-states rather than d-bands. In some intercalation compounds of V and Cr, a localized description is needed (eg Wiegers 1980, Murphy et al 1976).

RIGID-BAND MODELS OF INTERCALATION

In the simple picture of Fig.3, the outer electron states of atoms like Li lie several eV above the Fermi energy, so the electrons from those states enter the M-d states. The simplest approximation to the band structure of an intercalation compound is just that of the parent host compound with the Fermi level moved up to accommodate the extra electrons. This model, called the rigid band model, is frequently used in describing intercalation compounds.

There have unfortunately been few calculations to check to what degree the rigid-band model is valid in transition-metal chalcogenides. The only self-consistent calculations are for $LiTiS_2$ (Umrigar et al 1982). Those calculations show that the rigid-band model is a reasonable first approximation, but that it has limitations. Fig.6 shows the calculated density of states for TiS_2 and $LiTiS_2$, with the peaks in the lower d-band aligned. The two curves are similar, but the S-p bands are shifted to lower energy in $LiTiS_2$ than in TiS_2. This shift, of order 1eV, arises from the charge rearrangements that accompany intercalation. The Li atom, although it has formally given up its 2s electron, polarizes the S ions around it, drawing charge density towards itself; compared to TiS_2, there is actually more extra electron density between the Li and S layers than between the S and Ti layers (Umrigar 1982). This polarization lowers the energy of the S states more than that of the Ti states. This charge rearrangement is not surprising; in general, the charge distribution of atoms in solids corresponds more closely to that for neutral atoms than to ions with a charge equal to the formal

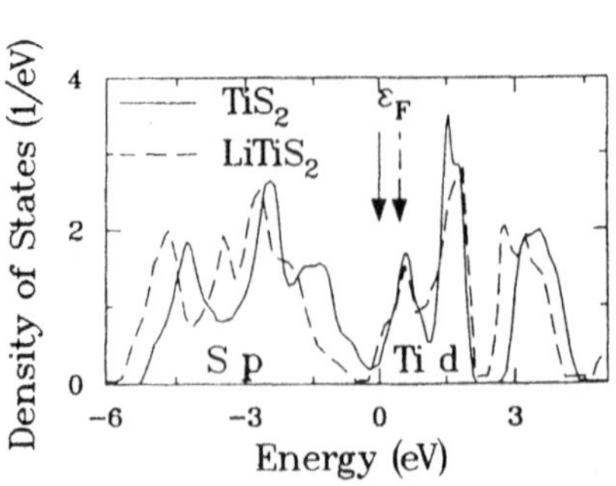

Fig.6. Density of states for TiS_2 and $LiTiS_2$, calculated by Umrigar et al (1982), shifted so the peaks in the lower d-bands are aligned.

oxidation state of the ion. Thus although Li is formally +1, the charge it polarizes from the S atoms makes it appear almost neutral (Umrigar 1982).

Except for the shifts just mentioned, the band structure of $LiTiS_2$ and TiS_2 are similar, so it is a reasonable approximation to look on the band structure of the compound as rigid, unchanging in shape, as the electrons are added. Unfortunately, the term "rigid-band model" is often misinterpreted to mean that the energy of the electrons in the band remains fixed. But most experiments cannot tell whether the bands move up or down on an absolute energy scale, because most measure either the density of states at the Fermi energy (the energy of the upmost-occupied state) or the difference in energy between two states. Sellmyer (1978) distinguishes two versions of the rigid-band model: the electron-gas version, where the bands are rigid in shape and fixed in energy, so the Fermi energy rises like water in a glass as electrons are added; and the screened-impurity version, where the bands are rigid in shape but move down as electrons are added.

The screened-impurity version is due to Friedel (1954). He argued that any charged impurity added to a solid will polarize the solid, and attract charge to itself. The field from the impurity will thus be screened over some distance L, called the screening length. Since the solid must remain neutral, a positively charged impurity (of charge $+Ze$, say) must be accompanied by Z electrons. Effectively those Z electrons will be added to the region within L of the impurity, even if they are being added to electron states of the solid. As long as the impurities are far apart compared to L, there will be regions of the solid where no extra electrons are present. Thus the chemical potential of the electrons, the Fermi energy, will remain constant, even though electrons are being added to the energy bands of the solid. The only way this can happen is if the bands move downward in energy to compensate. This lowering in energy is simply due to the Coulomb attraction between the ions and the screening electrons.

Friedel's arguments are based on a Thomas-Fermi model, where the potential from the impurity ions varies slowly with distance. In this limit, the electron energy averaged over the solid gives a density of states of the same shape as in the empty host, but shifted down in energy. Thus the rigid-band description holds no matter how the impurity ions are distributed. When the potential varies rapidly, however, a rigid-band description can break down even if the bands in the ordered limits (all sites full or all empty) have the same shape. Such a breakdown is observed in alloys, and must be described by modifications of band theory like the coherent-potential approximation (for a review, see Ehrenreich and Schwartz, 1976). Unfortunately, there has been little study of this problem in intercalation compounds.

Another consequence of the local nature of the charge rearrangements in intercalation is that the potential that attracts the electron near the intercalated ion could be strong enough to change the electron wavefunction from itinerant to localized. This can happen when the Fermi energy is near a gap in the density of states, for there the lowering of energy of the electron states can move states into the gap.

STUDIES OF THE ELECTRONIC STRUCTURE IN INTERCALATION COMPOUNDS

This section reviews some of the measurements of the electronic structure of intercalation compounds. A more extensive discussion is given by Friend and Yoffe (1987).

Charge transfer

The first point to be established is that charge really is transferred to the host during intercalation. The most often-cited evidence is the measurement of the Knight shift in Li_xTiS_2 (Silbernagel and Whittingham, 1977). There it was found that the NMR frequency of Li was shifted by less that 12 ppm from that for Li in solution, compared to 240 ppm for Li in Li metal. This indicates that the charge density at the Li nucleus is almost the same for Li in Li_xTiS_2 as for Li ions in solution.

This measurement indicates charge is transferred away from the Li; other measurements show charge is transferred to the M ions. Mossbauer studies of 2H-TaS_2 (Salomon et al 1985) show the Mossbauer frequency of the Ta nucleus shifts during intercalation, indicating an increase in charge density at the Ta nucleus.

Extensive measurements have been made of the transport properties of intercalation compounds. Except for cases where the d-states become localized, as in Cr or V dichalcogenides, the conductivity is generally as expected from the schematic band structure in Fig.5. As group 4 compounds are intercalated, they change from semiconducting to metallic. Moreover, where the Hall effect has been measured (Li_xTiS_2 and Li_xZrS_2), the number of carriers is close to that expected for complete charge transfer from the Li (Friend and Yoffe 1987).

"Charge transfer" and electrochemical cells

The quotation marks in the subheading above are intended to distinguish the charge transfer we have spoken of so far from the charge flow that accompanies intercalation in an electrochemical cell. In the simplest electrochemical cell, the host compound is the positive electrode and the pure guest (Li, for example) is the negative electrode. As electrons pass from the negative to positive electrode through the external circuit, corresponding to discharging the cell, atoms of the guest dissolve in solution, travel through the solution as ions of charge $Z_s e$, recombine with the electrons from the external circuit, and intercalate into the host. The change in free energy when Δn atoms intercalate (or $Z_s \Delta n$ electrons pass through the external circuit) is given both by $-Z_s eV$ and by $\Delta n(\mu - \mu_0)$, where V is the voltage difference between positive and negative electrodes, μ is the chemical potential of the guest in the host, and μ_0 the chemical potential of the guest in the negative electrode (McKinnon and Haering 1983). Thus because μ_0 is constant the voltage of the cell directly measures μ, according to

$$-Z_s eV = (\mu - \mu_0).$$

But the quantity Z_S in this equation has nothing to do with the state of charge of the guest in the host. It simply relates the number of electrons transferred to the number of atoms intercalated, and so is determined by the ionization of the guest in the solution connecting the two electrodes. In general, the intercalated atom is expected to have the same charge state as it would have in solution, but it need not. We expect, for example, Cu to have the same charge state in a solid whether it comes from an aqueous solution where it is $Cu^{+}2$ or from a nonaqueous one where it is $Cu^{+}1$.

Changes in host structure induced by intercalation

By adding electrons to the host bands, intercalation can change the stability of various polytypes of the host and change the structure of the compound. Thus, adding Li to group 6 compounds like MoS_2 or $MoSe_2$ changes the coordination of the Mo from trigonal prismatic to octahedral. This change is attributed to the extra energy it costs to add electrons across the band gap of the trigonal-prismatic form, compared to the energy to change the structure to the octahedral form and eliminate that gap. This structural change has practical implications. Naturally-ocurring MoS_2 has trigonal-prismatic coordination of S by Mo, and makes a poor electrode in a battery if the voltage of the battery is never allowed to go below 1.2V. At lower voltages, however, Li enters the host and the coordination changes to octahedral (Py and Haering 1983). This new form of MoS_2 is the basis of a commercial cell (Lamann et al 1986).

Measurements of the Fermi surface

Various experiments have probed the Fermi surface of intercalation compounds by measurements of quantities like magnetic susceptibility or heat capacity. In the trigonal-prismatic MX_2 compounds of group 5, the density of states is maximum when the band is half-full (Mattheiss 1973), and so the density of states should decrease as guests are intercalated. The magnetic susceptibility of various intercalation compounds of TaS_2 falls as expected from the calculated density of states (Johnston 1982). The term linear in T in the heat capacity of Li_xNbS_2 also decreases with increasing x, and appears to drop abruptly near the composition where the calculated density of states for NbS_2 does (Dahn et al 1986).

Localization by disorder

An example where disorder localizes the electrons is $Li_xTa_yW_{1-y}O_3$. In this material W is octahedrally coordinated by O, so the material should be metallic when $x \geq y$. But the material remains insulating until $x - y$ exceeds about 0.2 (Dubson and Holcomb 1985). Photoemission experiments confirm the localized nature of the states at small $x - y$ and suggest that the disorder responsible for the localization is due to the Na, not to the random distribution of W and Ta (Hollinger et al 1985).

Band-filling as observed in electrochemical cells

If the guest atom separates into an ion and its electrons in an intercalation compound, the chemical potential of the guest, μ, should have a contribution from both the ions and from the electrons. Thus when electrons completely fill one band and begin to fill the next one, the chemical potential should step up and the voltage should drop abruptly by roughly the band gap. In most cases, drops in the voltage of electrochemical cells can also be explained by the filling of a certain kind of site; the drop is then related to the energy difference between ions in that site and ions in the next site to be filled. The first observation of a step in V that was clearly related to electrons was in the intercalation compounds $Li_xNb_yMo_{1-y}S_2$ (Py and Haering 1984). As shown in Fig.7, the voltage drops near $x + y = 1$, as expected since these are trigonal-prismatic compounds, so the lower d-band (Fig.5) should be full at this point if each Li transfers one electron to the host d-bands. These experiments were not detailed

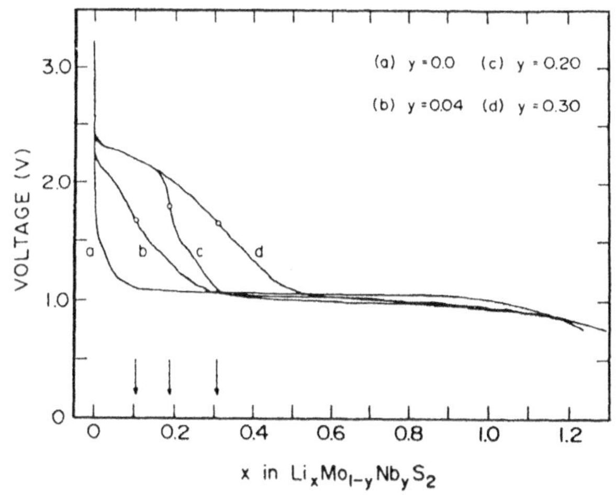

Fig.7. Voltage V versus x for 4 values of y in a $Li/Li_xNb_yMo_{1-y}S_2$ electrochemical cell at 20 ^{o}C. From Py and Haering (1984).

enough to distinguish the difference between the rigid-band models discussed above. Such experiments have been done in the Chevrel compounds $Li_xRu_zMo_{6-z}Se_8$, and are discussed below.

SEPARATION OF CHEMICAL POTENTIAL INTO ELECTRONIC AND IONIC PARTS

The above sections have established that the principle effect of intercalation on the band structure of transition-metal chalcogenides is to fill the bands of the host with electrons, but not to change the nature of the bands much. Most experiments are consistent with a rigid-band model, to the extent that they are accurate enough to test it. But what is not well appreciated is the difference between the two forms of rigid-band model discussed above. To distinguish these models, we need measurements of the absolute value of the chemical potential of electrons. Such measurements are in principle possible, through measurement of the work function, but have not been reported, and would be difficult to do to the accuracy required. The next-best thing is to measure chemical potentials of the guest atoms in electrochemical cells.

Such a measurement does not measure the chemical potential of electrons alone, because such thermodynamic measurements can only measure the chemical potential of neutral species. (Even the chemical potential or activity of ions in solution is really a formal definition; only sums of such chemical potentials corresponding to neutral entities can ever be measured.) Thus the separation of the measured chemical potential μ into an electronic and ionic part is really only a definition; because ions and electrons interact, there will always be an interaction term between them, and when we write $\mu = \mu_i + z\mu_e$ we must arbitrarily partition the interaction between the two terms. Nevertheless it is convenient to use that separation in what follows.

Once this separation of μ is made, it is tempting to suggest that the presence of the term μ_e, which is just the Fermi energy, means that μ will drop at a rate given by the density of electron states (Armand 1978; Molenda 1986). But that suggestion is based on the misinterpretation of the rigid-band model discussed above. Recently experiments on Chevrel compounds have confirmed that the more appropriate model is that where the bands drop as electrons are added. The next section reviews how well simple models can explain the ionic part of μ, and the following one examines the separation of μ into ionic and electronic parts.

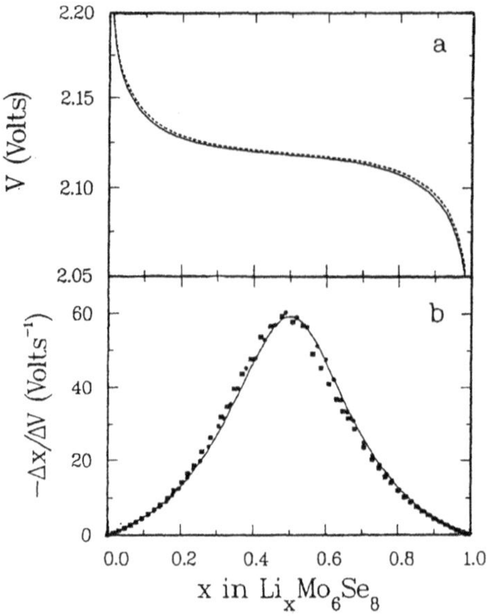

Fig.8. Voltage V and derivative $-\partial x/\partial V$ versus x in a $Li/Li_xMo_6Se_8$ electrochemical cell at 38 ^oC. From Dahn et al (1985a).

Lattice-gas models of intercalation in Chevrel compounds

The variation of the chemical potential of Li in $Li_xMo_6Se_8$ can be understood in terms of the ionic part μ_i only. For $x < 1$, Li enters a lattice of sites separated by over 6Å, and this region of x has been accurately modelled using a simple mean-field lattice-gas model. With one parameter, U, which describes the interaction of ions, such a model describes the derivative $\partial x/\partial \mu$ to a few percent between 5°C and 50°C. Fig.8 shows $V(x)$ and $-(\partial x/\partial V)_T$ at 38 ^oC. The same model explains the phase separation that occurs at lower temperature (Dahn and McKinnon 1985). With calorimetry, μ can be separated into entropy and energy contributions, and the composition-dependent part of the entropy contribution obeys a simple lattice-gas model with no adjustable parameters (Dahn et al 1985b). Moreover, the size of the interaction energy U is consistent with the expansion of the lattice by the ions. Thus all the variation in μ can be explained accurately with a model that completely ignores the electron contribution μ_e. This implies that μ_e is constant over this range of x. This in turn is consistent with the Friedel interpretation of screening, where the Fermi energy (μ_e) remains fixed as the electrons are added.

This fit alone does not prove the Friedel arguments, for the electronic contribution could be hidden in the parameters of the model. A variation in the energy part of μ due to a rise in the Fermi energy could add a term linear in x to μ, which could be absorbed into the parameter U. (That term, however, would be positive, whereas U is negative, so the elastic contributions would have to be even larger to compensate.) Moreover, since the electrons form a degenerate Fermi gas, their contribution to the entropy part of μ should be small. The next section discusses more convincing evidence that the electronic contribution really is constant, except near band edges.

$Li_xRu_zMo_{6-z}Se_8$

In analogy with $Li_xNb_yMo_{1-y}S_2$, electrons can be added to Mo_6Se_8 if Mo is replaced with Ru. Each Ru adds two electrons to the host, so, as noted above, the compound $Ru_2Mo_4Se_8$ is a semiconductor. Fig.9 shows the voltage versus x for $Li_xRu_zMo_{6-z}Se_8$, for four values of z. There are several points to note. First, it is possible to add Li beyond $x = 4 - 2z$, where the d-bonding band discussed above is full. Second, the voltage drops at that point. Third, the voltage of the first plateau, which is so well described by the lattice-gas model discussed above for $z = 0$, moves only by a few mV as z changes from 0 to 1, but then drops by about 1 volt between $z = 1$ and $z = 2$.

All these points are consistent with the arguments of Friedel discussed above. If the bands of Mo_6Se_8 were rigid in both energy and shape as Li intercalated, the first plateau would move down by 80mV between $z = 0$ and $z = 0.5$, according to the density of states (Bullett 1977); that it does not supports the argument that the band drops in energy as Li ions are added. The drop in voltage at $x = 4 - 2z$, just as in Fig.7 for $Li_xNb_yMo_{1-y}S_2$, indicates that the band is full. The shift in the plateau between $z \leq 1$ and $z = 2$ may be taken as a measure of the band gap, giving 1eV. Note this is probably a better measure of the gap than the drop in voltage at $x = 4 - 2z$, because adding Li (changing x) will cause more charge rearrangement than replacing Mo by Ru (changing z), so a rigid-band model is more likely to apply to differences in μ at fixed x than at fixed z.

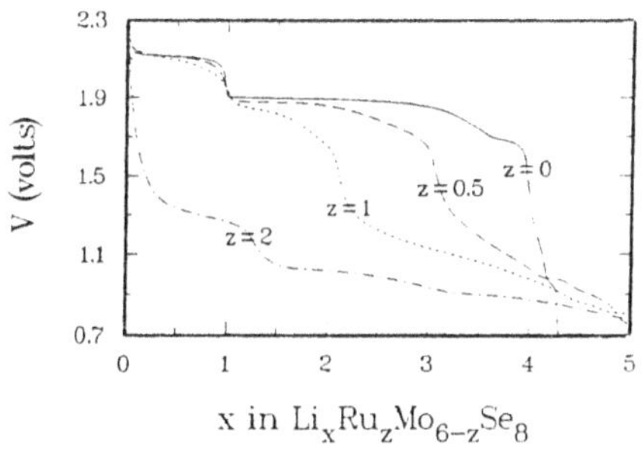

Fig.9. Voltage V versus x for 4 values of z in a $Li/Li_xRu_zMo_{6-z}Se_8$ electrochemical cell at 28 ^oC. From McKinnon and Selwyn (1987)

Fig.10 shows both the voltage and the resistance of the host powder versus x (McKinnon and Selwyn 1987). These resistance measurements were taken in-situ using a two-terminal technique, and so are not quantitative, but they do show an increase in resistance just where the bands fill at $x = 4 - 2z$ and the voltage drops. Thus we see evidence for charge transfer, band filling, and screening (by the small change in the first plateau) in this figure.

Conclusions

The band structure and charge transfer are well established in transition metal chalcogenides and their intercalation compounds. The rigid-band model is a useful first approximation to understanding the changes that occur in the electronic structure on intercalation, although there are cases where it is inadequate. But the rigid-band model does not imply that changes in chemical potential directly measure the density of states. The arguments of Friedel and the results of $Li_xRu_zMo_{6-z}Se_8$ show that the correct picture is one where the bands move down in energy as guest ions are added, compensating for the rise in the Fermi level relative to features in the band. This is shown schematically in Fig.11.

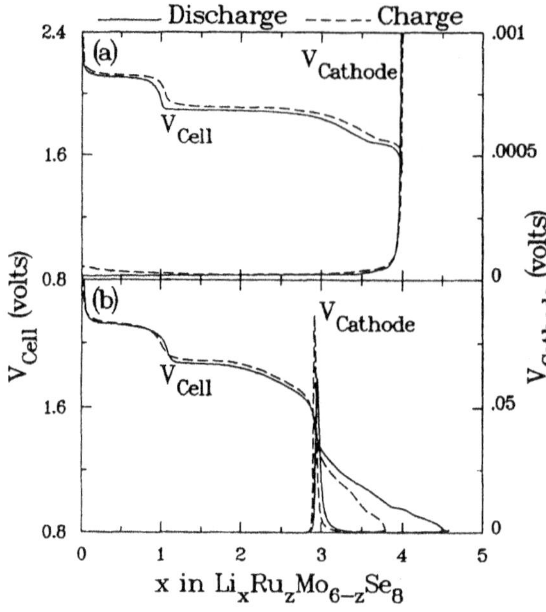

Fig.10. Voltage V and the voltage V_{cathode}, which is proportional to the resistance of the intercalation compound, versus x in a $Li/Li_xRu_zMo_{6-z}Se_8$ electrochemical cell at 28 ^oC. (a) $z = 0$; (b) $z = 0.5$. From McKinnon and Selwyn (1987) .

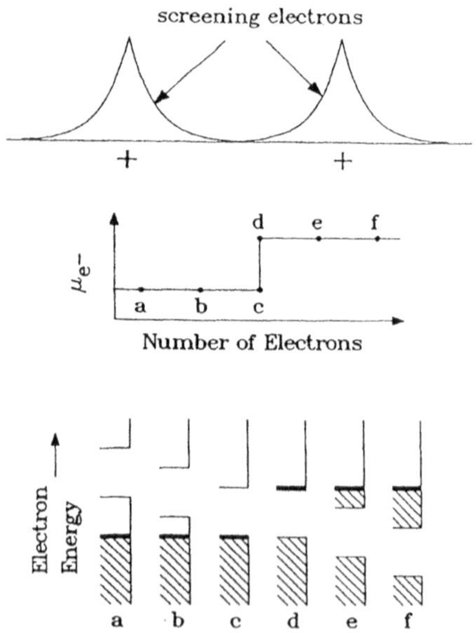

Fig.11. Schematic view of the screening charge around intercalated ions, the shift in the Fermi energy with intercalation, and the shift in the electron energy bands with intercalation, in the Friedel version of the rigid-band model.

References

Armand, M., 1978, Ph.D. Thesis. Université Scientifique Medical et Institut National Polytechnique de Grenoble. British Columbia.

Bullett, D. W., 1977, Phys. Rev. Lett., 39:664.

Chevrel, R., and Sergent, M., 1982. in: "Superconductivity in Ternary Compounds, Springer Topics in Current Physics", O Fischer and M.B. Maple, ed., Springer-Verlag, Berlin, 32:25.

Coehoorn, R., Haas, C., Dijkstra, J., Flipse, C. J. F., de Groot, R. A., and Wold. A., 1987, Phys. Rev. B, 35:6195.

Dahn. D. C., Carolan, J. F., and Haering, R. R., 1986, Phys. Rev. B, 33:5214.

Dahn. J. R., and McKinnon, W. R., 1985, Phys. Rev. B, 32:3003.

Dahn, J. R., McKinnon, W. R., and Coleman, S. T.. 1985a, Phys. Rev. B, 31:484.

Dahn, J. R., McKinnon, W. R., Murray, J. J., Haering, R. R., McMillan, R. S., and Rivers-Bowerman, A. H., 1985b, Phys. Rev. B, 32:3316.

Doni, E., and Girlanda, R., 1986, in: "Electronic Structure and Electronic Transitions in Layered Materials", V. Grasso, ed., Riedel, Dordrecht, pg. 1.

Dubson, M. A., and Holcomb, D. F., 1985, Phys. Rev. B, 32:1955.

Ehrenreich, H., and Schwartz, L. M., 1976, in: "Solid State Physics", H. Ehrenreich, F. Seitz, and D. Turnbull, ed., Academic Press, New York, 31:149.

Friedel, J., 1954, Adv. Phys., 3:446.

Friend, R. H., and Yoffe, A. D., 1987, Adv. Phys., 36:1

Goodenough, J. B., 1971, in: "Progress in Solid State Chemistry", H. Reiss, ed., Pergamon Press, New York, 5:145.

Haas, C., 1975, in: "Crystal Structure and Chemical Bonding in Inorganic Chemistry", C. J. M. Kooymans and A. Rabenan, eds., North-Holland, Amsterdam, pg. 103.

Haas, C., 1985, J. Solid State Chem., 57:82.

Heine, V., 1980, Solid State Phys., 35:1.

Hollinger, G., Pertosa, P., Doumerc, J. P., Himpsel, F. J., and Reihl, B., 1985, Phys. Rev. B, 32:1987.

Huisman, R., De Jonge, R., Haas, C., and Jellinek, F., 1971, J. Solid State Chem., 3:56.

Kertesz. M., and Hoffman, R., 1984, J. Am. Chem. Soc., 106:3453.

Johnston, D. C., 1982, Solid State Commun., 43:533.

Laman, F. C., Matsen, M. W., and Stiles. J. A. R., 1986, J. Electrochem. Soc., 133:2441.

Mattheiss, L. F., 1973. Phys. Rev. B, 8:3719.

McKinnon, W. R., and Haering, R. R., 1983, in: "Modern Aspects of Electrochemistry", R. E. White, J. O'M., Bockris, and B. E. Conway, ed., Plenum, New York, 15:235.

McKinnon, W. R., and Selwyn, L. S., 1987, Phys. Rev. B 35:7275.

Molenda, J., 1986, Solid State Ionics, 21:263.

Murphy, D. W., Di Salvo, F. J., Hull, G. W., and Waszczak, J. V., 1976, Inorg. Chem., 15:17.

Nohl, H., Klose, W., and Andersen, O. K., 1982, in: "Superconductivity in Ternary Compounds, Springer Topics in Current Physics", O Fischer and M.B. Maple, ed., Springer-Verlag, Berlin, 32:165.

Perrin, A., Chevrel, R., Sergent, M., and Fischer, O., 1980, J. Solid State Chem., 33:43.

Py, M. A., and Haering, R. R., 1983, Can. J. Phys., 61:76.

Py, M. A., and Haering, R. R., 1984, Can. J. Phys., 62:10.

Salomon, D., Lerf, A., Biberacher, W., Butz, T., and Saibene, S., 1985, Chem. Phys. Lett., 119:238.

Sellmyer, D. J., 1978, in: "Solid State Physics", H. Ehrenreich, F. Seitz, and D. Turnbull, ed., Academic Press, New York, 33:83.

Silbernagel, B. G., and Whittingham, M. S., 1976. J. Chem. Phys., 64:3670.

Umrigar, C., Ellis, D. E., Wang, D., Krakauer, H., and Posternak, M., 1982, Phys. Rev. B, 26:4935.

Vellinga, M. B., De Jonge, R. and Haas, C., 1970, J. Solid State Chem., 2:299.
Wiegers, G. A., 1980, Physica, 99B:151.
Wilson, J. A., and Yoffe, A. D., 1969, Adv. Phys., 18:193.
Yvon, K., 1979, in: "Current Topics in Materials Science",
E. Kaldis, ed., North-Holland, Amsterdam, 3:53.

EFFECTS OF INTERCALATION ON PROPERTIES OF TRANSITION METAL DICHALCOGENIDES

R.F. Frindt

Department of Physics
Simon Fraser University
Burnaby, B.C., Canada V5A 1S6

ABSTRACT

The effects of intercalation on the host structure and on the optical and transport properties of the transition metal dichalcogenides will be discussed. Modification of properties, for example optical absorption, can be used to monitor intercalation kinetics and staging in some systems. Some recent work on intercalation kinetics will be discussed.

INTRODUCTION

The transition metal dichalcogenides (TMDs) and their intercalates have attracted considerable interest in the past 20 years or so from a number of viewpoints, including optical and transport properties, superconductivity, electronic instabilities (charge density waves), order-disorder transitions and, of course, intercalation physics. The common thread that ties the work together is the highly anisotropic, pseudo two-dimensional aspect of the materials.

A very recent and useful review of the transition metal dichalcogenides and their intercalation complexes has been written by Friend and Yoffe.[1] The discussion here will be primarily concerned with intercalates of the alkali metals and Ag.

The TMD materials are structurally quite similar in that they are all made up of X-M-X layer sandwiches where X refers to a chalcogen (S,Se,Te) and M is one of the following transition metals: Group IV, Ti,Zr,Hf; Group V, V,Nb,Ta; Group VI, Cr,Mo,W. The sandwich layer or molecular layer has a hexagonally packed layer of M's with hexagonally packed planes X's on either side. The atoms in a X-M-X sandwich are covalently bonded into a relatively rigid framework but the sandwiches are in turn very weakly bound to each other, resulting in easy cleavage and pronounced anisotropy. The coordination of the chalcogen atoms occurs in only two ways for all the TMDs: either trigonal anti-prism (distorted octahedron) as is the case, for example, for 1T-TiS_2 and 1T-$TaSe_2$; or trigonal-prism, as is found for 2H-TaS_2, 2H-NbS_2 and 2H-MoS_2. By stacking the two basic sandwich units in various sequences, various polytypes of a given MX_2 can be generated.[1]

The TMDs can be intercalated by many different species, provided they are electron donors, from the alkali metals to large organic molecules. Intercalation can be achieved in a number of ways, for example for organic intercalation by immersion in a liquid or vapour; for alkali metals by electrolytic methods, immersion in metal-ammonia solution, and by use of lithiating and other agents. In addition, intercalated systems can be grown at high temperatures, a technique that has been used to make high quality single crystals. Another method, discussed below, involves the formation of single layers in a liquid, adsorbing species on the single layers and re-stacking the layers with the adsorbed species "included" between the layers.

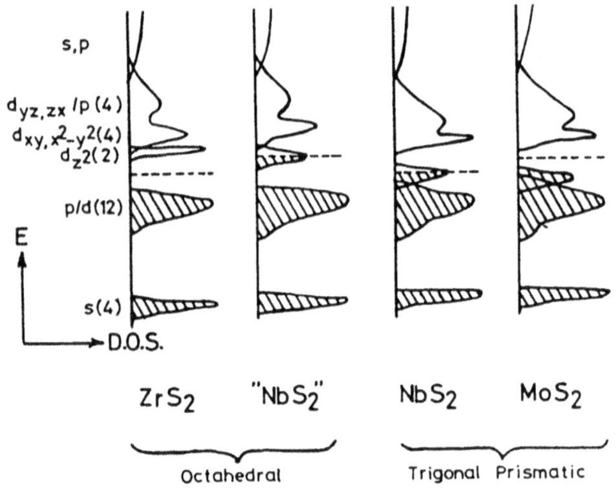

Fig. 1. Band scheme showing the density of states vs. energy for the octahedral and trigonal prism coordination. The shading indicates occupied states and the dashed line is the Fermi energy.[2]

Band structure calculations have been carried out on a number of the TMDs, and these, coupled with empirical evidence, lead to a band scheme proposed by Liang,[2] indicated in Fig. 1. The main feature of the band scheme is that the Group IV octahedrally coordinated materials (e.g. ZrS_2) are semiconductors or semimetals ($TaSe_2$), the Group V materials (either octahedral or trigonal prismatic) with one more electron and a half-filled dz^2 band (e.g. NbS_2) are metals and the Group VI materials (e.g. MoS_2) with a filled dz^2 band are semiconductors. When a TMD is intercalated, there is a great deal of evidence indicating that charge-donation occurs, that is, the intercalate donates electrons to the host layers, and that the main effect of charge transfer is to fill unoccupied states in Fig. 1, leaving the band scheme essentially unchanged. This so-called rigid band model has been used with considerable success in explaining and predicting properties of intercalated systems.

STRUCTURAL EFFECTS OF INTERCALATION

Diffraction studies on intercalated TMDs have revealed a variety of interesting structural changes in the host. The layer separation in general is observed to increase, from typically 5 to 10% for metal intercalates to a factor of ten for large organic molecules. Generally, the a spacing is observed to change by only a percent or so. The host stacking sequence can remain the same (e.g. $LiTiS_2$ and all the Group IV selenides have the 1T structure[1,3]) or a relative translation of host layers may be observed (e.g. $AgTaS_2$ [4]). There is also evidence that the coordination of the host layer can change: MoS_2 on intercalation with Li converts from trigonal prism to octahedral coordination.[5]

Staging, commonly observed in graphite, has also been observed in intercalated TMDs.[1,3] In a stage 1 system, each gallery between the host layers is occupied; in stage 2, every second gallery is occupied, and so on for higher order staging. With TMDs intercalated with alkali metals and Ag, stage 1 and stage 2 are most commonly observed. Over the years there has been much interest in the domain structure of intercalates and the mechanisms for stage transformations[6] as well as the energetics of staging and domain formation.[7,8,9] Recently there has been increased interest in non-equilibrium situations including 3-d modeling of the dynamics of intercalation and deintercalation.[10,11]

Order-disorder transitions of highly mobile intercalates such as Li and Ag are commonly observed. For example, in TiS_2 where both Li and Ag occupy octahedral sites, a $\sqrt{3}a \times \sqrt{3}a$ superlattice is observed for (stage 1) $Li_{1/3}TiS_2$ [12] and $Ag_{1/3}TiS_2$.[13-16] The X-ray work[15,16] shows 3-dimensional ordering for $Ag_{1/3}TiS_2$ with abab stacking for the Ag. The temperature for onset of ordering in $Ag_{1/3}TiS_2$ is, depending on the sample, in the range 250 to 300K. The Ag ordering is readily observed in TiS_2 by electron diffraction on cooling, as shown in Fig. 2. $AgTiS_2$ has provided a prototype system for lattice gas modeling of order-disorder transitions and Brillouin zone[14] and Fermi surface reconstruction.[17]

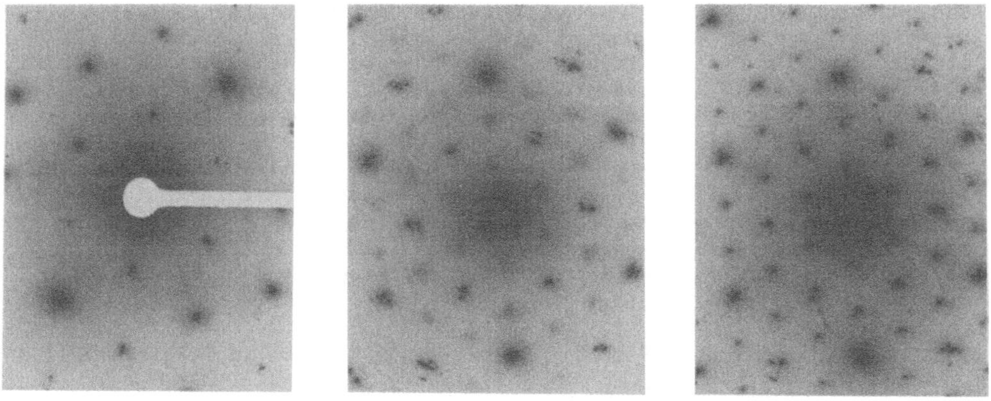

a b c

Fig. 2. Electron diffraction patterns showing Ag ordering in TiS_2 on cooling: (a) room temperature, (b) about 250K, (c) about 100K. (Curzon and Frindt, unpublished.)

The Ag_xTaS_2 (Scholz et al.,[18] Tatlock et al.,[1] Boebinger et al.[19]) and Ag_xNbSe_2 systems[20] are considerably more complex. Here the Ag occupies tetrahedral sites and on cooling a sequence of Ag ordering superlattice structures are observed. In $LiTaS_2$ it is found that the octahedral sites are occupied and the ordering is observed slightly above room temperature for x = 1/3 and 2/3.[21]

EFFECTS OF INTERCALATION ON TRANSPORT PROPERTIES

From the point of view of rigid band model, the main effect of intercalation on the transport properties of the TMDs occurs via a change in the number of carriers, the most dramatic effects of course occurring in the intercalated semiconductors. For example, the trigonal prism coordination semiconductor MoS_2 is converted to a metal (often superconducting) on intercalation with alkali metals. This is readily understood in terms of the rigid band model, where electrons donated to the host MoS_2 fill bands above the fully occupied dz^2 band (Fig. 1). On the other hand, for the metallic group V layered compounds, such as $2H-TaS_2$, intercalation can be expected to reduce the conductivity by electron donation to the half-filled dz^2 band, thus reducing the number of conducting carriers. In principle, with sufficient intercalation (x = 1) and electron donation, a conversion to a semiconducting state might be expected. The system Ag_xTaS_2 exhibits a nominal decrease in conductivity on intercalation to x = 2/3.[19,22]

The pure Group IV compound TiS_2 is a very small band gap semiconductor. In practice the compound is generally off stoichiometry and exhibits metallic behaviour.[1] Transport,[23] susceptibility,[24,25] and nmr[26] studies have been carried out on Li_xTiS_2. The results show that on intercalation there is an increase in carrier concentration which is consistent with increased d-band filling and a high degree of charge transfer. Transport and magnetic studies on $Ag_{1/3}TiS_2$ also demonstrate an increased carrier concentration and indicate transfer close to one electron per Ag.[28]

One of the most interesting effects observed in the Group V TMD's is the occurrence of charge density waves (CDW). These electronically driven transformations are accompanied by a periodic lattice distortion. New crystal periodicities break up the Fermi surface and have large effects on the conductivity of $1T-TaS_2$ and $1T-TaSe_2$, with more subtle effects seen in $2H-TaS_2$ and $2H-TaSe_2$.[1] Because the atomic displacements in CDW phases are relatively small, typically around 0.1Å, the effects on the intercalation process seem to be minimal. On the other hand, because of electron donation with resulting modification of the Fermi surface, possible host layer restacking and relatively high defect densities, intercalation can be expected to suppress CDW formation. In addition, a stiffening of the structure due to the intercalate will tend to oppose an electronically driven structural distortion.

An example of the effects of metal intercalation on CDW structure is given by the Ag_xTaS_2 system. Here single crystal resistivity measurements show that the CDW transformation is suppressed for a relatively low x value ($x \lesssim 0.13$).[22] In addition, first-order changes in the Ag superlattice for Ag_xTaS_2[19,22] and $Ag_{1/3}NbSe_2$[27] on cooling are detected in the resistivity. Conductivity effects due to the intercalant order-disorder transition in $Ag_{1/3}TiS_2$ are not observed because of the gradual (2nd order) nature of the transition.[28] However, the Hall coefficient displays a gradual change with temperature, reflecting the changes in the Fermi surface that occur as the new $\sqrt{3}a$ period develops on cooling.[28]

OPTICAL EFFECTS OF INTERCALATION

Although possibly the most convincing evidence for charge transfer on intercalation comes from nmr Knightshift measurements (for example, on Li_xTiS_2[26]), optical absorption studies have played a significant role in establishing the charge transfer concept. One of the earliest experiments in this regard is the work of Acrivos et al.[29] on Na_xMoS_2 (Fig. 3) which shows the disappearance of the A and B exciton peaks due to free carrier screening where the free carriers arise from band filling above the full dz^2 band. In addition, the onset of free carrier reflectivity is seen in region C of Fig. 3.

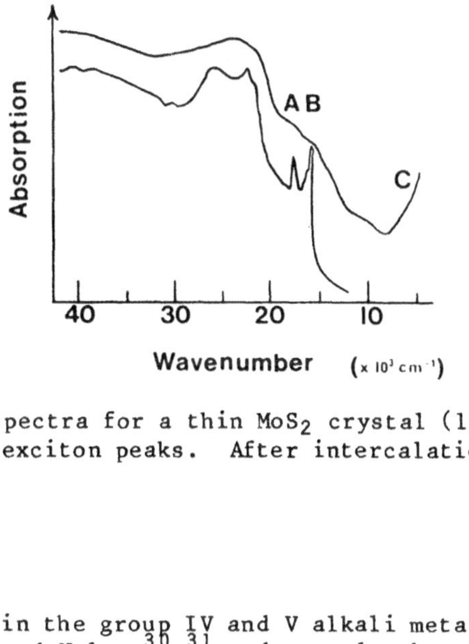

Fig. 3. Absorption spectra for a thin MoS_2 crystal (lower curve) showing the A and B exciton peaks. After intercalation with Na, upper curve.[29]

Optical effects in the group IV and V alkali metal intercalates have been studied by Beal and Nulsen[30,31] and can also be satisfactorily interpreted in terms of electron donation and a rigid band model. The optical absorption of Ag_xTaS_2 on intercalation is shown in Fig. 4. The effects are very similar to those for Li_xTaS_2.[30] The interband absorption edge around 2 eV shifts to lower energies with increasing x. This can be understood in terms of an increasing Fermi level relative to higher empty bands. The decrease in absorption at lower energies with increasing x can be attributed to a decrease in free carrier absorption as the half-filled dz^2 band fills via charge transfer. The decrease in transmission with increasing x in the 5500 Å region is more or less linear and to the eye, a darkening of thin crystals on intercalation is observed in transmitted light.

Very similar effects on the absorption spectrum of TiS_2 are observed on intercalation of Li[31] or Ag.[32] With Ag_xTiS_2, it is found that with increasing x up to stage 1 (x = 0.4), there is a steady decrease in absorption in the 1 to 2.5 eV range (independent of staging). When viewed by eye, intercalated thin TiS_2 crystals are thus more transparent than pure TiS_2.

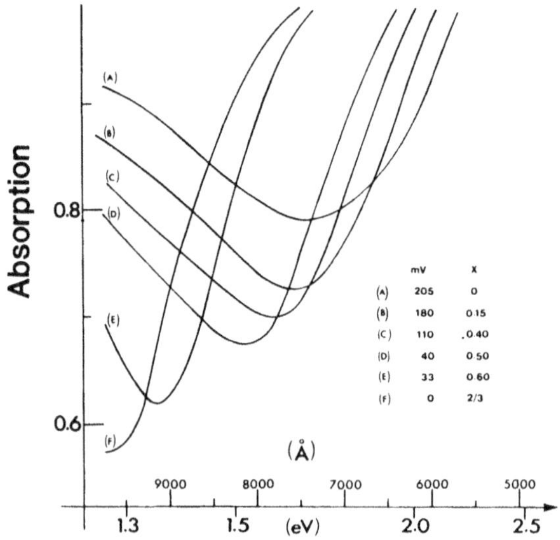

Fig. 4. Absorption spectra (A = 1-T) for a Ag_xTaS_2 crystal about 1500Å thick taken in situ. Open circuit potentials are also shown.[32]

It is of interest that we first saw the indications of staging in Ag_xTiS_2 and Ag_xTaS_2 by simply observing the electro-intercalation of thin crystals in an optical microscope. This led to structural studies of these systems, to studies on the distribution of intercalants within a crystal, and to studies of stage transformations on intercalation and deintercalation. Some of our visual observations on partially intercalated thin crystals were puzzling but subsequent work, both experimental and theoretical, as discussed below, have clarified many of the visual observations.

Photographs of a thin TiS_2 crystal partially intercalated with Ag and partially deintercalated are shown in Figs. 5a and 5b. In the intercalating crystal, two distinct regions are observed. As indicated by the effects of Ag on the optical absorption discussed above, the region closest to the crystal edge corresponds to the highest Ag concentration. In from the edge a lower Ag concentration region is seen, then empty crystal. Electron microprobe X-ray fluorescence work, weighing experiments and X-ray structural work, has shown that the region near the edge is stage 1 with a uniform Ag concentration of 0.4 and the inner region is stage 2 with a uniform Ag concentration of 0.2. With time it is observed that the two regions in Fig. 5a (for very thin crystals) disappear by diffusion into the unintercalated crystal. For thicker crystals, X-ray fluorescence and ^{110}Ag tracer measurements have shown that the stage 1 at the crystal edge converts to stage 2, with the stage 2 penetrating deeper into the crystal.[33,34]

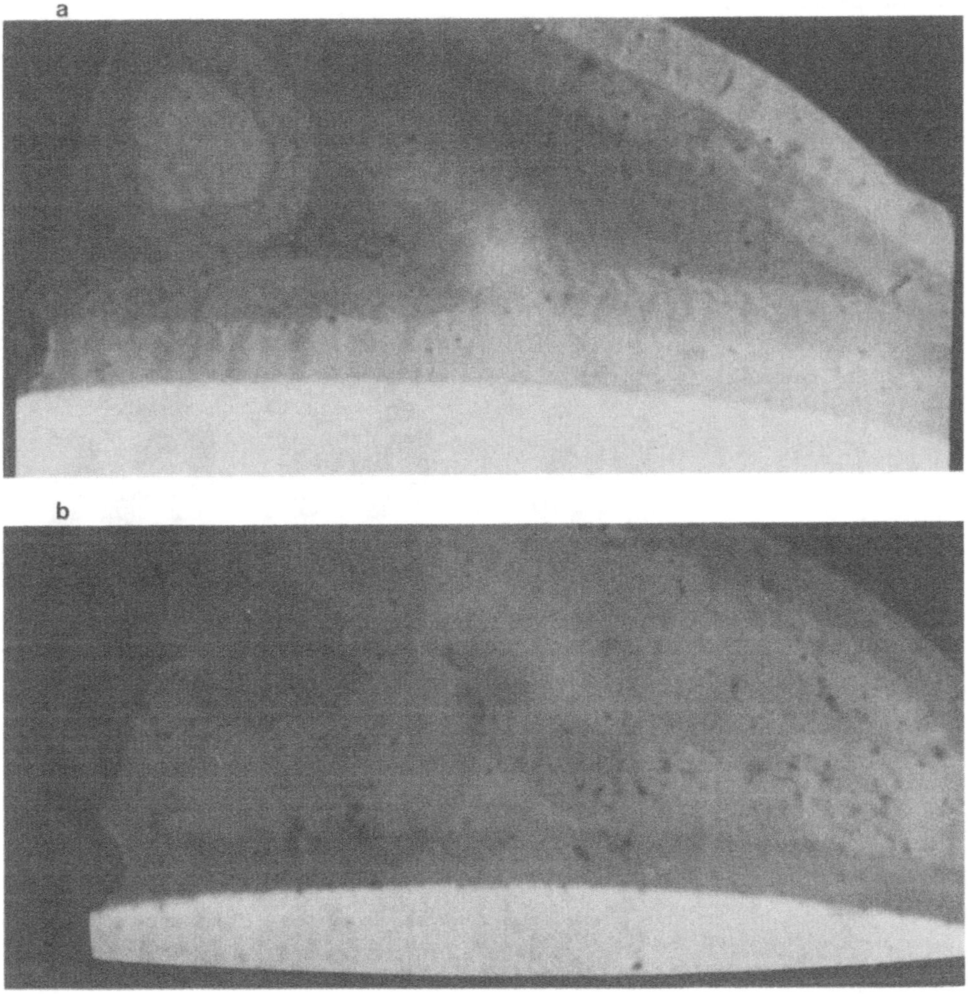

Fig. 5. (a) Optical transmission photograph of a TiS_2 crystal about 300Å thick partially intercalated with Ag. A light stage 1 region and a darker stage 2 region are observed propagating in from the crystal edges. The circular regions are due to pinholes in the sample. (b) Partial deintercalation of the stage 1 crystal showing a darker stage 2 region propagating in from the crystal edges. The crystal is in electrolyte. The long crystal edge is about 1 mm in length.

On deintercalation, a transparent Ag-depleted region is seen to propagate in from the crystal edge (Fig. 5b). XRF studies have shown that this region is stage 2 with x ≈ 0.2. We have not been able to deintercalate Ag from TiS_2 below stage 2 at room temperature with a liquid electrolyte, although we found that further deintercalation is possible at elevated temperatures (150C).[32] It appears that a lower x value can be attained on deintercalation using a solid electrolyte.[16,35]

The optical behaviour of thin TaS_2 crystals intercalated with Ag is very similar to the $AgTiS_2$ system (including observations of staging), except that, as indicated in Fig. 4, the crystal becomes less transmitting on intercalation, and deintercalation (more or less complete) proceeds much faster.

INTERCALATION KINETICS

Some recent interesting three-dimensional Monte Carlo computer simulations of the dynamics of intercalation and deintercalation have been carried out by G. Kirczenow.[10,11] In the model, intercalation proceeds via the motion of small "elementary islands" (EI's) of intercalate which interact via electrostatic and elastic forces. The EI size is of the order of the "healing length" of deformation introduced by an isolated intercalate atom (possibly a few Å to tens of Å in size). An abridged form of some of

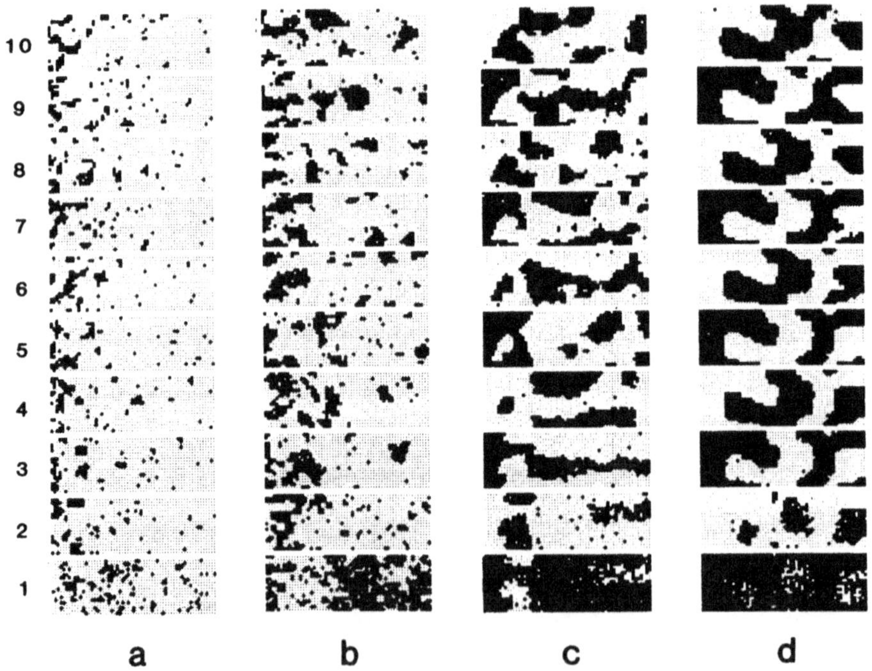

Fig. 6. Computer simulation of the growth of a stage 2 structure. Each column is a stack of galleries at a given time and time increases to the right. The dark regions are intercalate. Gallery 1 is just below the crystal surface. The reservoir is at the left in each frame. In (c), stage 3 is seen at the front of the stage 2 at the edge. From Kirczenow.[11]

Kirczenow's results are shown in Fig. 6, which displays in time the formation of a stage 2 structure on intercalation. Initially nucleation and growth of aggregates of EI's are seen; then stage 2 forms near the crystal edge with higher order stages (particularly stage 3) forming further in from the edge. Finally a complete stage 2 with well-developed Daumas-Hérold (DH) islands is found. In the model the size of the DH islands is determined to a large extent by the kinetics at the edge of the crystal in contact with the intercalate reservoir.

The above model with the well-formed lowest order stage at the crystal edge and higher-order regions penetrating into the unintercalated crystal appears to provide a reasonable general understanding of the initial phases of Ag intercalation in the TiS_2 and TaS_2 systems, as observed for example in Fig. 5a.

When TiS_2 crystals greater than a few thousand Angstroms thick are intercalated it is found (as for very thin crystals) that intercalation proceeds via propagation of a stage 2 region followed by a stage 1 region. If the intercalation is stopped, the stage 1 region converts to stage 2 as indicated in Fig. 7. This stage conversion is likely driven by Coulomb repulsive forces. The final stage 2 front is stationary at room temperature. Propagating fronts on intercalation have also been observed in $AgNbSe_2$ [36] and $LiTaS_2$.[37]

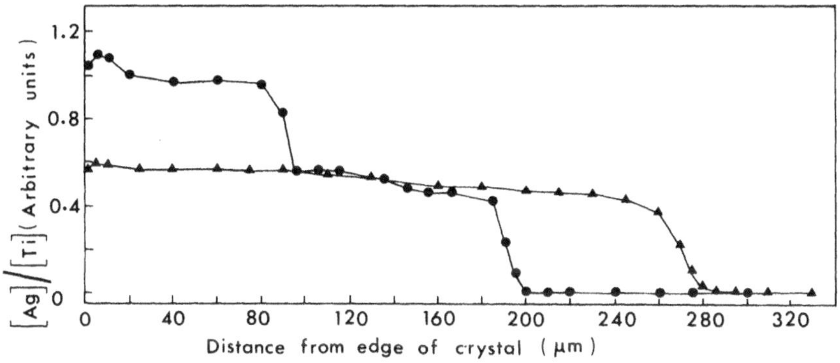

Fig. 7. Distribution of Ag in a partially intercalated TiS_2 crystal. Circles show initial stage 1 and stage 2 regions. Triangles: after conversion of stage 1 to stage 2.[33]

Results of calculations on the deintercalation of stage 1 are shown in Fig. 8. Deintercalation proceeds by the propagation of a stage 2/stage 1 "front" in from the crystal edge. In the stage 2 region, Daumas-Hérold islands similar to those for intercalation in Fig. 6 are seen. As the stage 1 retreats, stage 2 islands extend further into the crystal. The islands remain more or less stationary. Thus the intercalant EI's appear to flow through the island pattern as the crystal deintercalates. As deintercalation proceeds a residue type of situation is finally reached.

XRF data for the deintercalation of Ag from TiS_2 is shown in Fig. 9. The simulations in Fig. 8 appear to provide a good model for observations on the deintercalation of stage 1 $AgTiS_2$ and $AgTaS_2$, namely that the crystals deintercalate by the propagation of a stage 2 region and then a low x value region (for TaS_2) in from the edge of the crystal.

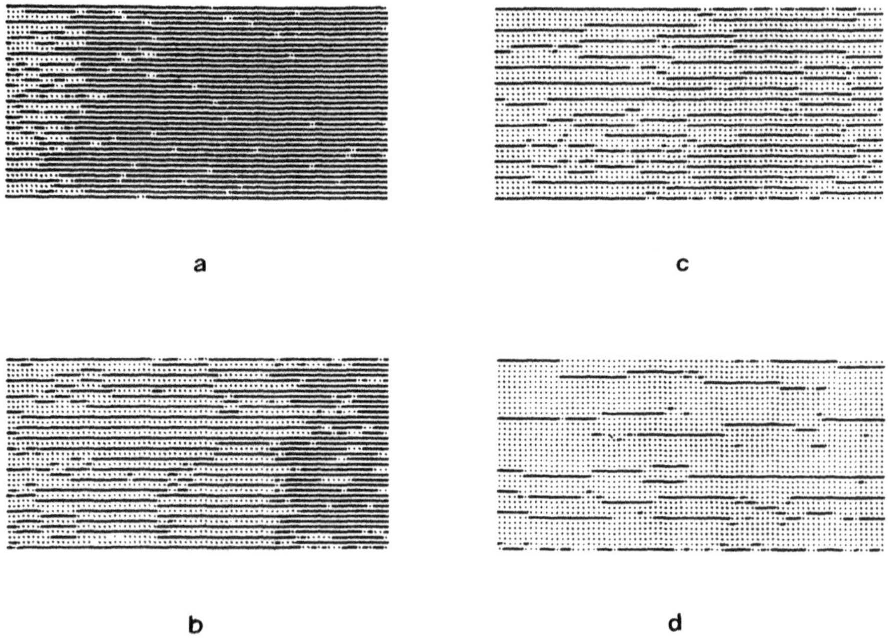

Fig. 8. Computer simulation of a stage 1 crystal, showing a slice through the crystal at right angles to the host. From Kirczenow.[11]

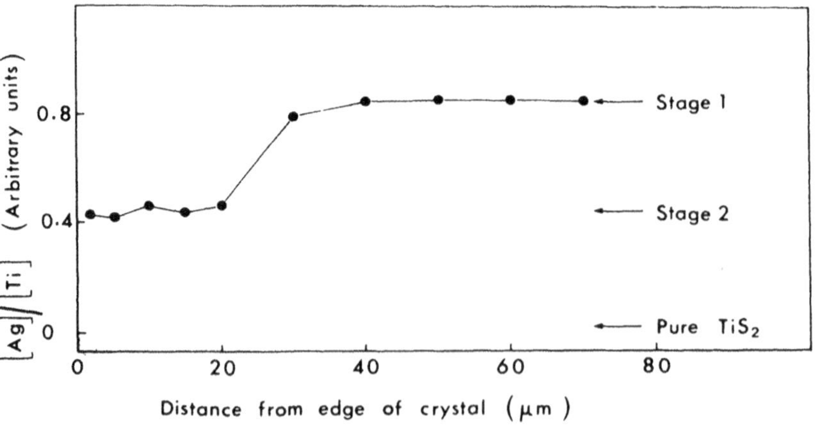

Fig. 9. Ag distribution showing partial deintercalation of stage 1 TiS_2. Crystal thickness between 0.5 and 1.0μm.[34]

For $AgTiS_2$ there is an increase in the stage 2 c-spacing of 6.3%, with an increase of about 1% in the a-spacing, so that macroscopic stress is built up as intercalation proceeds. Fig. 10 provides a model for how this stress might be relieved for a crystal in the micron thickness range. After the stage 1 to 2 conversion, the stage 2 front in Fig. 7 at room temperature is stationary with no significant diffusion from the front observed. On heating, diffusion is observed. A stationary front indicates that there is a point where the energy required to bend and possibly shear overlying layers is such that it is no longer possible for intercalant in a given layer to progress by thermal activation and diffusion essentially stops. Intercalant near the crystal basal plane is expected to be the least affected by such macroscopic strain effects and recent Auger work on $AgTiS_2$ shows that this is indeed the case. It is found that stationary fronts are not observed or are unstable near the crystal basal planes, down to a depth of around 100 to 150 crystal layers, and that there is an enhanced intercalation rate in this region. This is in agreement with optical observations of enhanced intercalation rates in crystals less than about 1000Å thick.[38] Such effects may have useful application. For example, it is often noted that intercalation is "slow" in certain materials (e.g. for alkali metals in the group V layered compounds). It is possible that in some cases intercalation may be hindered by macroscopic strain and that intercalation rates could be enhanced by the use of very small particles - that can be obtained, for example, by ultrasonication or vigorous blending.

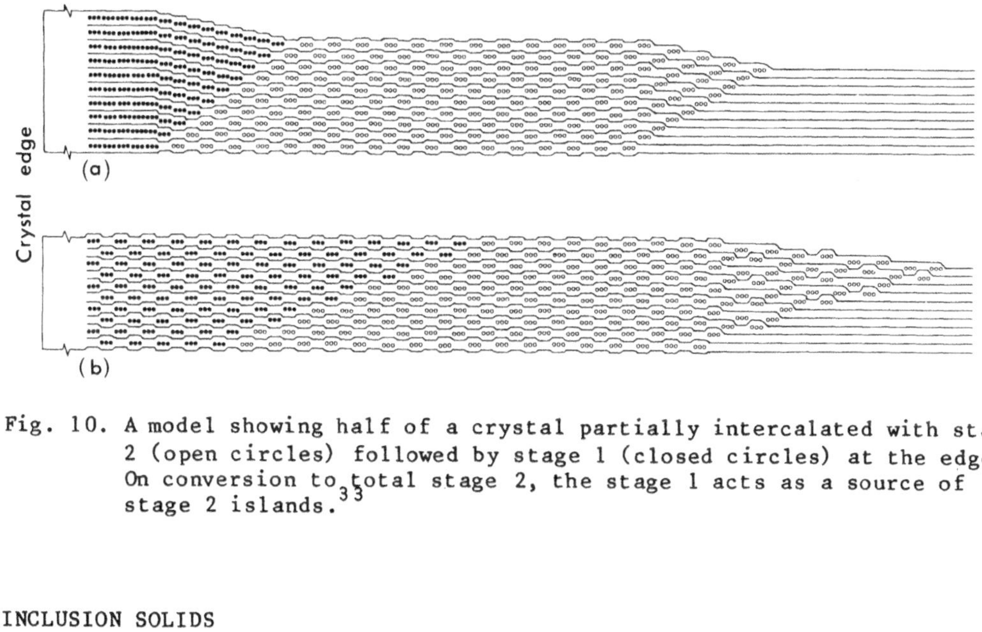

Fig. 10. A model showing half of a crystal partially intercalated with stage 2 (open circles) followed by stage 1 (closed circles) at the edge. On conversion to total stage 2, the stage 1 acts as a source of stage 2 islands.[33]

INCLUSION SOLIDS

It has been shown that an alternate method for producing intercalation-like systems is possible by exfoliating layered compounds, depositing metal atoms on both sides of single layers in suspension (usually as a hydroxide) and then restacking the layers by drying so that the deposited atoms are included between the host layers. If desired, the hydroxide can be reduced by heating or other treatment. Such "inclusion solids" are very similar in final form to intercalated systems. The included species can be in a form from sub-monolayer to monolayer or clusters of varying sizes.

Single layer suspensions of TaS_2 and NbS_2 have been produced by intercalation of hydrogen and water.[39,40] Recently, exfoliated MoS_2 has been produced intercalating Li and then exposing to water with suitable agitation.[41] It appears that the water penetrates the intercalated layers and reacts with the Li to produce hydrogen gas and LiOH. The hydrogen gas "blows" the crystal apart into single layers. Probably the best direct evidence for single layer production is X-ray diffraction. With no periodicity in the c-direction, all of the (h,k,ℓ) lines with $\ell \neq 0$ should be missing, in particular the very strong 002 line for MoS_2 should be missing.

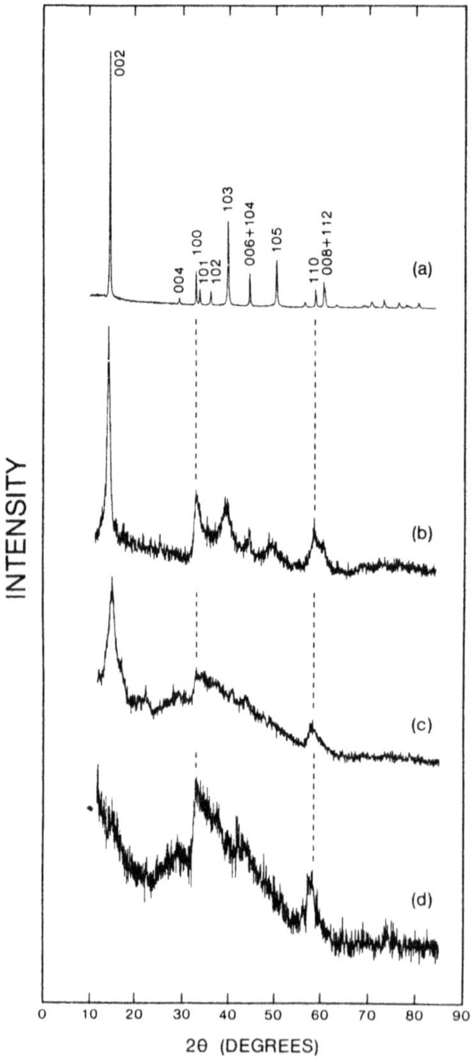

Fig. 11. X-ray diffraction patterns for (a) untreated MoS_2, (b) dried film of partially exfoliated MoS_2, (c) dried film of exfoliated MoS_2, (d) exfoliated MoS_2 in water.[41]

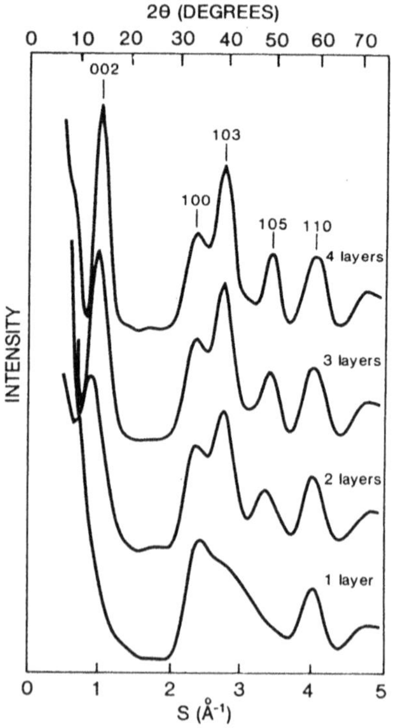

Fig. 12. Calculated X-ray diffraction patterns for MoS_2 microcrystallites one to four layers thick, with each layer containing 6×6 Mo atoms. After Chien et al.[42]

Fig. 11 shows the diffraction pattern obtained for exfoliated MoS_2. Basically only the asymmetric 100 line and the 110 line are observed. This pattern is to be compared with a calculation of diffraction intensity for single layer MoS_2 shown in Fig. 12. It is clear that the MoS_2 has been exfoliated.

To make an MoS_2 inclusion compound, a soluble metal salt can be added to a single layer suspension and the pH adjusted to give monolayer coverage of metal hydroxide on both sides of the MoS_2. After washing and drying, a powder inclusion solid is obtained. We have made inclusion solids using more than a dozen different metals, including Cu, Al, Cd, Ni, Co and Pb.[43] X-ray diffraction patterns show that the layer separation can be increased by the inclusion (up to 17% for Pb) or can decrease (7% for Ni, Co). The fundamental properties of such inclusion solids will be of considerable interest in both their monolayer and cluster forms.

ACKNOWLEDGEMENTS

I would like to thank P. Joensen, D. Kaluarachchi and G. Kirczenow for helpful discussions, and G. Kirczenow for some of his computer simulation diagrams. This work was supported in part by the Natural Sciences and Engineering Research Council of Canada.

REFERENCES

1. R.H. Friend and A.D. Yoffe, Adv. Phys. 36:1 (1987).
2. W.Y. Liang, Physics and Chemistry of Electrons and Ions in Condensed Matter, NATO ASI 130, J.V. Acrivos, N.F. Mott and A.D. Yoffe eds., Dordrecht: D. Reidel p. 459 (1984).
3. J. Rouxel, Intercalated Layered Materials, F. Lévy, ed., Dordrecht: D. Reidel, Vol. 6, p. 201 (1979).
4. G.A. Scholz and R.F. Frindt, Mater. Res. Bull. 15:1703 (1980).
5. M.A. Py and R.R. Haering, Can. J. Phys. 61:76 (1983).
6. N. Daumas and A. Hérold, C.R. Acad. Sci. Paris C268:373 (1969).
7. S.A. Safran and D.R. Hamann, Phys. Rev. B 22:606 (1980).
8. D.C. Dahn and R.R. Haering, Solid State Commun. 44:29 (1982).
9. S.E. Ulloa and G. Kirczenow, Comments Cond. Mat. Phys. 12:181 (1986).
10. G. Kirczenow, Phys. Rev. Lett. 55:2810 (1985).
11. G. Kirczenow, Can. J. Phys. (in press).
12. H.A. Hallak and P.A. Lee, Solid State Commun. 47:503 (1983).
13. W.K. Unger, J.M. Reyes, O. Singh, A.E. Curzon, J.C. Irwin and R.F. Frindt, Solid State Commun. 28:109 (1978).
14. R. Leonelli, M. Plischke and J.C. Irwin, Phys. Rev. Lett. 45:1291 (1980).
15. R.M. Suter, M.W. Shafer, P.M. Horn and P. Dimon, Phys. Rev. B 26:1495 (1982).
16. A.G. Gerhards, H. Roede, R.J. Haange, B.A. Boukamp and G.A. Wiegers, Synthetic Metals 10:51 (1984).
17. J.I. Meakin, P.C. Klipstein and R.H. Friend, J. Phys. C 20:271 (1987).
18. G.A. Scholz, R.F. Frindt and A.E. Curzon, Phys. Stat. Sol. 71:531 (1982).
19. G.S. Boebinger, N.I.F. Wakefield, E.A. Morseglia, R.H. Friend and G.J. Tatlock, Physica 117B:608 (1983).
20. A.E. Curzon and O.S. Rajora, Phys. Stat. Sol.(a) 87:157 (1985).
21. J.R. Dahn and W.R. McKinnon, J. Phys. C 17:4231 (1984).
22. G.A. Scholz and R.F. Frindt, Phys. Stat. Sol. 79:483 (1983).
23. P.C. Klipstein and R.H. Friend, J. Phys. C 17:2713 (1984).

24. D.W. Murphy, F.J. DiSalvo, G.W. Hull and J.V. Waszczak, Inorg. Chem. 15:17 (1976).
25. L. Bernard, W. Glausinger, and P. Colombet, Solid State Ionics 17:81 (1985).
26. B.G. Silbernagel, Solid State Commun. 17:361 (1975).
27. O.S. Rajora and A.E. Curzon, J. Less Common Metals 118:117 (1986).
28. J.I. Meakin, P.C. Klipstein and R.H. Friend, J. Phys. C 20:271 (1987).
29. J.V. Acrivos, W.Y. Liang, J.A. Wilson and A.D. Yoffe, J. Phys. C. 4:418 (1971).
30. A.R. Beal and S. Nulsen, Phil. Mag. B43:985 (1981).
31. A.R. Beal and S. Nulsen, Phil. Mag B 43:965 (1981).
32. G.A. Scholz and R.F. Frindt, Can. J. Phys. 61:965 (1983).
33. D. Kaluarachchi and R.F. Frindt, Phys. Rev. B 28:3663 (1983).
34. D. Kaluarachchi and R.F. Frindt, Phys. Rev. B 31:3648 (1985); also work to be published.
35. A. Bottini, M. Lazzari, G. Razzini, B. Rivolta, G. de Felici, M.A. Voso and B. Scrosatti, J. Electroanal. Chem. 96:165 (1979).
36. J.T. Folinsbee, M.H. Jericho, R.H. March and D.A. Tindall, Can. J. Phys. 59:1267 (1981).
37. T. Butz and A. Lerf, Ber. Bunsenges Phys. Chem. 90:638 (1986).
38. G.A. Scholz, P. Joensen, J.M. Reyes and R.F. Frindt, Physica 105B:214 (1981).
39. D.W. Murphy and G.W. Hull, Jr., J. Chem. Phys. 62:973 (1975).
40. C. Liu, O. Singh, P. Joensen, A.E. Curzon and R.F. Frindt, Thin Solid Films 113:165 (1984).
41. P. Joensen, R.F. Frindt and S.R. Morrison, Mater. Res. Bull. 21 457 (1986).
42. F.Z. Chien, S.C. Moss, K.S. Liang, and R.R. Chianelli, J. Phys., Colloq. (Orsay, Fr.) C4, Suppl. 10, Tome 42:273 (1981); J. Non-Crystalline Solids 79:251 (1986).
43. M.A. Gee, R.F. Frindt, P. Joensen and S.R. Morrison, Mater. Res. Bull. 21:543 (1986).

INTERCALATION IN OXIDES

FROM 2D TO 3D INTERCALATION

Claude Delmas

Laboratoire de Chimie du Solide du CNRS
Université de Bordeaux I
351, cours de la Libération - 33405 Talence CEDEX (France)

The intercalation properties of graphite, transition metal dichalcogenides (TMDC) and clays have been intensively studied for at least twenty years. Till the eighties, stoichiometric AMO_2 (A : alkali ion, M : transition element) sheet oxides have been mainly studied from a crystal chemistry point of view while only few works were devoted to A_xMO_2 layer oxides [1-6]. During the last few years, numerous new results have been reported in this chemical field, stimulated by the potential interest of some of these compounds as positive electrode for high energy density batteries. This lack of interest in oxides as intercalation compounds arises certainly from the non existence of the fully deintercalated form MO_2 with sheet structure. As it will be discussed in the following this property comes from the large electronegativity difference between the oxygen and the M element. Nevertheless, it must be pointed out that the positive plate of the Ni-Cd secondary battery, used for a century, runs from the topotactic hydrogen deintercalation reaction :

$$Ni(OH)_2 \longrightarrow NiOOH + H^+ + e^-$$

or in another way :

$$H_2NiO_2 \longrightarrow HNiO_2 + H^+ + e^-$$

GENERAL SURVEY OF THE CRYSTAL CHEMISTRY OF A_xMO_2 SHEET OXIDES

Their structure can be considered as the stacking of two dimensional slabs (O-M-O), containing a layer of metal atoms M sandwiched between two oxygen layers, these slabs being linked together by alkali ions.

In TMDC the lower electronegativity of the chalcogens ($\chi_S = 2.5$, $\chi_{Se} = 2.4$) in comparison with the oxygen one ($\chi_O = 3.5$) leads to high covalent character of the M-S bond. This gives rise to a small negative charge on the adjacent chalcogen layers so that the Van der Waals bonding can stabilize the 2D structure.

In oxides the M-O bonds are more ionic, then the electrostatic repulsions between highly negatively charged oxygen layers are very strong and prevent the existence of unintercalated layer oxides. Nevertheless if alkali ions are intercalated between the adjacent oxygen layers, strong ionic bonds appear and stabilize the A_xMO_2 layer oxides. In these materials x varies in the $x_\ell \leqslant x \leqslant 1$ range. The minimum intercalation rate x_ℓ is related to the nature of A and M elements but is always over 0.5^3.

This remark implies that only stage I compounds may be obtained in oxides. Nevertheless, the difference between the ionocovalent character of the M-O and A-O bonds is high enough to provide an important bonding anisotropy. It results that in the vicinity of RT a slab can be considered as an independent unit able to move with respect to other slabs. This behavior is well illustrated by the intersheet distance modifications and sheet glidings observed during the intercalation-deintercalation or exchange reactions.

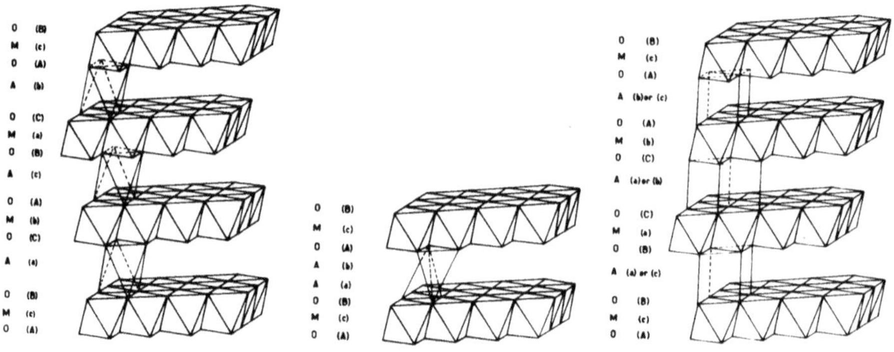

O3 T1 P3

Fig. 1 - Structures of sheet oxides.

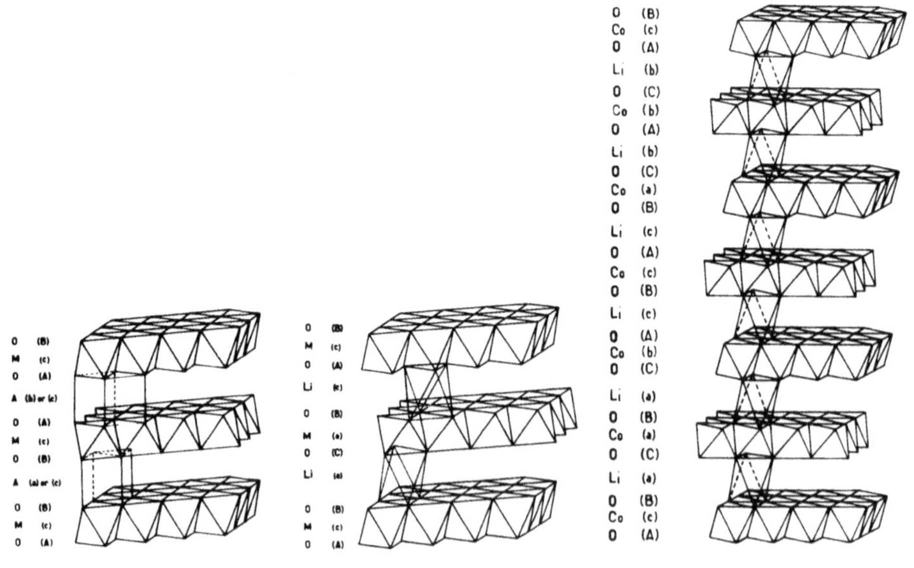

P2 O2 O6

Fig. 2 - Structures of sheet oxides.

Structure description

The M element stays generally in an octahedral surrounding. The various structural types differ by the oxygen packing and by the alkali ion surrounding : trigonal prismatic (P), tetrahedral (T) or octahedral (O). They are drawn on Figures 1 and 2. In the last case the structure anisotropy leads actually to a trigonal antiprismatic environment. To describe the various structural types the following notation is used. The letter P, T or O indicating the alkali ion surrounding is followed by a number indicating how many $(MO_2)_n$ sheets are in the unit cell. The ideal symmetry of these structures is always hexagonal or trigonal. When a crystallographic distortion occurs the symbol prime is added.

The (O3) oxygen packing. The (O3) type structure, generally related to the α-$NaFeO_2$ type, is the most frequently observed especially for stoichiometric compounds. As in this structure AO_6 and MO_6 octahedra share only edges Pauling's third rule is strictly obeyed. The intercationic electrostatic repulsions are then minimized and highly stable materials are obtained [7,8].

The (T1) oxygen packing. In the (T1) structural type only edges are shared but the alkali ion lies in a tetrahedron. As in an anionic close packing there are two tetrahedra for one octahedron, the formulation is A_2MO_2 (vacancies in the tetrahedral sites have never been found). In that case, the MO_2 sheets exhibit a CdI_2 type lattice. Three types of lithium compounds (summarized in Table 1) have been reported with this oxygen packing.

Table 1. Various types of A_2MO_2 phases

Formula	Crystallographic formula	Ref.
Li_8ZrO_6	$Li_2(Li_{2/3}Zr_{1/3})O_2$	9
Li_7NbO_6	$Li_2(Li_{1/3}\square_{1/3}Nb_{1/3})O_2$	10
Li_3InO_3	$Li_2(In_{2/3}\square_{1/3})O_2$	11

This MO_2 packing is also found in $M(OH)_2$ hydroxides derived from the brucite $Mg(OH)_2$.

It should be noticed that the (O1) structural type (same MO_2 packing but alkali ion in octaheral environment) which is very common for the lithium chalcogenides Li_xMX_2 does not appear in homologous lithium oxides. In such lattices related to NiAs-type, LiX_6 and MX_6 octahedra share faces along the c axis. Consequently it is observed only when the host lattice is predominantly covalent, which reduces the interface cation-cation repulsion. Furthermore ternary lithium oxides are generally observed for x = 1. Would the (O1) type exist for $LiMO_2$ oxides, the number of Li^+-M^{3+} interactions through common faces would be very high.

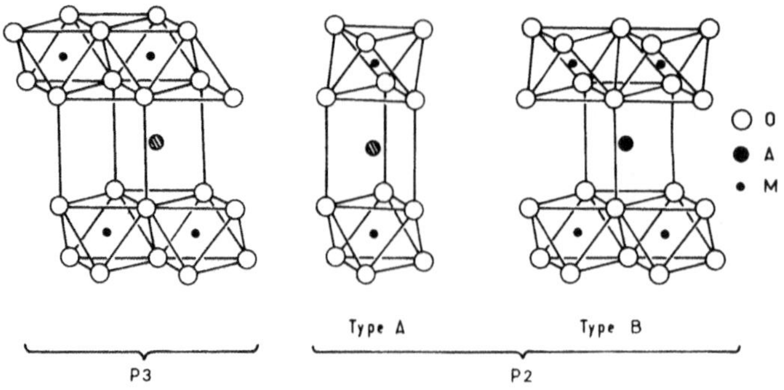

Fig. 3 - Alkali ion environment in P2 and P3 type phases.

<u>The (P3) oxygen packing</u>. In this structure alkali ions are inserted in trigonal prisms, which share one face with one MO_6 octahedron and three edges with three other MO_6 octahedra belonging to the other sheet (Fig. 1). The number of prisms is twice that of the formula units. All prisms are energetically equivalent, as a result the alkali ions are statistically distributed in them (Fig. 3). As in chalcogenides this structural type is always obtained for phases with high vacancy content, so that the repulsion between opposite anions of the prisms is limited by larger distance of the adjacent sheets [12,13].

<u>The (P2) oxygen packing</u>. As in the previous layer structure type, alkali ions occupy trigonal prismatic sites. There are two different types of prisms available which are schematically illustrated in Fig. 3. The a-type prisms share two faces with the MO_6 octahedra, whereas the b-type prisms share only edges with neighbouring MO_6 octahedra. This implies that the former type is less electrostatically favourable for the alkali ion than the second. Due to their proximity two neighbouring a- and b-sites cannot be occupied simultaneously.

These remarks explain why in the stoichiometric layer phases β-$RbScO_2$ [14], $CsLnO_2$ [15], the monovalent ions lie only in the b-sites. In non-stoichiometric phases (oxides or chalcogenides) both sites can be occupied [16,17].

<u>The (O2) and (O6) oxygen packings</u>. These two structural types have been respectively observed for the metastable variety of $LiCoO_2$ and $Li_{0.62}CoO_2$ obtained by soft chemistry technique [18,19]. The simultaneous presence of shared faces and small alkali ions destabilize these types of packing.

Classification

Because of the electrostatic repulsion between adjacent oxygen layers trigonal prismatic surrounding is generally observed for large size alkali ions, small fractional occupancy and prevailing covalent host lattices.

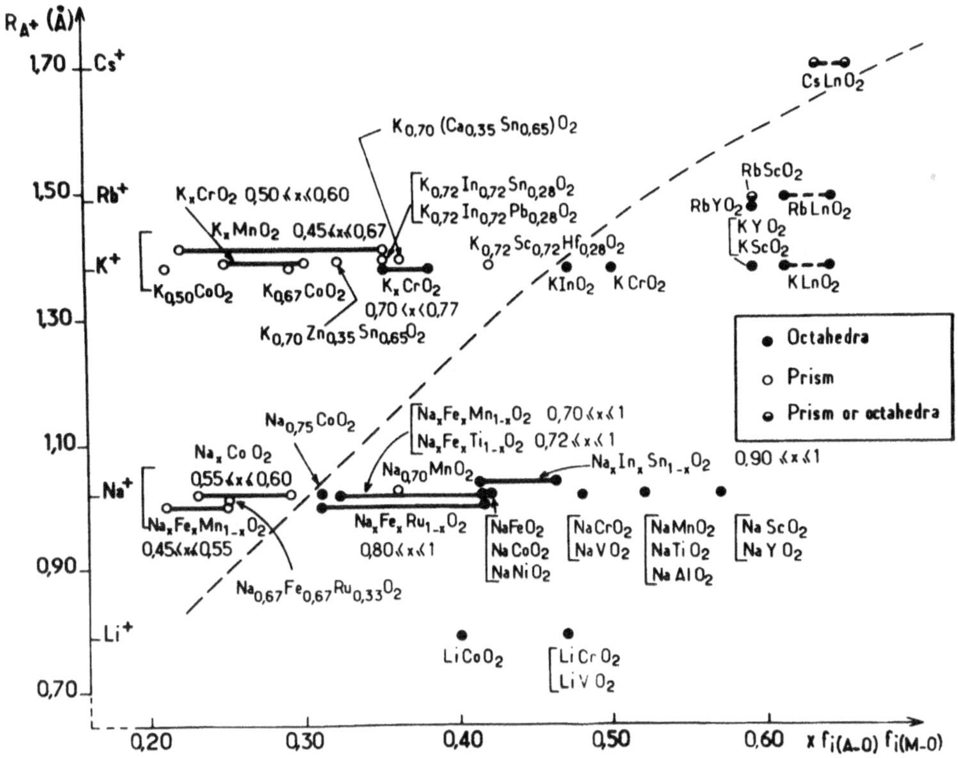

Fig. 4 - Classification of the A_xMO_2 phases in a R_{A^+} vs $xfi_{(A-O)}fi_{(M-O)}$ diagram.

From these criteria it is possible to classify the layer oxides in an ionicity-structure diagram analogous to ROUXEL's diagram[20]. The radius of the alkali ion (R_{A^+}) is plotted vs the $fi_{A-O}.fi_{M-O}$ product, in which x is the alkali content, fi_{A-O} the ionicity of the A-O bond and fi_{M-O} that of the M-O bond. Such a product characterizes the charge carried by the anionic layers. As shown in Fig. 4, both types of surroundings are well separated[12]. Every phase represented in this diagram is a stable variety obtained by high temperature solid state reactions.

This diagram can be very helpful from a dynamical point of view. Both exchange and intercalation/deintercalation reactions can be depicted on this figure : by vertical displacements for the former, by horizontal shifts for the latter ones. According to the position of the starting A_xMO_2 phase relatively to the border regions between both types of A^+ surroundings, structural changes may occur. Furthermore the two classes of reactions can be realized successively leading to a great number of new metastable phases, of high interest for their physical properties or their electrochemical behavior. Structural transition between phases with octahedral (O) or trigonal prismatic (P) environment and even between two octahedral environments can be observed.

Extension to copper and silver

Substitution of copper and silver for alkali ion in the intersheet space leads to the so-called delafossite type phases [21, 22, 23].

These materials exhibit the MO_2 slab packing of (P2) and (P3) type phases but the monovalent cations are situated on the edges of the trigonal prisms. This peculiar behavior results from the strong stabilization of Cu^+ and Ag^+ in linear coordination. Delafossite type phases are observed with a large number of M^{3+} cations with ionic radius situated between that of Al^{3+} ($r_{Al^{3+}}$ = 0.53Å) and that of La^{3+} ($r_{La^{3+}}$ = 1.03Å) [24]. Such structures are also observed for $PtCoO_2$, $PdCoO_2$ and $PdCrO_2$ in which the platinum and the palladium are in the monovalent state.

Physical properties

The purpose of this chapter does not consist in giving a complete review of the physical properties of these materials, but rather in showing their relation with the structure dimensionality.

Electronic properties. Considering a MO_2 layer, two types of orbital overlaps can be considered according to the number of d electrons and the HS or LS state :

. t_{2g}-t_{2g} direct overlap via common edge

. e_g-e_g indirect overlap via the anion orbitals

Interesting results can be derived from GOODENOUGH's formula which gives the value of the critical distance between M cations allowing an electronic delocalization in a t_{2g} band [25] :

$$R_c^{3d} = 3.05 - 0.03\ (Z-Z_{Ti}) - 0.04\ J(J+1) \quad (\text{Å})$$

$$R_c^{4d} = R_c^{3d} + 0.88\ ,\ R_c^{5d} = R_c^{3d} + 1.36\ (\text{Å})$$

For example, Na_xCoO_2 and K_xCoO_2 phases are metallic as the Co-Co distance is smaller than the R_c value. On the contrary, for chromium and manganese bronzes a semiconducting behavior is observed as a result of the larger M-M distance with respect to the orbital extension [26].

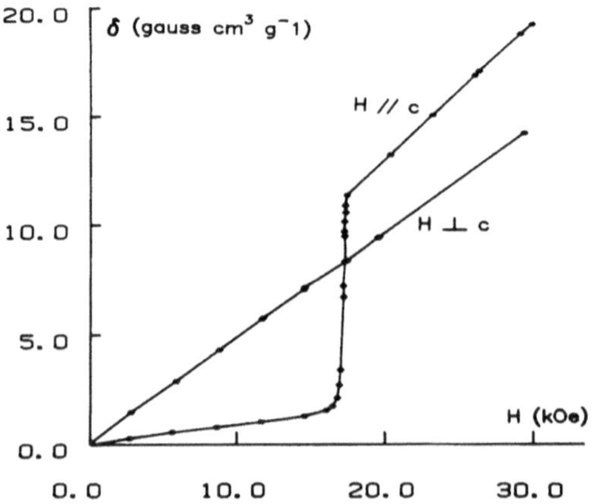

Fig. 5 - Magnetization δ per gram of a crystal of $NaNiO_2$ measured in a magnetic field parallel and perpendicular to $c_{hex.}$ (from ref. 27)

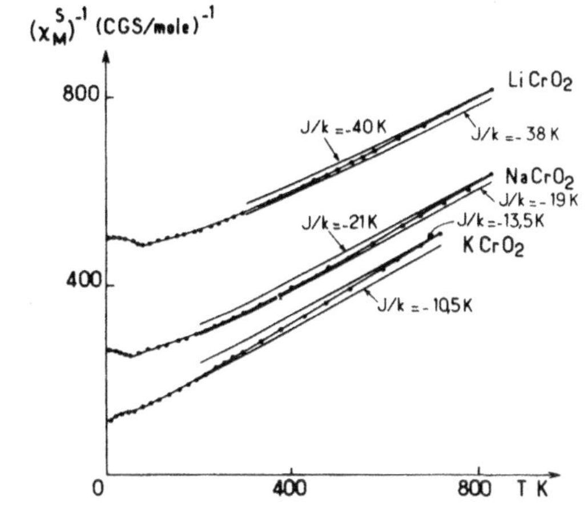

Fig. 6 - Variation of the reciprocal molar susceptibility vs temperature for $ACrO_2$ (A = Li, Na, K) phases.

The magnetic properties of these materials reflect clearly the structural anisotropy. $NaNiO_2$ shows a metamagnetic behavior as illustrated by Fig. 5. Within a layer the spins are ferromagnetically coupled parallel to the c-direction of the hexagonal lattice. Adjacent layers show an antiparallel arrangement. The magnetization vs magnetic field curve is typical for a spin flop in an antiferromagnet with uniaxial symmetry [27].

The prevailing 2D character of these structures is emphasized by the behavior of $ACrO_2$ (A = Li, Na, K) phases with (O3) type slab packing. The Cr-Cr intraplane distance increases slightly (4 %) while the intersheet Cr-Cr distance increases rapidly (33 %) from Li to K. Fig. 6 summarizes the thermal variation of the reciprocal susceptibility. Large deviations from the Curie-Weiss law are observed at low temperature. Exchange integrals were calculated from a 2D Heisenberg model using the expansion series method by RUSHBROOKE and WOOD for reciprocal magnetic susceptibility[28]. For $LiCrO_2$ a 3D magnetic order is observed at 61 K[29]. For $NaCrO_2$ and $KCrO_2$ not any 3D order has been found above 2 K by neutron diffraction study despite the maximum of susceptibility observed at 50 K for $NaCrO_2$ [30]. A similar behavior has been recently reported for chromium delafossites[31].

<u>Ionic conductivity</u>. All A_xMO_2 sheet structure phases (M = 3d) are mixed conductors. In order to study the ionic conductivity in these structures it was necessary to suppress the electronic conductivity by substitution of diamagnetic cations for transition elements. Fig. 7 shows plots of log σ vs reciprocal temperature for several layer oxides.
From these data two principal findings can be derived :
- the nature of the environment of the alkali ion (octahedral or trigonal prismatic) is the main parameter influencing the conductivity. When the mobile ion is in a trigonal prismatic site the conductivity is 3 or 4 orders of 10 higher than when it is in an octahedron. This can be related to the fact that in the trigonal prismatic environment the A-O distance variation during the diffusion process is much less pronounced than for the octahedral one.

- the mobility of the alkali ion can be correlated with the covalent character of the sheets : the greater the covalency of the sheets, the lower the electrostatic interactions between the mobile cation and its anionic environment will be and as a consequence the higher the mobility[32, 33]. In the case of layered sulfides, L. TRICHET and J. ROUXEL have observed such effects of covalency and alkali ion type of sites[34, 35].

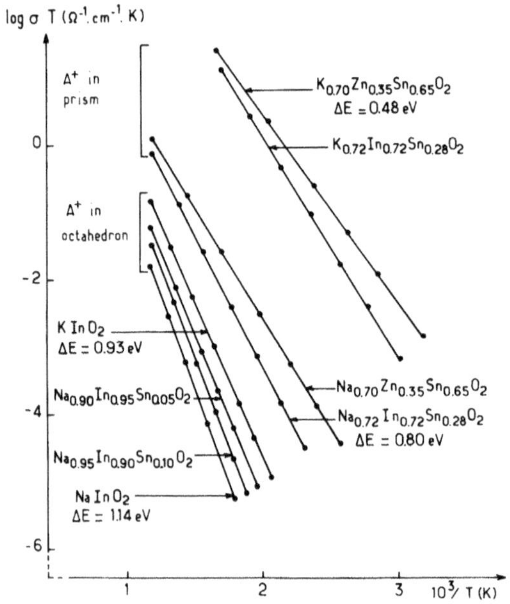

Fig. 7 - Variation of the logarithm of the σT product vs reciprocal temperature for $A_x(L,M)O_2$ phases.

SOFT CHEMISTRY IN SHEET OXIDES

The peculiar structure of sheet oxides with, as above mentioned, two main types of chemical bonds has given rise to a new chemistry realized in the vicinity of room temperature. At low or moderate temperature it is possible to modify the ionic bonds without modification of the starting slab, so new metastable materials are expected. Fig. 8 summarizes the various chemical reactions which can be realized from a layer A_xMO_2 oxide prepared at high temperature. According to the starting x value, intercalation or deintercalation reactions can be expected. Moreover new materials with B_xMO_2 formula can be obtained by exchange reactions. In some cases these materials cannot be directly obtained from classical solid state chemistry. These new materials can be used as starting phases for further intercalation or deintercalation studies. It results that each high temperature phase can be considered as the parent of an overall new family of materials.

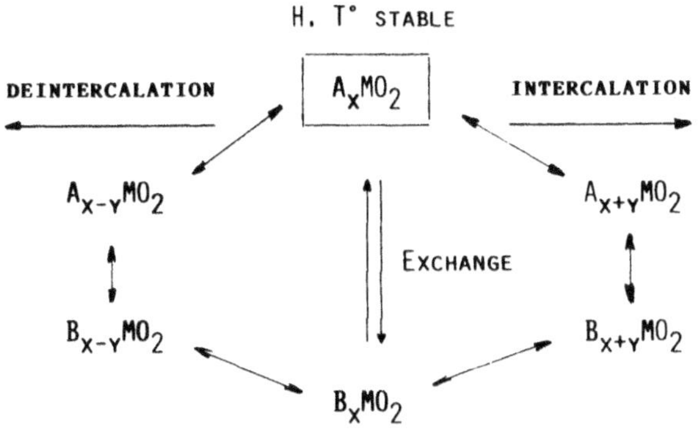

Fig. 8 - Soft chemistry reaction from a A_xMO_2 oxide.

Preparation of the starting phases

They are generally obtained by reaction of a 3d metal oxide with an alkali metal oxide, carbonate or nitrate in the temperature range 500°C-900°C. For lithium compounds only the stoichiometric phase $LiMO_2$ can be prepared in such conditions. As it has been recently reported by HEWSTON and CHAMBERLAND all these materials can be obtained with the lithium carbonate[36]. In the case of sodium or potassium compounds some troubles appear as the stability of AMO_2 phases decreases when the alkali ion size increases. It results that some of them can only be prepared at moderate temperature using Na_2O or K_2O as basic agent. This is particularly the case for $KCrO_2$ which is obtained at 450°C by direct synthesis from K_2O and Cr_2O_3[6]. The very reducing $NaTiO_2$ phase has been obtained by reduction of TiO_2 with Na metal under sodium pressure[37,38]. Some Na_xMO_2 and K_xMO_2 non stoichiometric phases can be obtained by high temperature chemistry[1,3,4,5,39]. It is well known that large size alkali ions stabilize the high oxidation levels of transition elements. The higher the ionicity of the A-O bond, the higher the covalency of the antagonist M-O bond will be and then the higher the stability of high oxidation states.

Intercalation-deintercalation reactions

These reactions can be realized either chemically or electrochemically. The chemical reactions are reviewed in MURPHY's chapter of this book. This author proposed a few years ago a redox scale which allows to select the more convenient reagent to get an intercalation or a deintercalation reaction[40].

In layer oxides the starting phase is generally fully intercalated (AMO_2 composition), oxidizing agents I_2 or Br_2 are used to obtain deintercalated phases. The deintercalation reaction stops when the potential of the A_xMO_2 phase is equal to the $I°/I^-$ (or $Br°/Br^-$) one. Most of the Na_xMO_2 phases can be obtained by this technique[41-44]. The high potential vs lithium metal of $LiMO_2$ phases hinders generally such a deintercalation reaction. Nevertheless, the reductive character of V^{3+} permitted to get Li_xVO_2 phases[45,46].

Electrochemical reactions are generally studied in a two electrode cell. An alkali metal foil is generally used as negative plate. The A_xMO_2 phase, often mixed with graphite as electronic agent and a binder, is used as positive electrode. Various salts ($LiClO_4$, $LiPF_6$, $LiAsF_6$, $LiBF_4$) dissolved in propylene carbonate are used as liquid electrolyte. The high cell voltage of the A_xMO_2 phases excludes utilization of other solvents like dioxolane or T.H.F.[47,48]. As the starting materials generally have the AMO_2 composition the studies begin with a deintercalation reaction :

$$AMO_2 \longrightarrow A_{1-z}MO_2 + zA^+ + ze^-$$

$$zA^+ + ze^- \longrightarrow zA$$

In lithium systems open circuit voltage vs Li exceeds generally 4v; this value being out of the electrochemical stability window of the electrolyte, there is a risk of electrolyte breakdown. To avoid this problem, GOODENOUGH et al. have proposed to use a three electrode cell with an intercalation phase ($Li_xV_2O_5$) as counter and reference electrodes[49]. $Li_{1+y}Mn_2O_4$ has been used as counter electrode in a two electrode cell design. As $Li_{1+y}Mn_2O_4$ operates as a two phase electrode over a wide range of y ($0.1 \leqslant y \leqslant 0.8$) it could be used as a constant voltage electrode to monitor lithium extraction from a $LiMO_2$ working electrode[50,51]. Though the potential seen by the electrolyte is not changed by this technique which does not either modify the behavior of the system from the point of view of thermodynamic, it nevertheless seems to suppress the problems of electrolyte decomposition.

Exchange reactions

Inter-alkali ions exchange reactions are generally realized in molten alkali nitrate salts[52,53]. Nevertheless as the reaction temperature is close to 250°C some very metastable phases cannot be obtained with this technique. Therefore if the final product is sensitive to moisture some difficulties appear to recover it. An improvement of the exchange technique using a solution of alkali ion chloride or iodide in an organic solvent (CH_3OH for example) has been recently reported[54]. The new $LiCoO_2$ (O2) variety has been thus obtained from $Na_{0.70}CoO_2$ (P2) [54].

Reversible intercalation-deintercalation reactions

Reversible deintercalation reactions are generally observed at the beginning of the deintercalation process. For high deintercalation amounts, irreversible reactions involving slab modifications are observed. They will be discussed in the next part. The first electrochemical studies were reported in 1980, they concerned the Li_xCoO_2 [49] and Na_xCoO_2 [55] systems. In the previous years several works were realized on non-stoichiometric layer oxides (Na_xMnO_2 [1], Na_xCoO_2 [3], K_xCoO_2 [4,5], K_xMnO_2 [39], K_xCrO_2 [6]), but all these materials were studied by high temperature chemistry. The preparation of K_xCrO_2 phases can be considered as a high temperature spontaneous deintercalation reaction as these materials are obtained from $KCrO_2$ according to the reaction :

$$KCrO_2 \xrightarrow[\text{vacuum}]{700°C} K_xCrO_2 + (1-x)\,K\nearrow$$

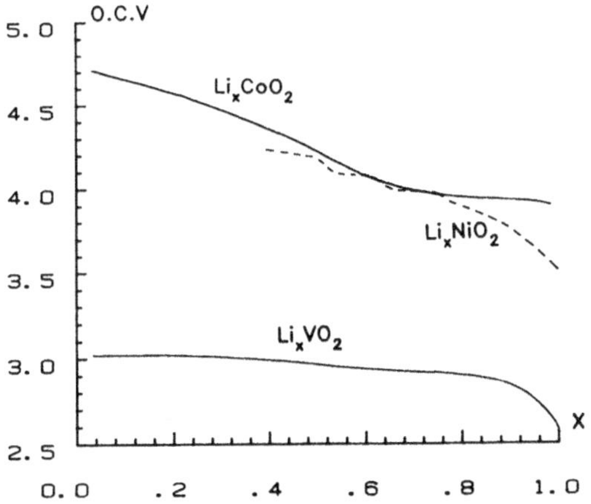

Fig. 9 - O.C.V. vs lithium content in Li_xVO_2(O3)[46], Li_xCoO_2(O3)[49] and Li_xNiO_2(O3)[56] phases.

The Li_xMO_2 phases. Lithium deintercalation from O3 type phases have been mostly studied by GOODENOUGH's group : $LiCoO_2$[49], $LiNiO_2$[56] and $LiVO_2$[46] have been successfully deintercalated. Table 2 summarizes the values of the intercalation range observed without irreversible modification. In these domains no structural modification has been seen. the O.C.V. curves are plotted in Fig. 9. For higher deintercalation amounts a part of the M atoms left the slab to the intersticial space. Such a behavior has not been seen in the Li_xCoO_2 system where a ferroelectric type Co^{4+} ion displacement occurs[57]. A cooperative ferroelectric c-axis displacement of Co^{4+} ions from the center of symmetry of an octahedral interstice can introduce dipole-dipole couplings between CoO_2 sandwich layers that alleviate the electrostatic repulsions between adjacent oxygen layers. Nevertheless, some interlayer cations are required for structural stability[58]. This behavior should be compared with that observed in $K_{0.72}(In_{0.72}Sn_{0.28})O_2$. The structure of this phase has been determined by X-ray diffraction profile analysis. In^{3+} and Sn^{4+} ions are at different z values within a $(In_{0.72}Sn_{0.28})O_2$ sheet.

Table 2. Reversible electrochemical domains of Li_xMO_2 layer oxides.

Li_xCoO_2	$0.5 \leqslant x \leqslant 1$	(49)
Li_xNiO_2	$0.6 \leqslant x \leqslant 1$	(56)
Li_xVO_2	$0.7 \leqslant x \leqslant 1$	(46)

The deintercalation of the metastable $LiCoO_2$ (O2) variety has been recently reported[60]. Fig. 10 shows the e.m.f. variation vs composition obtained from this material. For the composition $Li_{0.62}CoO_2$ a new layer type packing (O6) (Fig. 2) has been obtained.

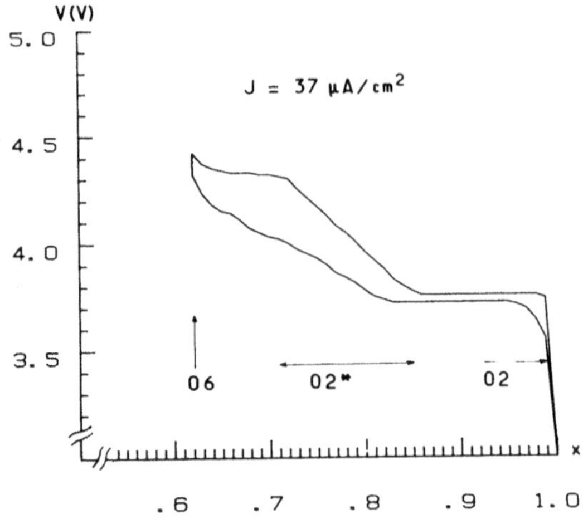

Fig. 10 - E.m.f. vs lithium content from $LiCoO_2$ (O2).

<u>The Na_xMO_2 phases</u>. The essential part of these studies comes from our group in Bordeaux. Sodium has been electrochemically deintercalated from $NaTiO_2$[38,61], $NaCrO_2$[62], $NaCoO_2$[55,63] and $NaNiO_2$[62]. Moreover intercalation deintercalation reactions were realized from the non stoichiometric Na_xMO_2 (M = Co, Mn) phases obtained at high temperatures[63,64].

Fig. 11 shows the voltage variation vs composition curves obtained from $NaMO_2$ (M = Ti, Cr, Co, Ni) phases.

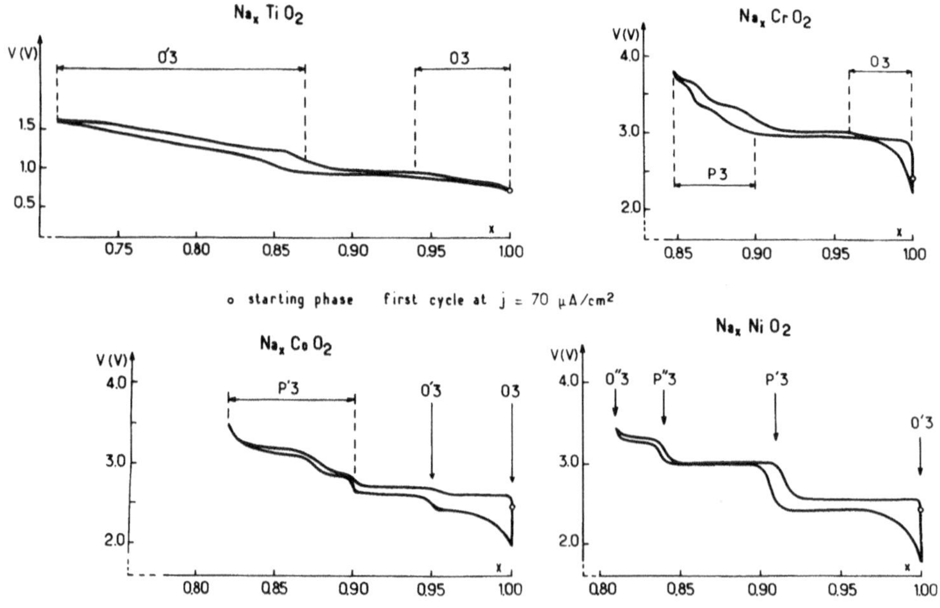

Fig. 11 - E.m.f. vs sodium content in Na_xMO_2 (M = Ti, Cr, Co, Ni) phases.

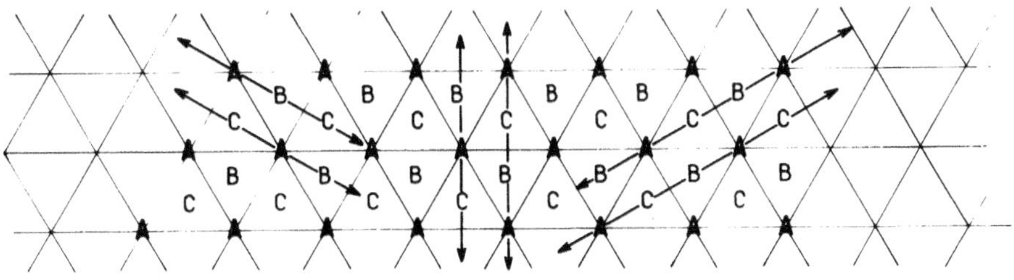

Fig. 12 - Gliding directions in a triangular lattice.

Deintercalation reactions from $NaMO_2$ allow to obtain new Na_xMO_2 materials. According to x values, solid solutions (continuous variation of voltage vs composition) or defined compounds separated from one another by biphased domains (constant voltage vs global composition) can be observed. The voltage values (vs Na/Na^+) clearly illustrate the oxidizing character of tetravalent chromium, cobalt or nickel on one hand and the reducing character of trivalent titanium on the other hand.

Reversible structural transitions between phases with octahedral (O) or trigonal prismatic (P) alkali ion environment can be observed.

The structures of layer oxides can be separated in two groups given respectively in Figures 1 and 2. Within each group a structural transition requires a sheet gliding. As in such a process the M-O bonds are not broken this transition is already possible at R.T. and is moreover completely reversible from a structural point of view.

As the integrity of the slab is maintained during its gliding, all packings cannot be obtained by sheet gliding from a given one. It is particularly the case of P2 type phases which cannot be obtained from O3 and P3 only by sheet gliding. In a more general point of view, group 2 structures (Fig. 2) differ from those of the first group (Fig. 1) not only by a sheet gliding but also by MO_6 octahedra rotations around the hexagonal axis. As a result the M-O-M bonds should have to be broken and the reaction does not occur at room temperature. On the contrary, these restrictions do not apply to GIC's as there is only one graphite layer per slab.

As shown on Fig. 12 in a triangular lattice with ABC oxygen positions, three equivalent gliding directions are possible. For each of them the new position can be deduced from the starting one by circular permutation of A, B, C.

This behavior is illustrated in Fig. 13 by the structural transformations between the P3 and O3 structural types. Such transformations occur for instance in the Na_xTiS_2 system studied by ROUXEL and coworkers[65,66]. Nevertheless it should be noted that the intermediate phase O'3 has never been observed in chalcogenides.

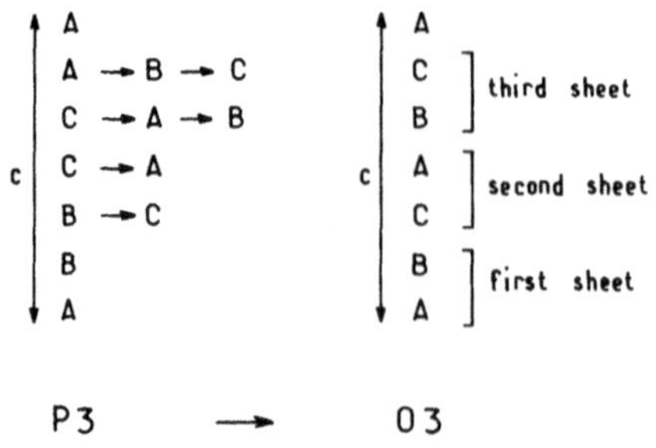

Fig. 13 - Sheet gliding in the P3 → O3 transition.

An interesting behavior has been found for intercalation-deintercalation reactions from $Na_{0.60}CoO_{1.92}$ (P'3) and $Na_{0.77}CoO_{1.96}$ (O'3) phases[63]. In Fig. 14 the emf composition curves (II and III) are compared with the diagram obtained from $NaCoO_2$ (O3) (I). In each cycle the previously described structural modifications occur. The comparison of the three curves shows that the x values for which a given structural type is observed are modified by the presence of oxygen vacancies. In each system the O3 type is obtained for the highest x values when the oxidation state of cobalt is close to +III. The strong voltage increase as x decreases for the P'3 phase is evidence of the instability in sheet structure oxides. For oxygen-deficient bronzes a smaller amount of Na^+ ions is necessary to stabilize a sheet structure due to smaller repulsion between oxygen layers. Each oxygen vacancy gives two electrons to the conduction band, so the t_{2g} band (Co^{3+} and Co^{4+} are in the LS state) is filled for an amount of intercalation smaller than 1.

A similar behavior has been observed during the sodium intercalation in $Na_{0.70}CoO_{2-y}$ (P2) phases[67,68]. Fig. 15 shows the variation of cell voltage versus sodium content for two phases having the same Na/Co atomic ratio (0.70), but different y values. The comparison of the two curves shows that they differ essentially by a horizontal shift. The two characteristic plateaus are obtained for each sample at exactly the same voltage : 2.54 V and 2.44 V. Nevertheless these voltage values are observed for both materials for different sodium contents. Indeed as a result

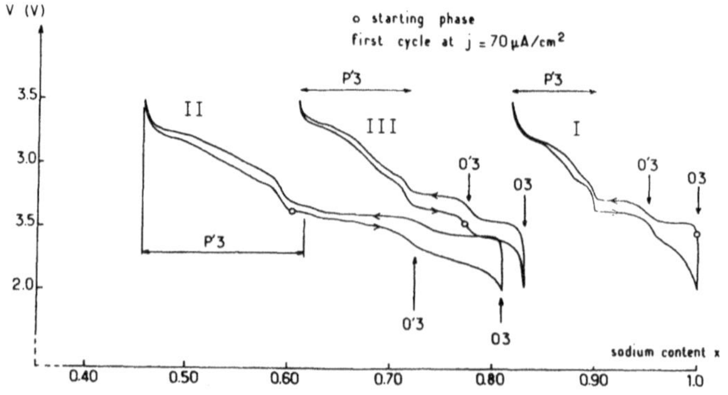

Fig. 14 - E.m.f. vs sodium content curves in the Na_xCoO_{2-y} systems.

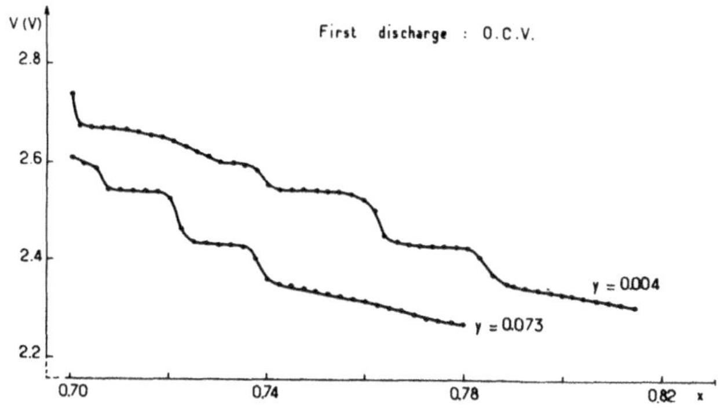

Fig. 15 - O.C.V. vs composition curves obtained from $Na_{0.70}CoO_{2-y}$ samples

the assumption of a succession of sodium-ordered phases fails. For the less oxygen-deficient phase (y = 0.004) a given value of the voltage is obtained for a bigger intercalation amount of sodium than for the more oxygen-deficient phase. This behavior emphasizes the results obtained by electrical measurements, the electrons arising from the oxygen vacancies compensate partially the original positive charge carriers. During the intercalation process the injected electrons raise the Fermi level by an energy which depends on the density of states in its neighborhood. When the electronic compensation resulting from the oxygen departure from stoichiometry decreases, a higher amount of sodium must be intercalated to obtain the same potential value (i.e. the same Fermi level).

A similar behavior has been recently observed by TARASCON and HULL for the $Na_xMo_2O_4$ system[69]. In this system five single phase domains were found at x = 0.55, 0.9, 1-1.4, 1.60 and 1.9. Moreover the maximum sodium content in $Na_xMo_2O_4$ is strongly related to the synthesis temperature of the starting material.

Chemical deintercalations of sodium from $NaMO_2$(O3) (M = Cr, Fe, Co, Ni) phases were reported by Koizumi and coll.[41-44]. These authors have used iodine and bromine as oxidizing agent. Table 3 summarizes the single phase compositions obtained in these systems.

Table 3. Sodium deintercalation phases obtained by chemical oxidation from $NaMO_2$ (O3) phases.

Starting phase (O3)	Deintercalated phases	
$NaCrO_2$	$Na_{0.75}CrO_2$ (O'3)	$Na_{0.50}CrO_2$ (P'3)
$NaFeO_2$	$Na_{0.90}FeO_2$ (O3)	
$NaCoO_2$	$Na_{0.60}CoO_2$ (P'3)	$Na_{0.50}CoO_2$ (P'3)
$NaNiO_2$		$Na_{0.50}NiO_2$ (P'3)

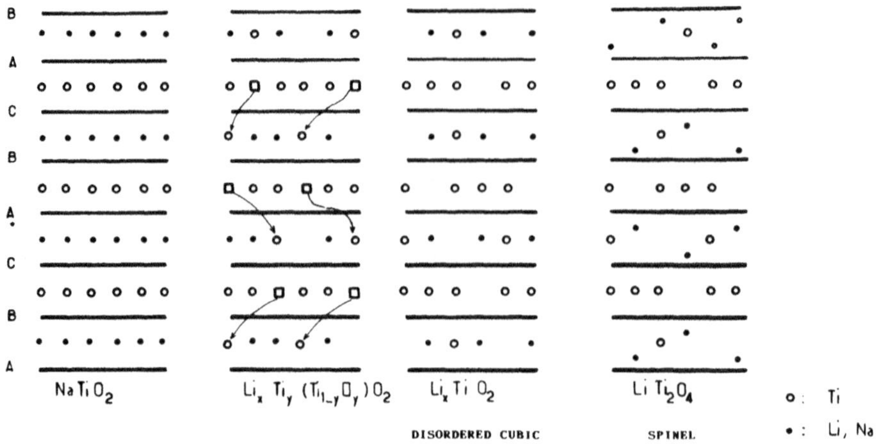

Fig. 16 - Cationic rearrangements occurring in the $NaTiO_2 \rightarrow Li_xTi_2O_4$ reaction.

Irreversible deintercalation reactions

As mentioned in the introduction, sheet oxides cannot exist without alkali ion inserted between adjacent MO_2 layers. For A_xMO_2 phases having a low cell voltage (i.e. for reductive M^{3+}) it is possible to deintercalate, by overcharging the cell or by using a strong oxidizing reagent, out of the reversible deintercalation domain. Then, to stabilize the highly deintercalated phases some M^{3+} may move from the MO_2 sheet to the intersheet space.

The first A_xMO_2 2D → spinel transition was reported by MAAZAZ in 1982[70]. $Li_{0.50}TiO_2$ with a disordered rock salt structure was first and foremost obtained by simultaneous Li^+/Na^+ exchange and oxidation reactions from $NaTiO_2$. The annealing of this phase has led to the spinel $LiTi_2O_4$. Fig. 16 describes the various steps observed during these reactions.

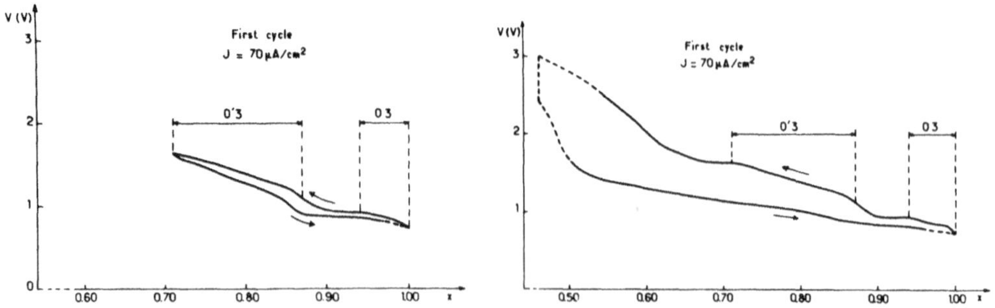

Fig. 17 - Electrochemical behavior of $Na/Na^+/NaTiO_2$ cells according to the cycling conditions.

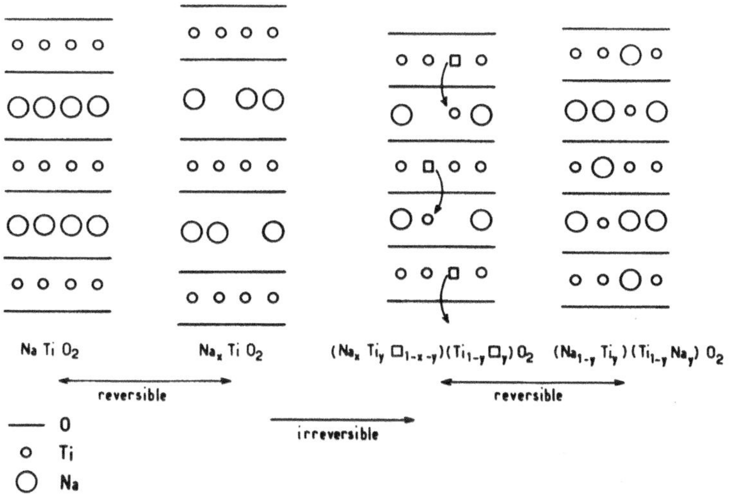

Fig. 18 - Cationic rearrangements occurring in the reversible and irreversible deintercalation of $NaTiO_2$.

A similar behavior has been observed by overcharging $Na/Na^+/Na_xTiO_2$ cells[61,38]. A reversible deintercalation occurs in the $0.71 \leqslant x \leqslant 1$ range (Fig. 17a). For higher deintercalation amounts some titanium ions migrate from the TiO_2 sheet to the intersheet space leading to $Na_xTi_y\square_{1-x-y}(Ti_{1-y}\square_y)O_2$ (Fig. 17b). During discharge from this phase Na^+ ions fill the two types of available sites and a composition close to $(Na_{1-y}Ti_y)(Ti_{1-y}Na_y)O_2$ is reached without modification of the new framework (Fig. 18).

2D → 3D transitions were reported in Li_xVO_2 [46,50,71] and Li_xNiO_2 systems[56]. In the vanadium one the 2D structure remains in the $0.7 \leqslant x \leqslant 1$ range. For $x < 0.7$ removal of lithium results in a rearrangement of the vanadium ions in the cubic close packed oxygen lattice. In $Li_{0.22}VO_2$ approximately one third of the vanadium ions are located in the octahedral sites left vacant in the lithium layer. The annealing at 300°C of a mixture of $Li_{0.50}VO_2$ composition leads to the spinel LiV_2O_4 [71].

A quite similar behavior has been observed by THOMAS et al. in the Li_xNiO_2 system. A metastable layered phase appears for $x < 0.6$. The normal spinel $Li[Ni_2]O_4$ has been obtained for the first time by a 200°C annealing of the metastable layered $Li_{0.50}NiO_2$ compound [56]. This spinel is metastable and disproportionates above 300°C into several phases.

The same transformation type occurs in TMDC as observed by DENIARD, CHEVALIER, TRICHET and ROUXEL in the Li_xZrS_2 system[72]. A (O1) type phase is observed for $0 \leqslant x \leqslant 0.20$, for higher intercalation amounts ($0.30 \leqslant x \leqslant 1$) a sheet gliding occurs and the O3 type phase is then obtained. The $0.30 \leqslant x \leqslant 0.50$ phase undergoes at 280°C a transition to spinel structure.

It should be noticed that the cfc oxygen packing of the O3 type starting phase remains unchanged during the 2D → spinel transition which requires only a displacement of a M cation to an adjoining layer through a tetrahedral site. The easiness of such a migration has been recently discussed by GOODENOUGH et al. according to the stability of the moving cation in tetrahedral environment[56]. From LiV_2O_4 [51], obtained as above mentioned, or other normal spinels $Li[M_2]O_4$ obtained by high temperature chemistry, Li intercalation or deintercalation reactions have been reported. In all cases the $[M_2]O_4$ framework remains unchanged. From $LiMn_2O_4$, $Li_{0.4}Mn_2O_4$ and $Li_2Mn_2O_4$ have been obtained electrochemically[73,74]. Using chemical oxydation by $S_2O_8^{=}$ anions TOURNOUX et al. have obtained $Li_{0.04}Mn_2O_4$ [75]. For a composition close to the rutile phase the $[Mn_2]O_4$ spinel framework is maintained. By lithium intercalation in the

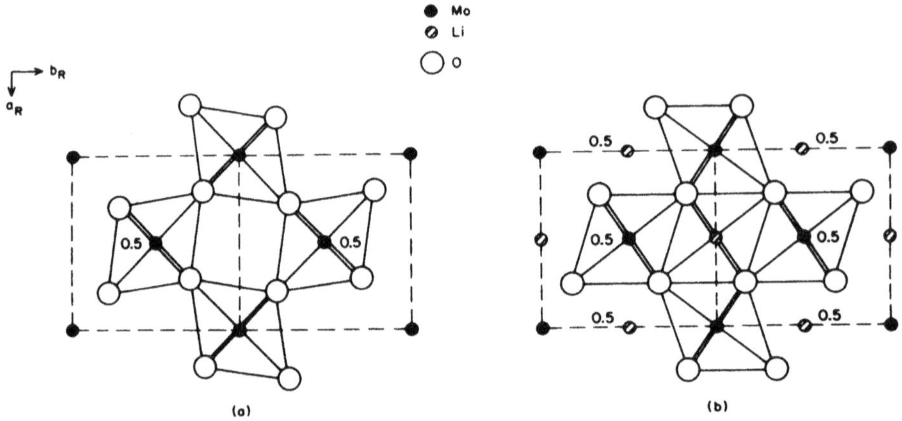

Fig. 19 - Structures of rutile MoO_2 (a) and NiAs-type $LiMoO_2$ (b) (from Ref. 78).

$Li[M_2]O_4$ structure a new type of ordered rock salt phase is obtained for $Li_2Mn_2O_4$ and $Li_2Ti_2O_4$ [75,74,76]. In order to accomodate the additional lithium the tetrahedrally coordinated lithium ions have been displaced to occupy octahedral sites.

Rutile type compounds have never been observed by delithiation reaction from AMO_2 layer oxides or AM_2O_4 spinels. An intensive study of lithium intercalation in rutile type phases has been realized by MURPHY et al.[77]. They have shown that when the intercalation occurs the rutile framework is preserved. Nevertheless the structure of $LiMoO_2$ obtained from MoO_2 can be regarded as having a NiAs type structure rather than a rutile type one [78]. It can be described from a (O1) type oxygen packing with lithium and molybdenum in the octahedral positions (Fig. 19).

Lithiation of the rutile phase β-MnO_2 results in a product $Li_{1-\delta}MnO_2$ isostructural with the tetragonal $Li_2[Mn_2]O_4$ phase obtained by room temperature lithiation of the spinel $Li[Mn_2]O_4$ [79]. This result shows that a rutile → spinel framework transformation occurs (Fig. 20). In a first step the rutile framework rearranges to an anionic hexagonal close packed (O1) then a (O1) → (O3) sheet gliding occurs with a cation displacement. The driving force of the rutile-spinel transformation is the electrostatic repulsion between Li^+ and $Mn^{4+/3+}$ ions.

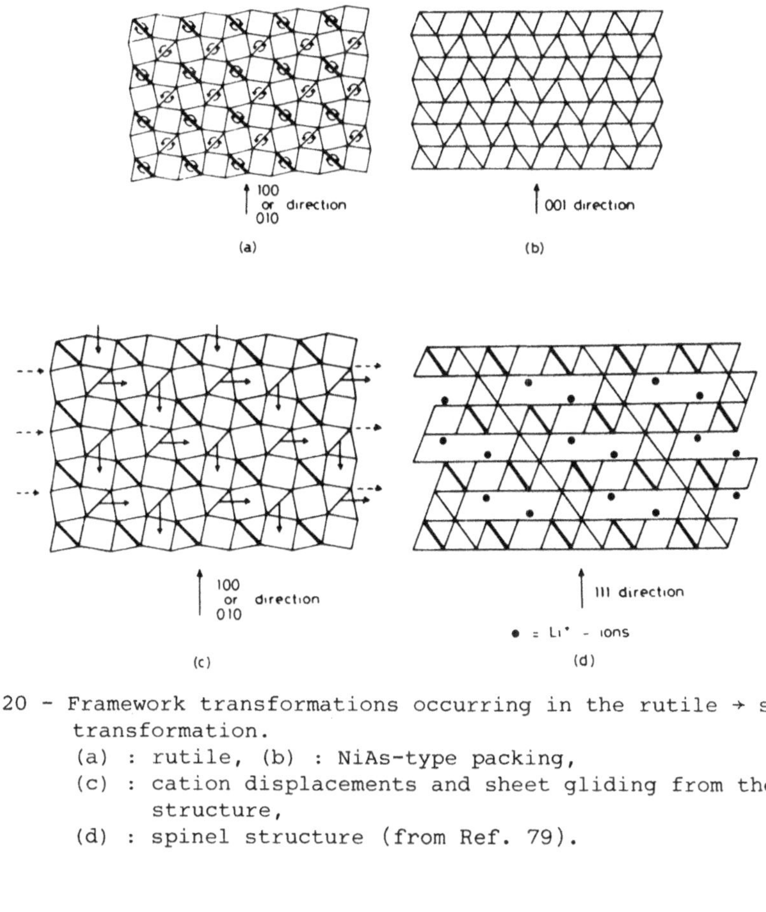

Fig. 20 - Framework transformations occurring in the rutile → spinel transformation.
(a) : rutile, (b) : NiAs-type packing,
(c) : cation displacements and sheet gliding from the rutile structure,
(d) : spinel structure (from Ref. 79).

APPLICATION TO THE NICKEL HYDROXIDE ELECTRODE

As mentioned in the introduction of this article, nickel hydroxide works in the positive electrode of alkaline secondary batteries (Ni-Cd, Ni-Zn, Ni-Fe and Ni-H_2) as an intercalation compound. The general reaction scheme was initially proposed by BODE et al.[80] :

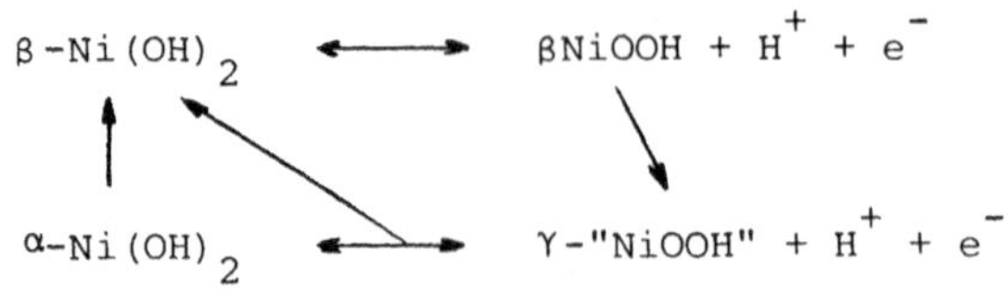

The understanding of this diagram requires the following comments :

- the α phase exhibits a turbostratic structure with a layer of water molecules intercalated between adjacent slabs (Fig. 21).
- the γ phase does not have a defined composition but a general formula : $H_xK_yNiO_2,zH_2O$ (Fig. 21) the nickel oxidation state varies between 3.3 and 3.7.

These materials were recently obtained by soft chemistry reactions (exchange and reduction) from the sodium nickelate $NaNiO_2$. A new variety of well crystallized nickel hydroxide $Ni(OH)_2,H_2O$ α* has been obtained and characterized[81]. Its chemical composition is close to that of the α-turbostratic variety. It differs nevertheless by its periodic packing (P3) and its large size cristallites which contrast with the imperfectly ordered character and the ultradivided texture of α.

Cobalt and iron substituted phases were obtained by exchange and reduction reaction from substituted sodium nickelates : $NaNi_{1-x}Co_xO_2$ and $NaNi_{1-x}Fe_xO_2$. Cobalt and iron stay in the trivalent state in the hydroxyde[82]. The excess of positive charges are compensated by CO_3^{2-} anions which are intercalated in the intersheet space[83]. The various intercalation-deintercalation reactions occurring in such processes are summarized in Fig. 22. Recently MENDIBOURE and SCHOLLHORN have reported the direct preparation by precipitation of the reevesite $Ni_{0.75}Fe_{0.25}(OH)_2(CO_3)_{0.125},0.38H_2O$ and related compounds[84]. They have studied the exchange properties of intercalated anions and have shown a selectivity for the stability of the intercalated compound in the following sequence $CO_3^{2-} \gg SO_4^{2-} > Cl^- > NO_3^- > CH_3COO^-$.

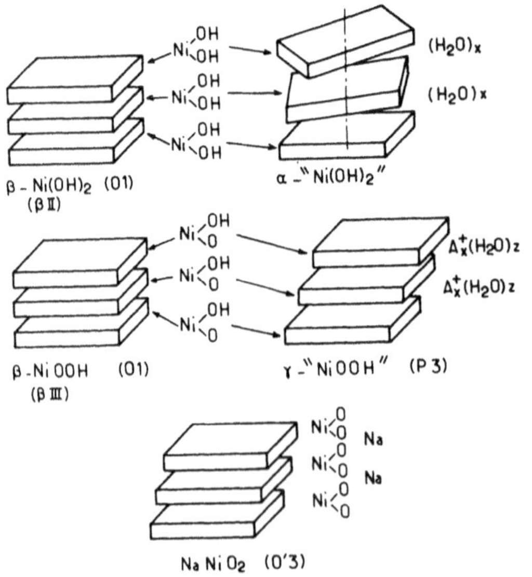

Fig. 21 - Stacking of NiO_2 slabs in nickel hydroxides, oxyhydroxides and $NaNiO_2$.

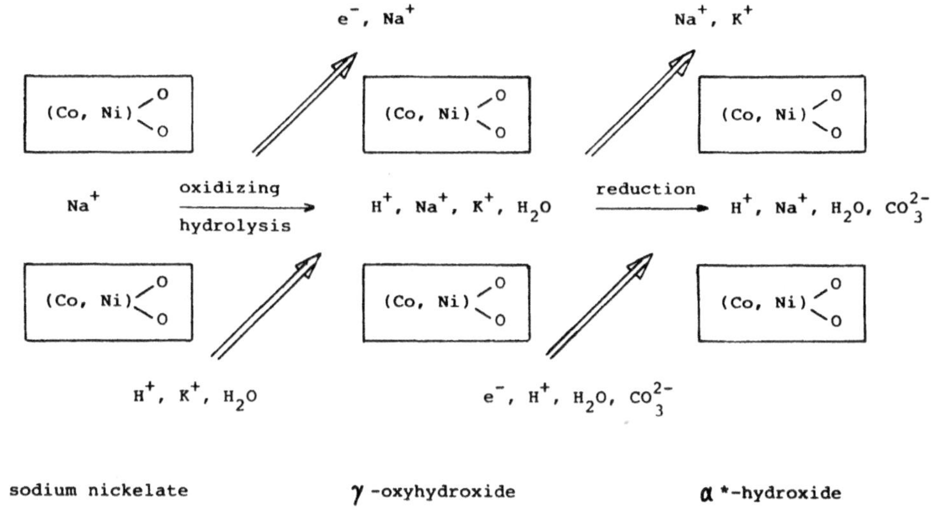

Fig. 22 - Schematic representation of the exchange reactions observed during oxidizing hydrolysis and reduction of cobalt substituted sodium nickelates.

CONCLUSION

Intercalation deintercalation properties of sheet structure oxides A_xMO_2 (A : alkali metal, M : 3d element) have been intensively studied during the last few years. As the high ionicity of the M-O bond prevents the existence of the fully deintercalated form, two types of behavior are observed according to x values.

For the highest alkali ion amounts (x 0.5) their crystal chemistry and physical properties are characteristic of 2D materials. The MO_2 sheets can be regarded as independent units :

- easy sheet gliding observed during reversible intercalation deintercalation reactions
- strong 2D character of the magnetic and transport properties.

For the lowest alkali ion amounts a structural transition from 2D to 3D is observed. Some M cations migrate irreversibly from the MO_2 sheet to the intersheet space. Disordered rock salt type phases are first obtained. Their annealing at medium temperature leads to the formation of spinel type phases. During all these transformations the oxygen cubic close packing is preserved.

REFERENCES

1. J.P. Parant, R. Olazcuaga, M. Devalette, C. Fouassier and P. Hagenmuller, J. Solid State Chem., 3:1 (1971).
2. C. Li, A.F. Reid and S. Saunders, J. Solid State Chem., 3:614 (1971).
3. C. Fouassier, G. Matejka, J.M. Réau and P. Hagenmuller, J. Solid State Chem., 6:532 (1973).

4. M. Jansen and R. Hoppe, Z. anorg. allg. Chem., 408:97 (1974)
5. C. Delmas, C. Fouassier and P. Hagenmuller, J. Solid State Chem., 13:165 (1975).
6. C. Delmas, M. Devalette, C. Fouassier and P. Hagenmuller, Mat. Res. Bull., 10:393 (1975).
7. R. Hoppe, Bull. Soc. Chim. Fr., 1165 (1965).
8. R. Clos, M. Devalette, P. Hagenmuller, R. Hoppe and E. Paletta, C.R. Acad. Sc. F., 265:801 (1967).
9. R. Scholder and H. Glaser, Z. anorg. allg. Chem., 364:113 (1969).
10. J. Hauck, Z. Naturforsch., 24b:252 (1969).
11. F. Stenner and R. Hoppe, Z. anorg. allg. Chem., 374:239 (1970).
12. C. Delmas, C. Fouassier and P. Hagenmuller, Mat. Res. Bull., 11:1483 (1976).
13. A. Leblanc-Soreau, M. Danot, L. Trichet and J. Rouxel, Mat. Res. Bull. 9:191 (1974).
14. R. Hoppe and H. Sabrowsky, Z. anorg. allg. Chem., 339:144 (1965).
15. R. Hoppe and H. Brunn, Rev. Chim. Min., 13:41 (1976).
16. C. Delmas, C. Fouassier and P. Hagenmuller, Mat. Res. Bull., 13:165 (1975).
17. P. Colombet, M. Danot and J. Rouxel, Mat. Res. Bull., 14:813 (1979).
18. C. Delmas, J.J. Braconnier and P. Hagenmuller, Mat. Res. Bull., 17:117 (1982).
19. A. Mendiboure, C. Delmas and P. Hagenmuller, Mat. Res. Bull., 19:1383 (1984).
20. J. Rouxel, J. Solid State Chem., 17:223 (1976).
21. R.D. Shannon, D.B. Rogers and C.T. Prewitt, Inorg. Chem., 10:713 (1971).
22. C.T. Prewitt, R.D. Shannon and D.B. Rogers, Inorg. Chem., 10:719 (1971).
23. D.B. Rogers, R.D. Shannon, C.T. Prewitt and J.L. Gillson, Inorg. Chem., 10:723 (1971).
24. J.P. Doumerc, A. Ammar, A. Wichainchai, M. Pouchard and P. Hagenmuller J. Phys. Chem. Solids, 48:37 (1987).
25. D.B. Rogers, R.J. Arnott, A. Wold and J.B. Goodenough, J. Phys. Chem. Solids, 24:347 (1963).
26. C. Fouassier, C. Delmas and P. Hagenmuller, Mat. Res. Bull., 10:443 (1975).
27. P.F. Bongers and U. Enz, J. Solid State Comm., 4:153 (1966).
28. G.S. Rushbrooke and P.J. Wood, Molec. Phys., 6:409 (1963).
29. C. Delmas, G. Le Flem, C. Fouassier and P. Hagenmuller, J. Phys. Chem. Solids, 39:55 (1978).
30. J.L. Soubeyroux, D. Fruchart, J.C. Marmeggi, W.J. Fitzerald, C. Delmas and G. Le Flem, Phys. Status Sol., 67:633 (1981).
31. J.P. Doumerc, A. Wichainchai, A. Ammar, M. Pouchard and P. Hagenmuller Mat. Res. Bull., 21:745 (1986).
32. C. Delmas, A. Maazaz, C. Fouassier, J.M. Réau and P. Hagenmuller in "Fast Ion Transport in Solids", ed. P. Vashista, C.N. Mundy and G.K. Shenoy, North Holland, New York (1979).
33. A. Maazaz, C. Delmas, C. Fouassier, J.M. Réau and P. Hagenmuller, Mat. Res. Bull., 14:329 (1979).
34. L. Trichet and J. Rouxel, Mat. Res. Bull., 12:345 (1977).
35. J. Rouxel, Mat. Res. Bull., 13:1425 (1978).
36. T.A. Hewston and B.L. Chamberland, J. Phys. Chem. Solids, 48:97 (1987)
37. P. Hagenmuller, A. Le Cerf and M. Onillon, C.R. Acad. Sci., 255:928 (1962).
38. A. Maazaz, C. Delmas and P. Hagenmuller, J. Incl. Phenom., 1:45 (1983)
39. C. Delmas and C. Fouassier, Z. anorg. allg. Chem., 420:184 (1976).
40. D.W. Murphy and P.A. Christian, Science, 205:651 (1979).
41. S. Miyazaki, S. Kikkawa and M. Koizumi, Rev. Chim. Min., 19:301 (1982)

42. S. Miyazaki, S. Kikkawa and M. Koizumi, Synth. Met., 6:211 (1983).
43. S. Kikkawa, S. Miyazaki and M. Koizumi, Mat. Res. Bull., 20:373 (1985)
44. S. Kikkawa, S. Miyazaki and M. Koizumi, J. Solid State Chem., 62:35 (1986).
45. K. Vidyasagar and J. Gopalakrishnan, J. Solid State Chem., 42:217 (1982).
46. L.A. de Picciotto and M.M. Thackeray, Mat. Res. Bull., 19:1497 (1984).
47. M.S. Whittingham, Prog. Solid State Chem., 12:1 (1978).
48. "Intercalation Chemistry", Ed. M.S. Whittingham and A.J. Jacobson, Academic Press, New York (1982).
49. K. Mizushima, P.C. Jones, P.J. Wiseman and J.B. Goodenough, Mat. Res. Bull., 15:783 (1980).
50. L.A. de Picciotto and M.M. Thackeray, Solid State Ionics, 18/19:773 (1986).
51. L.A. de Picciotto and M.M. Thackeray, Mat. Res. Bull., 20:1409 (1985).
52. W.A. England, J.E. Birkett, J.B. Goodenough and P.J. Wiseman, J. Solid State Chem., 49:300 (1983).
53. W.A. England, J.B. Goodenough and P.J. Wiseman, J. Solid State Chem., 49:289 (1983).
54. C. Delmas, J.J. Braconnier and P. Hagenmuller, Mat. Res. Bull., 17:117 (1982).
55. J.J. Braconnier, C. Delmas, C. Fouassier and P. Hagenmuller, Mat. Res. Bull., 15:1797 (1980).
56. M.G. Thomas, W.I. David, J.B. Goodenough and P. Groves, Mat. Res. Bull., 20:1137 (1985).
57. M.G. Thomas, P.G. Bruce and J.B. Goodenough, J. Electrochem. Soc., 132 (7) : 1521 (1985).
58. M.G. Thomas, P.G. Bruce and J.B. Goodenough, Solid State Ionics, 17:13 (1985).
59. C. Delmas and P.E. Werner, Acta Chem. Scand., A 32:329 (1978).
60. A. Mendiboure, C. Delmas and P. Hagenmuller, Mat. Res. Bull., 19:1383 (1984).
61. A. Maazaz and C. Delmas, C.R. Acad. Sci., 295:759 (1982).
62. J.J. Braconnier, C. Delmas and P. Hagenmuller, Mat. Res. Bull., 17:993 (1982).
63. C. Delmas, J.J. Braconnier, C. Fouassier and P. Hagenmuller, Mat. Res. Bull., 3/4:165 (1981).
64. A. Mendiboure, C. Delmas and P. Hagenmuller, J. Solid State Chem., 57:323 (1985).
65. J. Rouxel, M. Danot and J. Pichon, Bull. Soc. Chim. Fr., 3390 (1971).
66. J. Rouxel in "Intercalated Layered Materials", Ed. by F.A. Lévy, D. Reidel Publ. Comp. Dordrecht (1979).
67. J. Molenda, C. Delmas and P. Hagenmuller, Solid State Ionics, 9/10:431 (1983).
68. J. Molenda, C. Delmas, P. Dordor and A. Stoklosa, Solid State Ionics, 12:473 (1984).
69. J.M. Tarascon and G.W. Hull, Solid State Ionics, 22:85 (1986).
70. A. Maazaz, thesis, University of Bordeaux, France (1982).
71. L.A. de Picciotto and M.M. Thackeray, Mat. Res. Bull., 20:187 (1985).
72. P. Deniard, P. Chevalier, L. Trichet and J. Rouxel, Synth. Met., 5:141 (1983).
73. M.M. Thackeray, P.J. Johnson, L.A. de Picciotto, P.G. Bruce and J.B. Goodenough, Mat. Res. Bull., 19:179 (1984).
74. J.B. Goodenough, M.M. Thackeray, W.I. David and P.G. Bruce, Rev. Chim. Min., 21:435 (1984).
75. A. Mosbah, A. Verbaere and M. Tournoux, Mat. Res. Bull., 18:1375 (1983).
76. R.J. Cava, D.W. Murphy, S. Zahurak, A. Santoro and R.S. Roth, J. Solid State Chem., 53:64 (1984).

77. D.W. Murphy, F.J. Di Salvo, J.N. Carides and J.V. Waszczak, Mat. Res. Bull., 13:1395 (1978).
78. D.E. Cox, R.J. Cava, D.B. Mc Whan and D.W. Murphy, J. Phys. Chem. Solids, 43:657 (1982).
79. W.I. David, M.M. Thackeray, P.G. Bruce and J.B. Goodenough, Mat. Res. Bull., 19:99 (1984).
80. H. Bode, K. Dehmelt and J. Witte, Electrochim. Acta, 11:1079 (1966).
81. J.J. Braconnier, C. Delmas, C. Fouassier, M. FIGLARZ, B. Beaudouin and P. Hagenmuller, Rev. Chim. Min., 21:496 (1984).
82. J.J. Braconnier, Thesis University of Bordeaux (1983).
83. C. Delmas, J.J. Braconnier, Y. Borthomieu and P. Hagenmuller, Mat. Res. Bull., 1987 (in press).
84. A. Mendiboure and R. Schöllhorn, Rev. Chim. Min., 23:819 (1986).

SWELLING CLAYS AND RELATED

COMPLEX LAYERED OXIDES

Thomas J. Pinnavaia

Department of Chemistry and
Center for Fundamental Materials Research
Michigan State University
East Lansing, MI 48824

INTRODUCTION

The geological term "clay" refers to any naturally occuring material with a particle size less than 2μ. However, the term "clay mineral" refers to a specific composition of matter and a discrete structure, but with a particle size less than 2μ. Clay minerals are usually layered oxides. Included in this family are the layered silicate clays.

Since clay minerals are the products of the weathering of primary rocks, they are geologically recent materials found at the surface of the earth. They occur as mineralogically pure deposits and as components of many soils. The clay content of a soil contributes significantly to soil fertility and mechanical properties. Consequently, clay minerals have long been studied by the soil scientist. Much of our current understanding of the basic structures and properties of clay minerals is the result of early pioneering work by soil scientists and minerologists (1-5). It is only recently that chemists and physicists have become generally aware of the fascinating intercalation properties and guest-host chemistry of swelling clay minerals.

The object of this paper is to describe the structures and properties of swelling clays and to point out relationships to related complex layered oxides.

STRUCTURES OF LAYERED SILICATE CLAY MINERALS

Silicates constitute a very large family of minerals(6). With very rare exception, all of them contain the tetrahedral SiO_4 unit as a basic structural building block. The different families of silicate clays can be distinguished by the number of oxygens in a tetrahedral SiO_4 unit that are shared with like tetrahedra. For instance, the ortho silicates contain discreet SiO_4^{4-} anions in which none of the oxygen atoms of the tetrahedral unit are shared with other tetrahedra. Consequently, these are referred to as having zero tetrahedral connectivity and are designated Q^0-silicates. Examples of ortho silicates include an entire family of olivine minerals of the type M_2SiO_4, in which M can be any of a large number of divalent metal ions. If we were to link two SiO_4 tetrahedral units together by sharing a common bridging oxygen, the stoichiometry would be $Si_2O_7^{6-}$. Silicates with this type of Q^1 connectivity are designated pyrosilicates. An example is the scandium ore thortveitite, $Sc_2Si_2O_7$.

The sharing of two oxygens per tetrahedral unit results in the application of cyclic silicates with empirical formula SiO_3^{2-} and two chain silicates with the same empirical formula. An example of a Q^2 cyclic silicate is beryl, with the empirical formula $Be_3Al_2Si_6O_{18}$. Examples of chain silicates are enstatite, with the formula $MgSiO_3$, and diopside, $CaMg(SiO_3)_2$.

The sharing of three oxygens per SiO_4 unit generates two-dimensional sheet structures in which the empirical formula of the silicate is $Si_2O_5^{2-}$. However, examples of compounds containing single sheets are relatively rare. Single sheet structures exist in compositions of the type $KHSi_2O_5$ and $NaHSi_2O_5$, the latter example being found in nature in the form of the mineral kanemite(7). However, as we shall see shortly, the Si_2O_5 sheets with Q^3 linkages are found as components of layered silicate clay structures. The sharing of all four oxygen atoms per tetrahedral unit brings us to the framework silicates or tectosilicates. The various polytypes of silica contain such Q^4 linkages. In addition, the replacement of aluminum for silicon leads to other families such as the felspars, of which orthoclase, $KAlSi_3O_8$, and anorthite, $CaAl_2Si_2O_8$, are members. Also, the zeolites and ultramarines are familiar tectosilicates.

Let us consider in further detail the structures of layered silicates. The Q^3 Si_2O_5 sheets can be considered to be a structural unit of more complex layers in all layered silicate clays. For instance, in the clay mineral kaolinite, the layers are composed of a single Si_2O_5 sheet fused to an edge-shared, octahedral sheet of aluminum hydroxide (gibbsite). The apical oxygen atoms of the silicate sheet are commensurate with 2/3 of the hydroxyl groups of the gibbsite sheet. This commensurate relationship allows for condensation of the two sheets into a single layer consisting of three atomic planes of oxygens and hydroxyl groups. In crysotile and related asbestos minerals, the structure is similar to that of kaolinite, except that the silicic acid sheet is condensed with an octahedral sheet of magnesium hydroxide (brucite). The two sheets are not precisely commensurate, and this leads to an induced strain which causes the layer to curl on itself and adopt a fiber-like morphology. The layers found in kaolinite and crysotile are generally referred to as 1:1 layered silicate structures, in accord with the relative ratios of octahedral and tetrahedral sheets comprising the layers.

Figure 1 illustrates schematically the layered structure of kaolinite with a unit cell formula of $Al_4Si_4O_{10}(OH)_8$. In crysotile, where all of the octahedral positions are occupied by magnesium, the unit cell formula is $Mg_6Si_4O_{10}(OH)_8$. Since the structures consists of fused tetrahedral and octahedral sheets, the upper and lower atomic planes of atoms are non-equivalent. When the layers of kaolinite stack to form crystals, they stack in a head-to-tail fashion so that the hydrogen donor hydroxyl groups interact with the hydrogen acceptor oxygens of an adjacent layer. Thus, hydrogen bond interactions are important in the layer stacking of these structures.

Figure 1. A schematic illustration of the 1:1 layered silicate structure found in kaolinite. The layers consist of a single tetrahedral sheet (T) fused to an edge-shared (gibbsidic) octahedral sheet (O). The stacking of layers in crystals is head-to-tail allowing for H-bonding in the gallery region between layers.

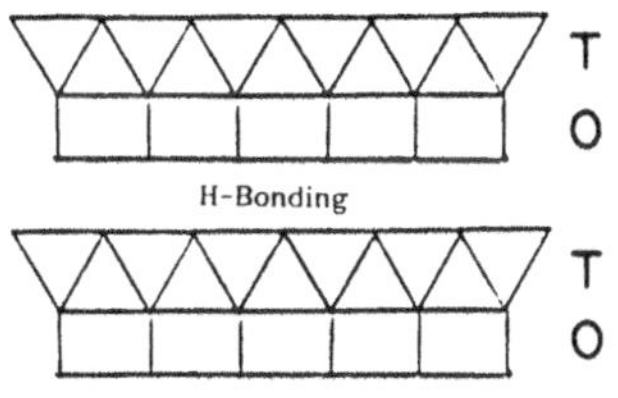

Since silicic acid sheets and edge-shared octahedral sheets are commensurate, or nearly so, it is reasonable to expect the formation of 2:1 layered structures in which two silicic acids sheets are condensed to a central octahedral sheet. Indeed, such structures are abundant and important in the family of layered silicate minerals. For instance, the condensation of two silicic acid sheets to a gisbbsitic aluminum hydroxide sheet leads to the structure found in the mineral pyrophyllite. An analogous condensation using brusitic magnesium hydroxide as the octahedral sheet gives rise to the structure of talc.

Figure 2 illustrates the 2:1 layered structure of talc. The stacking of these layers to form crystals leads to the formation of interlayers or "galleries" in which the layers are held together in van der Waals contact. Thus, talc and pyrophyllite differ from the 1:1 structures discussed earlier insofar as the layer stacking is governed by van der Waals interactions between the layers. Nevertheless, a useful structural relationship exists between the electrically neutral layers of 1:1 and 2:1 layered silicate clay mineral families.

Figure 2. A schematic illustration of the 2:1 layered structure found in talc. Note that there are two tetrahedral silicic acid sheets (T) fused to a central octahedral sheet (O). Also, the interlayer region defines a van der Waals gap.

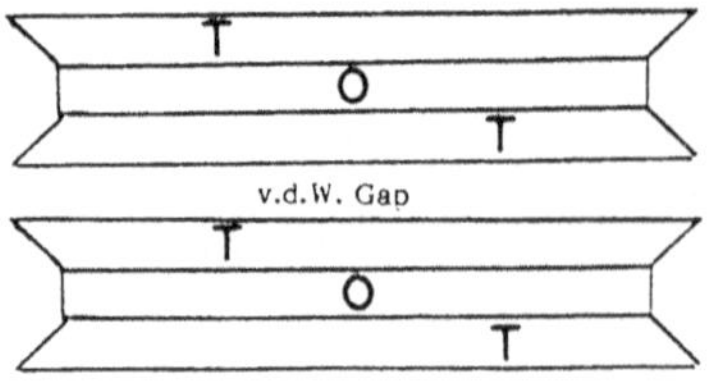

The same 2:1 oxide framework found in talc and pyrophyllite occurs also in the micas. We can think of the micas as being structurally derived from talc/pyrophyllite by replacing some of the tetrahedral silicon in the silicate sheets with trivalent aluminum ions. This isomorphous replacement of silicon by aluminum results in a positive charge deficiency and causes the layers to become two-dimensional polyoxoanions. To balance the layer charge caused by isomorphous replacement of silicon, potassium ions are incorporated in the gallery region between the layers of the 2:1 structure. Thus, in muscovite, where the unit cell formula is $K_2Al_4(Si_6Al_2)O_{20}(OH)_4$, two unit charges have been introduced per unit cell formula. In phlogopite, with a unit cell formula of $K_2Mg_6(Si_6Al_2)O_{20}(OH)_4$, the structure can be of thought of as being derived from talc as a result of replacing 1/4 of the tetrahedral silicons by aluminum. The potassium ions needed to maintain electrical neutrality are incorporated in the gallery region, as shown in Figure 3. The stoichiometry is such that the potassium ions occupy the hexagonal cavaties of the Kagome net of the layers. In muscovite, which can be thought of as being derived from pyrophyllite by replacement of 1/4 of the tetrahedral silicons by aluminum, also has potassium tightly bound to the oxygen atoms of the gallery. Thus, micas can be thought of as ionic layered compounds, or as intercalation compounds in which a permanent negative charge is fixed on the layers and potassium ions are sandwiched between the layers.

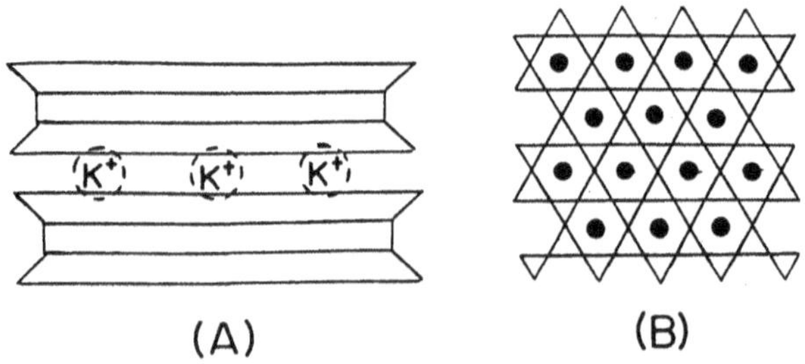

Figure 3. A schematic illustration of the structure of K^+-micas. (A) A side view showing the 2:1 structure and (B) a top view showing the position of the gallery K^+-ions (filled spheres) on the Kagome net.

There exists yet another class of 2:1 layered silicates, the swelling clays, which are ionic like the micas, but bear a substantially lower layer charge. These are the smectite clays, with charge densities in the range 0.4 to 1.2 charge units per unit cell formula. Montmorillonite, beidellite, hectorite and saponite are members of the smectite family of 2:1 layered silicates. Table 1 provides the unit cell formulas for these clays. Montmorillonite and beidellite both are dioctahedral minerals insofar as two-thirds of the interstices of the octahedral sheet are occupied by metal ions. However the former bears a layer charge due to octahedral substitution of magnesium for aluminum whereas the latter bears tetrahedral layer charge due to replacement of silicon by aluminum. Hectorite and saponite are both trioctahedral structures, with the former bearing octahedral layer charge and the latter tetrahedral charge. Note also that hectorite contains some fluorine which isomorphously replaces hydroxide in the structure. It is important to emphasize that the formulas in Table I are idealized, representative structures. True minerals contain other ions in isomorphous substitution such as iron. Also, the layer charge can vary substantially from the values illustrated in the Table.

Table I. Anhydrous Unit Cell Formulas for Typical Smectite Clay Minerals

Mineral	Unit Cell Formula
Montmorillonite	$Ca_{0.35}[Mg_{0.70}Al_{3.30}](Si_{8.0})O_{20}(OH)_4$
Beidellite	$Na_{0.7}[Al_{4.0}](Si_{7.3}Al_{0.7})O_{20}(OH)_4$
Hectorite	$Na_{0.6}[Li_{0.6}Mg_{5.4}](Si_{8.0})O_{20}(OH,F)_4$
Saponite	$Na_{0.9}[Mg_{6.0}](Si_{7.1}Al_{0.9})O_{20}(OH)_4$

LAYER CHARGE

The properties of 2:1 layered silicates depends critically on the magnitude of the layer charge. The end members in the talc-pyrophyllite family and the mica family exist as single crystals, with particle sizes that can be substantially greater than 2 microns. However, the smectite clays are always found in particle sizes less than 2 microns (clay fraction) and the crystals are turbostratic. In the latter case there is reasonable ordering due to face-to-face stacking of layers, but the orientation of the a & b axes of individual layers may be randomly oriented.

Also, unlike talc-pyrophyllite or the micas, the smectite clays are cation exchangers, with cation exchange capacities typically in the range 60-120meq per 100 grams, depending on the exact magnitude of the layered charge. The ion exchange capacity results from the replacement of the gallery ions in the native mineral. The hydrated ions found in the pristine minerals (alkali metals and alkaline earth) can be readily exchanged by simple ion exchange reaction with a very large number of hydrated cations. In addition, mono-and polynuclear transition metal complexes may be made to occupy the gallery region. There is virtually no limit to the range of cations which can be intercalated in smectite clays. Even organo cations such as alkylammonium and phosphonium ions and carbocations can be intercalated by ion exchange reaction. Perhaps the most sophisticated cations to be introduced to clay galleries are the protonated enzymes and related protein molecules.

HYDRATION AND SWELLING

A second feature which distinguishes the smectite clays from the talc/pyrophyllite and mica end members is their swellability. A large range of polar molecules can intercalate into smectite clays in multi-layer form and thereby swell the galleries to substantial heights. The guest molecules may also enter in stages, that is, one mono-layer at a time[8]. This is illustrated by the adsorption of water on sodium beidellite in Figure 4. At partial pressures from approximately 0.1 to 0.6, one mono-layer of water occupies the interlayer gallery. At partial pressures between 0.6 and 0.9, two mono-layers of water are formed. This staging phenomenon is the reverse of the staging found in graphite, insofar as two layers of guest water molecules are being accomodated between single layers of host clay. Sodium montmorillonite and hectorite can be swelled to the extent that literally hundreds of angstroms of water occupy the interlayer region and the layers are essentially exfoliated. In fact, a 3-7 wt% aqueous suspension of sodium montmorillonite will form a thixotropic gel due to the extensive exfoliation of the layers.

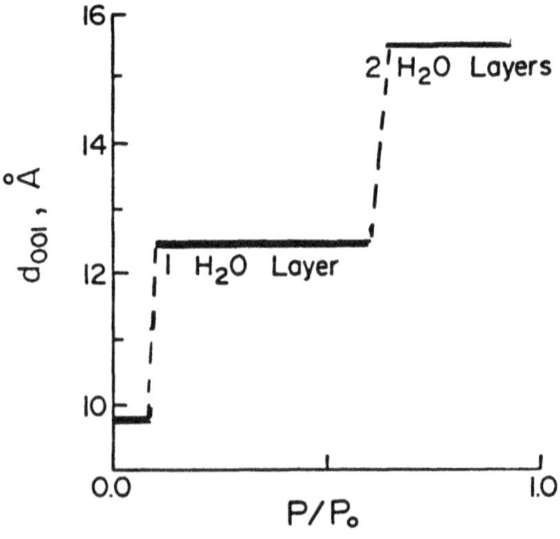

Figure 4. The step-wise expansion in the basal spacing of beidellite over the partial pressure range at 0 to 1.0.

The swelling of smectite clays, together with their ion exchange capacity, distinguishes them from all other 2:1 layer silicate structures. Swelling by water can be explained in terms of double-layer theory[9], but swelling by other molecules is not well understood. The extent of swelling is empirically observed to depend on the nature of the swelling solvent, the magnitude and origin of the layered charge, and the nature of the gallery cation. It is not possible to predict a priori the magnitude of swelling for a given solvent-clay system.

Smectite clays interlayered with hydrated, inorganic cations have been well recognized for their intrinsic Bronsted acidity[10]. The Bronsted acidity arises from the polarization and subsequent ionization of water molecules in the first coordination sphere of the intercalated ion, as illustrated in Equation 1, wherein the horizontal lines identify the intercalated species.

$$\overline{\underline{M(H_2O)_x^{n+}}} \longrightarrow \overline{\underline{M(OH)(H_2O)_{x-1}^{(n-1)+} \quad + \quad H^+}} \qquad (1)$$

The extent of ionization can be significantly greater in the gallery of the clay than in bulk aqueous solution, because of the greater polarizing ability of the partially hydrated gallery cation. The Bronsted acidity of smectite clays can contribute to the mechanisms available for the adsorption of molecules in the gallery regions of the clays. For instance, basic molecules can be adsorbed in the galleries via a protonation mechanism, where the proton is supplied by the hydrolysis of the inorganic cation. Figure 5 illustrates the uptake of three different bases, ammonia, pyridine, and urea by a series of montmorillonites exchanged with cations of differing charge-to-radius ratio. For a given gallery cation (for example, aluminum) the extent of binding of the base increases with increasing basicity (urea<pyridine<ammonia). For a given base, the extent of binding in the protonated form increases with increasing charge-to-radius ratio of the cation.

Figure 5. Protonation of ammonia, pyridine and urea in the gallery regions of montmorillonite containing exchange cations of different Pauling electronegativities. The polarizing ability, or charge-to-radius ratio, of the exchange cation increases with increasing electronegativity.

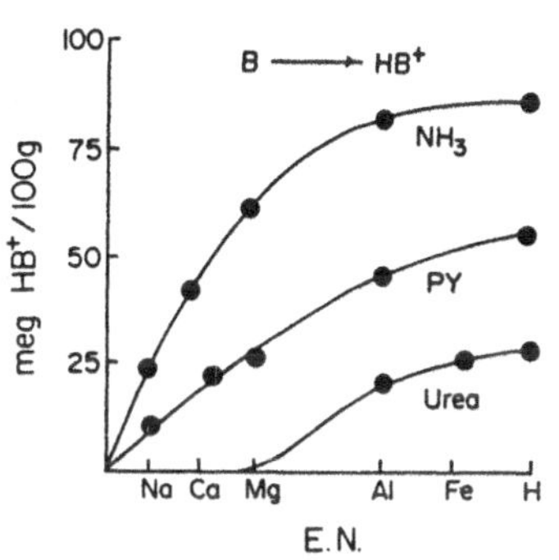

The Bronsted acidity of clays can also lead to the catalytic transformation of organic molecules on their internal and external surfaces. Such reactions are believed to be important in the soil chemistry of clays(10,11). Acid-base reactions on clay surfaces can also be used for the laboratory synthesis of a wide range of organic molecules(12-14).

The dynamics of water, solvated cations, and other molecules in the galleries of smectite clays have been investigated by magnetic resonance methods(15-23). In the limited space allowed here, only a few examples will be provided based on electron spin resonance studies of paramagnetic cations under different degrees of solvation. In Figure 6 is shown the electron spin resonance spectrum of copper (II) ions intercalated in a clay under conditions where the ion is solvated by three molecular layers of water molecules(21). When the applied magnetic field is positioned parallel to the clay layers, only the perpendicular component of the resonance is observed, indicating that the unpaired electron on copper is quantized along an axis perpendicular to the symmetry axis of the Jahn-Teller distorted ion. When the magnetic field is oriented perpendicular to the layers, the parallel resonance component is observed, split by hyperfine interaction with the I=3/2 nuclear spin of the copper nuclei. This spectral anisotropy clearly demonstrates that the C_4 symmetry axis of the tetragonally elongated $Cu(H_2O)_6^{2+}$ ion is oriented perpendicular to the silicate sheets. Analogous studies of copper ions solvated by one and two molecular layers of water, indicate that the $Cu(H_2O)_4^{2+}$ and $Cu(H_2O)_6^{2+}$ respectively adopt the orientations as shown in Figure 7.

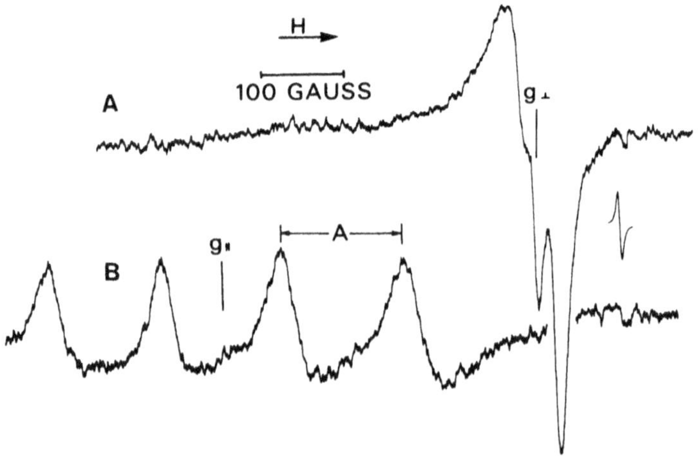

Figure 6. The electron spin resonance spectra for $Cu(H_2O)_6^{2+}$ in an oriented film sample of montmorillonite containing three molecular layers of gallery water. The spectral anisotropies observed when the applied magnetic field is oriented parallel (A) and perpendicular (B) to the silicate layers indicate that the symmetry axis of the Jahn-Teller distorted ion is orthogonal to the clay layers.

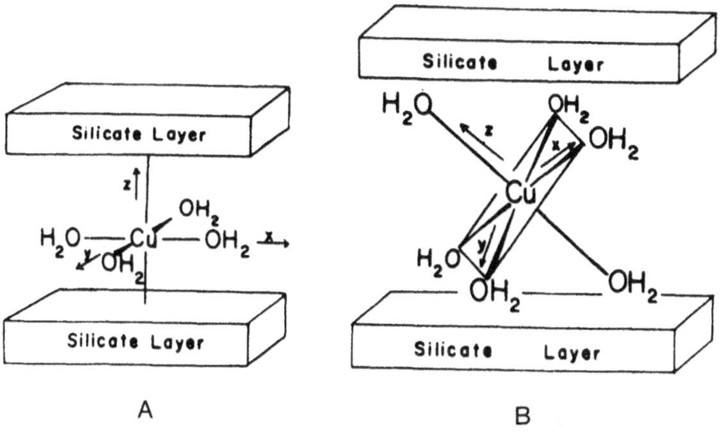

Figure 7. A schematic illustration of the orientations of the (A) $Cu(H_2O)_4^{2+}$ and (B) $Cu(H_2O)_6^{2+}$ in smectite clay galleries solvated by 1 and 2 molecular layers of water respectively. These orientations were deduced from the ESR spectra of oriented film samples.

The above electron spin resonance results for copper ions solvated with one, two, and three molecular layers of water indicate that the ion adopts highly oriented positions in the clay gallery. However, when the galleries are fully solvated with the maximum number of water molecules, then quite a different picture is provided by the electron spin results. As shown in Figure 8, when the ion is solvated in galleries which are swollen to 12Å in thickness by water, then isotropic spectra are observed independent of magnetic field orientation. This result is compatible with the presence of rapidly tumbling solvated ions in the gallery. Because of dynamic Jahn Teller effects, hydrated copper ions are not well suited for quantitative kinetic studies. Electron spin resonance line broadening studies of manganese ions in clay galleries, however, has allowed us to estimate the correlation time for tumbling[23]. Under conditions where the ion is solvated by 12Å of water, the correlation time is only about 50% larger, than the correlation time for the same ion in bulk aqueous solution. As the gallery is dehydrated, the ion becomes less and less mobile, and results analogous to those obtained for copper (II) are obtained under limited degrees of hydration.

Electron spin resonance studies have also been useful in determining the orientation and mobilities of organic cations in clay galleries. For instance, alkyl ammonium ions labeled with nitroxide spin probes, have shown that the ions are capable of undergoing solution-like motions when intercalated in fully wetted smectite clay galleries[24].

Figure 8. ESR spectra of $Cu(H_2O)_6^{2+}$ in an oriented film sample of smectite clay galleries swollen to 12Å by water. The $Cu(H_2O)_6^{2+}$ ions are solvent separated from the clay layers and tumble freely in a solution-like gallery environment.

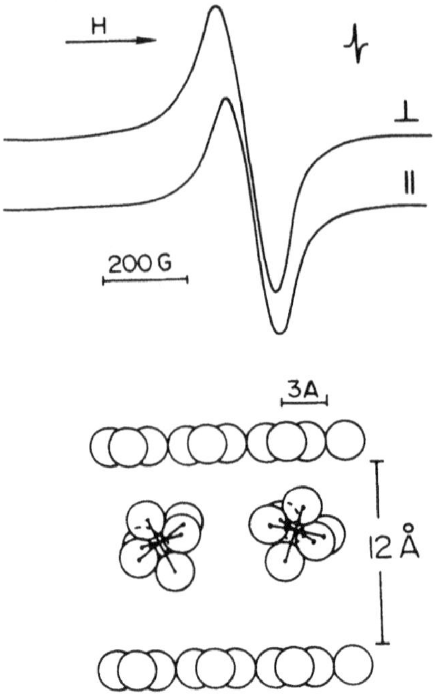

CATALYSIS

The above results for the dynamics of inorganic and organic cations in solvated clay galleries strongly implied that the galleries should be readily accessible under ordinary conditions of temperature and pressure. This prompted us to study metal complex catalyzed reactions in the gallery space of montmorillonite, which had been exchanged with rhodium phosphine complexes(25-30). Such species are known to be catalysts for the hydrogenation and hydroformylation of unsaturated carbon-carbon bonds. In an initial illustrative study, we intercalated complexes of the type RhL_2^+ where L is a triphenylphosphine ligand, and compared the reactivity of the intercalated complex with the reactivity of the complex in homogeneous solution(26). The sequence of steps involved in the hydrogenation of an olefin (ol) to an alkane is illustrated in Equation 2. An equivalent sequence of steps is involved in the conversion of alkynes to the corresponding cis-olefins with RhL_2^+ as the hydrogenation catalyst.

$$\underline{\overline{RhL_2^+}} \xrightarrow{H_2} \underline{\overline{RhH_2L_2^+}} \xrightarrow{ol} \underline{\overline{RhH_2(ol)L_2^+}} \longrightarrow$$

$$\underline{\overline{RhH(R)L_2^+}} \rightarrow \underline{\overline{RhL_2^+}} + \text{alkane} \qquad (2)$$

The rates of hydrogenation for a series of alkynes is shown in Table II(26). For a given swelling solvent (for example, methanol) the rate of the intercalated catalyst, relative to the homogeneous catalyst (R_I/R_H) decreases dramatically. These results imply that the size of the substrate dramatically influences reactivity under intercalated conditions. The importance of substrate size is further demonstrated in a complimentary experiment using 2-decyne as the substrate solvents of different swelling capacity. For instance, as the solvent is changed from methylene chloride (gallery height 10.0Å) to benzene (gallery height 5.7Å) the rate of reaction for the intercalated catalyst falls off regularly from a value of 0.85 to a value of less than 0.02.

Table II. Reduction of Alkynes to Cis Olefins by a Clay-Intercalated RhL_2^+ Complex

Substrate	Swelling Solvent	Gallery Height(Å)	R_I/R_H[a]
1-hexyne	MeOH	7.7	1.0
2-hexyne	MeOH	7.7	0.92
2-decyne	MeOH	7.7	0.48
3-decyne	MeOH	7.7	0.20
$C_6H_5CCC_6H_5$	MeOH	7.7	<0.01
2-decyne	Ch_2Cl_2	10.0	0.85
	MeOH	7.7	0.43
	Et_2O/MeOH	6.7	0.24
	C_6H_6	5.7	0.02

a/Reaction rate under clay-intercalated conditions relative to reaction rate in homogeneous solution.

The spatial dependence for the reactivity of the intercalated catalyst has been explained in terms of specific orientations for the transition states in the clay galleries[26]. For the substrate to react effectively with the $RhH_2L_2^+$ complex, the unsaturated carbon-carbon bond axis must be orthogonal to the trans LRhL axis. As is illustrated in Figure 9A, if the LRhL axis is parallel to the silicate sheets, then the substrate carbon-carbon unsaturated bond axis must be oriented orthogonally to the silicate sheets. If the gallery height (Δd_{001}) is less than the critical dimension for orienting the substrate in a reactive position, then reaction in the gallery will be inhibited. On the other hand, if the gallery height exceeds the critical dimension for reactivity, then reaction will be facile. As shown in Figure 9B, the critical dimension for 2-decyne is 7.7Å. The reactivity of this molecule is high in the intercalated state when the gallery height exceeds 7.7Å, but relatively low when the gallery height falls below this value (cf.,Table 2).

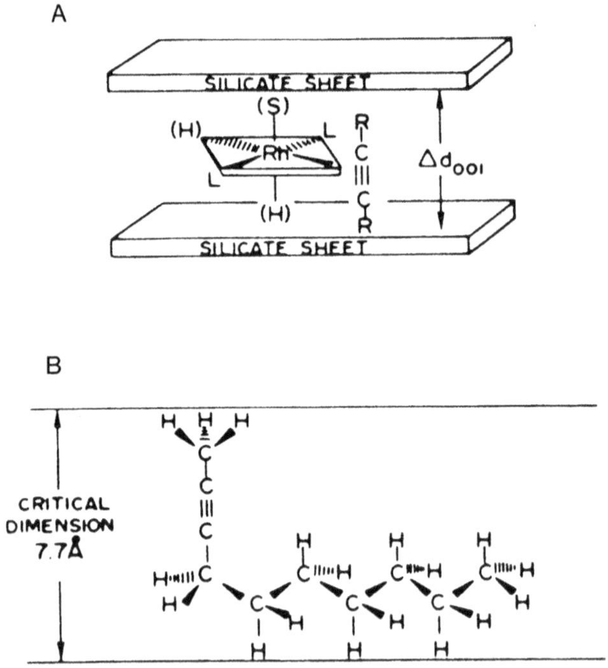

Figure 9. (A) The proposed orientation for $RhH_2L_2^+$-alkyne transition state in the gallery of a smectite clay. S represents a coordinated solvent molecule and Δd_{001} represents the gallery height (7.7A) when S is methanol. (B) The critical dimension needed to orient the triple bond in 2-decyne orthogonal to the clay layers.

PILLARING

The above magnetic resonance results and catalytic results demonstrate that the galleries of smectite clays can be quite solution-like when swollen with multiple layers of solvent. This also explains why it is possible to access the gallery region and to carry out solution-like catalytic reactions in the anisotropic intracrystal environment. However, the use of a solvent in heterogeneous catalysis is often not desireable because the solvent must be separated from the reaction products. To circumvent the limitations imposed by swelling solvents, current research efforts are focusing on the design of a new family of intercalated clay derivatives known as "pillared" clays.

Pillared clays are smectite derivatives which have been exchanged with large robust cations capable of acting as molecular props or pillars between the layers(31). Since the charge density of smectite clays is relatively low, the separation between gallery exchange ions is sufficiently large and the host layers are sufficiently rigid to allow for the existence of free void volume between the cations. This void volume then becomes available for adsorption and possible catalytic reaction of substrates sufficiently small to penetrate the intracrystalline pores.

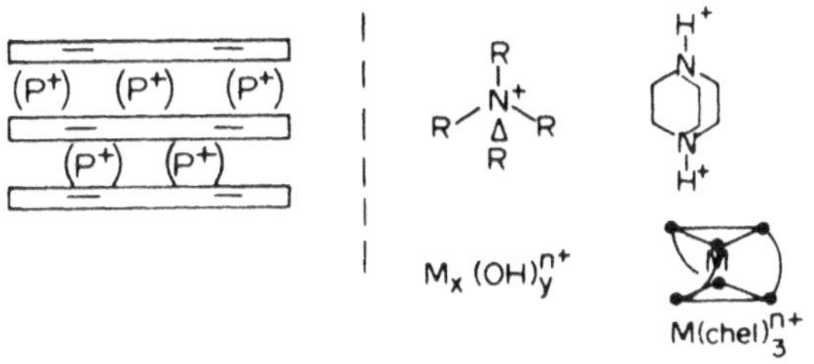

Figure 10. Various ions have been used as molecular propants or pillaring agents (P+) in smectite clay galleries. These include tetraalkylammonium ions, bicyclic amine cations, tris-metal chelates, and polyoxo metal cations.

The concept of pillaring clays extends back to the early work of R.M. Barrer, when he demonstrated that permanent microporosity could be introduced in montmorillonite by intercalation of tetraalkyl ammonium ions(32). As shown in Figure 10, several other robust cations have been utilized as pillars in smectite clay chemistry. These include bicyclic amine cations(33,34), tris metal chelates(35-37), and polyoxometal cations(30,38-44). Among the various pillaring agents which have been used, the derivatives pillared by polyoxo cations have received the greatest attention. When these derivatives are heated above their dehydration and dehydroxylation temperature, they generally form small metal oxide aggregates in the gallery. This imparts good thermal stabilities up to temperatures approaching 600 C. In contrast, organic pillaring agents generally decompose due to rupture of the carbon-carbon bond at temperatures below 350 C. These metal oxide pillared clays exhibit surface areas in the range 150-450m^2.

Alumina pillared clays have been developed by three independent groups(38-40). These derivatives are prepared by intercalative ion exchange of Al_{13} oligocations and their subsequent thermal conversion to oxide pillars, as illustrated in Equation 3. Alumina intercalated smectites exhibit gallery heights near 8Å. Moreover, they function as solid acids for a wide range of chemical conversions(41, 44-50), the most important of which is the cracking of petroleum to useful fuels(45-47). Since the microporosity of a pillared clay can exceed the porosity of a conventional three-dimensional zeolite, there is considerable promise that pillared clays may be tailored to refine high molecular weight low-grade petroleum fractions.

$$Na^{+} + Al_{13}O_4(OH)_{24+x}(H_2O)_{12-x}^{(7-x)+} \xrightarrow{-Na^{+}}$$

$$Al_{13}O_4(OH)_{24+x}(H_2O)_{12-x}^{(7-x)+} \xrightarrow{-H_2O} Al_2O_3 + H^{+} \qquad (3)$$

Most of the void volume of an alumina pillared clay is found in the microporous region where the pore radius is less than 25Å. The narrowness of the pore distribution arises because most of the clay layers are aggregated face-to-face with the oxide aggregates propping open the gallery regions. A schematic illustration of a pillared clay aggregate is shown in Figure 11A. Some defects are introduced due to layer folding and irregular platelet sizes. Nevertheless, the bulk of the void volume is in the microporous range.

We have recently developed a new family of microporous materials known as delaminated clays(51). These are metal oxide derivatives in which edge-to-face layer aggregation competes with face-to-face layer aggregation. The edge-to-face aggregation of layers is facilitated by very small particle size (<500Å) or by clays with lathe-like layer morphology as in hectorite. Some face-to-face aggregation allows for the presence of micropores as in conventional pillared clays, however the edge-to-face aggregation mechanism induces a card house structure similar to that originally proposed by Van Olphen for clay gels. The competitive edge-to-face and face-to-face aggregation mechanisms in a delaminated clay are qualitatively illustrated in Figure 11B. The extent of face-to-face aggregation is apparently limited to domains less than 70Å because these materials are typically X-ray amorphous. In contrast, a pillared clay will exhibit one to several orders of 00 X-ray reflections.

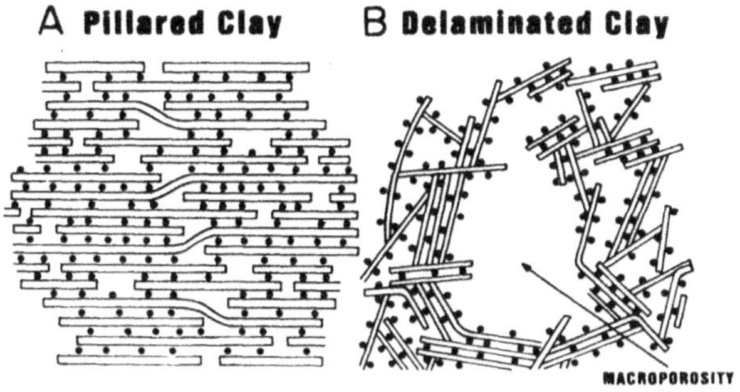

Figure 11. A schematic illustration of layer aggregation in (A) pillared clay in which the layers are stacked primarily face-to-face and (B) delaminated clay in which edge-to-face aggregation competes with face-to-face aggregation. The slabs represent the clay layers and the dark circles represent the robust pillaring cation. The latter aggregation model introduces macroporosity in addition to the microporosity typical of pillared clays.

Current interest in pillared and delaminated clays continue to focus on their utility as shape selective, heterogeneous catalysts. Their properties as petroleum cracking catalysts continues to be of major interest. Figure 12 illustrates the gasoline yields obtained with pillared and delaminated aluminum oxide derivatives[52,53]. Included in the figure for comparison purposes, are the gasoline yields obtained with an amorphous silica alumina catalyst. The pillared and delaminated clays are substantially larger than those afforded by the amorphous oxide. The shape selective performance of pillared and delaminated clays is competitive with zeolite catalysts in current commercial use.

The synthesis and structural characterization of pillared and delaminated clays are areas of current research activity. Recent MAS-NMR studies have demonstrated that a crosslinking reaction occurs between the pillaring metal oxide and the layers of certain smectite clays, such as beidellite and fluorohectorite[54,55]. These reactions lead to a structural rearrangement of the host layers. On the other hand, the pillaring of montmorillonite at temperatures below 500 C, involve clay layer pillar interactions which do not rearrange the layer structure[54-56]. Additional studies are needed to more fully understand the pillaring mechanisms, as well as the origin of the acidity which makes these materials useful as heterogeneous catalysts. In addition, experiments which will allow us to quantify the transverse layer rigidity of smectite clay layers(57), should help us to understand and to predict the limits of microporosity which can be designed into these systems and related complex layered oxides.

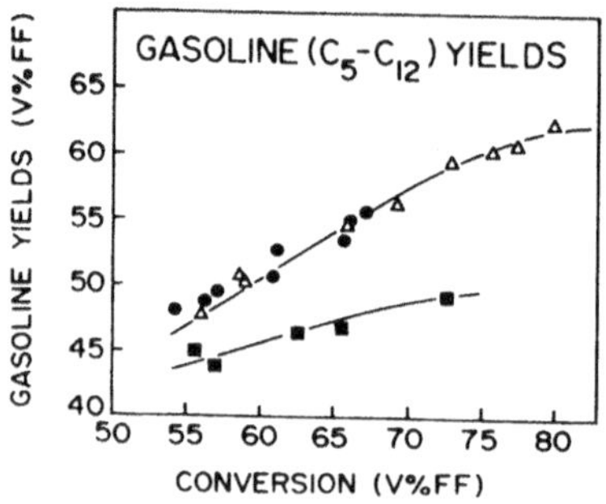

Figure 12. The shape selectivity toward gasoline yields in the catalytic cracking of petroleum by alumina pillared clay (triangles) and an alumina delaminated clay (dark circles). The gasoline yields increase with increasing conversion of fresh feed (FF). Included for comparison purposes are the yields observed with an amorphous aluminosilicate catalyst (dark squares).

ACKNOWLEDGMENT - The support of the National Science Foundation through the Division of Materials Research (Grant DMR-8514154) and the Michigan State University Center for Fundamental Materials Research is gratefully acknowledged.

REFERENCES

1. R.E. Grim, "Clay Mineralogy", McGraw-Hill, New York, 2nd Ed., **1968**.

2. "Crystal Structures of Clay Minerals and their X-Ray Identification", G.W. Brindley and G. Brown, Eds., Mineralogical Society, London, **1980**.

3. B.K.G. Theng, "The Chemistry of Clay-Organic Reactions", John Wiley and Sons, New York, **1974**.

4. "Minerals in the Soil Environment", J.B. Dixon and S.B. Weed, Eds., Soil Science Soc. Amer., Madison, **1977**.

5. "Chemistry of Clays and Clay Minerals", A.C.D. Newman, Ed., Mineralogical Society, London, **1987**.

6. A.F. Wells "Structural Inorganic Chemistry", 4th Ed., Oxford University Press, London, **1975**.

7. F. Liebau, "Structural Chemistry of Silicates: Structure, Bonding and Classification", Springer-Verlag, New York, **1985**.

8. R. Glaeser and J. Mering, C.r. hebd. Seanc. Acad. Sci. (Paris) **1968**, 267, 436.

9. H.Van Olphen "An Introduction to Clay Colloid Chemistry", 2nd Ed., Wiley, New York, **1977**.

10. M.M. Mortland, 9th Inter. Cong. Soil Sci. Trans. **1968**, 1, 691.

11. G.W. Bailey and J.L. White, Residue Rev. **1970**, 32, 29.

12. J.M. Thomas in "Intercalation Chemistry", M.S. Whittingham and A.J. Jacobson, Eds., Academic Press, New York, **1982**, Chapter 3.

13. J.A. Ballantine in "Chemical Reactions in Organic and Inorganic Constrained Systems", R. Setton, Ed., Reidel, Boston **1986**, 197-212.

14. P. Laszlo, Science **1987**, 235, 1473.

15. J.J. Fripiat, in "Advanced Chemical Methods for Soil and Clay Mineral Research", J.W. Stucki and W.L. Banwart, Eds., Reidel, Dordrecht, Netherlands **1980**, 245-316.

16. D.E. Woessner, J. Magn. Res. **1980**, 39, 297.

17. T.J. Pinnavaia, in "Advanced Techniques for Clay Mineral Analysis", J.J. Fripiat, Ed., Elsevier, New York **1981**, 139-161.

18. M.B. McBride, in "Advanced Chemical Methods for Soil and Clay Mineral Research", J.W. Stucki and W.L. Banwart, Eds., Reidel, Dordrecht **1980**, 423-450.

19. P.L. Hall, in "Advanced Chemical Methods for Soil and Clay Mineral Research", J.J. Stucki, Ed., 2nd Ed., Reidel, Dordrecht **1987**.

20. P.M. Clementz, T.J. Pinnavaia, and M.M. Mortland, J. Phys. Chem. **1973**, 77, 196.

21. M.B. McBride, T.J. Pinnavaia, and M.M. Mortland, J. Phys. Chem. **1975**, 79, 2430.

22. M.B. McBride, T.J. Pinnavaia and M.M. Mortland, Clays Clay Miner. **1975**, 23, 162.

23. M.B. McBride, T.J. Pinnavaia, and M.M. Mortland, AM. Mineral. **1975**, 60, 66.

24. M.B. McBride, in "Advanced Chemical Methods for Soil and Clay Mineral Research", J.W. Stucki, and W.L. Banward, Eds., Reidel, Dordrecht **1980**, 423-450.

25. T.J. Pinnavaia and P.K. Welty, J. Amer. Chem. Soc. **1975**, 97, 3819.

26. T.J. Pinnavaia, R. Raythatha, J.G.-S. Lee, L.J. Halloran, and J.F. Hoffman, J. Amer. Chem. Soc. **1979**, 101, 6891.

27. W.H. Quayle and T.J. Pinnavaia, Inorg. Chem. **1979**, 18, 2840.

28. R. Raythatha and T.J. Pinnavaia, J. Organometal. Chem. **1981**, 218, 115.

29. R. Raythatha and T.J. Pinnavaia, J. Catal. **1983**, 80, 47.

30. F. Farzaneh and T.J. Pinnavaia, Inorg. Chem. **1983**, 22, 2216.

31. T.J. Pinnavaia, Science **1983**, 220, 365.

32. R.M. Barrer, "Zeolites and Clay Minerals as Sorbents and Molecular Sieves", Academic Press, New York **1978**, 407-483.

33. J. Shabtai, N. Frydman, and R. Lazar, Proc. 6th Int. Congr. Catal. **1976**, B5, 1.

34. M.M. Mortland and V.E. Berkheiser, Clays Clay Miner. **1976**, 24, 60.

35. V.E. Berkheiser and M.M. Mortland, Clays Clay Miner. **1977**, 25, 105.

36. M.F. Traynor, M.M. Mortland, and T.J. Pinnavaia, Clays Clay Miner. **1978**, 26, 319.

37. M.I. Knudson and J.L. McAtee, Clays Clay Miner. **1973**, 21, 19.

38. D.E.W. Vaughan and R.J. Lussier, Preprints 5th Int. Conf. Zeol., Naples, June, **1980**.

39. N. Lahav, U. Shani, J. Shabtai, Clays Clay Miner. **1978**, 26, 107.

40. G.W. Brindley and R.E. Sempels, Clays Clay Miner. **1977**, 12, 229.

41. T.J. Pinnavaia, M.S. Tzou, and S.D. Landau, J. Amer. Chem. Soc. **1985**, 107, 2783.

42. J. Sterte, Clays Clay Miner. **1986**, 34, 658.

43. S.P. Christiano, J. Wang, T.J. Pinnavaia, Inorg. Chem. **1985**, 24, 1222.

44. J. P. Rupert, W.T. Granquist and T.J. Pinnavaia, in "Chemistry of Clays and Clay Minerals", A.C.D. Newman, Ed., Longman, London, **1987**, 275-318.

45. R.J. Lussier, J.S. Magee and D.E.W. Vaughan, Preprints, 7th Canadian Sym. Catal. Edmonton, October, **1980**.

46. M.L. Occelli, Ind. Eng. Chem. Prod. Res. Dev. **1983**, 22, 553.

47. J. Shabtai, R. Lazar, and A.G. Oblad, Stud. Surf. Sci. Catal. **1981**, 7, 828.

48. E. Kibuchi, T. Matsuda, J. Ueda, Y. Morita, Appl. Catal. **1985**, 16, 401.

49. M. Occelli, J.T. Hsu, L.G. Galya, J. Molec. Catal. **1985**, 33, 371.

50. J.A. Ballantine, in "Chemical Reactions in Organic and Inorganic Constrained Systems", R. Setton, Ed., D. Reidel, New York, **1986**, 197-212.

51. T.J. Pinnavaia, M.S. Tzou, S.D. Landau and R.H. Raythatha, J. Mol. Catal. **1984**, 27, 195.

52. M.L. Occelli, S.D. Landau, and T.J. Pinnavaia, J. Catal. **1984**, 90, 256.

53. M.L. Occelli, S.D. Landau, and T.J. Pinnavaia, J. Catal. **1987**, 104, 331.

54. D. Plee, A. Schutz, G. Poncelet, J. J. Fripiat, Stud. Surf. Sci. Catal. **1985**, 20, 343.

55. T.J. Pinnavaia, S.D. Landau, M.S. Tzou, I.D. Johnson, J. Amer. Chem. Soc. **1985**, 107, 7222.

56. D.T.B. Tennakoon, W. Jones, J.M. Thomas, J. Chem for Faraday Trans. I **1986**, 82, 3081.

57. B.R. York, S.A. Solin, N. Wada, R.H. Raythatha, I.D. Johnson, T.J. Pinnavaia, Solid State Commun. **1985**, 54, 475.

ELECTRONIC AND TRANSPORT PROPERTIES OF HIGHLY CONDUCTING POLYMERS

J.L. Brédas [(a)]

Laboratoire de Chimie Théorique Appliquée
Centre de Recherches sur les Matériaux Avancés
Facultés Universitaires Notre-Dame de la Paix
B-5000 Namur (Belgium)

INTRODUCTION

Conducting polymers! If, by the term "conducting", we mean a material displaying an electrical conductivity at the metallic level, i.e. a conductivity typically larger than 100 S/cm at room temperature, the possibility of labelling in this way an organic polymer was considered until the late sixties to be very remote. In the past fifteeen years, however, exciting discoveries have triggered a revolution in our knowledge and perception of the organic solid state. Nowadays, there exists a large number of electrically conducting polymers [1] and the steady improvement of their mechanical characteristics, environmental stability, and solubility properties opens up interesting technological applications. Conducting polymers can indeed offer to combine in a single material the electrical properties of metals with the plasticity, light weight, low cost, and synthetic engineering feasibility of polymers. Furthermore, in their pristine state, these polymers have been recently demonstrated to possess very promising nonlinear optical characteristics.

In this contribution, we mainly intend to describe from a chemical physics standpoint, the models that have been proposed to explain the mechanisms by which an intrinsically insulating organic polymer can be made highly conducting. We first review the key discoveries which are relevant to a discussion of these mechanisms. We then discuss the chemical and physical nature of charged defects such as solitons, polarons, and bipolarons, and explain their formation on conjugated polymer chains upon doping. We contrast the charge generation process proposed for the conducting organic polymers with those in traditional metals or doped inorganic semiconductors such as silicon or germanium. We describe the role played by the charged defects in the electronic and transport properties of doped conjugated polymers. Systems discussed in more detail include polyacetylene, polypyrrole, and polyaniline.

THE DEVELOPMENT OF CONDUCTING POLYMERS

A key discovery in the development of conducting polymers was the finding in 1973 that the inorganic polymer polysulfur nitride, $(SN)_x$, is a metal [2]. The room temperature conductivity of $(SN)_x$ is about 10^3 S/cm, to be compared with ≈ 6.10^5 S/cm for copper and ≈ 10^{-14} S/cm for polyethylene. $(SN)_x$ becomes even superconducting below a critical temperature of 0.3 K [3] and is still the only polymeric material known to date to possess that property. These discoveries were of particular importance because they provided an existence theorem for highly conducting polymers and stimulated the vast amount of work required to develop other polymeric conductors. The metallic character of $(SN)_x$ is an intrinsic property of the material due to the presence of an odd number of electrons per unit cell. As a result, there is no forbidden gap between the highest occupied and the lowest unoccupied electronic levels. Unlike $(SN)_x$, most other polymers have an even number of electrons per unit cell (closed-shell electronic configuration) and are dielectric.

$(SN)_x$ provided the field with further insight that eventually led to the discovery of an entirely new class of conducting polymers. In 1976-1977, it was observed that the room-temperature conductivity of $(SN)_x$ can be enhanced by an order of magnitude following exposure to bromine and similar oxidizing agents [4]. The conducting entity is no longer a neutral polymer but a polymeric cation, charge neutrality being preserved by incorporating into the material the reduced form of the oxidizing agent.

The major breakthrough in the area of conducting plastics occurred later in 1977 when the same redox chemistry was applied to an intrinsically insulating organic polymer, polyacetylene. It was discovered that polyacetylene, which has an intrinsic conductivity much lower than 10^{-5} S/cm, can be made highly conducting by exposing it to oxidizing or reducing agents [5]. The highest conductivities currently achieved reach values of the order of 10^{+5} S/cm [6]. The oxidizing or reducing process is often referred to in the physics litterature as "doping", by analogy with the doping of inorganic semiconductors. As we will show later, this is, however, a rather misleading analogy and the process is best viewed as a redox reaction. The insulating neutral polymer is converted into an ionic complex consisting of a polymeric cation (or anion) and a counterion which the reduced form of the oxidizing agent (or the oxidized form of the reducing agent). The intercalation process of the counterions between the polymer chains can sometimes induce significant structural modifications, as detailed elsewhere [7].

In solid-state physics terminology, the use of an oxidizing agent corresponds to p-type doping, and that of a reducing agent to n-type doping. An important criterion in selecting potentially conducting polymers is therefore the ease with which the system can be oxidized or reduced. This accounts in part for the choice of π-bonded unsaturated polymers which, like polyacetylene, have small ionization potentials (usually lower than about 6 eV in the solid state) and/or large electron affinities (generally larger than about 2 eV). Electrons of π character can be relatively easily removed or added to form a polymeric ion without much disruption of the σ bonds which are primarily responsible to hold the polymer together. In the case of saturated polymers such as polyethylene or polypropylene, for example the oxidation process would require higher energy and remove σ electrons, leading to σ bond breaking and chemical decomposition. Even in the case of conjugated polymers, it is important to choose quite mild oxidizing agents which do not cause any disruptive chemistry.

Following polyacetylene, this basic principle has been applied with success to an increasingly large number of other organic polymers [1]. The structures of some of the principal conducting polymers are displayed in Figure 1. The maximum dopant concentrations are typically of the order of several mole percent per repeat unit of the chain backbone. For instance, concentrations up to 20% are obtained in the case of polyacetylene doped by iodine and even larger than 50% in the case of polyparaphenylene doped by sodium.

In some cases, oxidation or reduction of the polymer can be achieved electrochemically by subjecting the neutral polymer to the appropriate oxidizing or reducing voltage in an electrochemical cell. The charge appearing on the polymer chain is then neutralized by a counterion from the electrolyte solution. This process constitutes the basis of the major efforts devoted worldwide to the development of conducting polymer batteries. An interesting group of conducting polymers consists of those prepared by the electrochemical oxidation and simultaneous polymerization of some monomer which reacts at the anode of the electrochemical cell. This group includes polypyrrole, polythiophene, and their derivatives [1].

RELATIONSHIP BETWEEN GEOMETRIC AND ELECTRONIC STRUCTURES IN CONJUGATED POLYMERS

Polyacetylene in its all-trans conformation is remarkable among conjugated polymers because it possesses a degenerate ground state, i.e. two geometric structures corresponding exactly to the same total energy in the case of the infinite chain [8]. This peculiarity, which has interesting consequences, is ilustrated in Figure 2. As we can observe, the two structures differ from one another by the exchange of the carbon-carbon single and double bonds. The total energy of the system takes the form of a double well potential as a function of the degree of bond-length alternation Δr (i.e. the difference between the length of a single and a double bond). Experimentally, the optimal degree of bond-length alternation Δr_0 is ≈ 0.08 Å [9].

Su, Schrieffer, and Heeger [8] have pointed out in 1979 that in such a degenerate ground state polymer, there can occur excitations of the system that lead from one potential well to another well <u>of the same energy</u> (from well A to well B or vice-versa in Figure 2) [10]. The geometric structure corresponding to such an excitation on a trans-polyacetylene chain is depicted in Figure 3. The excitation takes the form of a boundary between a segment of the chain adopting the geometric structure corresponding to one of the potential wells and a segment corresponding to the other well. Analogy can be made with a magnetic domain wall marking the separation between two Bloch domains in a ferromagnetic material [11]. In physics terminology, such an excitation associated with a boundary or domain wall is called <u>soliton</u>, because it has the properties of a solitary wave which can propagate without deformation and dissipation [10]. The soliton shown in Figure 3 corresponds to an unpaired π electron, a radical, which can be present in trans-polyacetylene when a chain contains an odd number of conjugated carbons. Such a soliton is thus a neutral species. In a long chain, it can be shown that the spin density corresponding to the neutral soliton is not localized on one carbon but is spread over several ($\approx$ 10-15) carbons [8, 12], which gives the soliton a width. Starting from one side of the soliton, the double bonds become <u>gradually</u> longer and the single bonds shorter, so that arriving at the other side, the bond-length alternation has completely reversed. Such an evolution implies that the bond-lengths do equalize in the center of the soliton.

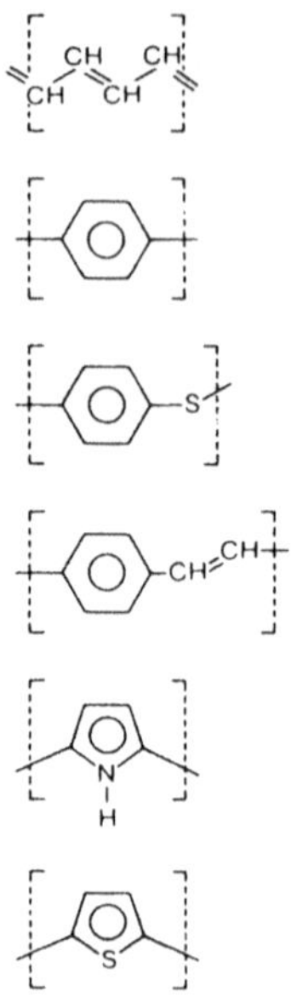

Figure 1: Structure of some of the more important conducting polymers. From top to bottom: trans-polyacetylene, polyparaphenylene, polyparaphenylene sulfide, polyparaphenylene vinylene, polypyrrole, polythiophene.

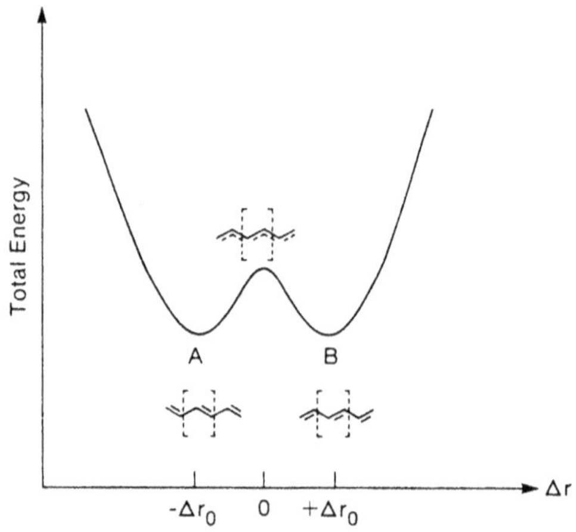

Figure 2: Total energy curve for an infinite trans-polyacetylene chain as a function of the degree of bond-length alternation Δr.

The presence of a soliton leads to the appearance of a localized (non-bonding) electronic level at mid-gap, see Figure 3. This level is half-occupied in the case of a neutral soliton (radical). As its associated electronic level is localized at the Fermi level, the neutral soliton can easily be ionized to produce either a positively charged soliton (in which case the soliton electronic level is empty) or a negatively charged soliton (the soliton level is then doubly occupied). It is interesting to realize that the soliton has unusual spin-charge relationships: a neutral soliton is a radical with spin 1/2 while a charged soliton is spinless [8].

Note that, within one-electron models, a trans-polyacetylene chain where all carbon-carbon bonds are equal is predicted to be metallic. Due to Peierls (pseudo Jahn-Teller) distortion effects, such a chain is, however, unstable towards bond-length alternation that leads to the appearance of a bandgap and semiconducting behavior. As we have pointed out above, bond-length equalization occurs in the middle of a soliton. The soliton can therefore be viewed as a kind of metallic island on an otherwise semiconducting chain.

Unlike trans-polyacetylene, compounds such as polyparaphenylene, polypyrrole, or polythiophene possess a nondegenerate ground state since their ground state corresponds to a single geometric structure which, in their case, is aromatic-like. A quinoid-like resonance structure can be envisioned but has a higher total energy (Figure 4). As a result, no soliton excitation can be produced and an isolated radical would tend to localize at a chain end.

The soliton example is illustrative of the intimate connection that exists in conjugated polymers between the geometric and electronic structures. This connection is representative of a strong electron-phonon coupling and will be further developed in the next section.

NATURE OF THE CHARGES APPEARING ON THE POLYMER CHAINS UPON DOPING

All the organic polymers we have discussed so far possess a bandgap larger than 1.4 eV, so that these materials are intrinsically semiconducting or insulating and require some "doping" process to achieve high conductivity. Initially, the conductivity increase observed upon doping was thought to result from the formation of unfilled electronic bands. It was simply assumed that upon p-type or n-type doping, electrons were, respectively, removed from the top of the valence band or added to the bottom of the conduction band, in analogy with the mechanism of generation of charge carriers in doped inorganic semiconductors. This assumption was, however, quickly challenged by the discovery that conducting polymers can display conductivity which does not seem to be associated with unpaired electrons (spin 1/2) but rather with spinless charge carriers.

To understand the origin of such a behavior, we have to address the following question: What happens on the polymer chain when it is oxidized or reduced, or in other words, when the chain becomes ionized by removing or adding an electron? (Note that in the rest of this contribution, we will mainly consider the process of removing electrons from the system, i.e. oxidation or p-type doping, but the concepts that will be developed apply equally well to reduction or n-type doping). To answer this question, it is informative to recall the situation in organic molecules.

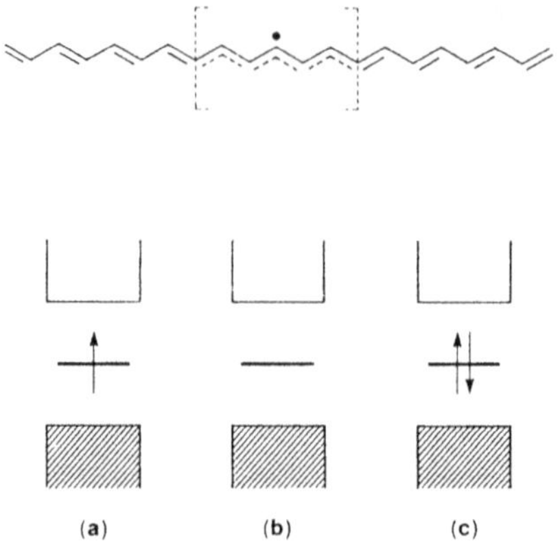

Figure 3: Illustration of the geometric structure of a (neutral) soliton on a trans-polyacetylene chain. At the bottom are displayed the schematic band structures of a trans-polyacetylene chain containing: (a) a neutral soliton; (b) a positively charged soliton; and (c) a negatively charged soliton.

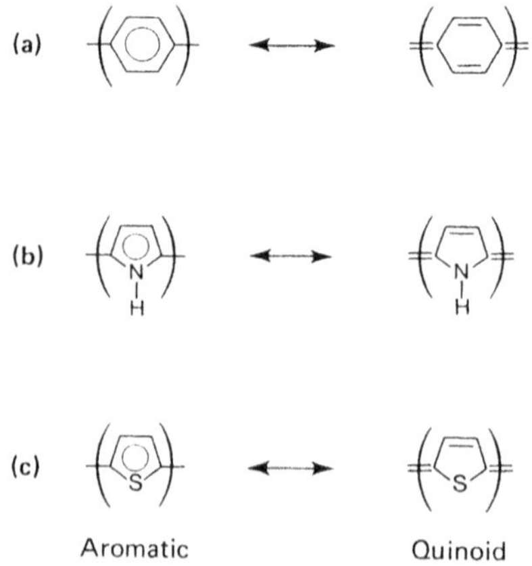

Figure 4: Aromatic (ground-state) and quinoidic geometric structures for: (a) polyparaphenylene; (b) polypyrrole; and (c) polythiophene.

In an organic molecule, it is usually the case that the equilibrium geometry in the ionized state is different from that in the ground state (e.g. the geometry of neutral biphenyl is benzenoid-like but becomes quinoid-like in the ionized state). We can depict the different energy terms relevant to the ionization process as sketched in Figure 5. A vertical, Franck-Condon-like ionization process costs an energy E_{IP-v}. If a geometry relaxation takes place in the ionized state, we gain back a relaxation energy E_{rel}. Conceptually, going from the minimum of the ground state total energy curve to the minimum of the ionized state curve, can also be viewed in the following way. First, the geometry of the molecule is distorted in the ground state in such a way that the molecule adopts the equilibrium geometry of the ionized state. This costs a distortion (elastic) energy E_{dis}. If we now turn our attention to the one-electron energy levels of the molecule, this distortion usually leads to an upward shift $\Delta\varepsilon$ of the highest occupied molecular orbital (HOMO) and a downward shift of the lowest unoccupied molecular orbital (LUMO), as illustrated in Figure 6. If we then proceed to the ionization of the distorted molecule, it requires an energy E_{IP-d}.

From Figure 5, we can observe that it is energetically favorable to have a geometry relaxation in the ionized state as soon as the quantity E_{IP-v}–E_{IP-d} (which actually corresponds to $\Delta\varepsilon$, as can be inferred from Figure 6) is larger than the distortion energy E_{dis}; or, in other words, when the reduction in ionization energy $\Delta\varepsilon$ upon distortion is larger than the energy E_{dis} required to make the distortion.

In a polymer, a vertical ionization (e.g. charge extraction) process E_{IP-v} results in creating a hole on top of the valence band, see Figure 7a. In this case, the positive charge on the chain becomes fully delocalized over the whole polymer chain and there is no geometry relaxation (by the very definition of the process). The presence of a hole (unfilled level) on top of the valence band leads, as we discussed above, to the appearance of a metallic character, a situation which would correspond to the initial assumption made about the conduction mechanism in doped organic polymers.

However, in an organic polymer chain, it can be energetically favorable to have the charge that appears on the chain localized and to have, around it, a local distortion (relaxation) of the lattice. This process causes the presence of localized electronic states in the gap due to a local upward shift $\Delta\varepsilon$ of the HOMO and a local downward shift of the LUMO (Figure 7b). Considering the case of oxidation, we lower the ionization energy by an amount $\Delta\varepsilon$. If $\Delta\varepsilon$ is larger than the energy E_{dis} necessary to distort the lattice locally around the charge, this localization process is favorable. We then obtain the formation of what condensed-matter physicists call a polaron [13]. In chemical terminology, a polaron is a radical-ion (spin 1/2) associated with a lattice distortion and the presence of localized electronic states in the gap referred to as polaron states. The ability of a charge to significantly deform the lattice around it, is another manifestation of the strong electron-phonon coupling that exists in organic polymers. It must be stressed that in the case of polaron formation, the valence band remains full and the conduction band empty. There is no appearance of any metallic character since the half-occupied electronic level is localized in the gap.

Next, we consider what happens when a second electron is removed from the polymer chain. Is it more favorable to have the two charges separated, in which case we have the formation of two polarons, or to have them coupled, in which case we have bipolaron formation [13, 14]. A bipolaron is defined as a pair of like charges (here a dication) associated with a strong local lattice distortion. The bipolaron can be viewed also as being the direct space

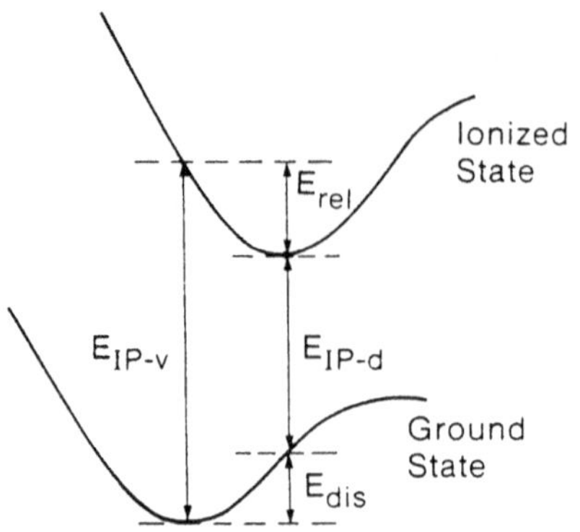

Figure 5: Illustration of the energies involved in an ionization process. The x-axis refers to deformation coordinates and the y-axis, to total energy.

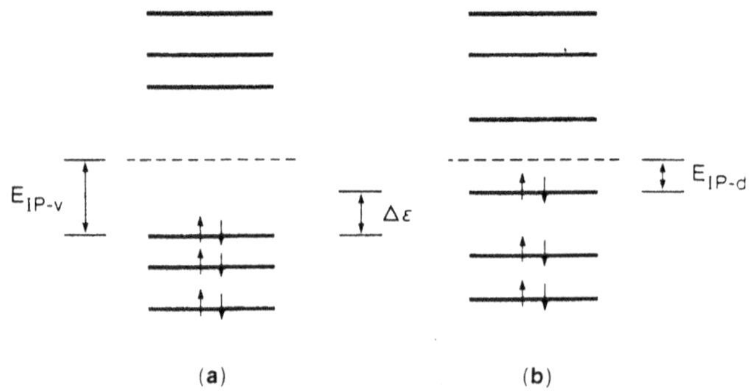

Figure 6: Illustration of the evolution of the one-electron energy levels of an organic molecule in its ground-state configuration, when going from the optimal geometry of the ground state (a) to the optimal geometry of the ionized state (b).

equivalent of the Cooper pair in the BCS theory of superconductivity [15] where the Cooper pair consists of two electrons coupled through a lattice vibration, i.e. a phonon. The formation of a bipolaron implies that the energy gained by the interaction with the lattice is larger than the Coulomb repulsion between the two charges of the same sign confined in the same location. Note that in the case of trans-polyacetylene, as a result of ground-state degeneracy, the two charges forming a would-be bipolaron can readily separate to form two isolated charged solitons, see Figure 8.

The electronic band structures corresponding to the presence of two polarons and that of one bipolaron are depicted in Figure 9. Since the lattice relaxation around two charges is stronger than around only one charge, E_{dis} for the bipolaron is larger than E_{dis} for the polaron. The electronic states appearing in the gap for a bipolaron are further away from the band edges than for a polaron, so that $\Delta\varepsilon^{bip}$ is larger than $\Delta\varepsilon^{pol}$. If we compare the creation energy of a bipolaron relative to that of two polarons, our calculations for polyacetylene, polyparaphenylene, polypyrrole, and polythiophene, indicate that the distortion energy E_{dis} to form one bipolaron (two solitons) is roughly equal to that to form two polarons [14, 16, 17]. On the other hand, the decrease in ionization energy is much more important in the bipolaron (soliton) case ($2 \times \Delta\varepsilon^{bip}$) than for two polarons ($2 \times \Delta\varepsilon^{pol}$), see Figure 9. This is the reason why, in these systems, one bipolaron (two solitons) is (are) thermodynamically more stable than two polarons. However, there can be compounds for which the equilibrium between bipolarons and polarons can go the other way around, such as in polyemeraldine, as will be discussed later; in polypyrrole, there also exist contradictory experimental reports [18]. Recent photoinduced absorption experiments in neutral polyacetylene [19], polythiophene [20], polyparaphenylene vinylene [21], and polydiacetylene [22] confirm that bipolarons (solitons) are indeed stable species in these materials, despite the Coulomb repulsion between charges of the same sign.

In case of p(n)-type doping, the bipolaron levels in the gap are empty (fully occupied), see Figure 9b. The bipolarons, like the charged solitons, are thus spinless. The presence of bipolarons on polymer chains result in the possibility of two optical transitions below the bandgap transition: for p-type doping, from the valence band to the lower bipolaron level and from the valence band to the upper bipolaron level. In the case of polarons, at least a third absorption is possible below the gap, corresponding to an optical transition between the two polaron levels (Figure 9a). The large modification of oscillator strengths upon bipolaron formation constitutes the basis for the promising nonlinear optical properties displayed by the pristine materials and for electrochromic applications (when switching between the doped and undoped forms).

The charge generation process that we have outlined for doped organic polymers can be contrasted to that in inorganic doped semiconductors. In the latter compounds, the atoms are covalently bonded in all three dimensions. The lattice is therefore much less deformable than along a polymer chain and rigid-band models usually provide a good description of their solid-state properties. The process of doping is achieved by introducing ppm quantities of atoms, such as boron or phosphorus atoms, which fit into the lattice either substitutionally or interstitially. In the example of boron-doped silicon, boron has one less valence electron (three) than silicon (four). This yields empty electronic levels associated with the boron slightly above the top of the silicon valence band. These empty dopant levels can be thermally populated with electrons from the silicon valence band, which results in the creation of holes on top of the valence band and the possibility of electronic conduction. By comparison, in polymer chains, the electronic levels coming into the gap are intrinsic levels of the polymer chains, the dopant levels lying usually well outside the polymer forbidden band.

EVOLUTION OF THE ELECTRONIC AND TRANSPORT PROPERTIES AT HIGH DOPANT CONCENTRATION: POLYACETYLENE, POLYPYRROLE, AND POLYANILINE

The evolutions of the transport properties on doping of trans-polyacetylene (as a prototypical example of a degenerate ground state system) and polypyrrole (as an example of a nondegenerate ground state system) have been reviewed in detail for instance by Epstein [23] and Brédas and Street [24], respectively. We therefore only briefly summarize here the major aspects. In trans-polyacetylene, at low doping levels, charged solitons are formed either directly from existing neutral solitons or as the result of recombination among polarons. Upon increasing the doping level, soliton states at midgap start overlapping and form a soliton band. Between ≈ 0.1-1% and ≈ 6-7% doping level, a high conductivity is observed which is associated with a vanishingly small Pauli susceptibility. This observation suggests that the charge carriers in that regime can be spinless charged solitons [25, 26]. However, there is still controversy as to whether the conduction process is really based on mobile charged solitons [27-29]. This is one of the most important aspects that still await a full understanding. Above 7% dopant concentration, Pauli susceptibility and metallic-like conductivity appear. It has been proposed that this phenomenon is the result of either (i) a first order phase transition, leading from a soliton lattice to a metal [30]; or (ii) the appearance of a disorder-induced gapless incommensurate Peierls insulator [29]; or (iii) the transition from a soliton lattice to a polaron lattice [31].

Theoretical studies of the evolution of the polypyrrole electronic band structure as a function of doping level have been performed using methods ranging from highly sophisticated ab initio techniques to simple Hückel theory with σ compressibility. They all converge towards the same picture [16, 32]. In the undoped state, the polypyrrole bandgap is about 3.2 eV. Taking an electron from the chain leads to the formation of a polaron whose associated quinoid-like geometry relaxation extends over about four pyrrole rings. When a second electron is taken out, the energetically favorable species, as we calculate it, is a bipolaron which also extends over four rings. This evolution from polarons to bipolarons is supported by ESR measurements on oxygen-doped polypyrrole [33]. At low doping, the ESR signal grows, in accord with the fact that polarons with 1/2 spins are formed. At intermediate doping, the ESR signal saturates and then decreases, consistent with polarons recombining to form bipolarons. At high doping, in electrochemically-cycled samples, no ESR signal is observed although the system is highly conducting, indicating that the charge carriers in that regime are spinless and may correspond to bipolarons. Analysis of the Pauli susceptibility shows that the density of states at the Fermi level is vanishingly small.

In poly-β,β'-dimethylpyrrole [34], polythiophene [35], polyparaphenylene [36], polyparaphenylene vinylene [37], polyparaphenylene sulfide [38], the evolution of the electronic and transport properties as a function of doping level is very similar to that presented above for polypyrrole. However, for instance in poly-3-methylthiophene [39] or in the case of the polypyrrole data reported by the Grenoble group [18], polarons seem to be the dominant species

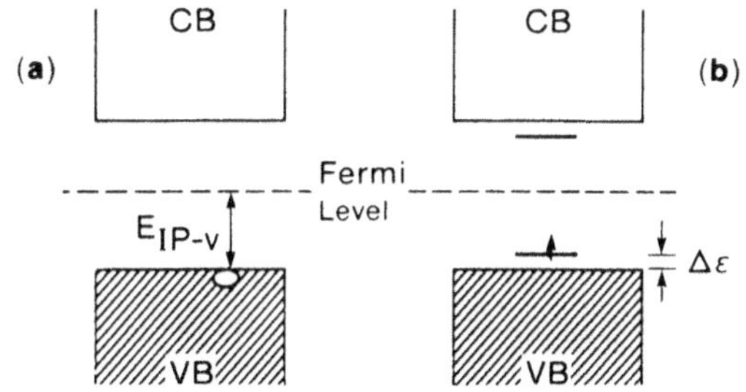

Figure 7: Schematic band structures of a polymer chain in the case of (a) a vertical ionization process and (b) the formation of a polaron.

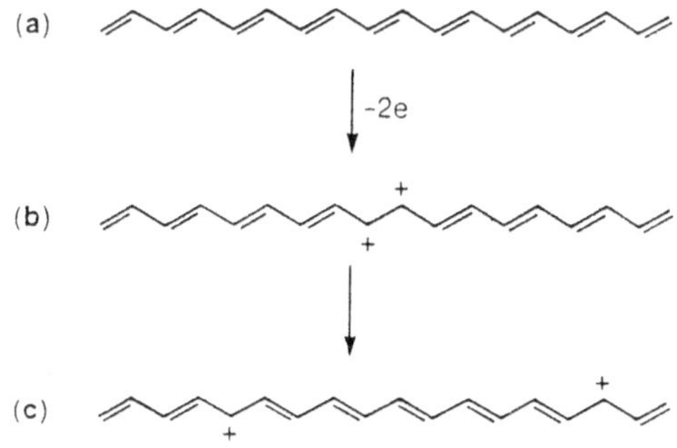

Figure 8: Illustration of the formation and separation of two charged solitons on a trans-polyacetylene chain.

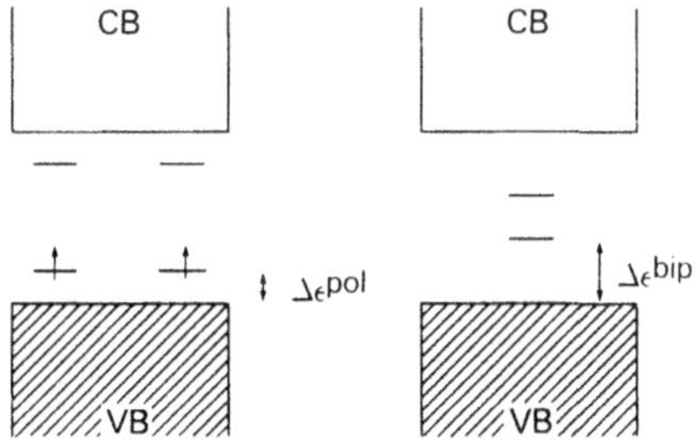

Figure 9: Band structures of a polymer chain in the case of two polarons (left) and one bipolaron (right).

at least at low doping levels. We now turn to a more detailed discussion of polyaniline, a material of high current interest and which presents some fascinating aspects.

Preparation techniques have by now been adequately optimized to produce, either chemically or electrochemically, several well-characterized forms of polyaniline [40, 41]. One of those polyaniline forms is polyemeraldine, $(-C_6H_4-NH-C_6H_4-NH-C_6H_4-N=C_6H_4=N-)_x$, see Figure 10a. Polyemeraldine is also called polyparaphenylene-amine-imine, since along the chains two amine-type (reduced) nitrogens, (–NH–), are followed by two imine-type (oxidized) nitrogens, (–N=). Polyemeraldine can be considered as being semioxidized with respect to the fully reduced form of polyaniline, polyparaphenylene-amine, $(-C_6H_4-NH-C_6H_4-NH-)_x$.

Having the structure depicted in Figure 10a, polyemeraldine is semiconducting and presents a first optical transition edge at about 1.4 eV. An important discovery has been to find that, upon protonation, the conductivity of polyemeraldine increases by about ten orders of magnitude to reach 5 S/cm [42]. A study of the temperature dependent static and dynamic magnetic susceptibility of polyemeraldine shows the presence of a Pauli susceptibility linearly proportional to the percent protonation [43, 44]. These results are in agreement with the idea that, upon protonation ("proton doping"), phase segregation and a first order transition to a metallic phase do occur [43]. A model based upon a two-step transition from a lattice of isolated doubly charged spinless bipolarons to a polaronic metal (Figure 10b to 10d) has therefore been proposed [43, 44]: (i) the first step relates to the unstability of a bipolaron on a polyemeraldine chain (Figure 10b) with respect to an internal redox reaction leading to two semiquinone radical-cations [45], i.e. polarons (Figure 10c) — a process which has been described for more than thirty years in the chemical litterature [46]; (ii) the second step implies the Coulomb separation of the so-produced polarons to yield a polaron lattice. This semiconductor-metal transition induced by protonation is most interesting since the protonation process does not modify the number of electrons on the polyemeraldine chains [43]. (We also stress that the electronic properties observed for 50% protonated polyemeraldine are totally equivalent to those observed after the first electrochemical oxidation of polyparaphenylene-amine, i.e. for $(-C_6H_4-NH-C_6H_4-NH-)_x^{+\cdot}$.)

The electronic band structures have been calculated for the bipolaron lattice and the polaron lattice in 50%-protonated polyemeraldine [47]. The polaron lattice band structure is presented in Figure 11. We only display the highest occupied valence bands and the first unoccupied conduction bands. It is important to mention that there are two extremely flat bands located around -0.1 atomic unit of energy in the band structure of the polaron lattice. (Those bands are so close to one another that they appear as a single band, band x, in the Figure.) Band a is only half-occupied and thus constitutes the polaron band.

It is interesting to realize that, both for the polaron and bipolaron lattices, there is only one defect (polaron or bipolaron) band which seems to be present in the gap. In all the other polymers investigated so far, there are always two distinct defect bands clearly appearing within the gap. As detailed before, this is due to a destabilization of the highest occupied band (leading to the lower defect band) and a stabilization of the lowest unoccupied band (leading to the higher defect band). The explanation for seemingly having only one defect band within the gap in 50% proton-doped polyemeraldine can be traced back via an analysis of the band characteristics for the highest occupied band and the lowest unoccupied band in fully reduced polyaniline, polyparaphenylene-amine [48]. In that case, relative to the polaron lattice band structure, the

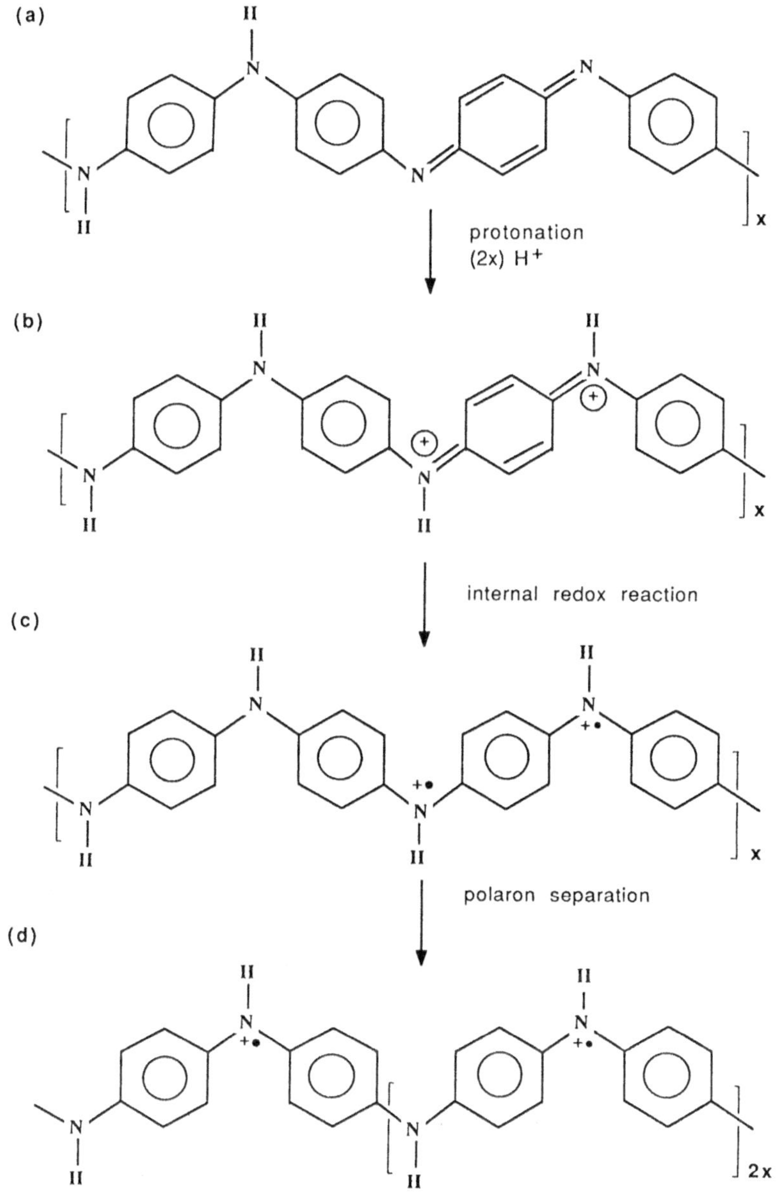

Figure 10: Sketch of the geometric structure of polyemeraldine: (a) before any protonation; (b) to (d) after 50% protonation. Protonation first results in the appearance of bipolarons, (b). Following an internal redox reaction leading to polysemiquinone radical-cations, the bipolarons split to form polarons (c). Due to Coulomb repulsion, the polarons separate, which results in a polaron lattice (d).

highest occupied (lowest unoccupied) band corresponds to a combination of bands a (x) and b (x'). As we pointed out, band x is extremely flat, its width being smaller than 0.05 eV. This narrowness originates from the fact that, for symmetry reasons, the atomic orbital contributions to this band coming from the nitrogens and the carbons connected to them are negligible. Actually, upon formation of the polaron or the bipolaron lattice, there is also an upper defect band, band x', which comes into the gap but is hardly stabilized with respect to band x because of the flatness of this band.

One of the crucial consequences of this particularity of the upper defect band in protonated polyemeraldine is to be found in the interpretation of the optical absorption spectra. In the other conducting polymers, the subgap absorptions that appear upon doping are related to electronic transitions involving both defect bands. In polyemeraldine, since the transitions to the upper defect band x' cannot be distinguished from those to the conduction band x, only one defect band, the lower defect band a, plays a role in the subgap absorptions. Thus, the detailed analysis of the polyemeraldine band structure establishes a modified picture for the transitions within the gap.

The polaron lattice band structure, which is displayed in Figure 11, yields electronic transitions which are in very good agreement with the optical absorption data for 50% proton-doped polyemeraldine [47]:

(i) we obtain a polaron band which is only half-occupied, allowing for intraband absorption and thus for far-infrared absorption, as is experimentally observed [47, 49, 50]. The calculated polaron bandwidth is 1.1 eV, in excellent agreement with the 1.0 eV estimate from the intraband optical absorption at low frequency;

(ii) direct optical transitions from band b to the polaron band are calculated to be at 1.8 eV, which compares favorably with the first optical absorption at about 1.5 eV;

(iii) direct optical transitions from band c to the polaron band are calculated to be at 2.6 eV, which is in very good agreement with the second absorption observed at 2.8 eV;

(iv) the first electronic transitions involving the upper defect band (x') and the conduction band (x) are predicted to occur at 4.1 eV, in agreement with the location of the third absorption. As a result of the excellent accord between the calculated band structure and the optical absorption data for 50% proton-doped polyemeraldine, the calculations offer a strong support of the polaronic metal model which has been proposed on the basis of the magnetic studies.

CONCLUSIONS

As is apparent from our discussion, conjugated polymers still present a strong fundamental challenge which has to be taken up by a diverse community of chemists, physicists, materials scientists, and theoreticians both in academia and in the industry. The close interaction between these scientists with very different backgrounds has been a significant factor in the discovery of much of the new exciting chemistry and physics in the field of conducting polymers and is still required in order to achieve further significant developments. In the light of the wealth of unanticipated discoveries that the field of conjugated polymers has witnessed in the last ten years, the scientific and technological future of conducting as well as nonlinear optical plastics has to be considered in a very positive way.

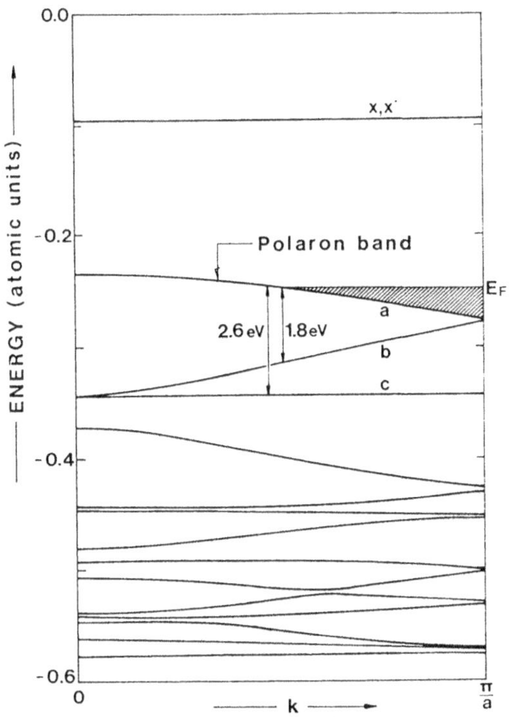

Figure 11: Band structure (as obtained by the Valence Effective Hamiltonian technique) for the polaron lattice in 50%-proton doped polyemeraldine.

ACKNOWLEDGEMENTS

I wish to thank all my colleagues that have at some point or another contributed to the works that have been presented here, in particular J.M. André, B. Thémans, and S. Stafström in Namur, G.B. Street at IBM San Jose, R.R. Chance at Exxon, R. Silbey at MIT, A.J. Epstein at OSU. This work has been supported by the Belgian National Fund for Scientific Research (FNRS). I am indebted to the Belgian National Fund for Scientific Research (FNRS), IBM Belgium, and the Facultés Universitaires Notre-Dame de la Paix for the use of the Namur Scientific Computing Facility (SCF).

REFERENCES

[a] Chercheur Qualifié of the Belgian National Fund for Scientific Research (FNRS).

[1] (a) Proceedings of the International Conference on Synthetic Metals –ICSM'86–, Synth. Met. **17, 18,** & **19** (1987); (b) "Handbook of Conducting Polymers", T.A. Skotheim, ed., Marcel Dekker, New York, 1986; (c) Proceedings of the International Winterschool on the Electronic Properties of Polymers –IWEPP'87–, Springer Verlag, Berlin, in press.
[2] V.V. Walatka, M.M. Labes, and J.H. Perlstein, Phys. Rev. Lett. **31**, 1139 (1973); M.M. Labes, P. Love, and L.F. Nichols, Chem. Rev. **79**, 1 (1979).
[3] R.L. Greene, G.B. Street, and L.J. Suter, Phys. Rev. Lett. **34**, 577 (1975).
[4] W.D. Gill, W. Bludau, R.H. Geiss, P.M. Grant, R.L. Greene, J.J. Mayerle, and G.B. Street, Phys. Rev. Lett. **38**, 1305 (1977); G.B. Street and W.D. Gill, in "Molecular Metals", W.E. Hatfield, ed., Plenum, New York, 1979, NATO Conference Series VI, vol. 1, p. 301.
[5] H. Shirakawa, E.J. Louis, A.G. MacDiarmid, C.K. Chiang, and A.J. Heeger, J. Chem. Soc. Chem. Commun. 578 (1977); C.K. Chiang, C.R. Fincher, Y.W. Park, A.J. Heeger, H. Shirakawa, E.J. Louis, S.C. Gau, and A.G. MacDiarmid, Phys. Rev. Lett. **39**, 1098 (1977).
[6] H. Naarman, Synth. Met. **17**, 223 (1987).
[7] P. Bernier, in this volume.
[8] W.P. Su, J.R. Schrieffer, and A.J. Heeger, Phys. Rev. Lett. **42**, 1698 (1979); Phys. Rev. B **22**, 2209 (1980).
[9] C.S. Yannoni and T.C. Clarke, Phys. Rev. Lett. **51**, 1191 (1983); C.R. Fincher, C.E. Chen, A.J. Heeger, A.G. MacDiarmid, and J.B. Hastings, Phys. Rev. Lett. **48**, 100 (1982).
[10] C. Rebbi, Scientific American **240**, 92 (1979).
[11] See for instance, M.A. Omar, "Elementary Solid State Physics", Addison-Wesley, Reading, 1978, ch. 9.11.
[12] D.S. Boudreaux, R.R. Chance, J.L. Brédas, and R. Silbey, Phys. Rev. B **28**, 6927 (1983).
[13] W.P Su and J.R. Schrieffer, Proc. Nat. Acad. Sci. USA **77**, 5626 (1980); K. Fesser, A.R. Bishop, and D.K. Campbell, Phys. Rev. B **27**, 4804 (1983); S.A. Brazovskii and N. Kirova, JETP Lett. **33**, 4 (1981).
[14] J.L. Brédas, R.R. Chance, and R. Silbey, Phys. Rev. B. **26**, 5843 (1982); Mol. Cryst. Liq. Cryst. **77**, 319 (1981).
[15] J. Bardeen, L.N. Cooper, and J.R. Schrieffer, Phys. Rev. **108**, 1175 (1957).
[16] J.L. Brédas, J.C. Scott, K. Yakushi, and G.B. Street, Phys. Rev. B **30**, 1023 (1984).
[17] J.L. Brédas, F. Wudl, and A.J. Heeger, Solid State Commun., in press.
[18] F. Genoud, M. Guglielmi, M. Nechtschein, E.M. Genies, and M. Salmon, Phys. Rev. Lett. **55**, 118 (1985).
[19] A.J. Heeger, in "Conducting Polymers (part II)", H. Sasabe, ed., CMC, Tokyo, 1987, pp. 5-54.
[20] Z. Vardeny, E. Ehrenfreund, O. Brafman, M. Nowak, H. Schaffer, A.J. Heeger, and F. Wudl, Phys. Rev. Lett. **56**, 671 (1986).
[21] D.D.C. Bradley, R.H. Friend, H. Linderberger, and S. Roth, in ref. 1c.
[22] F.L. Pratt, K.S. Wong, W. Hayes, and D. Bloor, in ref. 1c.
[23] A.J. Epstein, in ref 1b, pp. 1041-1097.
[24] J.L. Brédas and G.B. Street, Acc. Chem. Res. **18**, 309 (1985).
[25] D. Moses, A. Denenstein, J. Chen, A.J. Heeger, P. MacAndrew, T. Woerner, A.G. MacDiarmid et Y.W. Park, Phys. Rev. B **25**, 7652 (1982).

[26] T.C. Chung, F. Moraes, J.D. Flood et A.J. Heeger, Phys. Rev. B **29**, 2341 (1984); J. Chen, T.C. Chung, F. Moraes et A.J. Heeger, Solid State Commun. **53**, 757 (1985).

[27] E.J. Mele et M.J. Rice, Phys. Rev. B **15**, 5397 (1981); A.J. Epstein, H. Rommelmann, R. Bigelow, H.W. Gibson, D.M. Hoffmann et D.B. Tanner, Phys. Rev. Lett. **50**, 1866 (1983).

[28] K. Ehinger, S. Summerfield, W. Bauhofer, and S. Roth, J. Phys. C: Solid State Phys. **17**, 3753 (1984).

[29] X.Q. Yang, D.B. Tanner, A. Feldblum, H.W. Gibson, M.J. Rice, and A.J. Epstein, Mol. Cryst. Liq. Cryst. **117**, 267 (1985).

[30] J. Chen, T.C. Chung, F. Moraes, and A.J. Heeger, Solid State Commun. **53**, 757 (1985).

[31] S. Kivelson and A.J. Heeger, Phys. Rev. Lett. **55**, 308 (1985).

[32] J.L. Brédas, B. Thémans, J.G. Fripiat, J.M. André et R.R. Chance, Phys. Rev. B **29**, 6761 (1984).

[33] J.C. Scott, M. Krounbi, P. Pfluger, and G.B. Street, Phys. Rev. B **28**, 2140 (1983); J.C. Scott, J.L. Brédas, K. Yakushi, P. Pfluger et G.B. Street, Synth. Met. **9**, 165 (1984).

[34] K. Yakushi, L.J. Lauchlan, G.B. Street, and J.L. Brédas, J. Chem. Phys. **81**, 4133 (1984).

[35] T.C. Chung, J.H. Kaufman, A.J. Heeger et F. Wudl, Phys. Rev. B **30**, 702 (1984).

[36] G. Crecelius, M. Stamm, J. Fink, and J.J. Ritsko, Phys. Rev. Lett. **50**, 1498 (1983).

[37] D.D.C. Bradley, G.P. Evans, and R.H. Friend, Synth. Met. **17**, 651 (1987); L.D. Kispert, J. Joseph, J. Tang, M.K. Bowman, G.H. Van Brakel, and J.R. Norris, Synth. Met. **17**, 617 (1987).

[38] R.H. Friend and J.R.M. Giles, Synth. Met. **10**, 377 (1985).

[39] G. Harbeke, E. Meier, W. Kobel, M. Egli, H. Kiess, and E. Tosatti, Solid State Commun. **55**, 419 (1985).

[40] A.G. Green and A.E. Woodhead, J. Chem. Soc. 2388 (1910); R. de Surville, M. Josefowicz, L.T. Yu, J. Périchon, and R. Buvet, Electrochim. Acta 13, 1451 (1968); T. Kobayashi, H. Yoneyama, and H. Tamura, J. Electroanal. Chem. 161, 419 (1984); J.P. Travers, J. Chroboczek, F. Devreux, F. Genoud, M. Nechtschein, A. Syed, E.M. Genies, and C. Tsintsavis, Mol. Cryst. Liq. Cryst. 121, 195 (1985).

[41] J.C. Chiang and A.G. MacDiarmid, Synth. Met. 13, 193 (1986) and references therein.

[42] A.G. MacDiarmid, J.C. Chiang, A.F. Richter, and A.J. Epstein, Synth. Met. **18**, 285 (1987).

[43] J.M. Ginder, A.F. Richter, A.G. MacDiarmid, and A.J. Epstein, to be published.

[44] A.J. Epstein, J.M. Ginder, F. Zuo, R.W. Bigelow, H.S. Woo, D.B. Tanner, A.F. Richter, W.S. Huang, and A.G. MacDiarmid, Synth. Met.**18**, 303 (1987) .

[45] G. Wnek, Polymer Prepr. 27, 277 (1986); Synth. Met. 15, 213 (1986).

[46] H. Linschitz, J. Rennert, and T.M. Korn, J. Am. Chem. Soc. 76, 5839 (1954).

[47] S. Stafström, J.L. Brédas, A.J. Epstein, H.S. Woo, D.B. Tanner, W.S. Huang, and A.G. MacDiarmid, submitted for publication; J.L. Brédas, in ref. 1c.

[48] D.S. Boudreaux, R.R. Chance, J.F. Wolf, L.W. Shacklette, J.L. Brédas, B. Thémans, J.M. André, and R. Silbey, J. Chem. Phys. 85, 4584 (1986); S. Stafström and J.L. Brédas, Synth. Met. 14, 297 (1986).

[49] N.S. Sariciftci, H. Neugebauer, and H. Kuzmany, in ref. 1c.

[50] A.P. Monkman and D. Bloor, in ref. 1c.

EVOLUTION OF THE PHYSICAL PROPERTIES OF POLYACETYLENE DURING THE ELECTROCHEMICAL INTERCALATION WITH ELECTRON DONORS

P. Bernier, C. Fite, and A. El Khodary

Groupe de Dynamique des Phases Condensées
Université des Sciences et Techniques du Languedoc
Place E. Bataillon, 34060 Montpellier

ABSTRACT

A great number of conjugated polymers can be doped up to the metallic state with electron donating or accepting chemical species. Using electrochemical techniques, a reversible and controllable doping can be performed, which can be considered as an actual intercalation process. During such a process all the properties of the system experience marked changes. The main evolution occurs for the structural characteristics of the polymer/dopant system. For instance, in the case of polyacetylene, successive phase transformations are observed during doping with alkali metals, corresponding to various types of ordering.

Transport, optical and magnetic properties are also greatly influenced during the intercalation process. For instance the eight to ten orders of magnitude increase of the electronic conductivity observed during the doping of polyacetylene is in fact due to the transition of the system from the semiconducting to the metallic state. Such a transition can be studied continuously by measuring in-situ the optical or magnetic characteristics. The case of alkali intercalation is particularly interesting as the ion sizes agree reasonably well with the available space in the structure of linear conjugated polymers. In this case, the study of the magnetic properties of doped polyacetylene allow to clarify the role played by the dopant itself in interchain electronic transport.

Conjugated polymers such as polyacetylene $(CH)_x$ polyphenylene $(C_6H_4)_x$, polythiophene $(C_4H_2S)_x$, etc., which are insulators in their pristine state, can be brought to the metallic state after "doping" with chemical species which can be either electron donors or acceptors.[1]. This doping process involves a charge transfer between the dopant molecule and the polymer chain which are then supposed to be spatially close to each other. It follows that the mechanism of doping must be considered as an actual intercalation process, which will greatly affect the structural characteristics of the starting material, as well as its morphology, as has been observed during the intercalation of graphite and layered compounds[2]. In parallel with these modifications, the band structure of the system changes yielding a new set of electronic properties. It is evident therefore that the structural and electronic properties are intimately related, and must be studied simultaneously in the same system to give reliable information.

A great number of studies have been devoted to the structural and electronic properties of conjugated polymers after a chemical or electrochemical doping process[2]. Most of these concern the properties of the system for a given dopant concentration. With this approach a universal picture of the polymer/dopant system is very difficult to obtain, as a comparison between different experiments is very hazardous. On the other hand, only a small number of measurements have been performed during the continuous electrochemical doping of various polymers. The interest of such a doping process is to intercalate chosen chemical species, controllably at a given rate and as slowly as required, with the further possibility of investigating the undoping process which returns the system to its neutral state. Among the various properties which have been studied this way we find : the d.c. conductivity, the optical absorption, the infrared, Raman and electron spin resonance spectra, in parallel with the usual electrochemical measurements[3].

Of the various systems that have been investigated,

polyacetylene has been the most intensely studied, partly because it is easily produced as a film of controllable thickness, and partly because its degenerate molecular form leads to unusual electronic properties. For instance the doping process is supposed to proceed via the formation of charged solitons, polarons and bipolarons, [4] sometimes yielding magnetic properties which are not related to the existence of charge carriers. A common behaviour of all doped polyacetylene systems is the gradual transition during doping from a semi-conducting to a "metallic" state [5]. In the present paper we have studied the evolution of several physical properties during the electrochemical intercalation of polyacetylene with electron donors of the alkali metals series, focusing on the structural, transport and magnetic properties, with the aim of producing a unified description of the intercalation processes.

SOME EXPERIMENTAL DETAILS

As presented in the introduction the great advantage of electrochemical doping is that it permits : (i) a controllable intercalation, (ii) a controllable de-intercalation and (iii) a choice of rate for each process.

The electrochemical cell is generally made of a glass tubing specially designed to allow the specific measurements (5 mm cylindrical glass tube for ESR, 1 cm cylindrical glass tube for conductivity, cell with Lindeman glass for X-rays). One electrode consists of a $(CH)_x$ film gripped by a platinum grid. The counter electrode consists of a piece of the alkali metal attached to a Pt or NiCr wire. Various electrolytic solutions have been used, usually consisting of a salt ($LiClO_4$, $NaPF_6$, KCN + $(EtB)_3$, etc.) in a solvant (THF, MeTHF,...) A problem to be considered is the stability of the solvant in the range of voltages needed during the doping process (2 volts down to 0.1 volts). Another problem is the solubility of the salt in the solvent. The above mentionned compounds have proved to be good candidates for the intercalation of Li, Na and K in $(CH)_x$. The cell is generally sealed under vacuum to avoid any contact with air or moisture which will immediately degrade the $(CH)_x$ electrode [6,7,8].

The intercalation is performed by controlling the current flowing through the cell after the voltage has been lowered stepwise from 2V to a desired value which correspond to a given value of the dopant concentration y. This value y is determined from the open circuit voltage $V_{oc} = f(y)$ relationship, and by measuring the amount of charge passed through the cell (eventually corrected for non-ideal efficiency). More information concerning the experimental details of the various physical measurements can be found in the corresponding papers [1-8].

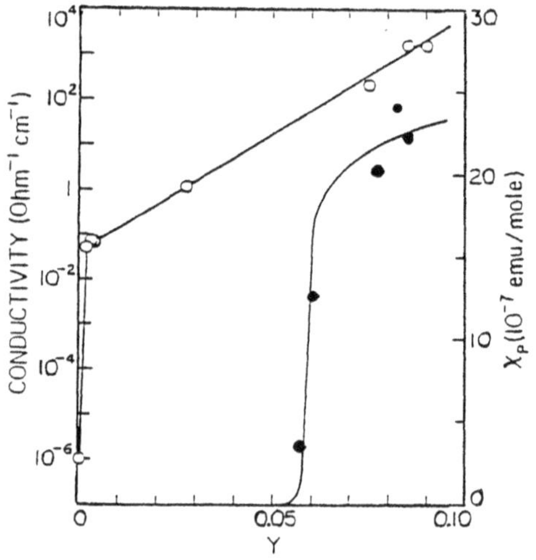

Figure 1 : Room temperature Conductivity σ (open circles) and Pauli susceptibility χ_P (full circles) versus dopant concentration y in $(Na_y CH)_x$. (from ref. 8).

The first strong evidence of an evolution from semiconducting to metallic state during doping comes from the variations of the d.c. conductivity (σ) with dopant concentration. The case of $(Na_y\ CH)_x$ as determined by Moraes et al[8], presented in Figure 1, shows that after a very sharp increase for the first percent, σ slowly increases up to ~ 1000 $(\Omega\ cm)^{-1}$ for higher concentrations. Such a behaviour suggests that an apparently "metallic" state has been reached at the low concentration of $\sim$ 0.5%. Nevertheless, the Pauli (temperature independent) contribution to the magnetic susceptibility has been shown to be negligible ($\chi_P < 10^{-8}$ emu/mole) for $y < 5\%$.

There follows a large range (0.5 to 5%) of concentrations where the d.c. transport does not correspond to the existence of extended electronic states at the Fermi level. Moraes et al[8] have suggested that in this range charges are transported by charged solitons or bipolarons which are intrinsically non-magnetic. The electrical transport would take place through a combination of correlated diffusive motion of mobile solitons and electron hopping through disorder-induced gap states. In the case of $(Na_y\ CH)_x$ the mobile soliton mechanism seems to be dominant, however, it is not possible to be conclusive on this point using transport data alone.

The situation for $(K_y\ CH)_x$ is quite different. As seen in Figure 2, the increase of σ with y at the onset of doping is more gradual than in the Na case, but nonetheless abrupt up to 5%, where the variation becomes very flat ($\sigma \sim 400\ \Omega^{-1}\ cm^{-1}$). It is important to note that no kinetic effects have been observed on a time scale greater than one minute, implying a very fast approach to equilibrium. Also, the very good reversibility of the behaviour of σ indicates that the dopant distribution is homogeneous throughout the whole sample. Unfortunately the same information is not available for Na doping, which makes interpretation difficult. We suggest that the significant differences observed between Na and K doped systems in the range 0 to 5% could be due to strongly inhomogeneous doping in the case of Na. In effect, it is possible to obtain artificially

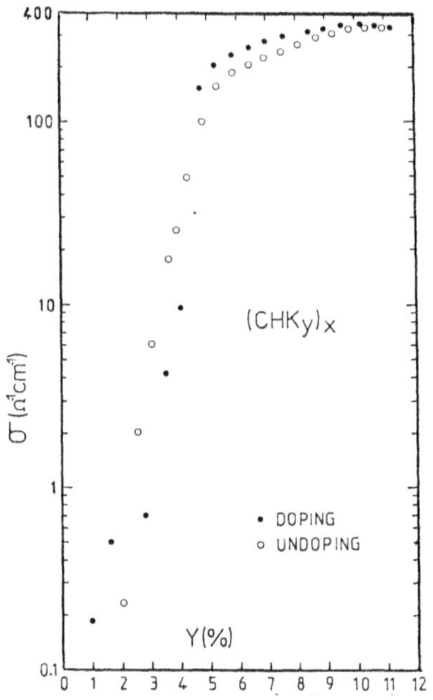

Figure 2 : Room temperature conductivity σ versus dopant concentration y during a full doping-undoping cycle in $(K_y\ CH)_x$.

high conductivities in film samples doped inhomogeneously as often occurs using chemical doping. In this case the measured conductivity is not characteristic of the <u>mean</u> y value, but of the <u>maximum</u> y value, which can be 2 to 10 times larger than the former value.

The situation is further confounded by the absence of magnetic susceptibility data for $(K_y CH)_x$, therefore preventing the complete comparison of Na and K doped systems.

STRUCTURE OF THE ALKALI DOPED (CH)x: THE QUESTION OF STAGING

Starting from the fact that the structure of the pristine trans-$(CH)_x$ has already available space for small species such as the alkali metal ions, Baughman and coworkers have proposed a channel structure for the doped systems[9]. In such a model the ion size plays the main role as it favors the formation of square channels (K^+ ions), or triangular channels (Na^+ ions) as proposed recently[10]. A challenge to this model is the intercalation model proposed by Flandrois et al, in which the alkali ions lattice is arranged in planes which are intercalated between planes of polymer chains[11].

The first consequence of these two models is the existence of stages. Baughman et al have proposed[9] to identify a given stage using the subscripts n and m in the formula $[(C_n H_n)_m M_y]_x$, where n represents the number of CH units per M^+ ion, m the number of CH chains per M^+ channel, and $y = (nm)^{-1}$ is the mean dopant concentration. Using X-ray diffraction techniques, these authors have identified in the case of K^+ and Na^+ doping, stages corresponding to n = 4, m = 3 and n = 3, m = 2. Nevertheless, the determination of the structure is not unequivocable, as a consequence of the very small number of diffraction lines which can be observed. Furthermore the planar intercalation model proposed by Flandrois[11] cannot be ruled out.

Staging also manifests itself in the evolution of the electrochemical potential V_{oc} of $(CH)_x$ during the intercalation of alkali ions[12]. The case of K^+ doping is presented on Figure 3[13]. The plateaus can be associated with the existence of various successive stages. For example, in the present case, two stages can be

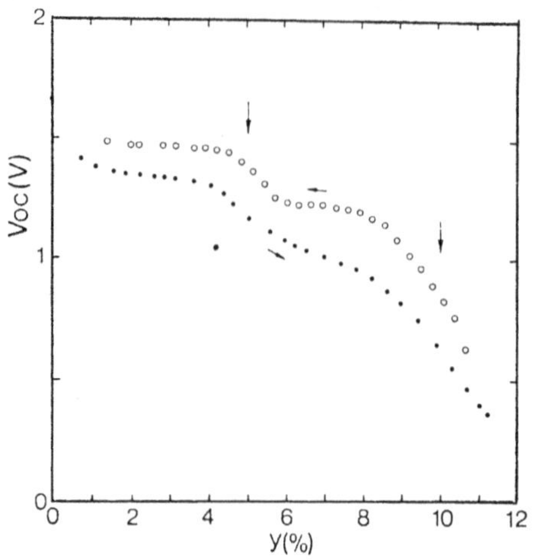

Figure 3 : Electrochemical potentiel V_{oc} of $(K_y\ CH)_x$ with respect to metallic potassium, versus the dopant concentration y. The arrows indicate the points at which the inverse potential derivative goes through a maximum.

identified at y ≅ 5% and y ≅ 10%. We note that these values correspond to the concentrations at which the slope of the conductivity σ versus doping level curve experiences dramatic changes (see Figure 2).

If only qualitatively, this correlation shows that the evolution of the transport properties depends on the evolution of the structure during doping. We finally note that the reversibility of both V_{oc} and σ during undoping guarantees that these behaviours are intrinsic properties of the polymer/dopant system itself. ESR spectroscopy also provides evidence of the existence of staging in K doped $(CH)_x$.

EVOLUTION OF THE ESR CHARACTERISTICS DURING DOPING: LINE-WIDTH

It has been clearly shown by Rachdi et al. that the ESR linewidth of doped $(CH)_x$ is a quantity very sensitive to the nature of the dopant species[14]. For example, for the alkali metal series, the linewidth ΔH_{pp} at the maximum doping level strongly increases with the atomic number of the element, from $\sim$ 0.1 Gauss for Li doping to more than 500 G for Cs doping. The case of K doping is intermediate, with ΔH_{pp} $\sim$ 20 - 40 Gauss. During the doping process the linewidth increases from $\sim$ 0.5 Gauss (characteristic of the undoped trans $(CH)_x$) to $\sim$ 40 Gauss in the highly doped state. This large window of values provides a favourable situation for the observation of fine details in the evolution of ΔH_{pp} during the intercalation process.

Figure 4 presents the variation of ΔH_{pp} versus the K^+ concentration y during one full doping-undoping cycle (the second cycle, as the first one is always slightly different from the others). Various features can be observed :

(i) three plateaus during the doping process, successively in the ranges 0-4%, 5-9% and 10-12%; four plateaus during the undoping process, successively in the ranges 12-10%, 9-4%, 3-2% and 1,5-0%.

(ii) the overall reproducibility is good : the linewidth can be cycled many times between 0.2-0.5 Gauss to $\sim$ 38 Gauss.

(iii) reversibility is observed only in the highly doped state (6-12%) or in the almost undoped state (0-1,5%). A reproducible hysteresis is observed between 1.5 and 6%.

We suggest that the various plateaus are associated with the existence of stages as evidenced by the dependence of V_{oc} versus y (figure 3) : the changes in the slope of the curves appear at roughly the same y values in both cases. The fact that the existence

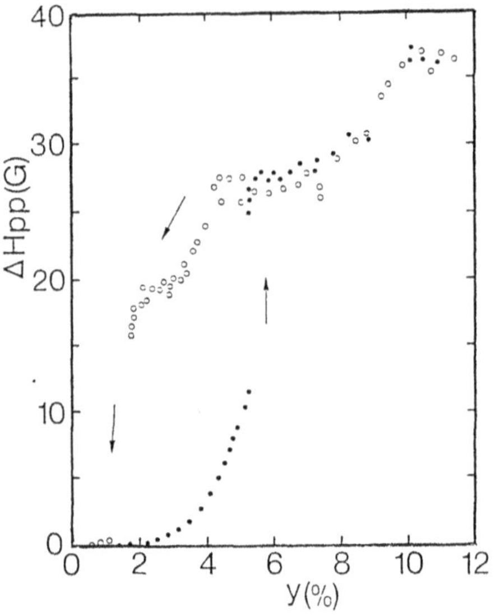

Figure 4 . ESR linewidth ΔH_{pp} (distance in Gauss between the peaks of the derivative absorption) versus dopant concentration y during a full doping-undoping cycle (T = 293 K) in $(K_y\ CH)_x$.
(• doping; o undoping).

of stages manifest itself as successive plateaus in the ΔH_{pp} behaviour can be understood as follows.

For a given stage the partitioning of K^+ ions in the polymer host is fixed. The strength of the spin orbit coupling with the delocalized electrons on the polymer chains depends on the <u>local</u> dopant concentration, and so depends on the stage order. Therefore, a phase A of the stage (n_A, m_A) will have a corresponding characteristic linewidth Δ_A, and a lower order (i.e. with a higher dopant concentration) phase B (n_B, m_B), a charactersitic linewidth $\Delta_B > \Delta_A$.

During intercalation the system can be considered as a mixture of various phases, whose relative importance changes with dopant concentration. Let us consider a simple biphasic model, with a growing phase B and a diminishing phase A. It can be shown that the experimentally measured ESR peak-to-peak linewidth is not a linear combination of Δ_A and Δ_B, but is in fact equal to Δ_A as long as phase B is not > $\sim$ 90% of the total system. Above $\sim$ 90% of phase B the linewidth increases sharply up to the value Δ_B.

We can then conclude that the linewidth dependence during doping, as shown on Figure 4, is consistent with the formation of two successive stages around 5% and 10% respectively. Again we note that these values agree well with those determined from the V_{oc} and σ variations (see above).

The evolution of linewidth during undoping can be interpreted in a similar manner. The new features, as compared with the doping process, are a longer plateau in the range 3% to 6%, and <u>a new plateau</u> appearing between 1.5% to 3%. While the first feature can be partially explained by kinetic and geometrical effects related to the shape of the sample, the second is an intrinsic and reproducible property of the system (it appears during all the successive undoping cycles). This third plateau corresponds to some high order stage (nm $\sim$ 50), which could arise from more concentrated stages by dilution of the K^+ ions along the channels. Such a suggestion has already been made by Shacklette and Toth for very similar systems[12].

It is important to note that the charactersitics discussed above are very specific to the K doped $(CH)_x$. The same type of experiments performed on Li or Na doped $(CH)_x$ produce very different results[7].

With Li, ΔH_{pp} <u>decreases</u> during doping from $\sim$ 0.8 Gauss to $\sim$ 0.12 Gauss for y $\sim$ 12%. The transition occurs gradually for concentrations higher than $\sim$ 5%. For Na doping, ΔH_{pp} <u>increases</u> from $\sim$ 0.8 Gauss to $\sim$ 6 Gauss for y $\sim$ 12%. The transition is also gradual around 5%. In both cases no successive plateaus have been observed for y > 5%. However we have to note that the rather small range of ΔH_{pp} values (0.8 to 0.12 G for Li,. 0.8 to 6 G for Na) do not favour the observation of fine details. Finally we can say that behaviour during K doping appears to be particularly exceptional, and enlightens somewhat the field of conducting polymers.

EVOLUTION OF THE ESR CHARACTERISTICS DURING DOPING: LINE-SHAPE

It has been known form a long time that the shape of the Conduction Electron Spin Resonance (CESR) depends on both the static and dynamic properties of the electronic system[15]. In most metallic materials, however, the characteristics are such that the CESR is unobservable, being much too broad. The advantage of organic conducting polymers is that they generally present narrow lines when compared with inorganic conductors. Another interesting property of these materials is the ability to vary continuously their conductivity. It is then possible to check in detail the validity of certain models developed to explain the CESR line shape.

In the Dyson theory, one of the quantities which can be expressed analitically is the asymmetry ratio A/B of the ESR spectrum, where A and B are amplitudes of the low field and high field first derivative absorption peak respectively (Figure 5). For an insulating material A/B = 1, i.e. the spectrum is symmetric. In a conductor we must take into account the skin depth δ which characterizes the penetration of the micowave field, and which is expressed as :

$$\delta = c\ (2\Pi\sigma\omega)^{-1/2}$$

with c = speed of light, σ = microwave conductivity of the material and f = $\omega/2\Pi$ is the frequency of the ESR spectrometer. In the simplest situation A/B is a function of the two quantities, d/δ and T_D/T_2, where d is the sample thickness, T_D is the time charge

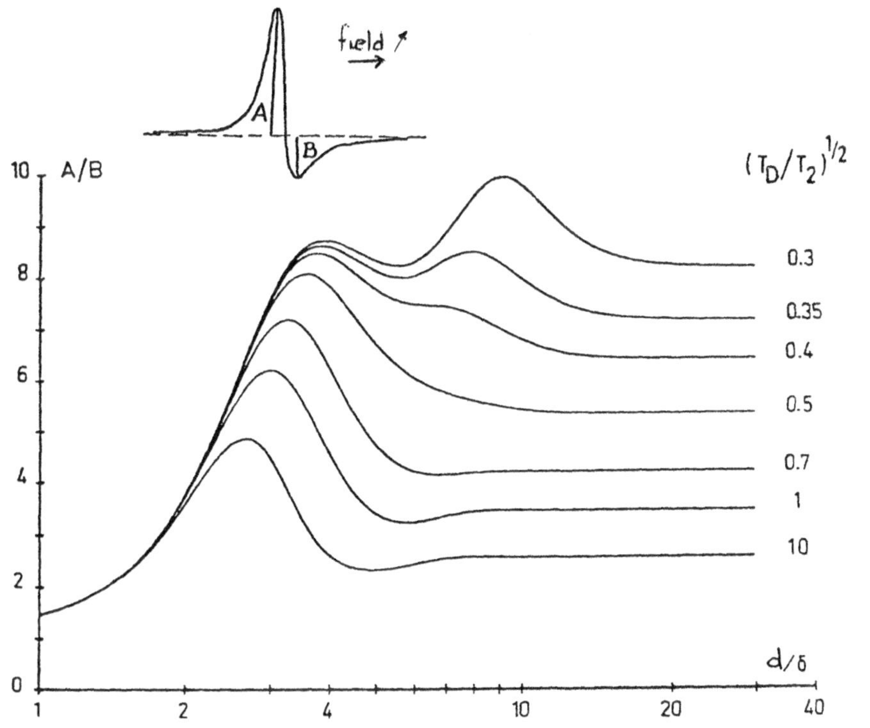

Figure 5 : Theoretical dependence of the CESR asymmetry ratio A/B versus the quantity d/δ as defined in the text. The quantity T_D/T_2 (see text) is introduced as a parameter.

carrier needs to diffuse through the skin depth, and T_2 is the homogeneous spin-spin relaxation time[16]. We note that d/δ is proportionnal to $\sigma^{1/2}$, and so increasing the conductivity by doping will result in an increase of d/δ. The variations of A/B versus d/δ are shown on Figure 5, where the quantity $(T_D/T_2)^{1/2}$ is introduced as a parameter. We note that in general A/B passes through a maximum (in some cases two maxima) and, after a decrease, stabilizes at a constant value. Quantitatively the maximum and the plateau values depend on T_D/T_2, a term which contains the electronic mobility. It follows that for a given T_2, and for a material reasonably conducting ($d/\delta > 5$), A/B increases as electronic mobility increases but stays in the range 2.6 to 8. In fact, for a simple sample geometry (thin slab of thickeness d) the above dependence can be investigated in detail.

Figure 6 presents the variation of A/B versus y in $(K_yCH)_x$ during a complete doping-undoping cycle [17](the second cycle ; the subsequent cycles being similar). It is clear that the general behaviour corresponds to the prediction of the Dyson model (Figure 5) : i.e. we observe a peak followed by a plateau these characteristics being reversible. Nevertheless some specific details should be pointed out.

(i) the d/δ value at which the maximum occurs is smaller during undoping than during doping. The conductivity σ is very reversible during the cycle (Figure 2) and so we suggest this shift to be due to the variation of sample thickness d (due to the intercalation). Francois et al[18] have shown that the apparent volume of a $(CH)_x$ film increases upon doping with Li and Na and experiences an hysteresis during the undoping process. This effect is supposed to be even move pronounced for K doping, for which an hysteresis of the staging process has been observed (see above) in the range 1.5 to 4%.

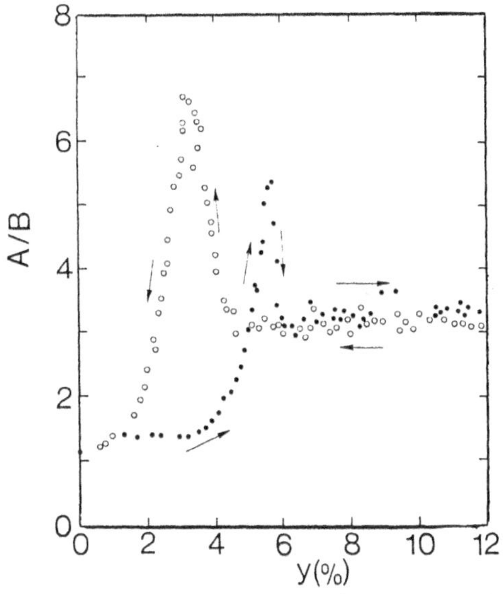

Figure 6 : Room temperature experimental variation of A/B versus dopant concentration y in $(K_y\ CH)_x$ during a full doping-undoping cycle.

(• doping, o undoping).

successive cycles, indicating that they are intrinsic characteristics of the polymer/dopant system itself.

(T_D/T_2) values should be different. Assuming that T_2 is not significantly changed (the change in linewidth is only 20% in the range 1.5 to 5%, see Figure 3) we can conclude that the electron mobility is higher during the undoping process compared with the doping one. This conclusion is reasonnable owing to the fact that the degree of long range organization of the doped system is supposed to be higher during undoping than during doping[12].

(iii) All the above features are almost totally reproducible over

(ii) The A/B value at the maximum is greater during undoping than during doping. This effect is due to the fact that the respective

It is important to note that other systems, such as Na or Li doped $(CH)_x$, obey the same theory. Nevertheless, the general aspect of the variation of A/B can be very different, as a consequence of the different values of T_D and T_2. For example for Na doping the linewidth ΔH_{pp} is much smaller ($\sim$ 6 Gauss) than for K doping[7]. The value of T_2 is therefore expected to be much larger, yielding smaller values for T_D/T_2. Consequently the A/B evolution has a different shape, increasing with d/δ up to $\sim$ 9, and stabilizing at a value close to 8. For Li doping also the A/B evolution has a specific shape which can be interpreted in a similar manner[7].

It is also important to note that, in spite of an infinite number of set values for d, δ, T_D, and T_2, all the above systems (and probably all conducting polymers) obey to the same theorical model. Nevertheless it is generally difficult to extract unequivocable information concerning the electronic properties of the system. Reliable results can only be obtained by following step-by-step the evolution of A/B where y (and therefore σ) is cycled back and forth between the neutral and the metallic states. For example, for $(K_yCH)_x$ a value of A/B $\sim$ 3.2 is reached after passing through the maximum. This constant value corresponds to values of $d/\delta > 5$ and $T_D/T_2 \sim 0.5$ (see figure 5). It is then possible to extract the value of T_D if one knows T_2. As we cannot determine exactly the anisotropic contribution to the CESR peak linewidth, only the lower limit of T_2 can be determined from ΔH_{pp}. For y > 8% we obtain $T_2 > 2\text{-}10^{-9}$S, which yields $T_D > 10^{-9}$S. At this point it is difficult to obtain more accurate information on the kinetic properties of the electrons. More data is needed, particularly on the role of the inhomogeneous broadening of the CESR peak and of the role of the anisotropic local transport properties.

CONCLUSION

The above results and analysis clearly show that there is a strong correlation between the structural characteristics and the electronic properties of alkali metal doped $(CH)_x$. This is probably a general conclusion concerning all conducting polymers, but is particularly spectacular in the case of alkali metal doped polyacetylene due to the good compatibility between the ionic sizes of the dopants and the host structure.

We have pointed out the fact that reproducibility of the observed behaviours over various doping-undoping cycles is necessary in order to give reliable information about the intrinsic properties of the polymer/dopant system. Reversibility is not a requisite as, for a given cycle, the structural properties generally present some hysteresis (at least in the K doping cas). With this in mind we have shown that the combination of in situ electrochemical process and of some physical measurements (dc conductivity, ESR) allow comparative studies to be made. Then it appears, principally on the K doped $(CH)_x$, that structural and electronic properties are intimately dependent on each other.

We have established that ESR can be used to investigate electronic as well as structural properties of intercalated polyacetylene. While the linewidth gives information on the successive stages which appear during doping and undoping, the shape of the ESR spectra is directly related qith the transport properties of the metallic or semiconductig system. Such a relation-ship is particularly spectacular in the case of K doping, but exists also for the other polymer.dopant systems. Nevertheless, if it is clear that all systems obey to the same general principles (existence of staging, transition to a metal for $y > 6\%$, applicability of the Dyson model, in particular), the details of the various behaviours can be extremely different from system to system.

ACKNOWLEDGEMENTS

We thank A. Whittaker for helpful discussion and comments on our manuscript. This work has been supported by CNRS under contract ATP "Chimie douce".

REFERENCES

1. "Handbook of conducting polymers", Ed. T.A. SKOTHEIM, Marcel Dekker, Inc., 1986.

2. Proceedings of the International Conference on Science and Technology of Synthetic Metals, Kyoto, Japan, June 1-6, 1986, Synth. Met. 17 (1987).

3. KAUFMAN, J.H., CHUNG, T.C., HEEGER, A.J. Sol. St. Comm., 1985, 47, p. 585.

4. CHANCE, R.R., BREDAS, J.L., SILBEY, R., Phys. Rev., 1984, 29, p.4491.

5. CHEN, J., CHUNG, T.C., MORES, F., HEEGER, A.J., Sol. St. Comm., 1985, 53, p. 757

6. EL KHODARY, A., BERNIER, P., J. Chem. Phys., 1986, 85, p.2243.

7. FITE, C., EL KHODARY, A., BERNIER, P., Solid St. Comm., 1987, 62, p.599.

8. MORAES, F., CHEN, J., CHUNG, T.C., HEEGER, A.J., Synth. Met., 1985, 11, p.271.

9. BAUGHMAN, R.H., MURTHY, N.S., MILLER, G.G., SHACKLETTE, L.W., METZGER, R.M., J. de Physique, Colloque C3, 1983,, 44, p. 53.

10. WINOKUR, M., MONN, Y.B., HEEGER, A.J., BARKER, J., BOTT, D.C., SHIRAKAWA, H., Phys. Rev. Lett., 58, 1987, p. 2329.

11. FLANDROIS, S., HAUW, C., FRANCOIS, B., J. de Physique, Colloque C3, 1983, 44, p. 523.

12. SHACKLETTE, L.W., TOTH, J.E., Phys. Rev., 32, 1985, p. 5892.

13. FITE, C., BERNIER, P., Phys. Rev., to be published.

14. RACHDI, F., BERNIER, P., Phys. Rev., 33, 1986, p.4000.

15. DYSON, F.J., Phys. Rev., 98, 1955, p.349.
FEHER, G., KIP, A.F., Phys. Rev., 98, 1955, p. 337.

16. KODERA, H., J. Phys. Soc. Jap., 1970, 28, p.89.

17. FITE, C., Thesis, 1987, Montpellier.

18. FRANCOIS, B., MERMILLIOD, N., ZUPPIROLI, L., Synth. Met., 1981, 4, p.131.

INTERCALATION COMPOUNDS :
COMPARATIVE POSSIBLE ELECTROCHEMICAL APPLICATIONS

Jean - François Fauvarque

Division Energétique, Laboratoires de Marcoussis,
Centre de Recherches de la Compagnie Générale d'Electricité,
Route de Nozay, F - 91460 Marcoussis (France)

I) - INTRODUCTION

Speaking about the applications of intercalation compounds is quite challenging since there are currently only a few examples of them, at least from an industrial point of view.
I suggest two ways of overcoming that difficulty :

One is by foretelling or dreaming about possible applications of intercalation compounds.

The second is by looking at materials actually used in industry and by choosing those which could be named intercalation compounds. Although this may appear as some kind of an imperialistic way of describing applications of intercalation compounds, borderlines in science are moving lines and this method may indeed suggest new possible applications of intercalation compounds.

There is another difficulty : Industrial research groups are presently involved in looking for possible industrial applications of intercalation compounds and I would like to discuss them. But information about industrial Research and Development is rather difficult to obtain, even inside my Company, specially when that information is economically sensitive. Also industrial work is not Science, Science may be necessary for it, but industrial work is mostly Technology and Marketting. These are not the subjects I shall talk about, I would just like to discuss with you some promising topics about intercalation compounds. However the capabilities of any material cannot be demonstrated unless a system has been built and tested. Every part of the system must be good and when a system fails it may be difficult to know why. In that sense technology is very important for obtaining the right information about intercalation compounds.

I - 1) What is intercalation?

Although intercalation can be exemplified in many chemical reactions such as alloying, hydriding, etc..., I shall restrict my paper to the following topic :

A solid may interact with dissolved cations and electrons from external sources, generating a new solid in which structural elements of the initial solid are maintained. Such elements may be tridimensional, bidimensional or unidimensional structures, in which the cations become intercalated, usually without their coordination sphere of solvent.

With such a restrictive definition, intercalation corresponds to an electrochemical reduction of the initial system. Conceptually at least, the process is reversible, but the intercalated product may also be unstable or metastable and transformed by irreversible subsequent processes into different kinds of products.

In order to proceed at reasonable rates, the intercalation solid must have sufficient electronic and ionic conductivity.

Most of the applications will be considered at room temperature with ionically conductive solids such as oxides (usually tridimensional structures with channels for ionic intercalation and conduction), sulfides or phosphosulfides (often layered structures), and conductive polymers (unidimensional). Graphite intercalation compounds will not be considered in this paper.

I - 2) Which applications for intercalation compounds?

In order to find widespread industrial utilization any material should meet stringent qualifications :

1) To have some specific properties at outstanding level.
Any other properties must be at a good level.
2) To be easily accessible and easily processable.
3) Must be reliable and have good ageing properties.
4) Must not have any bad property.
For example toxicity may impede any industrial application.
5) Must be economically profitable when substituted to other existing material.

Applications for intercalation compounds will be considered within the following fields :

Electricity conduction
Electrochemical generators
Electrochemical devices

II) - CONDUCTION OF ELECTRICITY

Electricity conduction is precisely one example of a mythic application of intercalation compounds. Such compounds are indeed conductive but their conductivities are usually **not** outstanding. One may just dream about it with the newly described oriented polyacetylene if other properties are sufficiently good.

However some companies are currently investigating the possibility of making processable conductive polymers with such applications as EMI shielding in mind.

Similarly silicium substitution by intercalation compounds, specially organic conductive polymers, is largely mythic. Of course any electronic device could be built with conductive polymers : diodes, field effect transistor, photodiodes, etc...; but to any such device would have poor or very poor performances because of the low mobility of the carriers (see Handbook of conductive polymers).

Only if compounds such as Chevrel Phases or mixed valence copper oxides are considered as intercalation compounds (and why not ?) then one may find some outstanding properties in electronic conduction.

Chevrel Phases such as $Pb\ Mo_6\ S_8$ are superconductive and show high critical fields. Critical currents of 100 Amp / mm^2 have been demonstrated at 20 Tesla at 4.2 K, which is outstanding.

$Ba_2\ Y\ Cu_3\ O_7$ oxides are supraconductive and have outstanding critical temperatures higher than 90 K. They may be considered as compounds of intercalated yttrium and baryum in layers of copper oxide.

Although their intercalation nature may be discussed, it suggests that their composition and properties could be modified by processes of "mild chemistry" or electrochemically.

Anyhow their possible industrial aplication in supraconductive devices is still far from being reached and will necessitate substantial work especially for obtaining good processability.

III) - ELECTROCHEMICAL GENERATORS

III - 1) Why intercalation compounds ?

Use of intercalation compounds in electrochemical generators may bring several advantages :

1) limitation of the amount of electrolyte.

For example in a lithium battery, during discharge, the concentration of lithium ion is kept constant, every lithium cation intercalated is replaced by a lithium cation coming from the lithium anode. Thus the salt concentration can be kept at its optimum value for the conductivity and the amount of electrolyte can be kept at a minimum value compatible with a good wetting contact of the electrolyte with the electrodes, thus saving weight for the whole battery. This behaviour of batteries with intercalation compounds contrasts with that of generators such as lead acid batteries in which sulfuric acid is consumed during discharge and with that of Leclanché batteries in which zinc salts accumulates in the electrolyte.

2) Intercalation compounds are ionically and electronically conductive.

This property allows a good utilization of the material. On the opposite sulfur has a high faradic capacity but no conductive properties. It is very difficult to use it in a solid electrode, even when mixed with electronically conductive additive such as carbon. Often intercalation compounds are poorly conductive and the power of the battery can be substantially increased by addition of conductive carbon, but anyhow, some electronic and ionic conductivity of the cathodic material is necessary for good performances.

3) Intercalation compounds may be completely insoluble.

This property avoids self-discharge of the battery, specially in organic solvent when they do not react at all with the materials of the electrode. It is also a good property for obtaining a large number of cycles, intercalation is a pure solid state chemical reaction and does not involve dissolution - recrystallization processes which are harmful to the length of life of rechargeable batteries. It must be emphasized that good ageing properties are dependent not only on the nature of the electrode material but also on the nature of the solvent which must be perfectly inert towards both electrode, sufficiently solvating for assuring a good conductivity and not too strongly bonded to the cation for allowing desolvation during the intercalation process.

III - 2) Drawbacks

As already mentioned intercalation compounds may suffer drawbacks and limitations. Unsufficient electronic conduction and ionic conduction may cause power limitations or energy losses. It is usually necessary to add electronically conductive materials and a binder for avoiding desagregation of the intercalation compound on cycling. This causes some losses on specific energy.

Since intercalation corresponds to the formation of a solid solution, theoretically without phase separation, the chemical potential of the material changes progressively during the process, and so does the cathode potential. It is usually considered that variable potential on discharge is a shortcoming, but sometimes it may also be looked as beneficial since this gives an indicator of the state of charge of a battery, a very challenging problem for such batteries as rechargeable nickel - cadmium.

III - 3) Theoretical energy densities

Table 1 shows some figures of theoretical energy densities for electrochemical couples which are used in batteries and may have something to do with intercalation.

Chromium oxide is a good potential intercalation cathodic material but still suffers from ageing problems.

Vanadium pentoxide is a good intercalation compound over the first Faraday and may be made reversible. It will be discussed with the rechargeable batteries. Discharge can be extended over 2 Faradays in primary batteries.

Manganese dioxide and carbon fluoride are typical intercalation compounds used in primary bateries. Noteworthy are their volumic energy densities.

Cupric oxide and cupric derivatives such as cupric oxyphosphate have ashonisihng ageing and conservation properties. Batteries made with those compounds can be kept at 70 °C during long period of time without losses of their capacities.

Metallic sulfides are famous for their safety properties. Lithium batteries with these cathodic materials have been shooted, burned, crushed, etc..., They never exploded. They are typical 1.5 V batteries, easily interchangeable with common zinc batteries without their conservation and leakage problems.

THEORETICAL ENERGY DENSITIES

Cell reaction	O.C.V.	Wh/kg	Wh/dm^3
$1.45\ Li + CrO_{2.86} \rightarrow CrO_{2.86}\ Li_{1.45}$	3.84	1420	2860
$Li + V_2O_5 \rightarrow V_2O_5Li$	3.4	482	1357
$Li + MnO_2 \rightarrow MnO_2Li$	3.5	1027	3185
$nLi + (CF)_n \rightarrow nLiF + C$	3.2	2260	3300
$2\ Li + CuO \rightarrow Cu + Li_2O$	2.24	1285	3140
$8\ Li + Cu_4O(PO_4)_2 \rightarrow Cu + Li_2O + Li_3PO_4$	2.7	1123	2817
$4\ Li + FeS_2 \rightarrow 2\ Li_2S + Fe$	1.75	1273	2474
$2\ Li + CuS \rightarrow Li_2S + Cu_2S$	2.12	557	1683

IV) PRIMARY BATTERIES

Detailed information concerning primary lithium batteries can be found in the book edited by Jean - Paul GABANO (1).

IV - 1) Lithium cupric oxide battery

These batteries have two outstanding properties : one is a very flat discharge plateau at 1.4 Volt at room temperature, the other is very good ageing properties even when stored at high temperature (70 °C).

Since cupric oxides lithium batteries have very flat discharge curves at 1. 4 Volt, it can ben questioned whether these coumpounds are intercalation compounds. The mechanism of running cupric oxide batteries is still controversial but several facts can be pointed out : flat potential indicates a multiphase system, it is certainly true that appearance of metallic copper helps the conduction properties of the cathodic material. But discharges at higher temperature may show a behaviour nearer those of intercalation compounds, and the same is true for discharges at low drain at low temperatures where dismutation of intercalation compounds may be slowed down by the the low temperature. It is thus possible to look at the cupric oxide cathode as an intercalation cathode with poor ionic migration properties and dismutation of the lithium intercalation compound as soon as it is formed. This is detrimental of course to the energy content of the battery which should have a theoretical OCV (Open Circuit Voltage) of more than 2 volts. Furthermore dismutation suppresses the possibility of recharge by a desintercalation process.

Better voltages are achieved when using copper oxyphosphate as cathodic material. Discharges can be typically achieved at 2.2 Volts. Several advantages remain such as the absence of self discharge and good ageing properties or even are enhanced such as running at higher temperature up to the melting point of lithium (180 °C).

IV - 2) Manganese dioxide

Lithium $Mn\ O_2$ primary batteries are widespread and have found many applications in electronics such as memory back-up. They may be considered as an actual example of industrial utilization of intercalation compound.

Outstanding is the energy content of MnO_2

300 Ah / Kg can be easily obtained at more than 3 volts versus lithium, and the specific gravity of $Mn\ O_2$ is 5, which means 1 500 Ah per liter of cathodic material !

This is an outstanding value specially for small volume batteries.

Manganese dioxide is known under several crystallographic varieties. It possesses both electronic and ionic conductivity. The gamma variety, obtained electrochemically, has larger channels and usually better discharge caracteristics. Powder X ray diffraction during electrochemical processes demonstrates the intercalation processes, with conservation of the diffraction pattern and a small increase of the lattice parameter. Incorporation of lithium ions in the lattice is attended with reduction of Mn^{4+} to Mn^{3+} and then partly to Mn^{2+}, this last step being electrochemically irreversible.

In actual D - size cells (R 20, external volume 50 ml) energy densities of 750 Wh / L or 330 Wh / Kg have been obtained (weight of the container included in the cell weight), power however is limited in the range of 25 to 40 W / Kg (60 to 80 W / L). These figures can be compared with those of a D size nickel - cadmium sealed rechargeable cell : capacity 4 Ah under 1.2 Volt, weight 150 g, 32 Wh / Kg, but any such cells can deliver 85 W power : 570 W / Kg or 1 700 W / L.

New varieties of manganese dioxide are still under industrial development for enhancing the energy content, but the great challenge is discovering the way of making rechargeable Li / Mn O_2 cell. The possibility of making rechargeable cells depends also on the possibility of finding appropriate solvents. Dioxolane, which is appropriate for lithium - copper oxide batteries or for rechargeable lithium - nickel phospho trisulfide cells, is oxidized by manganese dioxide and solvents like mixture of propylene carbonate and dimethoxy ethane (PC - DME), used in manganese dioxide lithium cells could be inadequate for lithium cyling.

IV - 3) Carbon fluoride cells

Carbon fluoride cells compete strongly with manganese dioxide cells for many applications in electronic devices. The lithium polycarbon monofluoride system was first introduced by MATSUSHITA in the early seventies, it was selected partly because of its lack of solubility in all existing organic electrolytes.

Considering carbon fluoride as an intercalation compound is somehow abusive. Carbon forms layers of hexagonal 6 carbon cycles, with the chair conformation of cyclohexane, whose equatorial bonds are C - C single bonds and axial bonds are covalent localized C - F bonds. Reduction of carbon fluoride in the presence of lithium cation causes the formation of insoluble lithium fluoride and electronically conductive graphite, the process is not electrochemically reversible and carbon fluoride, presently used in lithium cells, cannot strictly be considered as a genuine intercalation compound.

A typical $(C F_{0.8})_n$ material has a theoretical electrochemical capacity of 790 Ah / Kg, which is outstanding, For a lithium cell with 3 Volts OCV, that means about 2 KWh per Kg of theoretical energy density. But even at low drain voltage drops to 2.4 volts, less than the 3.0 Volts of manganese dioxide batteries, reducing the theoretical energy density to 1.6 KW / Kg, which is still an outstanding figure. Specific gravity of carbon fluoride is in the range of 2.5 which means about 2000 Ah / L theoretical compared with actual 1500 Ah / L for manganese dioxide. Precipitation of lithium fluoride limits the actual volumic energy density to figures close to that of manganese dioxide, in the range of 450 Wh / L in actual cells.

As many other lithium cells, the system is not very powerful. Its power depends on the specific surface of carbon fluoride, with a quite good figure of about 300 m^2 / g and on the conductivity of the electrolyte, usually lithium tetrafluoborate dissolved in pure gamma butyrolactone. Such an electrolyte allows good performances from - 20 to + 85 °C.

On a volume basis, manganese dioxide and carbon fluoride have very similar performances, depending strongly on the cell design, carbon fluoride being superior on a weight basis.

As for manganese dioxide, obtaining rechargeable carbon fluoride cells is a challenge. Hope can be put for success if true carbon fluoride intercalation compounds can be prepared and keep the good properties of polycarbon monofluoride, specially insolubility and inertness towards organic electrolyte.

V) RECHARGEABLE LITHIUM BATTERIES

V - 1) Present status

Presently there are no commercially available rechargeable lithium batteries. A few years ago, EXXON has sold Li / $Ti\ S_2$ cells but use of lithium perchlorate in the electrolyte was considered as unsafe and this activity was cut down. Since 1985 MOLI ENERGY, a Canadian Company, sells Li / $Mo\ S_2$ rechargeable cells for industrial evaluation and is believed to prepare industrialisation. Many other companies are reported to investigate possible development of lithium rechargeable compounds, for example (non exhaustive) :

Li - $S\ O_2$	Duracell, Amoco
Li - $S\ O_2$ - $Cu\ Cl_2$	Duracell

Lithium / Inorganic intercalation cathodes

$Ti\ S_2$	E I C , several japanese companies
$Mo\ S_2$	Moli Energy
$Ni\ P\ S_3$	Laboratoires de Marcoussis - CGE
$V_2\ O_5$	N T T (Japan)

Lithium / Organic conductive polymers

Polyacetylene	General Electric (abandoned)
Polypyrrole	BASF - Varta, French CEA
Polyaniline	Varta - CEA
Poly para phenylene	CNET

Information concerning these developmental studies is difficult to obtain or partial. So I would like to comment on the use of intercalation compounds in lithium rechargeable batteries on the bases of a few published data. The principles are still valid even if the informations do not reflect the most up to date state of the art.

Goals

For secondary lithium cells, the goals are in the range of 100 Wh / Kg and of 200 Wh / L at discharge rate of C / 5, with possible augmentation of discharge rate up to 2 C. Comparatively rechargeable Ni Cd cells have capacities of 30 Wh / Kg, 90 Wh / L at low drain and can be discharged up to 40 C.

V - 2) Ni P S_3 secondary lithium cells

This rechargeable system has been developped at Laboratoires de Marcoussis and more detailed information is available in 3). Possible applications are military and the work was supported by DRET.

The choice of the system was made on the following basis :

Availability of a good performance electrolyte : Li Cl O_4 , 2 Molar in dioxolanne, with a conductivity of 4 10^{-3} (ohm . cm)$^{-1}$ at room temperature working from - 30 to + 60 °C, perfectly inert towards lithium and the cathodic material.

Good cyclability of the cathodic material. In flooded electrolyte cells, nickel phospho trisuflfide can be cycled over 300 times on 1.5 faraday per mole at C / 5 charge and discharge rate. Although intercalation oxides may have higher energy content, they do not cycle as easily as sulfides or phosphosulfides.

Mean voltage over lithium is 1. 8 volt which demonstrate a theoretical capacity of 375 Wh / Kg. This figure compares favorably with similar figures obtained in similar conditions for different intercalation compounds such as Ti S_2 , Li_x Fe S_2, Mo S_3, etc...

Nickel phospho trisulfide can be reliably synthetized in a cristalline form stable in moist air, which is an advantage over titanium disulfide which is very sensitive to moist air. It is easy to make cathodic materials from it.

Lithium aluminium alloy was chosen as anodic material and was demonstrated to have good cycling capabilities in testing conditions.

For evaluating the performances of the material, a definite system was built and tested : The cell was a 2020 button cell (20 mm external diameter, 2 mm external thickness). Its nominal capacity was 10 mAh, with large excess of anodic capacity, discharged and charged at C / 5 rate between definite voltage limits. Every cell contained a limited amount of electrolyte : 100 microliters, for demonstrating the main advantage of intercalation compounds : the possibility to use limited amount of electrolyte. Standard charge and discharge potentials were respectively 2.3 and 1.05 volts. A cell was considered as failing when its capacity fell under 5 mAh under these conditions.

The results are convincing :

100 cycles can usually be obtained under standard conditions.
Failure is caused by progessive degradation of the electrolyte.
This degradation is dependent on the upper voltage limit at the end of charge (solvent oxydation).
Self discharge is low and ageing is very good.

Caracteristic charge and discharge curves are given in the figure. It should be pointed out that after 10 cycles the OCV of the cells are higher than that of the first cycle. This corresponds probably to some amorphisation of the material which cycles in a somehow metastable state.

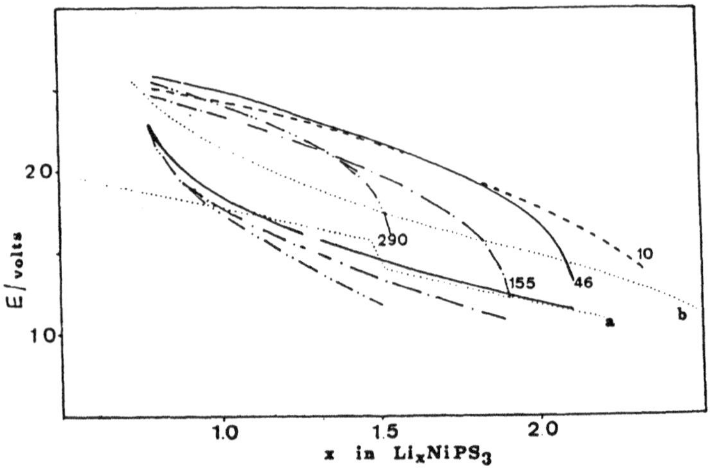

$NiPS_3$ cycling in experimental cells
(a) initial open circuit voltage; (b) after ten cycles.

Research and development is now on continuation in relation with the SAFT Company for enhancing the performances and developping more capacitive cells, specially obtaining more oxidation resisting electrolytes and substitution of lithium perchlorate. Results are very promising.

V - 3) Some other lithium - inorganic intercalation compounds systems

Among other systems currently under development, two at least deserve some special mention : Li - Mo S_2 from Moli Energy (4) and V_2O_5 - P_2O_5 glasses (5).

Mo S_2

The rechargeable cells are cycling upon the metastable intercalation compound Li_x Mo S_2, x varying from 0.2 to 0.7. This compound is obtained from easily accessible and cheap natural ore Mo S_2 in which Li is intercalated electrochemically at low potential up to one lithium per molybdenum. Electrochemical desintercalation yields the metastable phase.

Overcharge transforms the metastable phase into stable Mo S_2 and necessitates another electrochemical generation of the metastable phase before using the cell again.

Overdischarge forms irreversible reduction compounds.

Part of the success of the generator resides in the technology. Upon charging lithium anodes increase in volume more than the intercalation compounds decrease. An appropriate separator crushes down the growing dendrites of lithium allowing cycling without short circuit.

Perchlorate free electrolyte have been designed containing lithium hexafluoroarsenate, propylene carbonate and possibly other solvents which have been choosen as non- volatile for security aspects. Degradation of the solvents occurs when external voltage is put over 4.5 volt or put under 1 volt.

Standard cycling is performed on a C size cell (R 14) at 0.85 Amp on discharge and 0.28 Amp on charge between voltage limits of 2.4 volts on charge and 1.4 volt on discharge. Under these conditions C size cells have more than 2 Ah of cycling capacity and may work 500 cycles at more than 1Ah capacity. Furthermore pulse discharge can be obtained at 10 Amp without damage. For the same size Ni Cd batteries can deliver 2 Ah under 1.2 Volt.

Such a system clearly demonstrates the advantages of intercalation compounds : namely good retention of charge and good ageing properties, associated with good energy content, knowledge of the state of charge from OCV and good cyclability; and also their drawbacks : sensitivity to overcharge and overdischarge.

V_2O_5 - P_2O_5 glasses

Transition metal oxides have much higher reduction potentials than sulfides and thus higher theoretical energy content but their cyclability is not as good. Usually after several cycles, intercalation compounds show bad crystallinity, this suggests the possibility of cycling transition metal oxide glasses, an approach which was taken among others by Nippon Telegraph and Telephon (N T T) who published some results (5).

Vanadium pentoxide easily forms glasses when mixed with a few percent of phosphorous pentoxide (5 to 10 %). These cathodic materials have more than 250 Ah / Kg of proven cyclable capacity from 3.8 volts to 2.0 volts and can be overdischarged over 500 Ah / Kg with degraded cyclability.

The main advantage of adding phosphorous pentoxide is to enhance the performances at high drain.

At 2 volts cut off voltage on discharge, pure V_2O_5 gives 724 Wh / Kg of cathodic material at 0.5 mAmp / cm^2 and only 387 Wh / Kg at 5 mAmp / cm^2; with 10 % P_2O_5 the figures become 689 Wh /Kg at 0.5 mAmp / cm^2 and 661 Wh / Kg at 5 mAmp / cm^2. Several hundreds of cycles have been demonstrated at 0.5 mAmp / cm^2.

These figures demonstrate a new way for obtaining rechargeable lithium batteries with outstanding specific energy (100 Wh / Kg and 200 Wh / L) and good other properties of cyclability, power, conservation of charge, ageing properties and so on.

V - 4) Lithium - organic conductive polymers

Certainly being electronically conductive is an outstanding property for an organic material, but that is not a sufficient property for being used in rechargeable battery. Use of polyacetylene type polymers in lithium secondary batteries is still presently in the field of mythic proposals mostly based on false analogies. Organic materials does not necessarily mean thermoplastic processable materials. Polyacetylene, polyparaphenylene, polypyrrole, polythiophene, etc... are cross-linked, insoluble, non - fusible, untractable materials. Low specific gravity does not mean light -weight batteries, one kilogram of lead is not heavier than one kilogram of doped poly acetylene. Doped conductive polymers have usually less specific capacity and much less volumic capacity than inorganic electroactive materials. High specific area does not mean powerfull batteries when discharge rates are limited by ion diffusion velocities in the bulk of the material and ionic conductivities of the possible electrolytes. Low prices for acetylene or benzene do not mean cheap electroactive organic materials; etc...

Cyclability and self - discharge of several electroactive polymers have been checked by B. SCROSATI et al. (6) who write in conclusion " it now appears possible to obtain polymers characterized by fast and reversible electrochemical doping processes. Some questions on their effective applicability in lithium batteries, however, still remain unsolved. One of them is related to the electrochemical stability which, with the possible exception of poly - pyrrole, seems to be a drawback common to all polymers in their doped state. Another point of concern is the energy content, which is directly related to the doping level achievable with the electrochemical processes. At the moment, all of these aspects make it improbable that there will be a successfull utilization of conductive polymers as electrodes for high-rate, long-life lithium batteries. The field is very young, however, and new types of conductive polymer electrodes are being continuously characterized and proposed for electrochemical applications..."

More precisely polyacetylene, positively doped, has been demonstrated to be chemically unstable. Negative doping by potassium may however yield stable and reversible negative material (see the paper of BERNIER), but its energy content is quite low.

Poly - pyrrole is certainly the subject of many investigations, specially for enhancing its processability. It is claimed to be a positive material with 100 Ah / Kg of capacity, but this figure has never been demonstrated in a limited amount of electrolyte.

Poly - aniline is one of the oldest conductive polymer and is presently extensively studied. Its performances are claimed to be strongly dependent on the technique of preparation.

Poly para phenylene (PPP) is claimed to be easily cyclable both as positive or as negative material, but its performances seem to be limited by the electrolyte stability. Chemically modified PPP may bring a solution to this problem.

Another way of reshearch is looking at the conductive polymer as an electronic conductor. A Le MEHAUTE has described the possible use of sulfur - polyacetylene composite materials with 1000 Ah / Kg of capacity, looked as charge - transfer compounds and claimed to be more capacitive than any other sulfur electrode (7).

Another possibility is using these conductive polymers as indicators of charge end - points or discharge end - points.

In any event, use of conductive polymers for lithium intercalation batteries is still, at best, a dream.

VI) ELECTROCHROMIC DEVICES

Many compounds change color when changing oxidation state. Inter- calation compounds often do so. Since intercalation may be well reversible, such compounds could find use in electrochromic devices for large screens or flat screens.

Most studies have been devoted to proton intercalation in inorganic oxydes :

Reductive electrochromism : $W\,O_3$ - $V_2\,O_5$ - $Mo\,O_3$ - $Nb_2\,O_5$

Oxidative electrochromism : $Ir_2\,O_3$

Lutecium diphtalocyanines can give electrochromism both on reduction and oxydation.

Electrochromism can also be obtained with other cation intercalation, specially lithium which has been studied mainly with organic conductive polymers. For example GARNIER's group developped work on the electrochromism of polythiophene and derivatives (8) mainly in oxidative procedures with intercalation of different kinds of anions. Chemical derivatisation of thiophene opens the way of multichromic displays.

Potential advantages of electrochromic display over liquid crystal displays are :
Non directivity
Low current drain : 1 milli coulomb per square centimeter
Memory of state with very low energy consumption
Possibility of making integrated all-solid devices

For having good performances the response time must be fast, ten milli seconds, this appears to be achievable with electroactive polymers, at least in some instances, and the number of possible cycles must be very high, this appears to be possible with lutecium di phtalocyanine but still has to be confirmed with organic conductive polymers.

The main problem with electrochromic devices is marketting. One has to find some specific need which cannot be satisfied by the very common liquid crystals diplays.

VII) CONCLUSION

For finding some industrial application, any material must have at least one outstanding property and no poor ones. Lithium intercalation inorganic compounds have outstanding properties for electrochemical energy storage. They are used in primary lithium batteries and currently developped in lithium rechargeable batteries. Organic conductive polymers have not yet found industrial applications, although some are being investigated such as semi - conductive plastics in electrical engineering applications, rechargeable lithium batteries, and electrochromic devices. The advent of these applications also depends upon marketing and ecomomic arguments.

VIII) LITERATURE

1) Lithium batteries.
Book edited by J. P. GABANO, Academic Press 1983

2) Handbook of conducting polymers, vol 1 and 2;
T. A. SKOTHEIM Editor, Marcel Dekker 1986

3) Ageing of rechargeable lithium batteries
A. de GUIBERT and A. HERMELIN
International symposium on lithium batteries, Paris 15 April 1984
J. of Power Sources 1985

4) Performances of rechargeable Li / Mo S_2 cells
W. A. ADAMS, G. J. DONALDSON and J. A. R. STILES
14 th International Power Sources Symposium 1984

5) $V_2 O_5$ - $P_2 O_5$ glasses as cathode for lithium secondary battery
Y. SAKURAI and J. YAMAKI
J. Electrochem. Soc., **132,** 512 (1985)

6) Kinetics of semiconducting polymer electrodes in lithium cells
B. SCROSATI, S. PANERO, P. PROSPERI, A. CORRADINI and M. MASTRAGOSTINO
J. of Power Sources, **19**, (1987) 27 - 36

7) Charge transfer complexes between sulphur an polyacetylene like materials
A. LE MEHAUTE, A. PERICHAUD, P. BERNIER, TRAN VAN HOANG et A. GUYOT
Synthetic Metals, **11**, (1985) 373 - 375

8) Organic conducting polymers derived from substituted thiophenes as electro - chromic material
F. GARNIER, G. TOURILLON, M. GAZARD et J. C. DUBOIS
J. Electroanal. Chem., **148** (1983) 299 - 303

PART B. SEMINARS AND COMMUNICATIONS

STRUCTURAL PROPERTIES

OF DOPED POLYACETYLENES

Serge Flandrois and Ali Boukhari

Centre de Recherche Paul Pascal
Domaine Universitaire
33405 Talence Cedex, France

INTRODUCTION

Among polymers that become highly conducting upon doping, polyacetylene, $(CH)_x$, plays a special role: this simplest conjugated polymer can be obtained under high crystallinity, which gives the opportunity to study the relationship between structural and electrical properties. However polyacetylene cannot be obtained as single crystals, at least up to now, but under the form of films constituted of randomly distributed fibers with diameters ranging from about 200 A to 500 A. The fibers can be more or less aligned preferentially, for example by stretching the films. Nevertheless, X-ray spectra show only Debye-Scherrer diffraction rings, which in addition are limited in number (10 to 12). Thus the general approach consists in imagining a structural model a priori and then in comparing observed positions and intensities of diffraction peaks with those calculated from the model. Also of interest are lattice packing calculations and analysis of molecular packing in model compounds.

$(CH)_x$ can be prepared in two isomers, cis and trans, according to the polymerization temperature. The cis isomer is obtained at low temperature, whereas the trans form is stable at room temperature. The structure of the cis isomer has been studied by Baughman et al.,[1]. The unit cell was found orthorhombic with parameters a = 7.61 A, b = 4.47 A (chain axis) and c = 4.39 A. There are two chains per unit cell and the setting angle of the polymer chains with respect to the (001) plane is equal to 59°. The structure of the trans form has been determined by Fincher et al,[2]. The main results are a monoclinic unit cell with parameters a = 24 A, b = 7.32 A, c = 2.46 A (chain axis) and β = 91.5°. There are two chains per unit cell and the setting angle with respect to the (010) plane was found to be Φ = 55°. The chains lie in the (ac) plane and form layers perpendicular to b-axis with a ABA stacking, like the carbon layers in graphite,[3]. Fig. 1 gives a schematic representation of the structure projected along the chain axis. Using the Van der Waals radii of C and H atoms, a close packing of the polymer chains is observed.

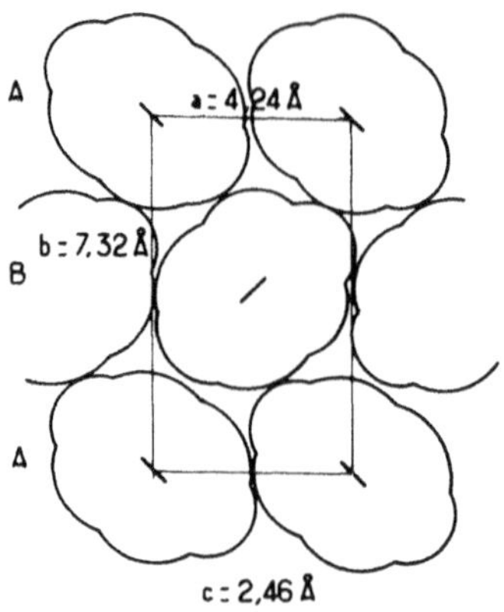

Fig. 1. Structure of pristine $(CH)_x$ projected along the chain axis.

$(CH)_x$ can be doped with either electron donors or acceptors. In both cases the main structural features are the appearance of new diffraction peaks and a decrease in crystallinity: the coherence lengths estimated with the Scherrer formula from the width of the reflections are generally half those (~ 100 A) estimated for pristine $(CH)_x$. Moreover the doping induces a cis trans isomerization,[4], even in the undoped regions. Finally the doping appears to be inhomogeneous: whatever the doping levels, there is always coexistence of the new structure with that of the pristine material. Thus, structure studies of doped $(CH)_x$ are still more difficult than for pristine $(CH)_x$. This explains that controversies can appear. In this contribution we will give the main proposals for the structure of $(CH)_x$ doped with acceptors and donors and we will conclude with a comparison with graphite intercalation compounds.

STRUCTURE OF (CH)x DOPED WITH ACCEPTORS

The first thoroughly investigated structure was that of $(CH)_x$ doped with iodine. Iodine is easily incorporated from the vapour phase into the polyacetylene lattice. Raman spectra[5] indicate the presence of I_3^- ions and also, for high doping levels, of I_5^- ions. X-ray diffraction spectra show the appearance of several new diffraction peaks, one of them corresponding to a large spacing of about 8.0 A. Baughman et al.,[1] noticed that this value is roughly the sum of the separation of close-packed planes in the parent polymer and the effective Van der Waals thickness of a dopant layer formed of an array of I_3^- ions. Thus, by analogy with graphite, they proposed an intercalation model, susceptible to be valid for other dopants. These results were corroborated by other authors, [3,6].Figure 2 shows the corresponding structure (of first stage, using the graphite terminology). Such a model leads to a composition of $(CHI_{0.33})_x$ for the fully-intercalated material, in good agreement with the highest observed composition, $(CHI_{0.3})_x$.

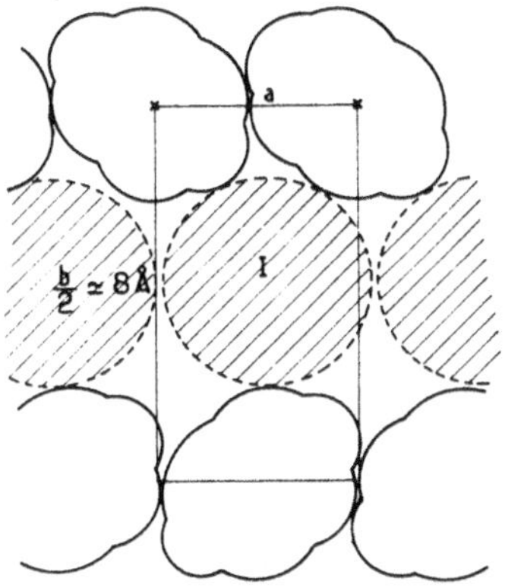

Fig. 2. Structure of iodine-doped $(CH)_x$ projected along the chain axis.

Of particular interest are the in situ X-ray experiments of Monkenbusch[6]. By recording the changes in diffraction pattern during doping with iodine, he was able to show that at low doping levels ($I/CH < 0.023$) iodine is not incorporated into the polymer lattice but rather adsorbed on the surface of fibrils. Moreover it seems that the formation of chainlike I_n^- ions occurs before intercalation. This doping mechanism is quite comparable with the intercalation of nickel chloride into graphite: nucleation of small $NiCl_2$ islands on the edges of graphite flakes before diffusion of these islands between the carbon planes,[7].

The same intercalation model was proven to be valid for polyacetylene doped with AsF_5 . From neutron diffraction studies, Hässlin et al.[8] showed that intensity calculations support a first-stage intercalation compound with a graphite structure.

A slightly different model has been proposed by Wieners et al.[9], for $(CH)_x$ doped with SbF_6^-. The counterions are still located between planes formed by polymer chains but, in addition, some $(CH)_x$ chains are present in the counterion plane, filling the space between the SbF_6^- ions. The authors compare this arrangement with the structures found in some radical-cation salts.

X-ray patterns of $(CH)_x$ doped with ClO_4^-, $FeCl_4^-$, $FeBr_4^-$, $GaCl_4^-$, and $InCl_4^-$ tetrahedral anions,[10] shows the appearance of diffraction rings corresponding to large spacings of 8.6 A, 9.3 A, 10.0 A, 9.2 A, respectively. These large spacings were interpreted as resulting from the intercalation of tetrahedral anions between (100) planes of $(CH)_x$.

More recently, however, the same authors proposed a quite different structural model for $(CH)_x$ doped with $FeCl_4^-$, based on X-ray, electron and neutron diffraction studies [11]. This model is shown in Fig. 3. Planes of $(CH)_x$ alternate with mixed planes of $FeCl_4^-$ and $(CH)_x$. In other words, $(CH)_x$ chains form channels in which columns of $FeCl_4^-$ tetrahedra can be introduced.

However no intensity calculations were performed. In addition this channel model assumes a total reorganization of the polymer structure, with not only the rotation of the $(CH)_x$ chains but also the suppression of one chain over two in the mixed planes $FeCl_4^-$ - $(CH)_x$. Finally, the model as-

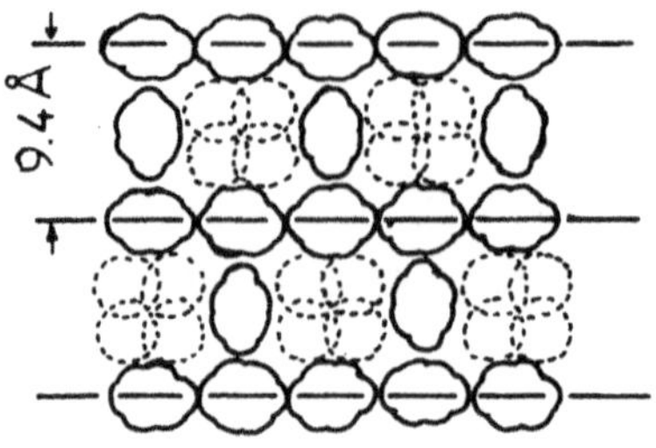

Fig. 3. Structure of $FeCl_4^-$ doped $(CH)_x$: channel model.

sumes a maximum $FeCl_4^-$ content y = 0.08 in $[CH(FeCl_4)_y]$, whereas our magnetic susceptibility experiments showed[13] that y_{max} = 0.055. The weight uptake above this value is only due to chlorine fixation on $(CH)_x$. Therefore, we reinvestigated the structure of these compouds by X-ray diffraction on a series of $(CH)_x$ doped with $FeCl_4^-$ either chemically or electrochemically[12].

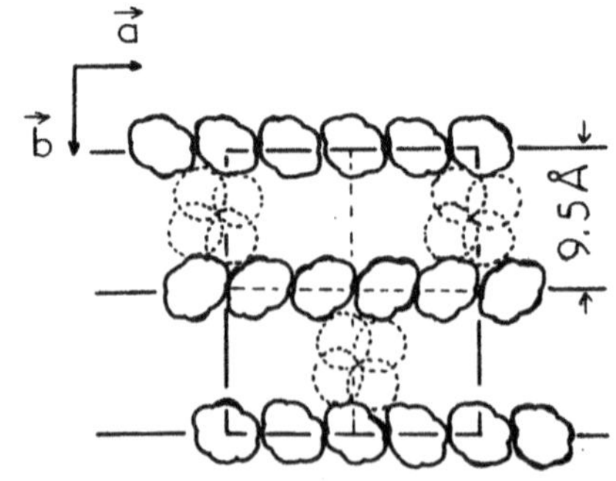

Fig. 4. Structure of $FeCl_4^-$ doped $(CH)_x$: intercalation model.

Our results were consistent with the model shown in Fig. 4. $FeCl_4^-$ tetrahedra are intercalated between (100) planes of polymer, in such a way that each $(CH)_x$ chain is on one side in contact with $FeCl_4^-$ columns. The unit cell parameters are a = 4a$(CH)_x$ = 17 A and b = 9.5 x 2 = 19 A. This model is in agreement with intensity calculations and with the maximum content in $FeCl_4^-$ which is found to be y_{max} = 0.06. Finally it accounts for the chlorine fixation which occurs at high doping level. When the content in $FeCl_4^-$ reachs 0.055, chlorine is fixed on the polymer chains, on the free side of the chains between $FeCl_4^-$ tetrahedra.

STRUCTURE OF $(CH)_x$ DOPED WITH DONORS

The structure of polyacetylene doped with alkali metals has been extensively studied. Using samples prepared from exposure of $(CH)_x$ films to the alkali metal naphtalide complex (Li, Na, K, Rb or Cs) in tetrahydrofuran, Baughman et al[13] found tetragonal unit cells in every case except for the lithium-doped samples which were observed to be amorphous. The a-parameters were 6.0 ± 0.1 A, 5.98 ± 0.05 A, 6.19 ± 0.07 A and 6.43 ± 0.07 A for Na, K, Rb, and Cs respectively. Then they proposed a channel model as shown in Fig. 5, where the host $(CH)_x$ structure is quite different from the pristine polymer, with $(CH)_x$ chains forming channels filled with columns of alkali metals. The size of the channels estimated for close packing of host molecules is very nearly equal to the diameter of K^+ ions, whereas for rubidium and cesium ions an expansion of the host lattice must occur. The observed amorphous structure of Li-doped samples was found to be reasonable due to the smallness of Li^+ ions. Finally it must be noticed that reasonably good agreement was obtained beween calculated and observed relative intensities.

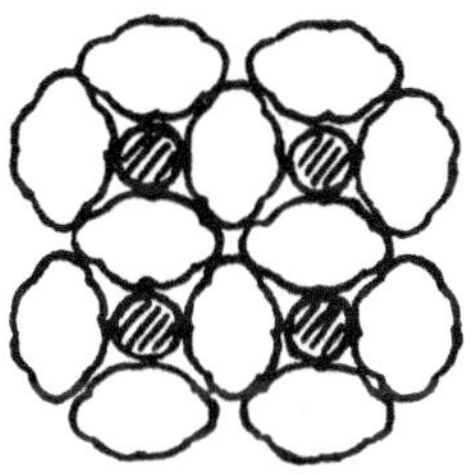

Fig. 5. Structure of alkali metal doped $(CH)_x$: channel model.

Neutron diffraction experiments were carried out[14] on a K-doped sample of deuterated polyacetylene $(CH)_x$, the doping being obtained as above from contact with a solution of K-naphtalide in THF. A monoclinic unit cell was proposed (slight distortion of the previous tetragonal cell) with a polymer setting angle of about 10 to 20°. Moreover it was found that annealing at 180°C, which increases the conductivity of the samples, produces noticeable changes in the neutron diffraction pattern. Besides a slight contraction of the unit cell, the order is improved, possibly indicating that the chains are slightly reoriented.

Flandrois et al.,[15] investigated the structure of Li, Na, K and Cs doped $(CH)_x$ by X-ray diffraction. Doping with sodium and cesium was realized by exposing the fibers to a solution of alkali metal naphtalide complex in THF. Potassium doped $(CH)_x$ was prepared by reaction with tetraphenyl 1, 1,4,4, butylene di-potassium in THF, whereas, for lithium doping, diphenylhexyllithium in cyclohexane was used.

All samples, even in the case of lithium, gave a number of diffraction rings (up to 17 in the case of sodium). As usual, some of them corresponded to undoped polymer, which means an inhomogeneous doping. Of particular interest is the first diffraction line which corresponds to the following spacings: 4.5 ± 0.1 A for Li, 5.98 ± 0.05 A for Na, 6.15 ± 0.05 A for K and 6.42 ± 0.05 A for Cs. These values are very nearly equal (for Li, K and Cs) to the repeat distance along b-axis ((020) peak) expected in an intercalation model (Fig. 6) for close packing of host molecules and guest cations, using the ionic diameter of alkali metals. In the case of Na, the value is larger than expected by about 0.6 A. This could be explained by a partial ionization of sodium (partial charge transfer) leading to a Na diameter intermediate between ionic and metal diameters. It is worth noticing that sodium has also a particular behavior compared to other alkali metals in graphite intercalation compounds. Finally, the intercalation model allowed to index the other diffraction peaks and calculation of diffraction intensities gave a rather good agreement with the observed intensities.

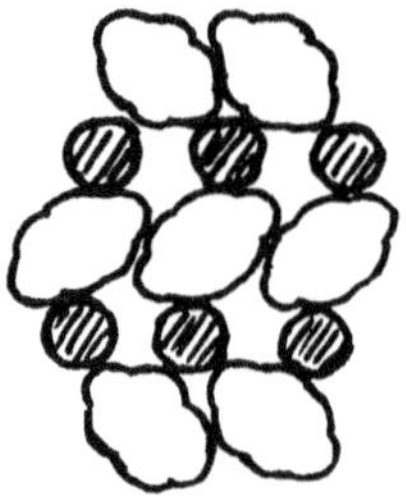

Fig. 6. Structure of alkali metal $(CH)_x$: intercalation model.

Further comparison with graphite intercalation compounds is useful. Firstly the intercalation model gives distances between alkali metals and nearest-neighbour carbon atoms identical for $(CH)_x$ and graphite. This suggests a similar type of bonding. Moreover, if the in-plane intercalate arrangement is assumed to be a triangular lattice with metal-metal distances of 4.9 - 5.0 A such as in graphite compounds, the maximum dopant concentration is 0.25. In addition, such metal-metal distances would give commensurate metal and $(CH)_x$ lattices. Another possibility is a honeycomb lattice which would lead to a maximum doping level of 0.17. The experimental maximum content is around 0.20. Note that the maximum doping level in the channel model would be 0.17.

CONCLUSION - COMPARISON WITH GRAPHITE INTERCALATION COMPOUNDS

Two main models for the structure of doped polyacetylene are in confrontation: the channel model and the intercalation model. There seems to be much more evidence for the intercalation model in the case of acceptor compounds. For donor doped $(CH)_x$, the difference in structure could come from the difference in sample preparation. The fact that Li compounds can be crystalline or amorphous according to the doping method is an indication. Nevertheless the intercalated model has the advantage of not changing drastically the host structure. In addition, NMR studies support this model[16].

In the course of this paper we have underlined several striking analogies with graphite intercalation compounds. Many others can be made. For instance, both materials, graphite and $(CH)_x$, can be doped with the same electron donors and acceptors. There are only a few exceptions: iodine cannot intercalate graphite, whereas chlorine, which is intercalated into graphite at low temperature, reacts with the double bonds of $(CH)_x$. In both cases three types of doping methods can be used: from vapor phase, in solution or electrochemically. The kinetics and the mechanism of doping are similar: the kinetics is mainly governed by the diffusion of dopant inside the host structure. As a result, the products are often inhomogeneous. Another common feature is the coexistence of residue compounds: the loss of dopant is not complete by deintercalation. In most cases the nature of intercalated species is the same: e.g. ions such as PF_6^-, ClO_4^- products of the disproportionation reaction of AsF_5 in AsF_6^- and AsF_3. However the case of metal halides is an exception: iron chloride is intercalated as $FeCl_3^-$ in graphite and $FeCl_4^-$ in $(CH)_x$. Finally the amount of charge transfer has the same order of magnitude in both families: 0.01 to 0.05 electronic charge per carbon atom. Such analogies can only be continued in the structural properties.

REFERENCES

1. R. H. Baughman, S. L. Hsu, L. R. Anderson, G. P. Pez and A. J. Signorelli, in "Molecular Metals", NATO Conf. Sci., W. E. Hartfield, ed., Plenum, N.Y. (1979) p. 187.
2. C. R. Fincher, C. E. Chen, A. J. Heeger, A. G. MacDiarmid and J. B. Hastings, Phys. Rev. Lett. 48:100 (1982).
3. S. Flandrois, C. Hauw and B. François, J. Phys. C3 44:523 (1983).
4. J. P. Pouget, P. Robin, R. Comes, H. W. Gibson, A. J. Epstein and D. Billaud, Physica B 127:158 (1984).
5. S. L. Hsu, A. J. Signorelli, G. P. Pey and R. H. Baughman, J. Chem. Phys. 69:106 (1978).
6. M. Monkenbusch, Makromol. Chem. Rap. Comm. 3:601 (1982).
7. S. Flandrois, J. M. masson, J. C. Rouillon, J. Gaultier and C. Hauw, Synth. Met. 3:1 (1981).
8. H. W. Hässlin, C. Riekel, K. Menke and S. Roth, Makromol. Chem. 185:397 (1984).
9. G. Wieners, R. Weizenhöfer, M. Monkenbusch, M. Stamm, G. Lieser, V. Enkelmann and G. Wegner, Makromol. Chem. Rap. Comm. 6:425 (1985).
10. J. P. Pouget, J. C. Pouxviel, P. Robin, R. Comes, D. Begin, D. Billaud, A. Feldblum, H. W. Gibson and A. J. Epstein, Mol. Cryst. Liq. Cryst. 117:75 (1985).
11. J. P. Pouget, A. Pron, A. Murasik, D. Billaud, J. C. Pouxviel, P. Robin, I. Kulszewicz, D. Begin, J. J. Demai and S. Lefrant, Sol. State Comm. 57: (1986).
12. A. Boukhari, S. Flandrois and A. Pron, To be published.
13. R. H. Baughman, N. S. Murthy and G. G. Miller, J. Chem. Phys. 79:515 (1983).
14. J. P. Aimé, M. Bertault, P. Delannoy, R. L. Elsenbaumer, G. G. Miller and M. Schott, J. Phys. Lett. 46:379 (1985).
15. S. Flandrois, C. Hauw and B. François, Mol. Cryst. Liq. Cryst. 117:91 (1985).
16. M. Ziliox, Thesis Strasbourg (1985).

X-RAY ABSORPTION IN INTERCALATED LAMELLAR COMPOUNDS

D. Bonnin and P. Kaiser

Laboratoire de Physique Quantique C.N.R.S. U.A. 421
E.S.P.C.I. 10 rue Vauquelin 75231 Paris Cedex 05

INTRODUCTION

This paper explains the possibilities of the X-ray absorption spectroscopies, particularly when they are used to study 2D-compounds. Different materials have been studied and the published results can be classified in several classes, the most usual are:
graphite and related intercalation compounds / dichalcogenides and related intercalation compounds / III-V layered compounds / FeOCl, MPS_3 / Some organic and biological compounds (for example: ferrocene and heme in myoglobin ...) / Some mineral compounds (micas, clays, pillared clays, several oxides and hydroxides..)

These spectroscopies are able to give some results about the structural local properties (examples concerning alkali-GIC are given) and about the electronic properties (examples of intercalated dichalcogenides).

X-RAY ABSORBTION

In the range of a discontinuity in a X-ray absorption spectrum, three parts can be distinguished. The first one called "edge region" or "threshold region" extends over about 10 eV. "XANES region" is the second one and the third one which corresponds to the range of highest energies is the "EXAFS region". XANES and EXAFS means respectively X-ray Absorption Near Edge Structure and Extended X-ray Absorbtion Fine Structure.

Each region can provide different informations. To understand the physical meaning of the first one, let us consider the notion of "threshold" [1]. There is usually a large confusion between the three following definitions :

a/ The "absorption threshold" Ea : the energy of the lowest energy core transition.

b/ The "absorption jump edge" Ej : the energy at half height of the principal absorption jump.

c/ The "ionization threshold" Ei : the energy required to eject the electron in the continuum ie

- the vacuum level in atoms and molecules
- the Fermi level (Ef) in metals
- the bottom of the conduction band in insulators.

Ei can de determined from XPS and optical data. Ej is a very used definition, however it is the one which has not a general physical sense.

The edge region

In non metals, there are bound states below the ionization threshold and therefore Ea is lower than Ei. In the edge region (below Ei), it is possible to observe different states according to the type of the compound.

-- In atoms : Series of discrete peaks appear, the Rydberg series. Each peak corresponds to a dipole transition between the core state and a bound state of allowed symmetry (according to dipole selection rules). For instance, for a K-edge, the components of the serie are 1s --> np transitions with $n = n',...,\infty$ and n' is the first vacant state of symmetry p. The absorption cross section for the first transition 1s --> n'p is very strong then the first spike is very intense. This peak is called "white line".

-- In molecules : Rydberg states are also present but difficult to be detected because broad intense peaks corresponding to transitions to bound valence states dominate the spectrum. These states are due to molecular orbitals (unoccupied in the ground state) which are deepen by the core hole coulomb attraction. They are often the most intense because the overlap of the inner shell is greater with these molecular valence states which are localized on the absorbing atom than with Rydberg states which extend far beyond the ligands in real space.

-- In insulators : "molecular-like" bound valence states occur also below Ei. They exist because the core hole is not completely screened by valence electrons, then the orbitals are pulled below Ei and form bound excited states or "core excitons". Given the atomic local character of the hole, the Coulomb attraction mainly acts on the *local* atomic or molecular vacant orbitals. Therefore the edge region of insulators is probing only the *local* structure of the system through the excited electronic states confined in a small cluster around the absorbing atom.

Because of the chemical bond, symmetry of an electronic state is a mixing of symmetries of the states of the absorbing atom and ligands. Then, transitions which are not allowed by dipole selection rules in isolated atom can be observed in some compounds. This phenomenon has a great importance for k-edges in transition metal in insulator compounds : maxima of absorption coefficient occur at lower energy than the principle jump (due to the first dipole permited transition) and have the name of "pre-edges". Their positions and magnitudes can give informations about oxidation state of the absorbing atom, coordinence and symmetry of its site.

-- In metals : The core hole is completely screened by conduction electrons in the valence band and no vacant states occur below Ei. Then Ei and Ea are the same and equal to Ef. The edge region extends over about 10 eV *above* Ef. Generally electrons of conduction are not completely delocalized except in ideal metals. The edge region probes the local and partial density of unoccupied electronic states near the Fermi level. Only "local" density because the measure is at the point of the excited atom. Only "partial" density because the features depend on symmetry of the initial state ie the core hole : according to selection rules, absorption cross section can be enhanced or suppressed for excitations close to the Fermi level.

Xanes and Exafs regions

When the photon energy E is only a few eV higher than Ei, the edge region is still concerned but when E is higher, the photoelectron is really ejected from the "central" (absorbing) atom with a certain kinetic energy Ekin. Then, processus of scattering of the photoelectron with the others electrons become important and the photoelectron mean free path λ decreases drastically [Fig. 1].

Sharp Xanes peaks and Exafs oscillations are respectively strong and weak modulations of the *atomic* absorption coefficient due to interference effects on the photoelctron wavefunction determined by scattering with neigbour atoms. In the Xanes region, λ is great enough to describe the processus in term of multiple scattering within a small atomic cluster and in the Exafs region, λ is smaller and that permits the description only in term of single scattering. The multiple scattering pathway are classified according with the number of scattering events. This number indicates the order of correlation functions. Exafs give only the first order pair correlation function while multiple scattering gives the higher orders (infinite in principle but because of inelastic scattering, the contribution of the longest pathways is reduced).

It is important to note that calculations in multiple scattering (in the real space) and cross section calculations starting from band structure of metals (in the wave vector space) are two equivalent ways to study the same unoccupied density of states.

Exafs gives informations about local structures in term of atomic radial distribution around the central atom within a range of about 4 Å and Xanes contains informations on coordination geometry and bond angles. However, in "2D-powder" compounds, it is possible to extract bond angles from Exafs as it is shown above.

Xanes theory is not so simple than Exafs theory and calculation procedures are rather long. Nevertheless, it is a powerful tool for local structure determination especially in complex disordered systems. Only Exafs theory will be developed here. Its advantage is to contain an analytical formulation.

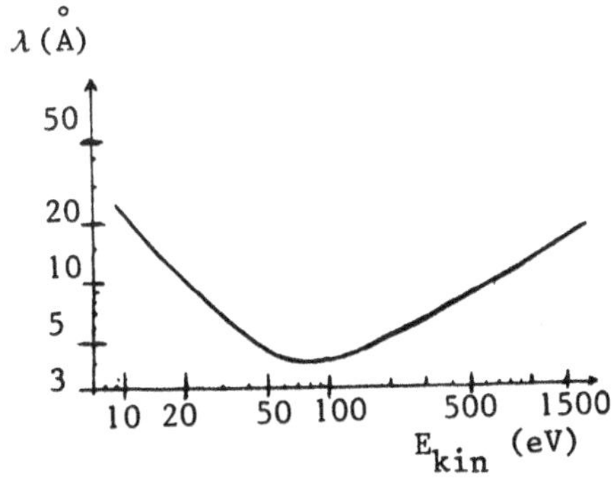

Fig. 1
The photoelectron mean free path.

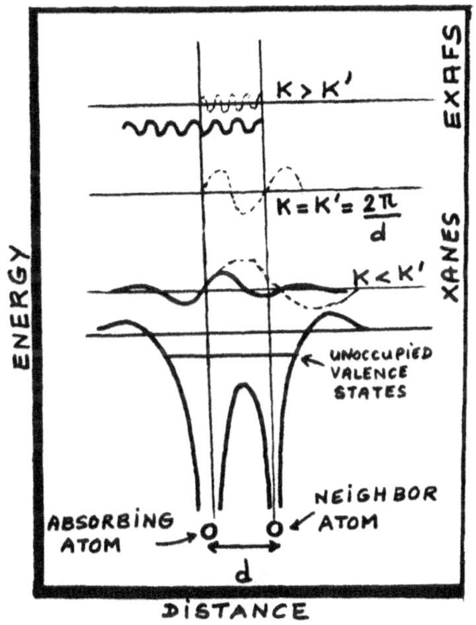

Fig. 2
Pictorial view of the final state wave function in a core excitation. dotted curves are the wave functions of the emitted photoelectron. From A. Bianconi [1].

The criterion for that single scattering theory becomes available is that the photoelectron wavelength has to be smaller than the smallest interatomic distance (central atom / first neighbour) [Fig. 2]. In this energy range, the kinetic energy of the photoelectron is sufficiently high to consider it as a free electron. Then the relation between its wave vector k and its kinetic energy Ekin is :

$$Ekin = (\hbar k)^2 / 2m \qquad (i)$$

Exafs oscillations are due to neighbour atoms so they are described by the quantity

$$\chi(E) = \frac{\mu(E) - \mu^\circ(E)}{\mu^\circ(E)}$$

μ is the linear absorption coefficient of the absorbing atom in the compound and μ° is the hypothetic coefficient of the same atom if it was isolated. According to the relation (i), $\chi(E)$ can be expressed in function of k. Its space averaged formulation for an initial state 1s (K-edge) is given :

$$\chi(k) = -\sum_j \frac{N_j \; |f_j(\pi,k)|}{kR_j^2} \exp[-2R_j/\lambda(k)] \quad \exp[-2\sigma_j^2 k^2] \quad \langle 3\cos^2\theta_j \rangle$$

$$* \; \mathrm{SIN}\,[2kR_j + 2\delta(k) + \phi_j(k)]$$

N_j is the number of scattering atoms of type j (given chemical element localized at a given distance R_j).

δ is the dephasing due to the electronic cloud of the absorber-emitter. The factor 2 takes into account the two crossing of the photoelectron through the cloud (ejection and return after backscattering).

ϕ_j is the dephasing due to the electronic cloud of scatterer atoms of type j with a backscattering power f_j. Calculated values of δ and ϕ_j are given in the litterature or experimental values extracted from reference spectra can be used.

The first exponential term accounts for the probability of the inelastic processus and λ is assimilable to the mean free path of the electron.

The second exponential contribution accounts for fluctuations of distances around their average value R_j. That comes from a dynamic disorder (thermal agitation) and from a static disorder (structure). σ_j is the *relative* mean displacement (in comparison with R_j) between the absorber and the backscatterer. This damping term is related to the Debye-Waller term in X-ray diffraction study. This formulation supposes a weak disorder and a gaussian distribution of distances around R_j.

θ_j is the angle between the polarisation vector of the X-ray beam and the axis joining the central atom and the jth neighbour.

Polarization dependence of Xanes and Exafs spectra

Several investigations have been done by angular studies or by modifying the intercalated species. Heald and Stern [2] have shown that for a lamellar compound the K-edge spectrum recorded with the polarization of the beam in the plane corresponds to a p_x or p_y-like final state, and to a p_z-like state if it is perpendicular.

As pointed out in ref.[2], materials with symmetry lower than cubic, and threefold or higher about the normal to the layers, present absorption coefficient μ with no dependence with the polarization in the layered planes and which vary as

$$\mu(\theta) = (\mu_{\parallel} - \mu_{\perp}) \cos^2\theta + \mu_{\perp}$$

It is always the case with HOPG or any 2D-powder which own an infinite axis of symmetry.

This dependence holds *at each energy* and can be used to determine, *via* linear regression and extrapolation, the spectrum which could have been measured if the polarization vector was perpendicular to the layers [Fig. 4]. With that kind of materials, it should be impossible to obtain samples thin enough to make possible transmission experiments along the planes.

XANES

The polarization effect is clearly demonstrated in the polarization dependence of the K-edge of carbon atoms in graphite (Rosenberg et al.[3]) for which dramatic changes in the intensity of the first peak is observed. This peak disappears for polarization in the plane and corresponds to a transition toward a maximum of the density of states of the first empty π-band (p_z-like band) [Fig. 4]. Several π and σ band peaks agree well with the band structure of graphite.

Frétigny et al. [4] present XANES spectra of alkali first and second stage compounds: MC_8 and MC_{24} (M = K, Rb, Cs) - K-edge for potassium and rubidium and L-edges for cesium. The spectra were taken at several angles θ between the photon beam and the normal to the basal planes of the samples, the rotation axis being normal both to the beam and to the polarization vector. An angular analysis of the data is presented and the results are discussed in connection with published band-structures.

Figure 3 presents angular dependence of alkali XANES spectra of first stage K compound.

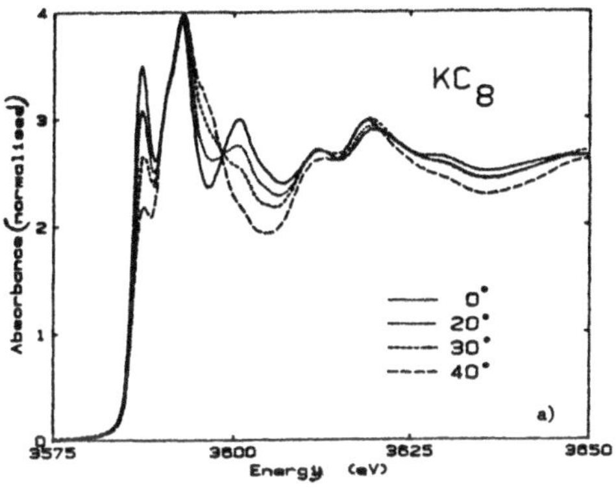

Fig. 3
Thickness normalized K-edge XANES spectra of KC_8 for several orientations.

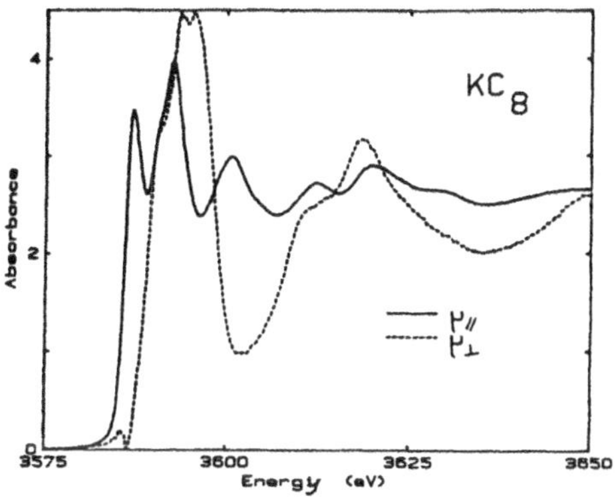

Fig. 4
Absorption $\mu_{\parallel}$ and $\mu_{\perp}$ obtained after linear extrapolation of the data of fig. 3.

From Fretigny and al. [4].

Angular effects in XANES may be used to determine symmetry of the empty states [3], in the limit of long elastic mean free path for the photoelectron. It is interesting to notice that, in pure graphite, the first peak in C-XANES is a p_z-like [3], due to empty π high density of states. This is not the case in the experiments on intercalation compounds [4]. Even in a rigid band model, this can be explained since the absorbing atom is not carbon, so the p_z character is not relevant.

For a given stage, the three studied compounds are expected to exhibit similar electronic and structural properties. In the spectra , the similarity is not clear. The Cesium and Rubidium spectra appear to be very smooth. On the opposite, the potassium ones exhibit many structures. This is due to the lifetime effect. The core hole lifetime broadening is about 0.6 eV for potassium and is above 3 or 4 eV for heavier alkali atoms. A convolution of the potassium spectra leads to results similar to the rubidium K-edge [Fig. 5] and cesium L_1-edge near the threshold.

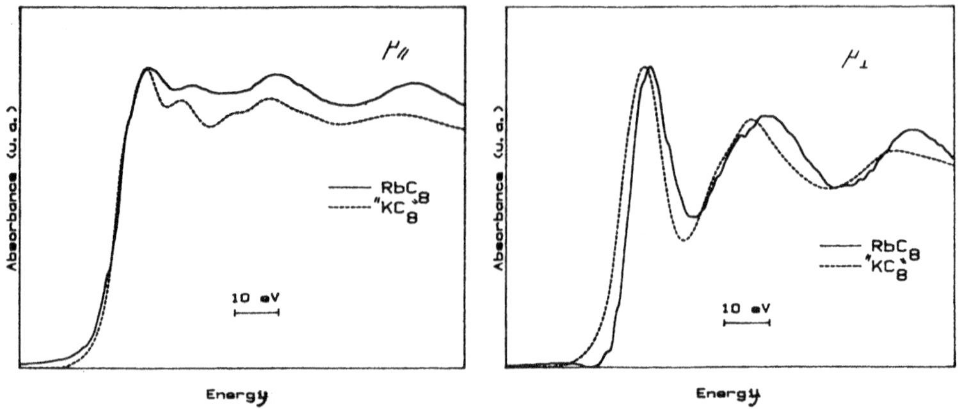

Fig. 5 Comparison between Rb and K K-edges in Rb_8 and KC_8. The potassium experimental spectra are smoothed to take into account the core hole life time effect. From Fretigny and al. [4].

Layered transition-metal dichalcogenides TX_2 are composed with metals belonging to the groups IVa, Va and VIa. Octahedral coordination is adopted by group IV compounds (Ti, Zr, Hf) and some group V compounds (V), and the trigonal prismatic coordination by group VI (Mo) and the other group V (Nb) compounds. Ohno et al. [8,13] and Davies and Brown [9] suggested that the rigid band model is valid for compounds with the same coordination. The host materials band structures are discussed in [7,13], they correspond to the band diagrams of the figure 6. Then, as group V compounds exhibit metallic properties and most of group IV and VI compounds are semiconductors, the authors expect that the addition of an extra electron from an intercalated atom to group IV (V) compounds leads to the same electronic structure as that of group V (VI) compounds, resulting in the semiconductor-metal (metal-semiconductor) transition.

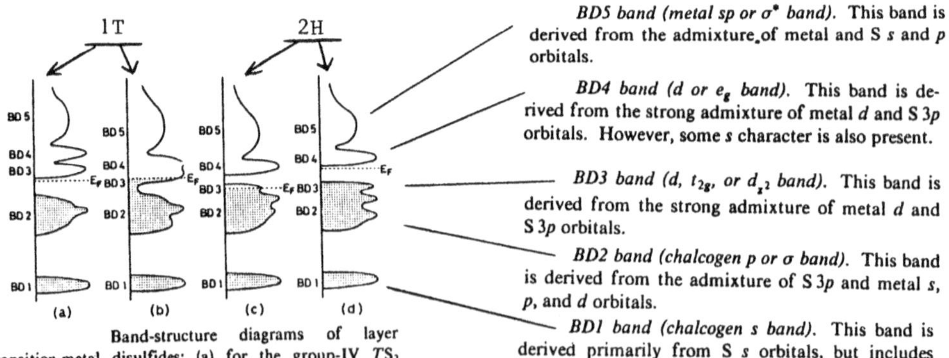

Band-structure diagrams of layer transition-metal disulfides; (a) for the group-IV TS_2 compounds (b) for the group-V $1T$-TS_2 compounds, (c) for the group-V $2H(3R)$-TS_2 compounds, and (d) for the group-VI TS_2 compounds.

Fig. 6 Structure group and representative material From [8].

Spectra agree well with these expectations. Figure 7a represents S K-edges of group IV host TiS_2 and intercalated TiS_3 and group V VS_2 and some intercalated species, all in octahedral coordination. It is clear that the edges have the same shape. A similarity is obtained for L_{III}-edge of Nb in group V NbS_2 and L_{III}-edge of Mo in MoS_2 [Fig. 7b]. Moreover, Ohno et al. [8] show that the intercalation effects on X-ray absorption spectra do not depend so much on the host material studied, as on the intercalated species. It means that the transfer of electron from the intercalated atom to the conduction band of the host layers does not change the two-dimensional behavior of the material and confirms that the rigid band model can be used in the first approximation.

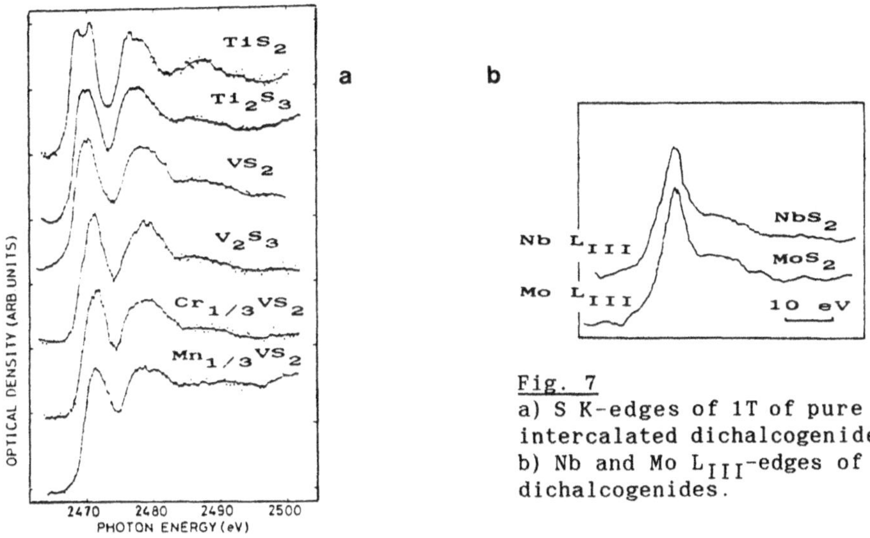

Fig. 7
a) S K-edges of 1T of pure and intercalated dichalcogenides.
b) Nb and Mo L_{III}-edges of dichalcogenides.

EXAFS

Structures of graphite and dichalcogenides are well known so these materials are used in EXAFS to check features of the spectra [6,9]. By contrast, the interest is evident in the case of intercalation of these compounds, because EXAFS permits to examine the location of the intercalated atom and the bond lengths in the host layer. For example, Bourdillon et al. [10] studied the intercalation compound $Rb_{0.28}NbSe_2$ and Thulke et al. [11] $Cu_{0.5}NbS_2$. Spectra were compared respectively to $NbSe_2$ and NbS_2. Essentially no change in the Nb-Se (Nb-S) near neighbor distance was found, while the Nb-Nb distance was observed to increase by 0.03 Å upon intercalation. The strongest bonding in the crystal is between Nb and Se (S) atoms. The corresponding σ-band, referenced BD2 in the figure 6, lie well below the Fermi level. That band remains unaffected by the charge transfer which fills the BD3 band. This BD3 band have a considerable metal d-character, this explain the repulsion between neighboring Nb atoms.

Heald and Stern [6] gave a general formulation of the orientation dependence of the EXAFS both at K and L-edges. They observed the anisotropy of 2H-WSe_2 both on Se K-edge and on W L_{III}-edge, taking into account two directions of the electric field with respect to the c-axis, they measured on the Fourier transform the variations of amplitudes of the peaks . The amplitude ratios of these peaks agree well with the known structure of the 2H-WSe_2. Moreover, these results confirm that the p-s transitions are small and the p-d transitions dominate.

Several investigations of intercalated graphite have been

performed on Br_2, alkali atoms and ternary compounds with alkali and tetrahydrofuran. Br K-edge was investigated both in graphite and Grafoil [12,14]. It was demonstrated that Br is present as Br_2 molecule in all the cases. Nevertheless, several species were found to present different stretching of the Br-Br distance which have different angular dependence spectra. In the intercalated form, the stretching is about .03 Å (2.31 Å instead of 2.28 for Br_2 molecule) and the orientation of the molecule is found to be about 20° with respect to the plane owing three angle measurements 0°, 45° and 90° using graphite fibers. The same kind of investigation was done for Rb-graphite with tetrahydrofuran [15], for which three structural models have been proposed. As each one leads to different angular variation spectra, the angular study permits to choose one, which is consistent with the Rb-C distance measured. The most striking behavior is obtained for the binary compounds alkali-graphite. A previous study of Caswell et al.[16] have shown that a stage one KC_8 have not the same spectrum than a stage 2 KC_{24}. Spectra of RbC_{24} were recorded above and below the temperature of the commensurate-incommensurate transition [17]. At low temperature - commensurate phase - only one Rb-Rb in plane distance was observed. Above the transition two distances appear, one smaller and the other one larger than the low temperature distance.

Samples intercalated with alkali metal, from HOPG, were investigated for the two first stages: RbC_8 and RbC_{24} [5].

Bond angle determination

The absorption coefficient is treated as described in the XANES part, this permits to obtain the 90° spectrum from a set measurement angles from 0° up to 40°. Figure 8 corresponds to the EXAFS part, a very large increase can be observed in the amplitude of oscillations from 0° to 90°.

The value of $\chi(k)$ varying with $\cos^2 \theta_j$, it for one kind of neighbor, where θ_j is the angle between the polarization vector and the binding, it is necessary to calculate the mean value $< \cos^2 \theta_j >$ because of the disorientation in the plane, the following formula is obtained:

$$< \cos^2 \theta_j > = \cos^2 \alpha \sin^2 \theta + (\sin^2 \alpha \cos^2 \theta)/2$$

where α_j is the binding angle with respect to the **c**-axis and θ is the angle between the polarization vector and the ab plane.

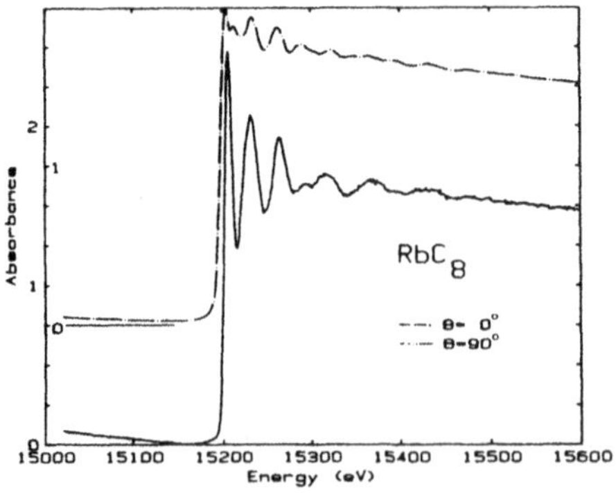

Fig. 8 EXAFS spectra of RbC_8 for θ = 0° and 90°.

From BONNIN and al. [5]

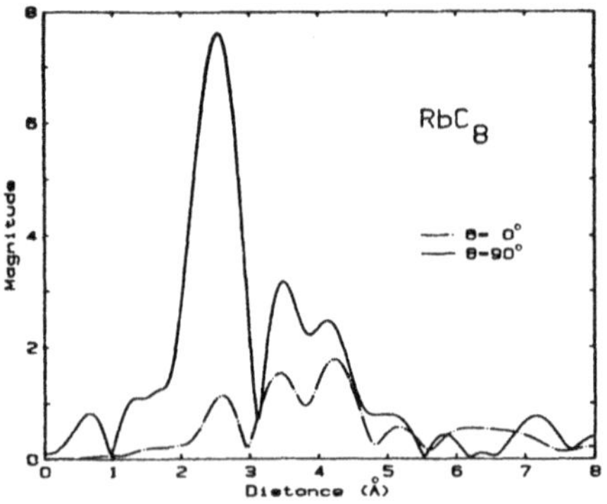

Fig. 9 Fourier transform of spectra of Fig. 8.

The radial distribution for the two perpendicular directions are shown in figure 9. For each peak of the Fourier transform, the 0° to 90° amplitude ratio permits to calculate the angle of the corresponding $\vec{R}$ vector with respect to **c**-axis, according to the formula:

$$I_{0°} \;/\; I_{90°} = (\, tg^2\, \alpha \,)/2$$

where I corresponds to the height of a peak of the Fourier transform. The obtained results agree well with the known structure of the first stage [Fig. 10]. Moreover, that analysis does not need the knowledge of the type of neighbour.

	Distances Å	Expected angles	Experimental angles
Rb - C1	3.18	27°	31°
Rb - C2	4.02	45°	44°
Rb - C3	4.72	53°	50°
Rb - Rb	4.92	90°	/
Rb - C4	5.90	61°	/

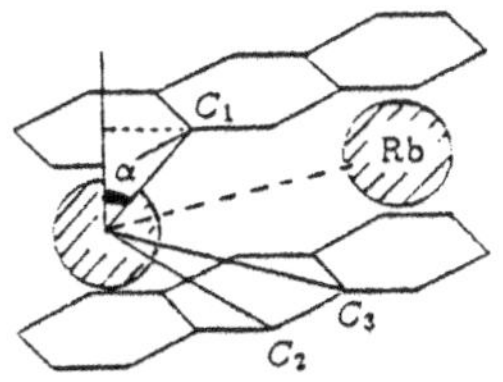

Fig. 10 Local structure of RbC_8. Comparison between experimental and expected angles.

ANISOTROPY AND TEMPERATURE DEPENDENCE OF THE DEBYE-WALLER FACTOR

EXAFS spectra were recorded for a set of temperature 45 K, 70 K, 140 K and 300 K. The experimental angle used is 30° in order to have comparable intensities of the first three peaks of the Fourier transform of the low temperature spectrum. Figure 11 shows the temperature dependence of the Fourier transforms, higher is the order of the peak, higher the decrease of the amplitude. Taking into account the angles of the bonds, it appears clearly that the Debye-Waller factor is much more higher in the plane than perpendicularly. For each temperature and for each peak, we calculate the inverse Fourier tranform and obtain the quantities $\chi_j(T)$ relative to the peak j at the temperature T. Then, the ratio $\chi_j(T)/\chi_j(45\ K)$ depends only on the Debye-Waller factor :

$$\text{Log}\ (\ \chi_j(T)\ \ /\ \ \chi_j(45)\ ) = 2\ k^2\ (\sigma^2_{45} - \sigma^2_T) = 2\ k^2\ \Delta\sigma^2$$

The variations of $\Delta\sigma^2$ are given on figure 12, they permit the calculation of $\Delta\sigma^2_\perp$ and $\Delta\sigma^2_\parallel$.
A very high anisotropy of vibrations is found due to the very large in plane value of $\Delta\sigma^2$. Because of the large Debye temperatures, the carbon atoms belong to a very rigid network, their vibrations are neglected. So, there is no correlation effect in the σ^2 values, measured for the three first peaks, which correspond to the movements of Rb atoms.

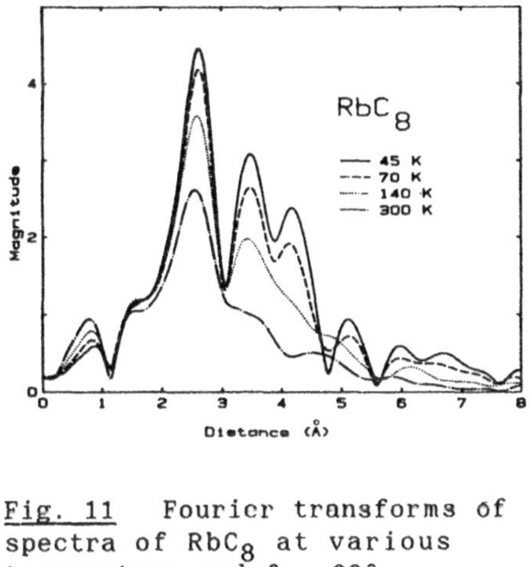

Fig. 11 Fourier transforms of spectra of RbC_8 at various temperature and θ = 30°.

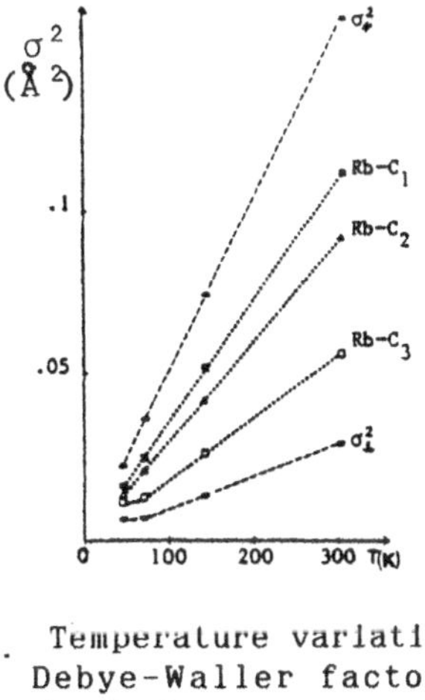

Fig. 12 Temperature variations of the Debye-Waller factors in and out of plane.

COMPARISON BETWEEN THE STAGE 1 (RBC8) AND STAGE 2 (RBC24)

On the figure 13, FT of 3 compounds are shown (they were calculated with the same window and for the same k-range in view of an easier comparison). The compound [a] is a first stage one (RbC_8), [b] and [c] are of second stage (RbC_{24}). The compound [b] is saturated with Rb while the compound [c] is not.

The magnitude of the FT of the low temperature spectrum (40 K)at $\theta = 0°$ of [b] and [c] is smaller (especially for the three first peaks) than those of [a]. Moreover, the third peak disappears in [c]. The decrease is not due to a smaller number of scattering Rb atoms seeing that the quantity $\chi(k)$ is per scatterer unit. The cause cannot be only a dynamic disorder because of the low experimental temperature. The difference unit. The cause cannot be only a dynamic disorder because of the low experimental temperature. The difference is attributed to a static disorder in RbC_{24}. This disorder is large enough to make impossible its description only with the Debye-Waller factor. It suggests that the local structure of a second stage compound is different than those of a first stage one.

Starting from the hypothesis that Rb atoms remain in the plane at half-way of two carbon layers, simulations are tentatively made for different positions that Rb is supposed to be able to occupy on the whole hexagon of carbon (in RbC_8, the alkali has only the central position) [5]. Taking into account the three first neighbour shells, simulations are calculated from amplitudes and phases extracted from the known compound RbC_8. It is attempted to give weights to positions.

In a first step, if equal-weighted positions are considered, only positions within a circle (centered on the center of the hexagon) of diameter about 0.35 a (with a = 1.42 Å) is needed to make the best description.

With a gaussian form of the weights W according to the distance R (Rb-C) $W = \exp(-R^2 / S^2) / N$ with N a constant as the sum of weights is 1. The best value of s is about 0.5 Å for the compound [c] and about 0.6 Å for [b]. Moreover, for the compound [b], others simulations are made with two gaussian constants for the two axes X and Y and it exists six equivalent pairs of axes X Y [Fig. 14]. The distribution of weights is not isotrope : the extension along the X direction is the most important.

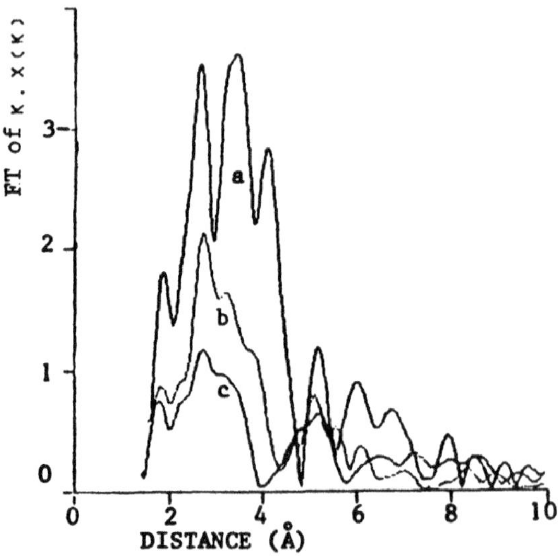

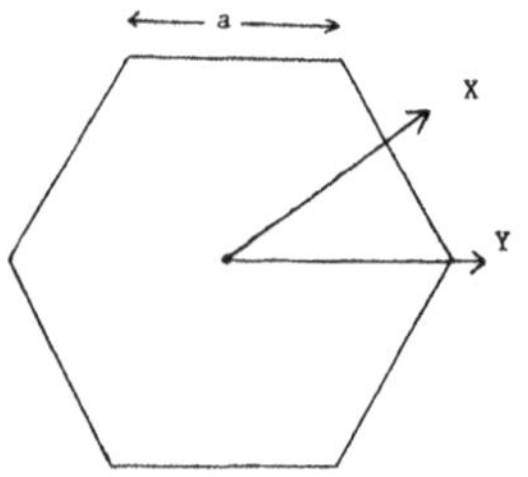

Fig. 13 FT for [a], [b] and [c] compounds.

Fig. 14 The hexagon of carbon.

REFERENCES

[1] BIANCONI A. Exafs and Near Edge Structure III. Springer Verlag proc. conf. Stanford (1984) 167/74

[2] HEALD S.M. STERN E.A. Phys. Rev. B16 (1977) 5549

[3] ROSENBERG R.A. LOVE P.J. REHN V. Phys. Rev. B33 (1986) 4034 EXAFS and Near Edge Structure III Springer Verlag (1984)

[4] FRETIGNY C. BONNIN D. CORTES R. Exafs and Near Edge Structure IV J. de Phys. proc. conf. Fontevreau (1986) C8 869

[5] BONNIN D. BOUAT J. KAISER P. FRETIGNY C. BEGUIN F. Exafs and Near Edge Structure IV J. de Phys. proc. conf. Fontevreau (1986) C8 865

[6] HEALD S.M. STERN E.A. Phys. Rev. **B 16** (1977) 5549

[7] OHNO Y. HIRAMA K. NAKAI S. SUGIURA C. OKADA S. Phys. Rev. **B 27** (1982) 3811

[8] OHNO Y. HIRAMA K. NAKAI S. SUGIURA C. OKADA S. Synth. Met. **6** (1983) 149

[9] DAVIES B.M. BROWN F.C. Phys. Rev. **B 25** (1982) 2997

[10] BOURDILLON A.J. PETTIFER R.F. MARSEGLIA E.A. J. Phys. **C 12** (1979) 3889

[11] THULKE W. FRAHM R. HAENSEL R. RABE P. Phys. Stat. Sol. **A 75** (1983) 501

[12] HEALD S.M. STERN E.A. Synth. Met. **1** (1980) 249

[13] OHNO Y. WATANABE H. KAWATA A. NAKAI S. SUGIURA C. Phys. Rev. **B 25** (1982) 815

[14] FELDMAN J.L. ELAM W.T. EHRLICH A.C. SKELTON E.F. DOMINGEZ D.D. QUADRI S.B. CHUNG D.D.L. LYTLE F.W. in EXAFS and Near Edge Structure III (1984) Springer Verlag and Solid State Commun. **49** (1984) 1023
FELDMAN J.L. ELAM W.T. EHRLICH A.C. SKELTON E.F. DOMINGEZ D.D. CHUNG D.D.L. LYTLE F.W. Phys. Rev. **B33** (1986) 7961

[15] BONNIN D. BOUAT J. Progress in X-ray studies by Synchrotron Radiation. Strasbourg (1985) // LEGRAND A.P. FACCHINI L. BONNIN D. BOUAT J. QUINTON M.F. BEGUIN F. Synth. Met. **12** (1985) 175

[16] CASWELL N. SOLIN S.A. HAYES T.M. HUNTER S.J. Physica **B99** (1980) 463

[17] BOUAT J. BONNIN D. FACCHINI L. Synth. Met. **7** (1983) 233

SURFACE EFFECTS IN GRAPHITE INTERCALATION COMPOUNDS

D.Marchand, M.Lagues and C.Fretigny

Laboratoire de Physique Quantique, CNRS Unité Associée 421
ESPCI 10 rue Vauquelin 75231 Paris Cédex 05 France

INTRODUCTION

The theoretical description of the band structure of these compounds has drastically changed during the last ten years [1-5]. The so called ONK description [1] first proposed the picture of a nearly rigid band structure, with the **s** band of the alkali atom partially filled. One of the main points was to calculate the charge transfer of the alkali atom (≈0.6 according to ONK calculations). Subsequently, Posternak et al. published a calculation exhibiting a completely new graphite band, the *interlayer band* [2]. They suggested that this level, located at 4 eV above the Fermi level in graphite, could play an important role in the intercalation mechanism and should be taken into account in the GIC band structure calculations. Later, this was clearly established for example by the calculations of Holzwarth et al. for LiC_6 [3]. At the same time, a new calculation for the band structure of KC_8 and CsC_8 was presented by R.C.Tatar and S.Rabii (TR) [4], showing complete charge transfer from the alkali atoms, the corresponding **s** level lying respectively 1.5 and 3.4 eV above the Fermi level. Moreover, according to TR the Fermi surface presents a pocket in the center of the Brillouin Zone (BZ) which relates only to the π bands and not to the alkali metal. Recently, H.Kamimura has given a new interpretation of ONK calculations [5], according to which the alkali atom also yields a unit charge transfer. Following this interpretation, the Fermi pocket at the BZ center should be related mainly to the interlayer band. Now, both the revised ONK (rONK) and the TR descriptions agree on the full ionization of the alkali atoms , and on the graphitic origin of the central Fermi pocket. The main difference between these descriptions relies on the location of the corresponding states in real space: they are located on the graphite layers for TR and between them for rONK.

The experimental determination of the band structure for first stages alkali metal GIC is thus extremely useful in clarifying the present picture. The careful study of the Carbon KVV Auger transition provides very useful information on the valence band density of states [6-9]. We show below that the main conclusion of this study is the presence of a sharp peak near E_F, which does not behave as the other levels of the valence band. The whole shape of the C KVV transition is now well described [5], except the origin of this peak which was extremely controversial [10-11]. We show here that there is strong evidences favouring the presence of a Charge Density Wave near the surface of the compounds, giving rise to the features observed both by

Auger Electron Spectroscopy (AES) and by Angle Resolved Photoemission Spectroscopy (ARPES).

ARPES is presently one of the most powerful tools to determine experimentally the band structure of a solid. On the other hand, the surface may induce specific effects which have to be recognized and studied as "surface driven", and separated from the true bulk properties. This technique is indeed a *surface technique* in the sense that the escape depth of the emitted electrons is of the order of 5 to 10 Å. This implies that the information comes from the first layers of the studied material. Nevertheless, these measurements provide accurate values of bulk properties. One can mainly distinguish three families of surface effects: effects from the intrinsic limits of the technique, effects on the equilibrium structure and composition of the material (i.e. reconstruction, segregation, ...) and effects on the electronic distribution (i.e. surface states, resonances,...).

We [12] and others [13-14] have observed, near the Fermi level, an unusual feature which does not correspond to any of the calculated bulk energy bands. New detailed ARPES data on CsC_8 convince us that we are seeing a new kind of surface-driven charge density wave (CDW). CDWs are expected to be stable mainly in low dimensional solids. Many cases are now well described in 1D solids such as the TTF-TCNQ, or 2D solids such as dicalchogenides. Graphite intercalation compounds (GIC) are *a priori* very good candidates for CDW, as they are 2D compounds with a rather large Fermi surface area. However, no CDW is observed in GIC despite numerous studies of the electronic properties for different compounds.

JENSEN and PLUMMER recently suggested a new kind of CDW in Na [15].It is localized at the surface because the electron-lattice interaction is stabilized by relaxation of the first few atomic layers, i.e. the CDW vector is normal to the surface. In this paper we present evidence for a surface-stabilized cdw with $q_{\parallel}$ in the plane, as in the bulk transition metal dichalcogenides. Here, the stabilization mechanism relies on relatively weak interactions between two sublattices, namely graphite and intercalated alkali metal layers, whereby the local 2D Cs structure need not be the same in the outermost layers as in the bulk. Our evidence is based on new, very detailed low temperature angle resolved photoemission spectroscopy (ARPES) of the first stage heavy alkali compound CsC_8. We [12] and others [13-14] have observed valence band features at low binding energy, <2 ev, which do not correspond to any of the calculated bulk energy bands, and we show that these non-bulk features are readily explained by the proposed CDW mechanism.

Both AES and ARPES spectroscopies sample an extremely thin slice of the material studied. In the case of GIC's, we expect a possible segregation, i.e. a different equilibrium composition at the surface. The GIC preparation temperatures are often as low as room temperature, and this implies very rapid intercalant diffusion. The main point leading to the consideration of possible surface segregation is the existence of the **staging phenomenon**. Indeed, this shows the sensitivity of the intercalated composition in a definite layer, to the composition of other distant layers in the bulk. Consequently, we expect a large sensitivity to the absence of neighbouring layers at the surface. Auger [16], XPS and ARPES [9] measurements show a considerable segregation of intercalant, towards the surface layers in alkali metal GIC, and towards the bulk in acceptor compounds. In donor compounds, this leads to a high intercalant concentration in the uppermost layers even with a low bulk concentration

[16]. This *surface intercalated compound* should not be classified as a stage. In the case of cesium GIC's, we show that the different surface compounds observed for bulk stages 1 and 2 agree largely with theoretical calculations for CsC_8, but not with CsC_{24} calculations. This is a typical case where ARPES measurements do not account only of the bulk properties.

AUGER ELECTRON SPECTROSCOPY (AES)

A study of the Auger spectrum in GICs provides information about both the intercalant and the carbon bonding. The Auger process is initiated by an incident electron (or photon) which creates a hole in the 1s carbon K level (or intercalate core level). The carbon KVV signal corresponds to the recombination of this 1s carbon K level hole with a valence electron and the simultaneous emission of an electron from a valence state. Studies of the carbon KVV Auger lineshape have been carried out in first stage C_8Rb and C_8Cs by Oelhafen et al.[17] and $MnCl_2$-GICs by Laguës et al.[7].

Figure 1 shows the carbon KVV Auger lineshape of graphite, $MnCl_2$-GIC and Cs-GIC, where we see that the lineshape are nearly identical for graphite and $MnCl_2$-GIC. This observation and the quantitative interpretation of the Mn and Cl Auger peak intensities [16] show that the intercalate concentration on the surface and between the top two graphite planes is very low, so that the surface layers for stage 1 can be approximed by GGIGIG..., where G and I represent graphite and intercalate layers, respectively. The surface depletion of the intercalant in acceptor compounds is in contrast to the intercalant surface enrichment observed in donor compounds [9], which is implied by Auger spectroscopy, by ARPES (angle resolved photoemission spectroscopy) studies and by XPS studies on C_8K [18].

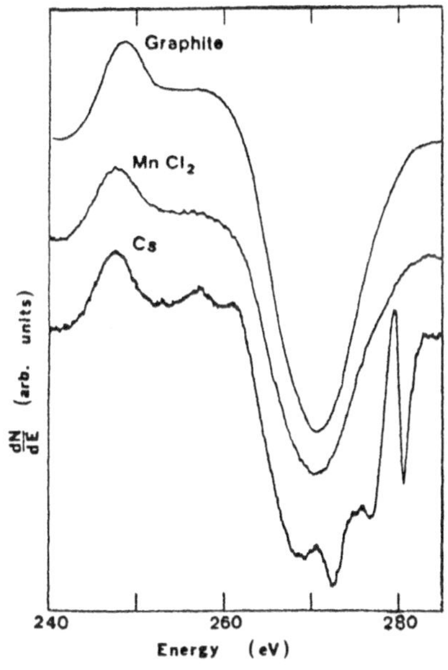

Figure 1. The Auger spectra of pristine graphite, stage 1 $MnCl_2$-GICs and stage 1 Cs-GICs [16]

We [16] have suggested that the surface intercalant depletion in acceptor-GICs and the surface intercalant enrichment in donor GICs, arises from the electrostatic screening associated with a surface state. The electrostatic energy associated with this screening (≈1 eV per intercalant atom) is larger than the typical chemical potential difference between the various GIC stages. The intercalant concentration is thus expected to change at least in the uppermost intercalated layer, in order to reach a free-energy minimum. This description applies not only to GICs but to any compound. However binding energies and diffusion activation energies are usually much too large so that the equilibrium situation at the surface is not very different from the bulk. The occupation of a surface state level results in a local electron enrichment. This picture is consistent with dangling π bonds at the surface and the large value of the work function. Thus in the case of GICs, this favors a surface enrichment of positive ions (donors) and depletion of negative ions (acceptors).

We now apply these concepts to discuss the carbon KVV lineshape and its relation to the valence density of states in donors GICs. Referring to Fig.1, we see that the Auger spectrum for the Cs-GIC exhibits a modified line shape relative to graphite, with an additional sharp structure not present in the graphite spectrum, as well as several weaker structures. This sharp feature arises from the increased occupation of the graphite π^* band conduction states as the Fermi level rises due to donor intercalation. An explanation for the lineshape of the sharp feature is provided by the Lander model for the Auger lineshape. According to the Lander model [19], the Auger intensity I(E) is given by the sum over all possible transitions leading to the final state energy E. Lineshape effects are particularly important for KVV transitions where the density of states (DOS) of the valence band n(E) enters the calculation. Neglecting the energy dependence of the matrix elements, the lineshape I(E) is simply the self-convolution of n(E) given by

$$I(E) \propto \int n(x)n(E-x)dx \equiv n*n \qquad (1)$$

This simple description assumes that the Auger transition does not involve localized states and thus is a true band structure effect. Large screening effects of the initial K state hole or final valence state hole would modify the lineshape. The energy dependence of the transition matrix element would also modify the lineshape. for example, in a tight binding description, we write

$$n(E)=n_s(E)+n_p(E) \qquad (2)$$

where $n_s(E)$ and n_p respectively represent the density of states associated with s and p bands. In this description n_s*n_s, n_s*n_p and n_p*n_p terms occur in the self convolution of Eq.(1) and should have different matrix elements. finally, the Lander model which works well for pristine graphite,assumes direct emission of the Auger electron to the vacuum. Each Auger electron reaches the surface with a definite probability of experiencing an energy loss,which can be determined by an EELS experiment (energy electron loss spectroscopy), and must be included in a detailed analysis of the lineshape.

The application of the Lander model to GICs makes use of the well established graphitic nature of the graphite derived bands in GICs and the model is able to explain the observed lineshape. The density of states n(E) is written as a sum of the graphite DOS g(E) and of the sharp feature in the DOS p(E) induced by intercalation. We note that p(E) also corresponds to a sharp peak in the UPS (ultraviolet photoelectron spectroscopy) spectra at the Fermi level [7]. Following the Lander model, we write the convolution for the GICs as

$$I_{GIC} \propto g*g+2g*p+p*p \quad (3)$$

where $I_G(E) \propto g*g$ is the KVV lineshape of pure graphite. Thus, it is possible to simplify the analysis of $I_{GIC}(E)$ by substracting the experimental Auger spectrum of graphite $I_G(E)$. Since $I_{GIC} \approx I_G(E)$ in the energy range E<260 eV, the contribution of the sharp feature p(E) can be found by substraction, noting that the second term 2g*p in Eq.(3) corresponds to the graphite DOS, slightly broadened by convolution with p(E).

According to the quantitative modeling of the Auger lineshape by Laguës et al.[7], the relative weights of the integral of g and p are respectively 4 electrons and f_C electrons, where f_C is the charge transfer per carbon atom. Thus, the integral in Eq.(3) has a weight of 16 for the g*g term, a weight of $8f_C$ for the 2g*p term and a weight of f_C^2 for the p*p term. After normalization to the graphite peak weight g*g, the relation

$$T^2 = 4S \quad (4)$$

is obtained where T and S are respectively the normalized integrals of 2g*p and p*p. Thus, the charge transfer per carbon atom f_C can be obtained either by convolutionof g with p to obtaind T, or the convolution of p with itself to obtain S. A detailed analysis shows that the graphite Auger spectrum can be substracted from the corresponding GIC spectrum to obtain the density of states g and p appropriate for the GIC. The functions g and p thus obtained successfully reproduce the feature in the Cs-GIC Auger lineshape with regard to the energies of the various peaks in the spectrum. The model explains the magnitude of T, but fails to explains the magnitude of S or of the sharp intercalation-induced peak near E_F, which is 5 to 10 times larger than calculated on the basis of complete charge transfer. An attempt to modify the Lander model [11] by taking into account the local perturbation of the band structure due to the K hole, enhances the magnitude of the expected sharp peak, but is inconsistent with the experimentally observed lineshape. A possible explanation of this discrepancy is that the Auger feature near E_F is related to a different electronic level which should not be convoluted with the valence band in Lander's model. Though this level could be a bulk state, it is probably a surface-driven charge density wave which is observed at the Fermi level by ARPES [20] and described below.

ANGLE RESOLVED PHOTOEMISSION SPECTROSCOPY (ARPES)

The new experiments were performed on the undulator line of the UV storage ring ACO located at LURE in ORSAY, FRANCE. It was possible to scan a wide range of experimental parameters: photon energy from 17 to 120 eV, photon incidence angle from 5° to 85°, analyzer orientation from -55° to 75°, permitting detailed exploration of $k_\perp$ and $k_\parallel$ dispersions with an overall energy resolution (photon plus electron) of about 200 meV and 1° angular resolution. Natural single crystals of graphite were bulk-intercalated in the usual way, transferred to a prechamber under flowing argon, then into the analyzer chamber under a residual pressure of 10^{-10} Torr. After azimuthal orientation via LEED observation of the graphite 1x1 pattern, the samples were cooled to 130 K and then cleaved with an adhesive tape. Significantly, we did not observe any spots from the bulk 2x2 Cs superlattice or from surface reconstruction.

The complete band structure observed for alkali metals is discussed elsewhere [20]. All measurements are presented for the graphite ΓK crystalline direction, except when the direction is specified. Figure 2 shows a sampling of energy distribution curves (EDC's) taken with 50 eV photons at 85° incidence along the trajectory shown. Figure 3 presents the bands observed in a range of 5 eV below E_F (experimental points for a photon energy of 50 eV, and 85° incidence), compared with Tatar and Rabii

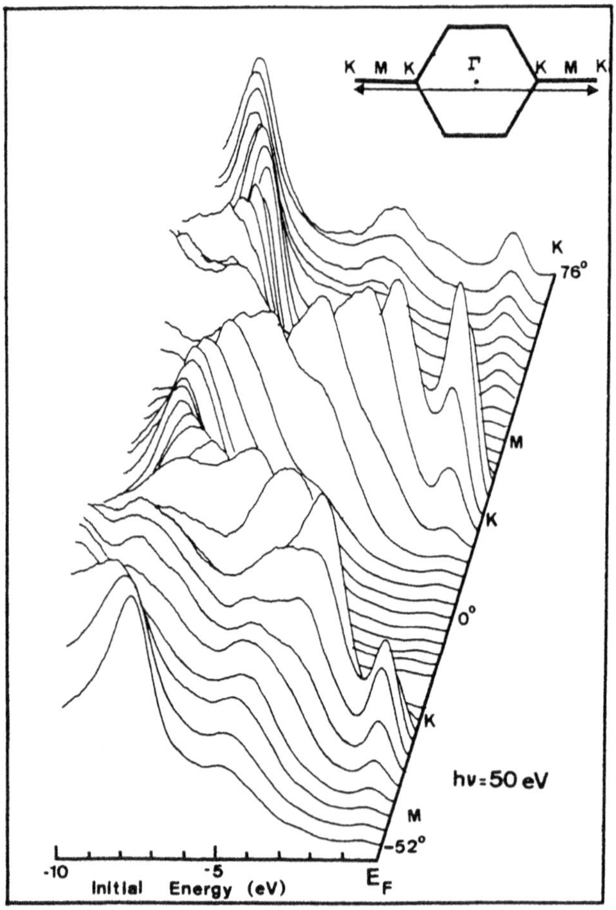

Figure 2 . ARPES EDC's for bulk CsC_8 cleaved and measured at -130 K, 85° incidence with 50 eV photons. The graphite BZ shown in inset describes the corresponding path explored in the reciprocal space. Note the sharp maxima of the Fermi level band, at points near but clearly different from the point K. Note also the very low intensity of this level near the point Γ.

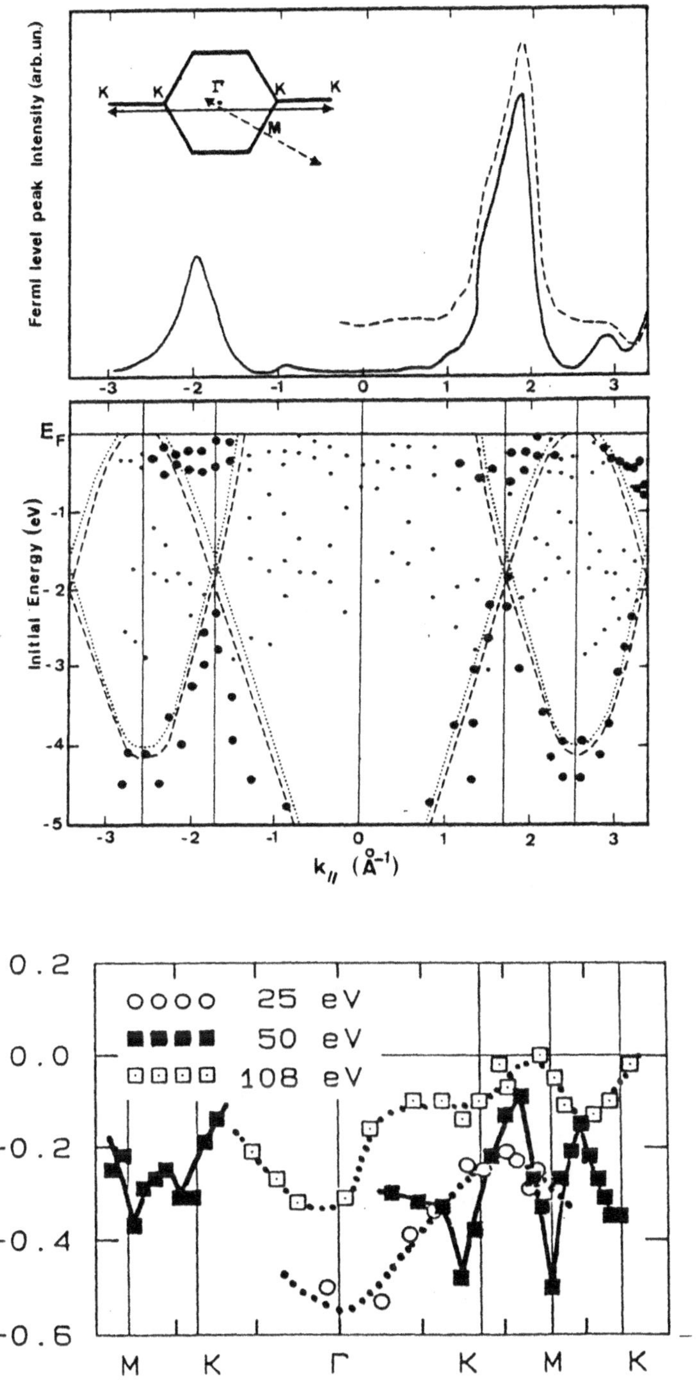

Figure 3 : ARPES measurements for CsC_8 at $h\nu$ = 50 eV. Experimental points : large dots for intense emission, small dots for low emission. dashed line : TR calculations, dotted line : ONK calculations.

calculations for CsC_8 [4] (dashed line) and ONK calculations for KC_8 [1] (dotted line) for π bands. Below -2 eV, the unfolded π bands agree very well with both calculations which are very close to each other. On the other hand, the 2x2 folding is not clearly observed at all: the points which are observed do not agree with the 2x2 folding, and the corresponding photoemission intensity is more than one order of magnitude lower than for unfolded π bands. Between ≃ -0.5 and -2 eV, no point is observed with a high intensity even around the K point. Between E_F and -0.5 eV, a low dispersion band is present in the whole BZ. The corresponding photoemission intensity is high around the K' point, and much lower at other points. Maxima are observed for two symmetrical values $Q = \pm 1.93$ Å^{-1}, clearly different from the K point wave vector of 1.7 Å^{-1}. By changing the photon energy from 17 to 110 eV,a $k_\perp$ dispersion of the same order as the $k_\parallel$ dispersion (≃ 500 meV) is observed both for $k_\parallel = 0$ and $k_\parallel = 1.7$ Å^{-1}.

The main explanations for such a band are, i) an indirect photoemission transition (with loss of momentum information), ii) a surface state, iii) a CDW. Clearly the first possibility is inconsistent with the experimental observations. The *surface state* explanation is inconsistent with three observations: the low sensitivity of this feature to the surface pollution and the $k_\perp$ dispersion, both implying a bulk extent of the corresponding electronic levels, and the incommensurability of the dispersion. In the following, we show that the CDW description is supported by many direct observations.

The structural and electronical properties of the first intercalated layers are completely different from the properties of bulk intercalated layers. A simple model explains this by the screening of a surface state [17]. The LEED pattern does not show any contrasted superstructure. Therefore we think that the sharp maxima are related to the inplane structure of the first intercalated layer. The same kind of pattern is observed for liquid state of first [21] and second [22] stage of alkali metals CIG, but here the structure is specially dense. The corrugation energy of such an hexagonal Cs layer is practically independent of its orientation relative to the graphite layer [13], and this explains that no orientation locking is observed.

The Fermi surface boundaries are accurately measured because of the steep variation of the intensity around the K point. Using the trigonal symmetry around this point leads to a good picture of the 2D Fermi surface, in the graphite symmetry (Figure4). The ratio of the Fermi surface area divided by the BZ area is simply the charge transfer f_c relative to this band, per carbon atom (here $f_c \simeq 0.13$, corresponding roughly to one electron transferred by Cs atom for the bulk stoechiometry CsC_8). This value means also that the intercalate structure leads to a folding which tiles almost perfectly the reciprocal space with the Fermi surface. This is the primary feature of a CDW: a perturbation of the crystalline potential leads to nesting at least in one direction- of the FS, and to gaps appearing at the FS boundaries. This tiling involves here two sheets: the band observed near E_F fits exactly the expected upper sheet obtained by this tiling. On the other hand, the lower sheet could be hidden by cross section effects.

CONCLUSION

We have presented here a new kind of surface-driven charge density wave stable only close to the surface of the compound. The perturbation of the crystalline potential arises from the dense and incommensurate structure of the uppermost intercalated layer. Plummer et al. [15,24] recently published similar results for the sodium surface.

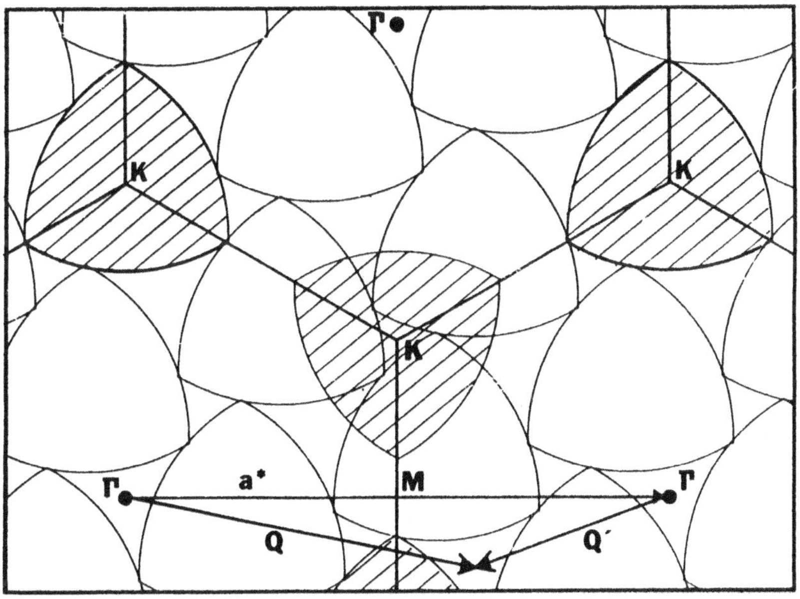

Figure 4 : Fermi surface tiling of the 2D graphite lattice, (shaded sections), accounting approximately for trigonal warping via the triangular shapes. The vector Q assumes a triangular Cs lattice with 1.93 $^{-1}$ fundamental, rotated -10°. The same point is reached by a* + Q', where a* = G(100) and Q' is interpreted as a modulation satellite of locally-commensurate √7 x √7 R 19.1° domains in a discommensuration lattice. The open sections result from either option with a FS area corresponding to 1/8 electron per C atom (see text).

However in the present case the dispersion of the Fermi level band is much larger, and this allows us to check that its periodicity is consistent with $2k_F$ and the intercalant structure. In contrast to the proposed surface CDW in metallic Na [9,12], our CDW wave vector lies in the surface plane and hence its periodicity in k can be traced across the surface Brillouin zone.

However, in our discussion of Auger electron spectroscopy, evidence was presented favoring an enriched intercalate concentration near the surface for donor GICs and a depleted intercalate concentration for acceptor GICs. Since the emitted electron originates from approximately the same depth for both Auger and ARPES techniques, acceptor GICs are expected to exhibit ARPES spectra very similar to those for pristine graphite; this may account for the absence of ARPES studies for acceptor GICs in the literature.

REFERENCES

1-T.OHNO, K.NAKAO and H.KAMIMURA J.Phys.Soc.Jpn.**47**(1979)1125
2-M.POSTERNAK, A.BALDERESCHI, A.J.FREEMAN, E.WIMMER and M.WEINERT Phys.Rev.Lett.**50**(1983)761
3-N.A.W.HOLZWARTH, S.G.LOUIE and S.RABII Phys.Rev.**B28**(1983)1013
4-R.C.TATAR and S.RABII "Graphite Intercalation Compounds" Symposium 1 of the **Mat.Res.Soc.** (BOSTON November 1984)p 71
5-H.KAMIMURA Annales de Physique Coll.2 **11**(1986)39
6-P.PFLUGER, P.OELHAFEN, H.U.KUNZI, R.JEKER, G.HAUSER, K.P.ACKERMANN M.MULLER and H.J.GUNTHERRODT Physica **99B**(1980) 395
7-M.LAGUES, D.MARCHAND, C.FRETIGNY and A.P.LEGRAND Solid State Commun. **49**(1984)739
8-M.LAGUES, D.MARCHAND and A.P.LEGRAND Solid State Sciences **38** (1981) 49 (Ed. Pietronero and Tossatti) Springer Verlag N-Y
9-M.LAGUES, D.MARCHAND and C.FRETIGNY Annales de Physique Coll.2 **11**(1986)49
10-J.G.MURDAY, B.I.DUNLAP, F.L.HUSTON and P.OELHAFEN Phys.Rev.**B24** (1981)4764
11-B.I.DUNLAP, D.E.RAMAKER and J.S.MURDAY Phys.Rev.**B25**(1982)6439
12-M.LAGUES, D.MARCHAND and C.FRETIGNY Synth. Metals **12**(1985)263 and "Graphite Intercalation Compounds" Symposium K of the 1986 Fall Meeting of the Mat.Res.Soc.(BOSTON December 1986) p 80
13-N.GUNASEKARA, T.TAKAHASHI, F.MAEDA, T.SAGAWA and H.SUEMATSU J.Phys.Soc.Jpn. **56**(1987)2581
14-M.T.JOHNSON, H.I.STARNBERG and H.P.HUGHES Solid State Commun.**57** (1986)545
15-E.JENSEN and E.W.PLUMMER Phys.Rev.Lett. **55**(1985)1912
16-M.LAGUES, D.MARCHAND and C.FRETIGNY Solid State Commun.**59**(1986)583
17-P.OELHAFEN, P.PFLUGER and H.J.GUNTHERODT Solid State Commun.**32** (1979)885
18-S.B.DI CENZO Synth.Metals **12**(1985)251
19-J.J.LANDER Phys.Rev.**91**(1953)1382
20-C.FRETIGNY, D.MARCHAND and M.LAGUES Phys.Rev.**B32**(1985)8462
21-D.S.ROBINSON, T.C.McGLINN, J.L.ZARESTKY and M.B.SALAMON "Graphite Intercalation Compounds" Symposium K of the 1986 Fall Meeting of the Mat.Res.Soc. (BOSTON December 1986)p 16
22-N.WADA, S.MINOMURA and J.PLUTH Solid State Commun.**55**(1985)393
23-This is clear when the corrugation energy is deduced from the substrate potentiel as described for instance by :
S.C.MOSS, G.REITER, C.THOMPSON, J.L.ROBERTSON, J.D.FAN and K.OSHIMA "Graphite Intercalation Compounds" Symposium K of the 1986 Fall Meeting of the Mat.Res.Soc.(BOSTON December 1986) p 20
24-A.W.OVERHAUSER Phys.Rev.Lett. **55**(1985)1916

INTERCALATION MECHANISMS IN THE POLYLAYERED TERNARY COMPOUNDS OF GRAPHITE WITH ELECTRON DONORS

Daniel Guérard and Philippe Lagrange

Laboratoire de Chimie du Solide Minéral (UA 158)
Université de Nancy I
B.P. 239 - 54506 Vandoeuvre les Nancy Cédex, France

INTRODUCTION

When a metal intercalates into graphite, two different questions arise :

- where is the intercalation coming from ? (parallel or perpendicular to the planes)[1]
- what are the successive stages from graphite to the suitable stage ?[2-4]

If one has two species, like, for instance, two alkali metals, the situation becomes more complicated since the composition of the intercalated sheet can be far from that of the starting alloy[5]. But, even in this case, the intercalated layer remains simple.

When the intercalate is an alloy MM'_x of a heavy alkali metal (M = K, Rb or Cs) and a heavy metal (M' = Hg, Tl, Bi) or an alkali hydride (KH or NaH), the intercalated layer is made of three different sheets M-M' (or H)-M and the process of intercalation and building of this threelayered intercalate is not obvious[6,7].

SYNTHESIS

The general method of preparation of such ternaries consists of the immersion of graphite in a large excess of the reactant. The large excess is used in order to avoid a drastic change of the composition during the intercalation process. Such a change could involve a variation of the obtained compound : for instance, if one starts from a KHg amalgam, one obtains a first stage ternary whereas a KHg_2 alloy leads to a second stage compound[7].

The variation in the composition of the intercalant can be due either to a partial decomposition (hydrides) or to a difference in the composition of the intercalated species compare to that of the free one.

The reaction temperatures and lengths are in the ranges 180°C (KHg) to 500°C (KBi) and a few minutes to a few weeks.

MECHANISM OF THE INTERCALATION

The reactions are followed by X-Rays, chemical analysis and the color of the compounds at different rates of intercalation.

The general behaviour is as follows :

- in a first step, there is a quasi-selective intercalation of the alkali metal, with decreasing stage from graphite (considered as infinite stage) down to a stage s, according to the reaction :

$$M + nsC \longrightarrow MC_{ns}$$

- in a second step, there is a simultaneous cooperative intercalation of both elements, without any change of stage (isostage process) :

$$MC_{sn} + M + 2xM' \longrightarrow 2MM'_xC_{n/2}s$$

during this step, there is the building of the threelayered sheets with a large increase of the interplanar distance.

In fact, a small amount of the second element has already been intercalated in the alkali metal planes at the end of the first step. In the case of potassium hydride, the two previous reactions are actually :

$$8\ C \xrightarrow[400\ C]{KH} KH_xC_8$$

where x << 0.1 and :

$$KH_xC_8 + \frac{1.6-x}{2} H_2 + K \longrightarrow 2\ KH_{0.8}C_4$$

According to the nature of the reactant, one has three different behaviours :

- with potassium hydride, the mercury and thallium alloys with potassium or rubidium, the reactions occur exactly as described before.

According to the temperature, the first step ends when graphite has become first or second stage binary. During the second step, the cooperative intercalation of both elements leads to the ternarization without any stage change[6-8].

- the bismuth alloys intercalate even when the alkali metal is cesium. The end of the first step leads to a binary compound of stage 4 (cesium) or stage 5 (potassium or rubidium). The second step is the ternarization of the binary compound and leads to the ternary of the corresponding stage.

In order to get a ternary lower stage, there is, in that case, a third step which consists of decreasing stage from the ternary 4th or 5th stage down to stage 1[9].

- with sodium hydride, there is only one step during the intercalation : NaH intercalates directly into graphite. In fact, the H/Na ratio is equal to one, whereas, with potassium hydride, the value of H/K is equal to .8.

On the other hand, this direct intercalation of NaH accounts for the fact that pure sodium leads only to high stages by intercalation into graphite[10].

There are also other possibilities to prepare ternary polylayered compounds with mercury or hydrogen, by ternarization of the binary first stages by calculated quantities of hydrogen or mercury. In these cases, the compound is a second stage and the reaction is longer than when starting from the amalgam or the hydride[7,11].

An other interesting method for the preparation of ternaries is to start from potassium hydride and a binary compound of potassium according to the following reactions :

$$KC_{24} + 2\ KH \longrightarrow 3\ KH_{2/3}C_8 \qquad (1)$$

$$KC_{24} + 5\ KH \longrightarrow 6\ KH_{5/6}C_4 \qquad (2)$$

$$KC_8 + KH \longrightarrow 2\ KH_{0.5}C_4 \qquad (3)$$

The reactions (1) and (3) are isostage and lead directly to the ternary phase. The reaction (2) involves a change of stage : starting from the binary second stage, one obtains the ternary first stage. In fact, during that reaction, one can observe firstly a binary first stage before the preparation of the ternary first stage. This behaviour confirms the mechanism described before : the ternarization is an isostage process.

The different reaction mentioned previously involve the presence of both elements during the whole process of intercalation. There are also other possibilities to prepare ternaries, for instance the ternarization of second stages of metals by organic molecules, like, for instance benzene[12] or THF[13].

$$\underset{\text{stage 2}}{KC_{24}} + {}_xTHF \longrightarrow \underset{\text{stage 1}}{K(THF)_xC_{24}}$$

A similar behaviour occurs with other organic molecules[14] or ammonia[15], but there are two differencies :

- the intercalated layer is formed by a single alkali metal and molecule layer

- the conditions of the ternarization are such that there is no external source of metal. The reaction occurs like when one prepares a second stage potassium-hydrogen, according to the reaction :

$$\underset{\text{stage 1}}{KC_8} + 1/3\ H_2 \longrightarrow \underset{\text{stage 2}}{KH_{2/3}C_8}$$

This ternarization by hydrogen involves also a change of stage and takes a long time (one or two weeks) compare to the direct intercalation of KH (one day at most).

The different steps necessary to obtain a ternary compound have been described. In fact, in most cases, with the heavy metals, there are, for a given stage, two different forms, called α and β which differ according to the thickness of the intercalated layers, the α phase being the thinner one. The α and β phases are very often mixed together and there is, then, a last step in the intercalation process : the thermodynamically stable phase is obtained in a pure form by annealing of the mixture[16].

There is no general rule about the relative thermal stability of forms α and β.

Since the transformation can be done generally at low temperature and without free reactant, the formula of α and β phase remains the same, as shown by chemical analysis, except for the cesium-bismuth stage one α and β whose formulae are respectively $CsBi_{.53}C_4$ and $CsBiC_4$. But, in this case, the transformation occurs only in the presence of the free alloy.

APPLICATIONS OF THIS STUDY

In the case of the potassium-hydrogen ternaries, there can be a problem : the repeat distances along the $\vec{c}$ axis for ternary first stage and binary second stage on one side, the ternary second stage and the binary third stage on the other are very close. The values of I_c are respectively 8.53 Å and 8.70 Å ; 11,88 Å and 12.05 Å. The compounds are generally identified by their 001 reflexions and there can be a confusion since the corresponding angles are very close.

Table 1. Comparison between the 001 Reflexions of binary and ternary compounds of potassium

Compound	I_c (Å)	Stage	main 001 peak		
			001	d (Å)	θ° (MoK$_\alpha$)
$KH_{.8}C_4$	8.53	1	003	2.84	7.17
KC_{24}	8.70	2	003	2.90	7.03
$KH_{.8}C_8$	11.88	2	004	2.97	6.88
KC_{36}	12.05	3	005	3.01	6.76

Since one knows that binary and ternary compounds whose stages are different cannot coexist during the intercalation process, the confusion is not possible anymore.

On the other hand, the compounds KC_{24} and $KH_{.8}C_8$ could also be mistaken, however, in this case, the repeat distances along the $\vec{c}$ axis are quite different.

An other application of this study consists of the determination of the relative affinities of the heavy metals for the alkali metals.

If one considers the ability of the different binaries of the heavy alkali metals to lead, under the action of hydrogen, to ternaries, the easier hydrogenation is for the compound KC_8 (100°C, hydrogen pressure : one atmosphere or less). The compounds KC_{24} and RbC_8 react with more difficulty (pressures of 100 atmospheres or more). The other binaries are impossible to hydrogenate.

The preparation of a ternary compound proceeds of an equilibrium between the affinity of the alkali metal towards the graphite and the third component (hydrogen in the present case). When the affinity of the alkali metal is too high for hydrogen (case of LiC_6 and hydrogen which leads to a mixture of LiH and graphite[17]) or for graphite (no hydrogenation), one cannot obtain a ternary compound.

From these considerations, one can deduce that the stability of the graphite-alkali metal bond increases when the stage increases (KC_8 reacts easily, KC_{24} with difficulty, no reaction at all with higher stages) and as the electropositivity of the alkali metal increases (KC_8 reacts easily, RbC_8 with difficulty, no reaction with CsC_8).

If one considers now the reaction of the mercury, thallium and bismuth alloys with the heavy alkali metals on graphite (see table 2), one has two different beheviours :

- Mercury and thallium alloys :
  - react only when the alkali is potassium or rubidium
  - the heavy metal starts to intercalate when the stage is low (stage 1 or 2). The intercalation occurs, then, with the less stable binaries.

- Bismuth alloys :
  - react also when cesium is the alkali metal
  - bismuth starts to intercalate in the binary stage 4 (Cs) or 5 (K, Rb), in very stable binaries.

From these considerations, one can say that the affinity of bismuth for the alkali metals is higher than that of mercury or thallium.

Finally, it is necessary to balance the affinity of graphite for potassium or rubidium by 1.5 atoms of bismuth (general formula $MTl_{1.5}C_{4s}$), by one mercury atom ($MHgC_{4s}$) or 0.6 atom of bismuth ($MBi_{.6}C_{4s}$). Thus, the affinity of thallium for the alkali metals is weaker than that of mercury and one can classify the affinity of the considered heavy metals for the heavy alkali metals : it increases from thallium to mercury to bismuth.

Table 2. Comparative Behaviour of the three Alloys during the Intercalation into Graphite

Heavy metal	Alkali metal	End of 1rst step	Final stage	M/M'
Tl	K, Rb	Stage 1 or 2	1 or 2	1.5
Hg	K, Rb	Stage 1 or 2	1 or 2	1
Bi	K, Rb or Cs	Stage 4 (Cs) Stage 5 (K, Rb)	1 → 7	.6

CONCLUSION

The intercalation into graphite of binary species (alkali metals hydrides or alloys with heavy metals) does not occur directly, except in the case of sodium hydride. The composition of the intercalated species can be different of that of the free reactants.

From the knowledge of the intercalation mechanisms, one can :

- follow easily the reaction,
- determine the relative affinities of the heavy metals for the heavy alkali metals.

REFERENCES

1. J.G. Hooley : Mat. Sc. Eng. 31 (1977) 17.
2. A. Hamwi, P. Touzain and C. Riekel : Synth. Met. 7 (1983) 23.
3. H. Suematsu : Ann. Phys. 11 (1986) 75.
4. J.E. Fischer, C.D. Fuerst and K.C. Woo : Synth. Met. 7 (1983) 1 and F. Rousseaux, G. Besson, C. Tchoubar and D. Guérard : MRS Symp. K (1986) 13.
5. D. Billaud, D. Balesdent et A. Hérold : Bull. Soc. Chim. Fr. 11 (1974) 2407.

6. N.E. Elalem, D. Guérard, F. Rousseaux, G. Loupias, H. Estrade-Szwarckopf, P. Lauginie and J. Conard : Proc. Carbon'86 - Baden-Baden (1986) 433.
7. P. Lagrange, M. El Makrini, D. Guérard and A. Hérold : Synth. Met. 2 (1980) 191.
8. M. El Makrini : Thèse Nancy 1980.
9. A. Bendriss-Rherhrhaye:Thèse Nancy 1986.
10. A. Métrot, D. Guérard, D. Billaud and A. Hérold : Synth. Met. 1 (1980) 363.
11. D. Guérard, P. Lagrange and A. Hérold : Mat. Sc. Eng. 31 (1977) 29.
12. R. Setton, F. Beguin, J. Jegoudez et C. Mazières : Rev. Chim. Min. 19 (1982) 360.
13. R. Setton : J. Mol. Cat. 27 (1984) 263.
14. B.R. York, S.K. Hark and S.A. Solin : Phys. Rev. Lett. 50 (1983) 1470 and J. Chem. Phys. 82 (1985) 921.
15. H. Miyazaki, C. Horie and T. Watanabe : J. Phys. Soc. Jap. 53 (1984) 3843.
16. A. Bendriss-Rerhrhaye, P. Lagrange and F. Rousseaux : to appear in Synthetic Metals.
17. D. Guérard : Thèse Nancy 1974.

VIBRATIONAL MODES OF A TWO DIMENTIONAL SUPERLATTICE STRUCTURE

Neil D. Shrimpton

Ottawa-Carleton Institute for Physics
University of Ottawa
Ottawa, Ontario, Canada

A simple crystal structure with small unit cells has vibrational modes that must be understood from the motion of the individual atoms in the unit cell. Superlattice structures have a large number of atoms per unit cell and modes of vibration are possible which involve collective motion of the atoms. This collective motion can be described using renormalized models which do no require a detailed knowledge of the microscopic interactions of the systems. Vibrational modes associated with the collective motion of the atoms are known as phasons.

Superlattice structures have a large number of atoms per unit cell. Rare gas monolayers adsorbed on substrates are systems that have superlattice structures which have been studied extensively. The example being considered in this paper is a krypton monolayer adsorbed on graphite. The interaction between the krypton atoms and the graphite substrate is not uniform and certain adsorption sites minimize the adsorption energy. These adsorption sites form a triangular lattice with spacing 2.46A and are generated by the primitive vectors $\vec{B}_1$ and $\vec{B}_2$ shown in fig. 1. The krypton atoms will fill a $\sqrt{3} \times \sqrt{3}$ sublattice of the adsorption sites, the spacing 4.26A of this lattice is close to the spacing of 4.03A which minimizes the krypton pair interactions. In fig. 1, $\vec{D}_1$ and $\vec{D}_2$ are the primitive vectors which generate the $\sqrt{3} \times \sqrt{3}$ sublattice. This solid is known as being commensurate or registered with the substrate. The monolayer can also have an incommensurate solid phase. It is noted above that the krypton pair interaction is minimized for a spacing different from 4.26A. Such mismatched systems have been considered by Frank and van der Merwe[1] and configurations are possible where the positions of the adatoms have soliton like displacements. For krypton on graphite, this corresponds to the monolayer having regions where the adatoms are registered with the substrate. These regions are separated by domain walls of higher density (fig. 2). The monolayer will form either a commensurate or an incommensurate monolayer depending on the stoichiometry of the system.

The width of the domain walls depends on two quantities; the strength by which the substrate pulls the adatoms to an adsorption site, and the strength of the interactions between the adatoms. The first effect arises because the adsorption energy is not constant over the surface of the substrate. This variation in energy is known as the substrate corrugation. If the substrate is highly corrugated the adatoms will be

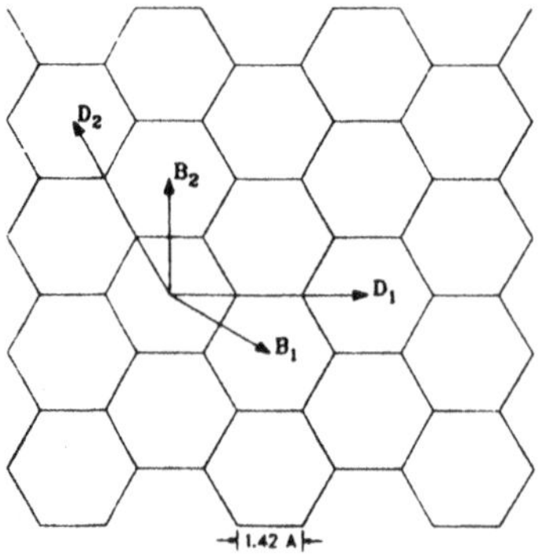

Fig. 1 Graphite substrate, carbon bonds represented by honeycomb network. B_1 and B_2 generate the set of adsorption sites, D_1 and D_2 $\sqrt{3} \times \sqrt{3}$ sublattice.

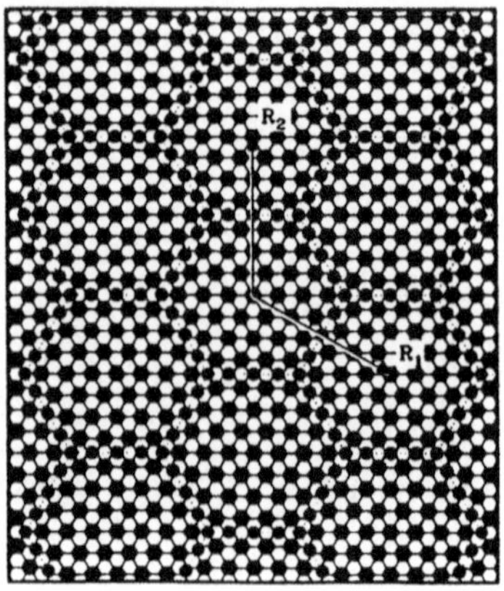

Fig. 2 Incommensurate monolayer with hexagonal symmetry R_1 and R_2 indicate the periodicity.

pulled strongly towards adsorption sites. The second effect depends on the strength of the interactions between adatoms. If this interaction is strong the adatoms will favour a triangular lattice. The interplay between these two influences results in the domain walls being sharp when the substrate corrugation is dominant: More adatoms are fixed at adsorption sites. On the other hand, when the adatom interaction is dominant, the domain walls will be very broad.

An example of a typical incommensurate configuration is shown in fig. 2. The vectors $\vec{R}_1$ and $\vec{R}_2$ identify the periodicity of the superlattice. The size of a given registered region, or domain depends on the length of $\vec{R}_1$ and $\vec{R}_2$. The configuration of the monolayer can also be specified from the averaged positions of the adatoms in the monolayer. These averaged positions form a triangular lattice and because the incommensurate monolayer is denser than the commensurate monolayer the primitive vectors $\vec{d}_1$ and $\vec{d}_2$ of this lattice are shorter than the vectors $\vec{D}_1$ and $\vec{D}_2$ of fig. 1. The difference between these lengths is known as the misfit M%

$$M\% = \frac{\vec{D}_1 - \vec{d}_1}{\vec{D}_1} \times 100\%$$

It is sometimes energetically favourable for the monolayer to be rotated with respect to the substrate. This has been observed experimentally[2] and predicted theoretically[3,4]. The difference between the angles of the vectors $\vec{d}_1$ and $\vec{D}_1$ specifies the rotation of the monolayer.

Although fig. 2 shows an incommensurate phase with hexagonal periodicity, it is also possible to form a uniaxial phase where the domain walls form stripes across the monolayer. For krypton on graphite this type of configuration has not been observed experimentally[5]. Calculations which consider only the static properties of the monolayer have, however, predicted that the striped phase may be more energeticalloy favourable when the domain size is large[6]. However, as shown by Villain[7], the hexagonal phase has a higher entropy due to a breathing mode of the domain walls and even for large domains at low temperatures the striped phase will not be formed. An excellent review of the general properties of rare gas monolayers adsorbed on graphite is provided by Villain and Gordon[8].

The dynamic properties of krypton monolayers on graphite have been considered in a thesis by this author[9].

The low energy vibrational behaviour of the commensurate and incommensurate monolayers are quite different. The commensurate phase has every adatom locked into adsorption sites and vibrational modes are not possible with energies less than a given value known as the commensurate band gap. The incommensurate phase, on the other hand, has domain walls that can move easily[10]. This motion is analogous to a bulge in a carpet, it may be difficult to move the carpet as a unit, but it is easy to shift the position of the bulge. Thus the incommensurate monolayer can have vibrational modes associated with the motion of the domain walls that are lower in energy than the value of the commensurate band gap. In fact, for krypton on graphite the domain walls translate freely and the monolayer has accoustic modes[10].

A detailed calculation of the dynamics of the monolayer based on microscopic interactions of the krypton atoms and graphite substrate has been performed[9]. For a given misfit and orientation of the monolayer the static minimum energy configuration is determined. Within the harmonic approximation, the dynamical matrix of this configuration is calculated. The frequency and form of the vibrational modes are determined from the eigenvalues and eigenvectors of the dynamical matrix. An analysis of the eigenvectors reveals that the low energy vibrational modes involve only the adatoms on the domain walls and the motion of the adatoms can be ascribed collectively to motion of the domain walls. Fig. 3 shows the

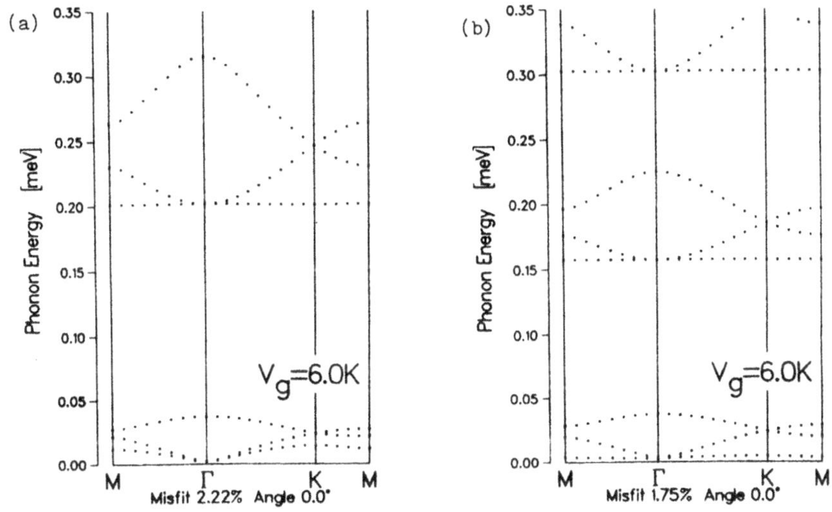

Fig. 3 Phonon energies as a function of wavevector for an incommensurate monolayer when the substrate corrugation is 6.0K. Γ is at the brillouin zone centre, K is at a corner and M is at the middle of an edge (see fig. 7).

dispersion curves of a typical incommensurate configuration. The misfit value of 2.22% corresponds to a domain size of 110A across. The misfit value of 1.75% corresponds to the domains being 140A across. The substrate corrugation of Vg = 6.0K causes the domain walls to be 35A thick. The commensurate configuration with this value of substrate corrugation has a band gap value of 0.8 meV, as the energy of the incommensurate vibrational modes approaches this value, the adatoms in the registered regions can start to vibrate and the motion of the adatoms cannot be ascribed to that of the domain walls. However, for the range of energies shown in fig. 3 the vibrational modes do correspond to domain wall motion. Fig. 4 shows the form of the domain wall motion for the 6 lowest energy modes of fig. 3a. The wavevector of the motion is at the centre of the brillouin zone. The dashed lines show the deformation of the domain walls about their equilibrium positions (solid lines). A scale is provided with each figure to indicate the size of the domains although the magnitudes of the deformation have been scaled arbitrarily to make the modes clearly visible. The value used for the scaling is, however, consistent in all of figures 4. Fig. 5 is similar to fig. 4 except the wavevector of the motion is at the middle of a brillouin zone edge.

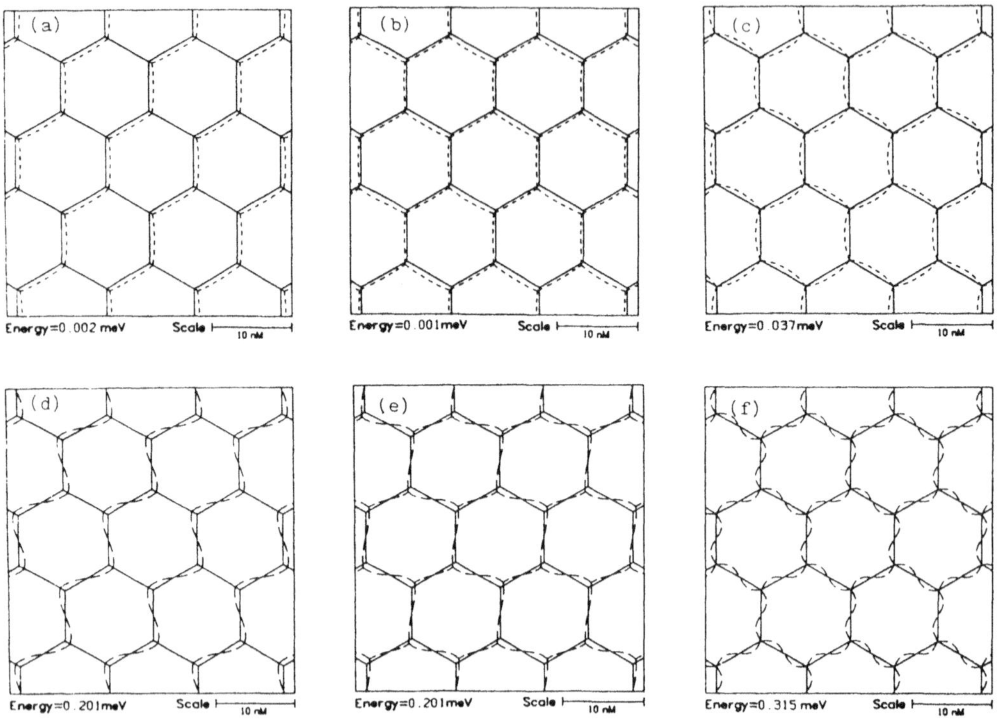

Fig. 4 Domain wall modes for fig. 3a wavevector is at the Γ point (see fig. 7).

The lowest energy mode fig. 4a and fig. 5a is the breathing mode predicted by Villain[7]. This mode has the characteristic that the total length of the domain walls is constant. The fact that this mode becomes increasingly soft (fig. 3b) as the misfit decreases is a consequence of the energy of the monolayer at small misfits being dominated by the energy per length associated with the domain walls. Under such circumstances, configurations which have the same total length of domain walls will be degenerate in energy. The modes which generate these alternate configurations will thus be completely soft. The next two higher energy modes correspond to shearing motions of the domain walls. The energy of these modes appears to be relatively insensitive to the misfit of the monolayer (fig. 3a and 3b). The next higher energy group of three modes are of interest because of the relationship that they have to the lowest energy group of three. If fig. 4a is compared to fig. 4d and fig. 5a is compared to fig. 5d the fourth lowest energy mode corresponds to a flexing of the domain wall. This can be compared to the vibrations of a string. If the lowest energy mode corresponds to the perpendicular translation of the string, the fourth mode corresponds to a vibration of the string with wavelength twice the length of the string. An analysis of the seventh lowest energy mode (not shown) reveals a wave with wavelength equal to the length of the string. This mode has twice the energy of the fourth mode

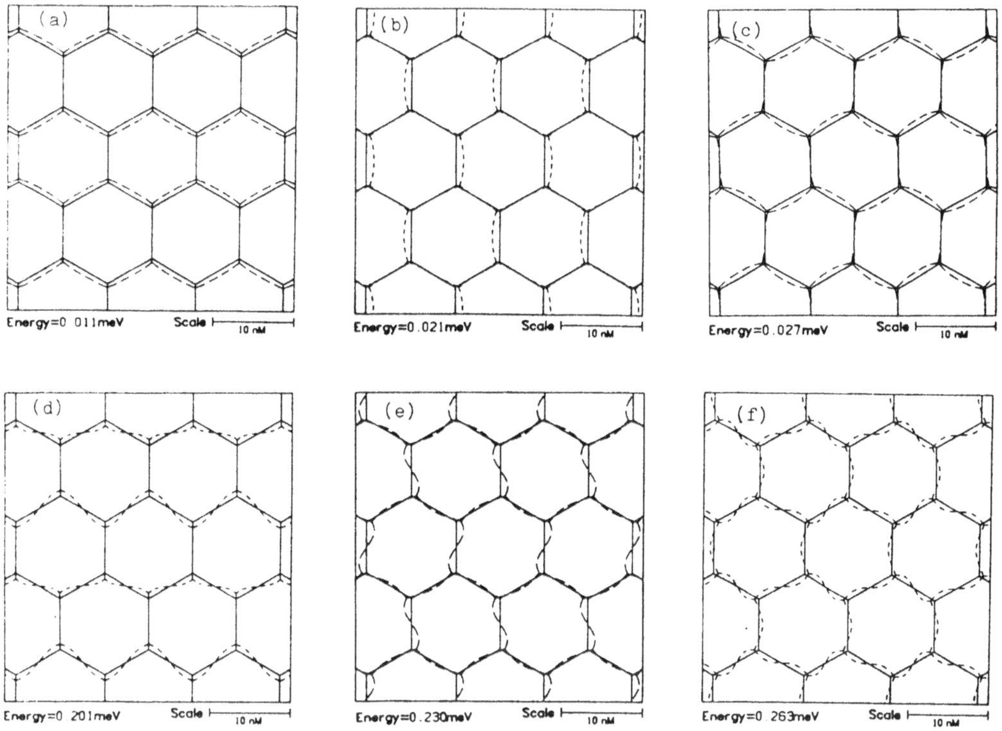

Fig. 5 Domain wall modes for fig. 3a wavevector is an M point vertical to Γ

(fig. 3b). From the frequency of the mode, and the length of the domain wall segments, the wavevelocity is calculated to be 100m/sec for the substrate corrugation Vg = 6.0K of fig. 3. This correspondence between the first three modes and the next three modes is also noted in fig. 4c and fig. 4f, the higher energy mode is clearly a more energetic form of the lower energy mode.

These results inspired a renormalized model based on string vibrations. The concept that domain walls should have string like qualities has already been considered for uniaxial systems[11], however, this type of behaviour has not been considered for the more complex system of a domain wall network. If the domain wall dynamics are dominated by the flexing of the domain walls, a simple model can be constructed to explain their motion. The domain walls are treated as string segments; these segments have a given wavevelocity c and are connected at the vertices (fig. 6). Given the equations of motion for a string, the displacement of the string y perpendicular to its length must satisfy

$$y(x, t) = f(x + ct) + g(x - ct)$$

where f and g are arbitrary functions. Given that the strings are connected at the vertices, the displacements y of the three strings at the vertex must total to zero. From fig. 6, if the values of the string segments at the central vertex are given as $y_1(0, t)$, $y_2(0, t)$ and $y_3(0, t)$ then

$$y_1(0, t) + y_2(0, t) + y_3(0, t) = 0$$

Also the strings exert a force at the vertex that is proportional to the derivatives dy/dx. Because the vertices are free to move in this model,

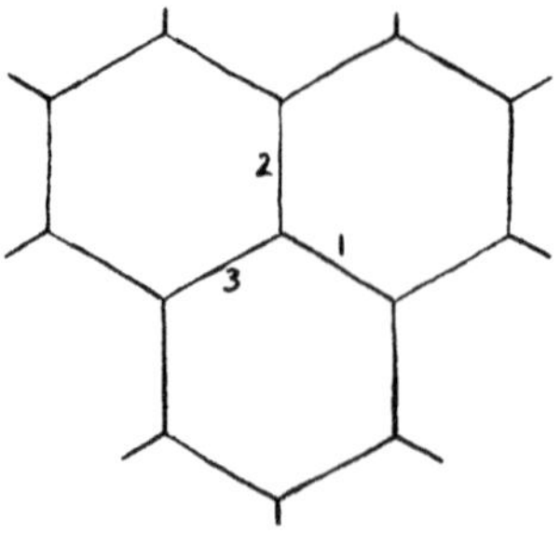

Fig. 6 String network model, the string segments connected to the central vertex are labelled.

the total force exerted by the strings must be zero. This provides the additional condition that

$$\frac{\partial y_1(0,\ t)}{\partial x} = \frac{\partial y_2(0,\ t)}{\partial x} = \frac{\partial y_3(0,\ t)}{\partial x}$$

These conditions must be met at every vertex in the network.

Corresponding to the normal mode solutions based on the dynamical matrix of the monolayer, solutions are sought so that the central vertex moves about its equilibrium position as

$$\vec{u}(0,\ t) = \vec{u}\ e^{i\omega t}$$

and the corresponding equivalent vertices of the surrounding domains move about their equilibrium positions as

$$\vec{u}(\vec{R},\ t) = \vec{u}\ e^{i(\vec{q}.\vec{R}\ +\ \omega t)}$$

$\vec{R}$ ranges over the triangular lattice formed by the primitive superlattice vectors $\vec{R}_1$ and $\vec{R}_2$ (see fig. 2). This network of strings will then have vibrational frequencies for wavevectors along the line from the centre of the brillouin zone Γ to the middle of a zone edge M (see fig. 7) of

$$\omega = \frac{c}{L}\left[\pi n \pm \frac{1}{2}\cos^{-1}\left(\frac{1}{9}(1 + 8\cos\theta)\right)\right] \quad \text{or} = \frac{c}{L}\pi n$$

where n = 0, 1, 2 ... , and θ = 0 at the Γ point and π at the M point. Along the line from the Γ point to a corner of the brillouin zone K, to the middle of the next zone edge M (fig. 7)

$$\omega = \frac{c}{L}\left[\pi n \pm \frac{1}{2}\cos^{-1}\left(\frac{2}{9}(1 + 2\cos\theta)^2 - 1\right)\right] \quad \text{or}\ \frac{c}{L}\pi n$$

where n = 0, 1, 2 ... , and θ = 0 at Γ, θ = 2/3π at K and θ = π at M. The dispersion curves produced are shown in fig. 8.

It is clear that the successive nature of the modes has been modelled. However, a discrepancy arises because the third and fourth modes meet at the Γ point, a fact not found in fig. 3: The model lacks effects due to the vertices. The vertices can interact with each other and they can interact with the substrate. They can have a mass and perhaps an inertia. To modify the model the vertex is described as a rigid rotator of fixed radius and the strings are connected to its edge.

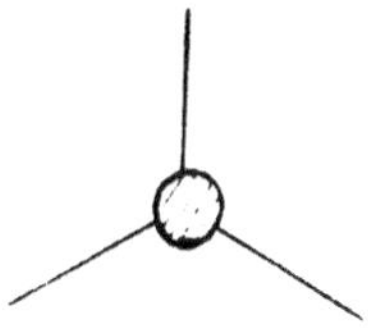

This introduces extra parameters to the model, the effects of which can be predicted.

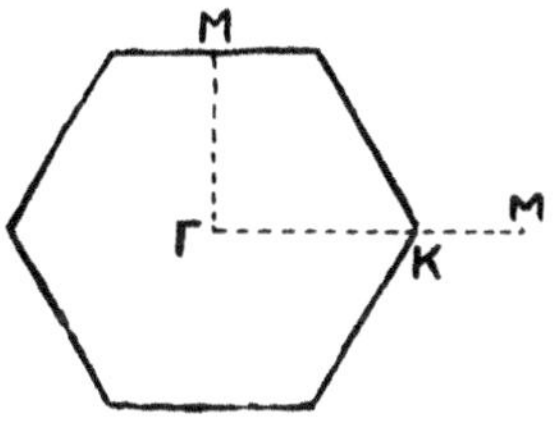

Fig. 7 Diagram of the brillouin zone showing the wavevector paths.

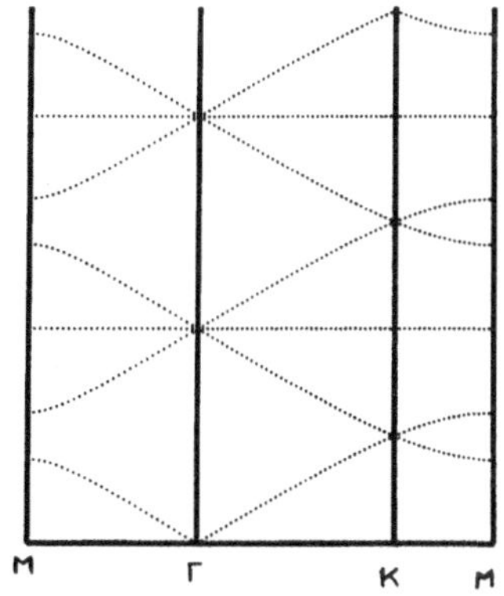

Fig. 8 Vibrational frequencies (arbitrary units) as a function of wavevector for the string model.

The interaction between the vertices causes the lowest energy breathing mode to have dispersion. This is found in fig. 3a. Increasing the size of the domains decreases the vertex interaction strength and the mode becomes softer (fig. 3b) the vertex can also have an interaction with the substrate. Although the substrate cannot exert a force on the vertex[10], the substrate can exert what amounts to a torque. This torque arises from the domain walls preferring to intersect each other in a staggered configuration.

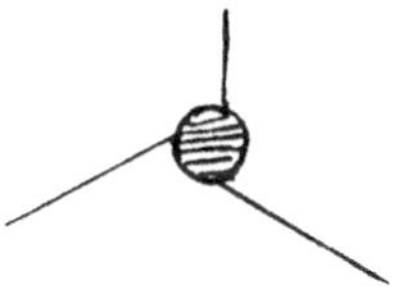

The advantage of the staggered configuration is that the increased density of the domain walls when intersected at the vertex can cause an overcompression of the adatoms[9]. This overcompression is eased if the vertex is rotated. From figs. 4 and 5, the second, third, fifth and sixth modes involve rotations of the vertices. If the substrate influence favours rotated configurations, the energy of the modes will be decreased. The fourth and seventh modes on the other hand, do not involve rotations of the vertices and their energies will not be affected.

Work is in progress on the more complete model that includes not only the vertex pair and substrate interactions, but also the mass and inertia of the vertices. The results of this will be published elsewhere. The general conclusion can be made, however, that a simple model based on the vibrational properties of string segments can explain the dynamic behaviour of a two dimensional superlattice configuration. The general requirements of this model are that the superlattice has domain walls or boundary segments that are independently free to translate but resist flexing.

Acknowledgements

I would like to thank Béla Joos and Birger Bergersen for their helpful discussions and comments. I would also like to acknowledge the financial support of N.S.E.R.C.

References

1. F.C. Frank and J.H. van de Merwe, One-Dimensional Dislocations I. Static Theory, Proc. R. Soc. London A198:205 (1949).
2. S.C. Fain, M.D. Chinn and R.D. Diehl, Commensurate-Incommensurate Transition of Solid Krypton Monolayers on Graphite, Phys. Rev. B21:4170 (1980).
3. A.D. Novaco and J.P. McTague, Substrate-Induced Strain and Orientational Ordering in Adsorbed Monolayers, Phys. Rev. B19:5299 (1979).
4. H. Shiba, The Nonregistered-Registered Phase Transition of Rare Gas Monolayers Adsorbed on Graphite. II, J. Phys. Soc. Japan 48:211 (1980).
5. P.W. Stephens, P.A. Heiney, R.J. Birgeneau, P.M. Horn, D.E. Moncton and G.S. Brown, High-Resolution X-Ray-Scattering Study of the Commensurate-Incommensurate Transition of Monolayer Kr on Graphite, Phys. Rev. B29:3512 (1984).
6. R.J. Gooding, B. Joos and B. Bergersen, Krypton on Graphite: Microstructure at Zero Temperature, Phys. Rev. B27:7669 (1983).
7. J. Villain, Two-Dimensional Solids and Their Interaction wih Substrates, Ordering in Strongly Fluctuating Condensed Matter Systems, Ed. T. Riste: 221 (Plenum, New York, 1980).
8. J. Villain and M.B. Gordon, Static Properties of Physisorbed Monolayers in their Solid Phases, Surf. Sci. 125:1 (1983).
9. N.D. Shrimpton, Low Temperature Behaviour of Krypton Monolayers on Graphite, Ph.D. Thesis University of British Columbia, Canada (May, 1987).
10. N.D. Shrimpton, B. Bergersen and B. Joos, Domain Wall Modes for Krypton Monolayers on Graphite, Phys. Rev. B34:7334 (1986).
11. V.L. Pokrovsky and A.L. Talapov, Ground State, and Phase Diagrams of Two-Dimensional Incommensurate Crystals, Phys. Rev. Lett. 42:65 (1979).

MAGNETIC RESONANCE

AS A TOOL FOR G.I.C. STUDIES

Jacques Conard

CRSOCI CNRS
45071 Cedex
Orléans France

INTRODUCTION

The graphite intercalation compounds conductivity is of great interest among many other properties as staging and bidimensionnal character. II^{d} stage A_sF_5 and NO_3H graphite compounds obtain conductivities not far from copper and aluminium respectively while the donor potassium GIC get only half this values and only for the 4th to 5th stage.[1]

Then the electronic structure of these compounds is a challenge if considered either as alloys or molecular solids. If we remember that graphite has carriers with the highest known mobility but not enough density, it seems that the general effect of intercalation is to rise this carrier density but with a reduction of the mobility which reduces the hoped gain in conductivity.

So the purpose of this lecture is to answer the questions :

1 What is the "atomic struture" of the free electron wave?
2 Does the "free electron" visit the inserted species?
3 Is the inserted species responsible for shortening the carrier lifetime?

Which are in fact three aspects of the same question.

And with magnetic resonances, we have two kinds of tools :

1 E.S.R. which is, in conductors, a non local tool which sees some average along the electron orbit.
2 N.M.R. which is a local tool in two ways :
  - in the direct space where the nucleus measures a local density of orbitals
  - in the reciprocal energy space where it is sensitive to the atomic structure of the orbital at the Fermi level through the Knight shift.

But the nucleus is also sensitive in aromatics to changes in the deep part of the valence band which act on the chemical shift.

RESULTS OBTAINED BY ELECTRONS SPIN RESONANCE

In conductors E.S.R. line has a specific lineshape known as Dysonian. It results from a mixture of absorption and dispersion shapes depending on the dephasing during the penetration and the output of the microwave in and from the sample respectively. Line intensity is related to the free electron paramagnetism, proportional to the density of states at the Fermi level N_{EF}. In good conductors the free electron averages rapid fluctuations of its coupling with the environment, during its trajectory, giving a $T_1 = T_2$ situation (T_1, T_2 longitudinal and transversal relaxation times). One limiting source of broadening is the lifetime of the observed electronic state with a magnetic correlation time τ . I may be related to the resistivity correlation time τ_r if we are able to define a yield for the spin flip during an electronic collision. But specific relaxation mechanisms are also able to act on T_1 and T_2 values.[2]

In fig 1 we give a plot of the reciprocal line-width log T_2 VS stage written as $1/x$ in MC_x for two acceptor and four donor compounds. The minimum of linewidth (max of T_2) seems to be related to the minimum of resistivity but with some shift : in Li and K compounds the minimum of linewidth occurs for the II^d stage, meanwhile in acceptors the linewidth is almost stage insensitive at the beginning. This comparison supports the idea that the 1st neighbour plane in donor GICs has some source of line broadening which does not exist in acceptor ones. We propose that partial covalency with alkaline atom is responsible for this effect because of a small presence of the free electron on the inserted alkaline atom. Then we can think that the alkaline atomic spin-orbit coupling could be responsible for this relaxation, as we see in donor compounds an atomic number (Z) dependence (fig 2).

The Eliott's formula : $T_2 = \alpha\tau/(\Delta g)^2$ $(1/30 < \alpha < 1)$ gives a relation between the linewidth $1/T_2$ and the spin orbit coupling Λ through the relation $\delta g \;\alpha\; \Lambda/\Delta E$, where the lineshift is related to the atomic Λ value.

After Mc Clure, the Λ value is proportionnal to Z^5 for the lighter elements and evolutes towards Z for the heavier ones. Then we look for a $T_2 = a/Z^n$ relation.

Acceptor studies of Davidov[3], where the central atom M of Cx MFy series has been changed, is also interesting as it shows a clear but limited role of the Z_M value in the line widening. The reduction factor for II^d stage compounds is about 10 relatively to donor ones.

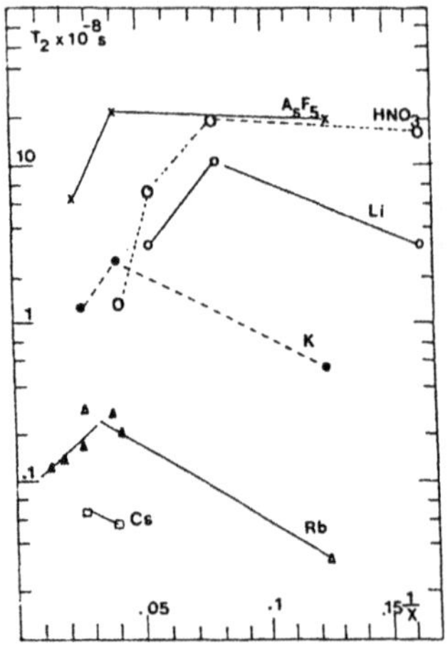

Fig. 1 $T_2 = 1/\gamma\Delta H$ VS stage

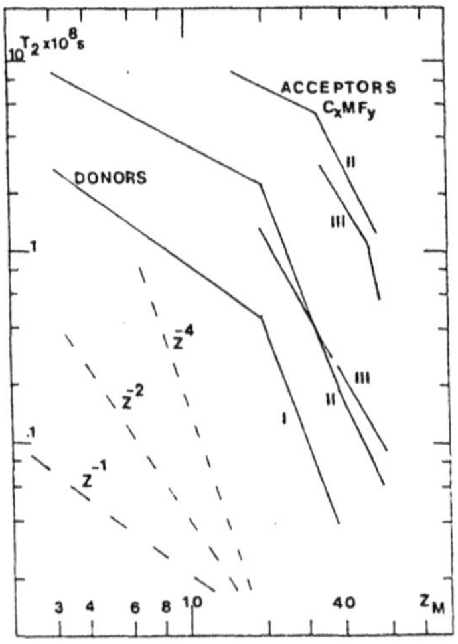

Fig. 2 T_2 VS Z (atomic number)[2,3]

So these ESR experiments are decisive for the role of the inserted species in the relaxation of the free electron. All the details are not perfectly understood but we can agree with a small presence of the free electron upon alkaline atoms in the first stage donor GIC meanwhile a light magnetic coupling, through exchange mechanism, is enough to explain the limited role of spin orbit coupling in acceptor GICs.

ELECTRON-NUCLEI INTERACTION IN G.I.C.

If we compare N.M.R. in molecules and in a conductive solid, we can propose a list of similarities going from the bottom to the top of the energy levels.

1 The core electrons are assumed to give a constant isotropic screening effect.

2 The paired electrons of the higher levels are equivalent to the main part of the valence band : their screening diamagnetic effect is the chemical shift σ or orbital one. As no averaging mechanism occurs, like in liquids, this shift is generally anisotropic with a non zero trace. That means that it exists a non zero average value which is observed in liquid state N.M.R. Typical of this effect is the ^{13}C spectrum of an isolated graphene plane with an axial symmetry and the (Ho//c) line position in a field higher than the (Ho//a,b) one, because of the diamagnetic screening of the in plane ring currents.

3 At the top of the conduction band, it exists a small amount of unpaired electrons, proportionnal to the density of states at the Fermi level and responsible, as in paramagnetic liquids, of a paramagnetic shift. It depends on the detailed structure of the free electron wave function and has to be described in a more accurate way.

a) The contact interaction results in a paramagnetic Knight shift $\frac{\Delta H}{H} = \frac{8\pi}{3} P_a \xi \mu_B^2 N_{EF}$ which is related to the probability of presence $P_a \xi$ of the s electron at the observed nucleus, with the magnetization $\mu_B^2 N_{EF}$ characteristic of the Pauli electrons. $P_a \xi = |\psi o|^2_N$ is compared to the atomic value P_a by the reduction factor ξ. This effect characteristic of an s electron which wave function has a non zero value at the nucleus, is called the direct contact. It can be partially cancelled by the indirect one resulting from the s core magnetic polarization produced by the paramagnetism of another kind of orbital e.g p type in carbon atoms.

b) The dipolar interaction is specific of non s orbitals and gives a variety of effects. For a p_z orbital with axial symmetry it gives an additionnal local field when Ho is along the z axis meanwhile the Ho//a,b position deduces a local field, half the preceding one but with twice the probability value. Then this dipolar paramagnetic shift is traceless, anisotropic and acts against the orbital shift. In cubic sites it reduces to zero.[4]

ALKALINE NUCLEI N.M.R.

These nuclei have clearly an s character in I^{st} stage compounds whereas II^{d} stage ones offer more diversified situations. Next table gives an idea of the experimental results. The components of the shift can be analyzed in various ways as indicated.

$$\frac{\Delta H}{H} \sim \frac{8\pi}{3} \overbrace{\underbrace{P_{at}}_{\substack{\text{atomic}\\ \text{hyperfine}\\ \text{field}}} \times |a_s^2|}^{\langle\psi o^2\rangle_{E_F}} \times \overbrace{\underbrace{|a_s^2| \times N_{EF}}_{\substack{\text{local density}\\ \text{of states}}} \mu_B^2}^{\chi\ \text{Pauli}}$$

Table I

	LiC_6	KC_8	RbC_8	CsC8	LiC_{12}	CsC_{24}	Unit
$P_{at,M}$	12,5	58	120	200			Tesla
Metal shift	260		6600	15600			ppm
σ range	5		150	630			ppm
GIC shift	40		1950	2550	$\sim$ 40	f(T)	ppm
$a_s{}^2$	0,07		0,13	0,09			

$a_s 2 = \xi$ is known as the s character in a simplified LÇAO wave function at E_F like $\psi = a_s \phi s_M + b\phi p_C + \ldots$

This s character, partially related to the charge transfer, as defined in molecular chemistry, is very dependent on the wave structure near the Fermi level. Theoretical calculations using a correct model for the potential near the nucleus[5,6] are able to find good $a_s{}^2$ values, but for Li at less, the contribution of 2p Li orbitals is important. That means that the wave function' structure near E_F is in fact more complicated in relation with a non spherical sharing of charges around the inserted nucleus.[7,8].

^{13}C N.M.R.

This technique offers a lot of interesting informations on GIC : it is a general method using the nucleus primarily concerned with the π free electron wave function. It sees all the contributions yet described and the problem now is to separate the various contributions and their physical content. The poor natural ^{13}C richness(1%) limits the perturbation from dipolar effects but needs some perseverance to get good signal over noise ratio with relaxation times T_1 up to 90 sec. in pure graphite.

The shift can be written, if we assume axial symmetry as :

$\delta_{a,b} = (A + B)\, N_{EF} + \delta_{a,b}$ orb. A contact interaction

$\delta_c = (A - 2B)\, N_{EF} + \delta_c$ orb. B dipolar interaction **Figure 7**

where four values are involved with only two measurements ; more, no purely ionic reference is available to define the shift. We propose to use boronated graphite where $N_{EF} \sim 0$ and δ orb. ~ 0.

Then if we plot $\delta_{a,b}$ VS N_{EF} as obtained from other techniques, we check a good proportionnality except for LiC_6. The observed absence of any orbital component for Ho//a,b is related to the difficulty to build very large electron orbits, crossing from C plane to C plane. Except for LiC_6 where a reduced anisotropy is generally observed.[9] (Fig. 3).

If we plot the same $\delta_{a,b}$ value VS the Fermi level shift, we are able to show that the donor compounds of first and second stage leave the rigid band model whereas acceptor ones, until now, stay very close to this model. (Fig.4).

Now if we look at higher stage donor compounds, ^{13}C N.M.R. reveals quite different spectra for each kind of C plane. Essentially in donor compounds the 2^d and 3^d carbon planes appears to be charged and slightly coupled C planes with, now, an important orbital δ_c term. And HOPG spectra[10] show ^{13}C doublet for each kind of planes which can be described by a spin

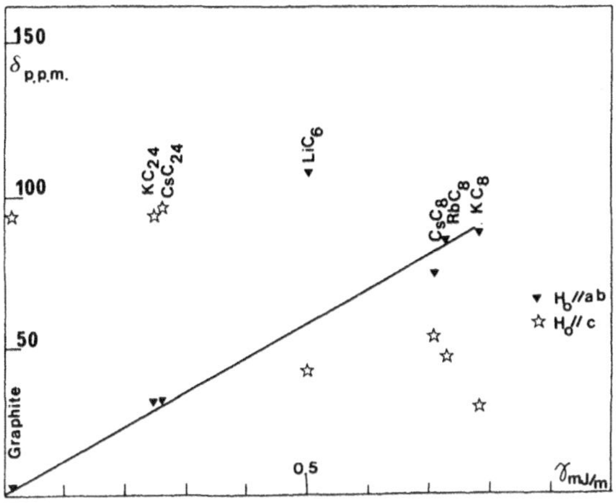

Fig. 3 ^{13}C N.M.R. ; $\delta_{a,b}$ VS γ (low temperature specific heat) shows that $\delta_{a,b} = (A + B)N_{EF}$ relation is valid except for LiC_6.

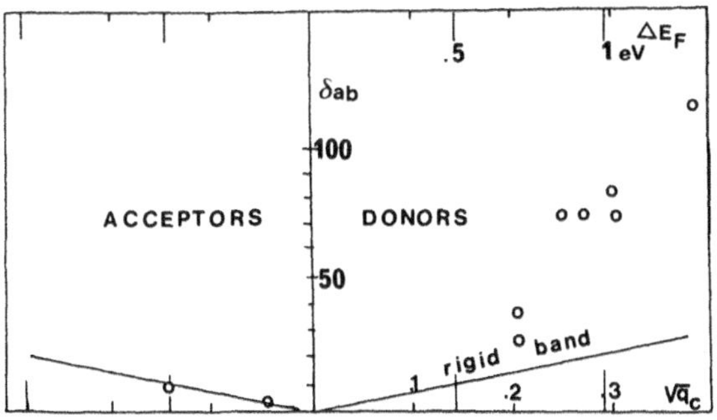

Fig. 4 ^{13}C N.M.R. : $\delta_{a,b}$ VS Fermi level shift shows an excess of density of states at E_F relatively to the rigid band model for donors and not for acceptor GICs density like in paramagnetic aromatic free radicals. It shows the oscillation of the charge and its magnetization along the aromatic lattice. The use of measured T_1 values permits to identify a chain of relaxation towards the alkaline atom acting as a relaxation well. Along this chain the ferromagnetic coupling is the rule in the plane or from plane to plane through the superimposed carbon atoms (Fig. 5).

In acceptor GICs all the C planes appear like doped graphene ones with a low magnetization coming from the relatively low density of states at E_F. We see very important orbital effects. Donor ternarization restores more or less this doped graphene state, from the first carbon plane.

Then it is possible to observe purely orbital (i.e. chemical) effects in acceptor compounds. Two interesting cases yet studied are the NO_3H and $SbCl_5$ II^d stage compounds where two lines with different δ_C values are observed at room temperature.[11] These compounds start from a lattice liquid organisation of the inserted layer which evolutes towards a non commensurable solid in the 200 K range. In NO_3H GIC (α,II) the transition as seen by ^{13}C N.M.R., shows the appearance of a continuous distribution of chemical shifts, probably related to a continuous distribution of bond lengths between the carbon atoms and the oxygen ones.

CONCLUSIONS

As a whole a residual s character of the order of 10% is seen in I stage donor GICS by the alkaline N.M.R.. The π character near the Fermi level necessarily raises much higher than 0,4, the rigid band limit. Then non rigid band is the rule in donors for the sandwich of the inserted species with the two joining C planes. The remote C planes and the acceptor C ones suffer more easily the approximation of separated charged graphene planes. So the layer by layer model[4] we proposed is a good working tool for very local measurements as given by N.M.R.; meanwhile ESR teaches us that the free electron feels more or less the inserted alkaline atom (Fig.6).

The problem of the relation between these results and the charge transfer is, first, a conceptual problem. In a slightly covalent solid, let us say an alloy, it cannot be defined, as in molecules, by a number only, but the new charge localization has to be described in direct space and energy values.

As seen by our techniques, it is split essentially in two parts :
- One, near the Fermi level, can be related to the Knight shift, if we are able to give a good description of the new bandshape in that energy region.
- The other one results from the creation of new states in the deep part of the valence band and acts on orbital properties.

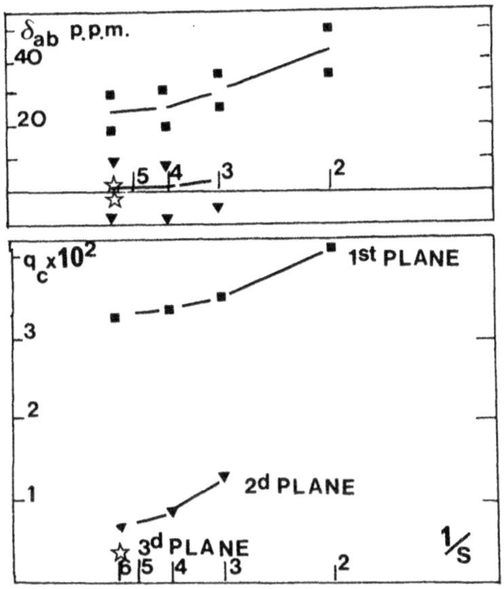

Fig. 5 comparison of graphene plane charge as proposed by Safran to the ^{13}C $\delta_{a,b}$ shifts in KC_x HOPG[10],VS reciprocal stage.

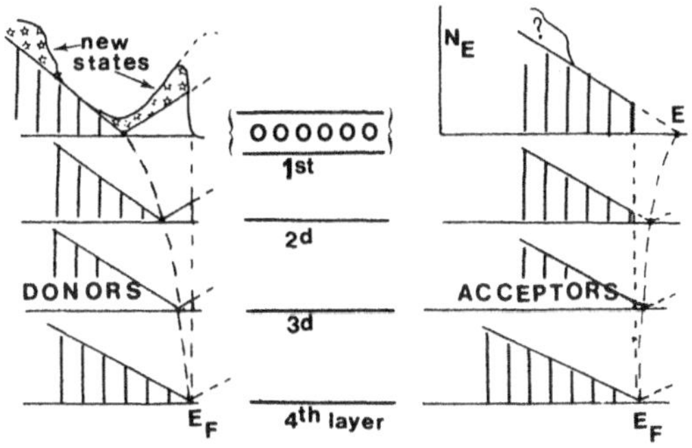

Fig. 6 Model of density of states layer per layer, able to explain local properties seen by ^{13}C N.M.R. It explains also γ values[9]

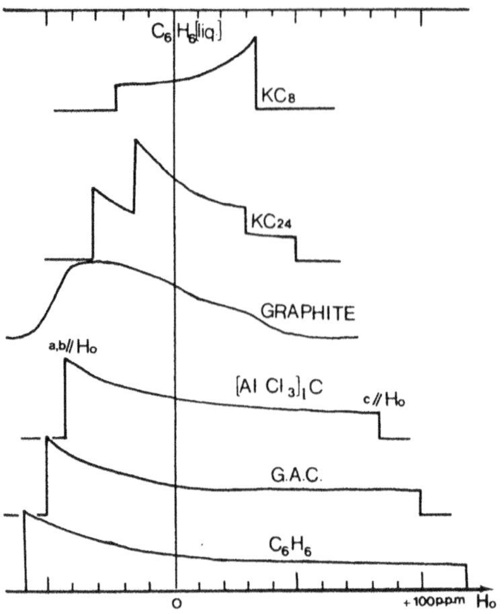

Fig. 7 Schematic presentation of the change in lineshape with the kind of graphite compounds. Axial anisotropy is a common feature of these spectra. The peak is corresponding to a, b, crystalline axes $\|H_0$ and the end of the plateau to $c\|H_0$ axis. The a, b, peak position is mainly governed by the density of states at E_F. For all compounds in which $N(E_F)\simeq 0$ (C_6H_6, GAC, graphite) the c= position is depending only on orbital magnetism of valence electrons. For donor compounds ($N(E_F)\neq 0$), it depends moreover on the paramagnetism of Fermi electrons.

REFERENCES

1. E. Mc. Rae, Thesis, Nancy (1982).
2. P. Lauginie, H. Estrade-Szwarckopf, J. Conard, D. Guerard, P. Lagrange, M. El Makrini, Physica 99 B (1980) 514.
3. R. Davidov, O. Milo, I. Palchan, H. Selig, Synth. Metals 8 (1983) 83.
4. J. Conard, P. Lauginie, H. Estrade-Szwarckopf, G. Hermann, D. Guerard, P. Lagrange, Physica 105 B (1981) 285-89.
5. N.A. Hollzwarth, S. Rabii, L.A. Girifalco, Phys. Rev. B 18 (1978) 5198-216
6. G. Volpilhac, F. Achard, L. Ducasse, J. Hoarau, J. of Physics C 18 (1985) 1613.
7. N.A. Hollzwarth, S.G. Louie, S. Rabii, Phys. Rev. B 28 (1983) 1013.
8. H. Kamimura, Ann. Phys. Col 2 11 (1986) 39-48.
9. J. Conard, H. Estrade-Swarckopf, P. Lauginie, G. Hermann, Springer Series Solid State Science 38 (1981) 264.
10. K. Kume, Ann. Phys. Col. 2 11 (1986) 245.
11. J. Conard, V GIC conference Jerusalem to be published in Synthetic Metals.
12. P. Lauginie, same references.

NEW METHOD TO DETERMINE THE DIELECTRIC CONSTANTS OF ACCEPTOR TYPE GIC

F. Jost, S. Roth, and Y. Yacoby[1]

MPI für Festkörperfoschung
Heisenbergstr.1, D-7-- Stuttgart 80

ABSTRACT

The usual way to determine the dielectric constants is to measure the reflectivity at normal incidence in a large energy range to perform than a Kramers-Kronig analysis. We present a novel method based on a reflectivity measurement at fixed energy in a Fresnel configuration. The reflectivity is measured as a function of angle and polarization in the infrared region. We can simultaneously determine the value of the dielectric constants parallel and perpendicular to the graphitic planes. This method allows for in-situ intercalation and studies of the intercalation kinetics. The first results for AsF_5 stage 1 intercalated graphite are reported and discussed in the framework of the 2D-Blinowski-Rigaux model.

INTRODUCTION

The electrical conductivity of graphite is very anisotropic. This anisotropy increases in acceptor type compounds. Graphite intercalated with AsF_5 is one of the most anisotropic materials [1]. In this paper we present an in-situ method which permits to measure optically the change in anisotropy from pure graphite to stage 1 intercalated graphite. This method is based on angle dependent reflectivity measurements. The determination of the dielectric constants from the Fresnel relations is clearly not a new technique. But such measurements have been largely restricted to isotropic materials and in addition such measurements are often insensitive to small changes in the optical constants. Therefore we have used this technique in an special configuration - that of a internal reflection experiment (Fig. 1) and thus we obtain a very large sensitivity. The high dielectric constant of the germanium crystal provides a better optical index matching. Taft and Philipp [2] derived the complex dielectric constant parallel to the graphitic planes by application of the Kramers-Kronig relation to reflectance data for graphite. Hanlon et al. [3] have carried out room temperature measurements of the 0.07-2.0 eV optical reflectance of stage 1-4 AsF_5-graphite intercalation compound. Blinowski et al. [4] developed a 2D band structure model which they used to analyse reflectivity results in the region of the plasma edge for various GAC including AsF_5. From the theoretical fits they deduced the charge transfer coefficient. There are only a few measurments of the dielectric function of graphite in the c-direction [5],[6],[7] and non for intercalated graphite. The main difficulty for such measurements is to obtain a good optical a-face. With this consideration in mind we have used the Fresnel relations for our experimental situation and developed a best

[1]Racah Institute of Physics, Jerusalem, Israel

fit method to yield simultaneously the four dielectric constants from reflectivity data as function of angle and polarization.

REFLECTIVITY AT NON-NORMAL INCIDENCE

The following formulas are valid for the reflectivity of a uniaxial crystal, where the c-axis is perpendicular to the reflecting surface and the infrared light is polarized with the electric vector parallel or perpendicular to the plane of incidence. The formulas take in to account the internal reflection geometry with three different optical media: 1. germanium, 2. germanium oxyde layer (an inavoidable side reaction of in-situ intercalation), 3. graphite or graphite intercalated with AsF_5.

Vertical polarization

$R_V = | GGE |^2$
$GGE = (ZZGE - 1)/(1 + ZZGE)$
$ZZGE = ZGD/ZGE$
$ZGD = ZA * (1 + GG_{Ge})/(1 - GG_{Ge})$
$GG_{Ge} = GAG * e^{-4\pi i * D * \sqrt{\varepsilon_a^{\parallel} - A0}}$
$GAG = (ZZA - 1)/(1 + ZZA)$
$ZZA = ZG/ZA$
$ZA = 1/\sqrt{\varepsilon_a^{\parallel} - A0}$
$ZG = 1/\sqrt{AS}$
$ZGE = 1/\sqrt{\varepsilon_{Ge}^{\parallel} - A0}$
$AS = (\varepsilon_G^{\parallel} - A0)$
$A0 = \varepsilon_{Ge}^{\parallel} * sin^2\vartheta$

Horizontal polarization

$R_H = | GGE |^2$
$GGE = (ZZGE - 1)/(1 + ZZGE)$
$ZZGE = ZGD/ZGE$
$ZGD = ZA * (1 + GG_{Ge})/(1 - GG_{Ge})$
$GG_{Ge} = GAG * e^{-4*\pi*i*D*\sqrt{\varepsilon_a^{\parallel} - A0}}$
$GAG = (ZZA - 1)/(1 + ZZA)$
$ZZA = ZG/ZA$
$ZA = 1/\sqrt{\varepsilon_a^{\parallel} - A0}/\varepsilon_a^{\parallel}$
$ZG = \sqrt{AS}/\varepsilon_G^{\parallel}$
$ZGE = \sqrt{\varepsilon_{Ge}^{\parallel} - A0}/\varepsilon_{Ge}^{\parallel}$
$AS = (\varepsilon_G^{\perp} - A0) * (\varepsilon_G^{\parallel}/\varepsilon_G^{\perp})$
$A0 = \varepsilon_{Ge}^{\parallel} * sin^2\vartheta$

R_V= reflectivity for vertical polarisation
R_H= reflectivity for horizontal polarisation
$\varepsilon_{Ge}^{\parallel}$= dielectric constant of germanium
$\varepsilon_G^{\parallel} = \varepsilon_1^{\parallel} + i\varepsilon_2^{\parallel}$ - complex dielectric constant of graphite or GIC parallel to the graphitic planes
$\varepsilon_G^{\perp} = \varepsilon_1^{\perp} + i\varepsilon_2^{\perp}$ - complex dielectric constant of graphite or GIC perpendicular to the graphitic planes
$\varepsilon_a^{\parallel}$ = dielectric constant of germanium oxyde ($\varepsilon_a^{\parallel} = 1$)
D = oxyd thickness in units of the wavelength of radiation
ϑ = angle of incidence

The parameters $\varepsilon_G^{\parallel}$, $\varepsilon_G^{\perp}$ and D are determined by least square fits to the measured reflectivity. For graphit D=0 because the oxyd layer appears only as a interaction of AsF_5 with the germanium crystal.

EXPERIMENTAL METHOD

The measurements were carried out using a glow-bar as infrared source. After passing a monochromator, a filter, a polarizer and the germanium halfcylinder the light beam hits the graphit sample, which is encapsulated in a stainless steel box tightly sealed to the germanium crystal. A Golay cell is used as IR detector. In fig. 1 we present the arrangement of sample and detector and in fig. 2 the details of sample position. The HOPG sample can be moved inside the box so that it touches the germanium crystal during the recording of the reflectivity data, and can be removed in between to allow for swelling during intercalation. When the intercalation starts, the optical measurement can be made and also the stage can be controlled.

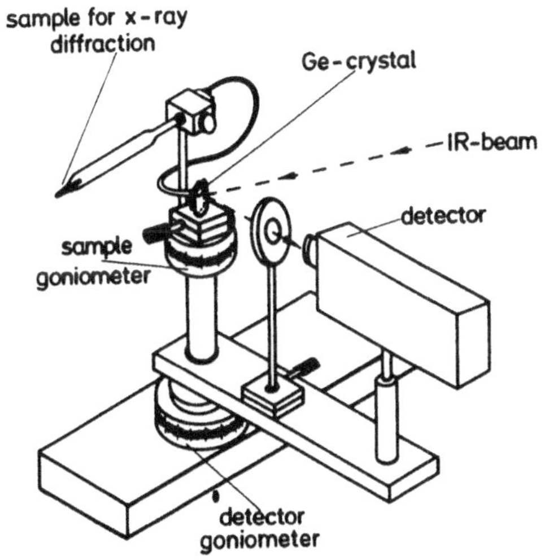

Fig. 1 sample and detector goniometers

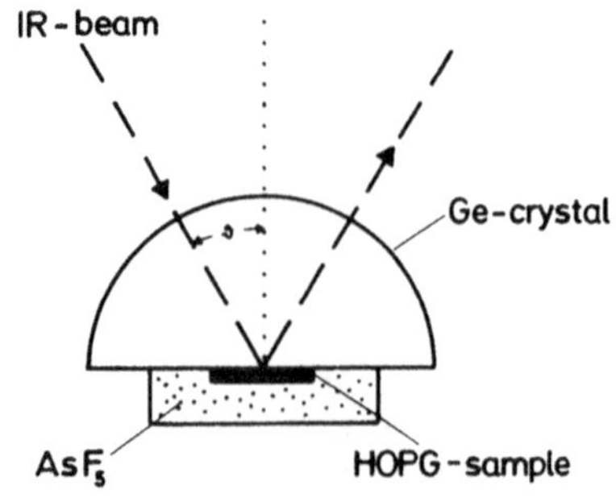

Fig. 2 Ge-Halfcylinder
with HOPG

EXPERIMENTAL RESULTS

In figs. 3 and 4 the intensity of the IR beam, reflected on the backside of the germanium crystal without sample is shown as a reference both for horizontal and vertical polarization. The reflectivity data for HOPG and stage 1 intercalated graphit is given in figs. 5,6. In all figures the reflected intensity of a $\lambda=5\ \mu m$ beam is plotted versus the angle of incidence.

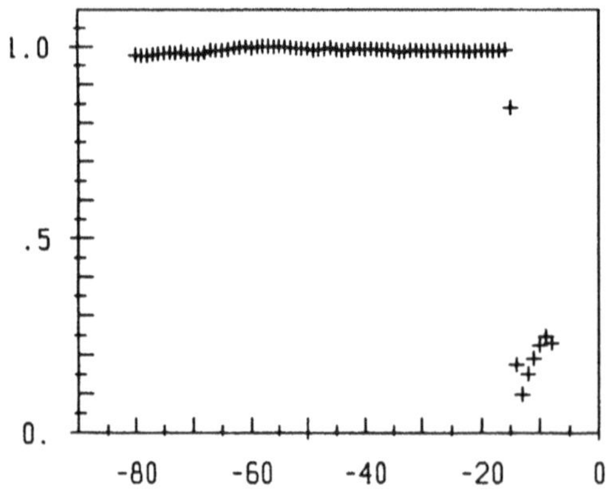

Fig. 3 Reference for
horizontal polarization
(reflected without sample)

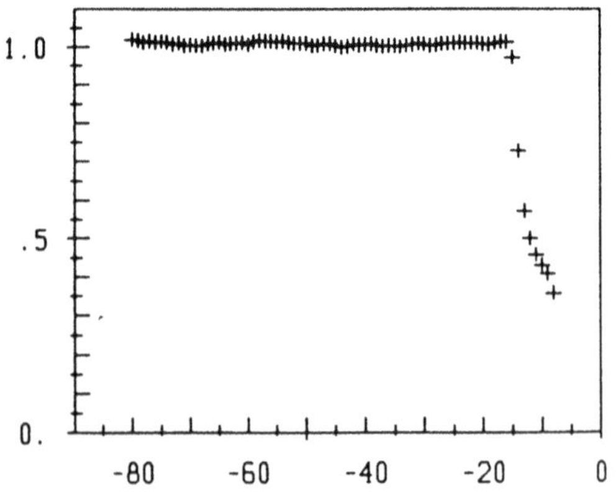

Fig. 4 Reference for
vertical polarization
(reflected without sample)

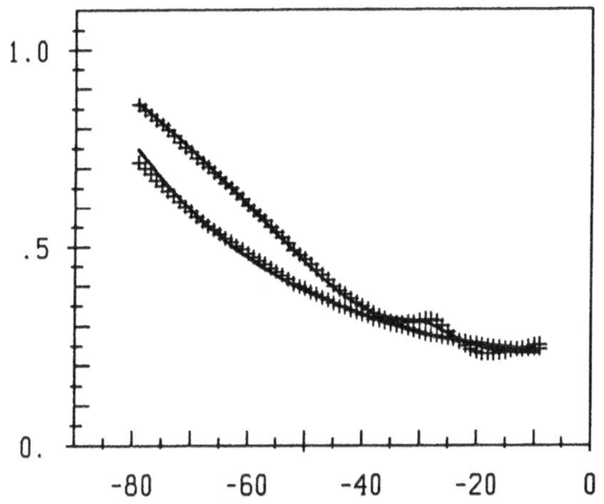

Fig. 5 reflectivity data for HOPG
horizontal polarization – lower curve
vertical polarization – upper curve
lines are the best fit

After normalization and the fitting procedure discribed above we obtain for the dielectric constants of HOPG

$$\varepsilon_1^{\parallel} = 4.17 \qquad \varepsilon_2^{\parallel} = 16.17$$
$$\varepsilon_1^{\perp} = 2.22 \qquad \varepsilon_2^{\perp} = 0.0$$

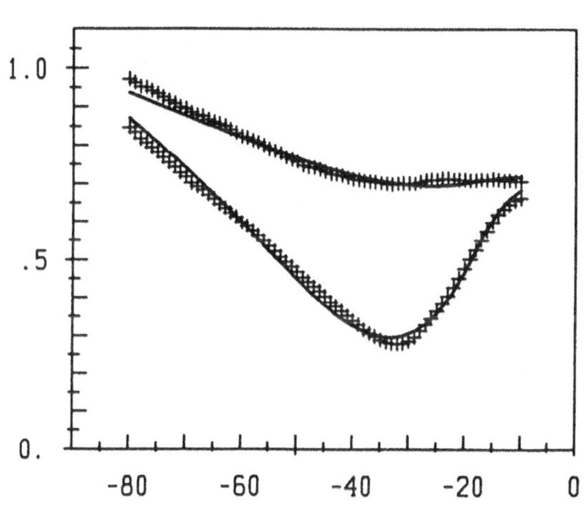

Fig. 6 reflectivity data for stage 1 AsF_5 GIC
horizontal polarization – lower curve
vertical polarization – upper curve
lines are the best fit

and for the dielectric constants for stage 1 AsF_5 GIC

$$\varepsilon_1^{\parallel} = -123.61 \qquad \varepsilon_2^{\parallel} = 82.73$$
$$\varepsilon_1^{\perp} = 1.66 \qquad \varepsilon_2^{\perp} = 1.60$$

Applying the theory of Blinowski et al. the shift of the Fermi level due to intercalation can be calculated from $\varepsilon_1^{\parallel}$ and $\varepsilon_2^{\parallel}$. Using the relation $E_F = 3/2\gamma_0 b K_F$ with $\gamma_0 = 3eV$ [8] we obtain for the carrier concentration of stage 1 the value $N = 1.27 * 10^{14} cm^{-2}$.

DISCUSSION

Our values of the dielectric constants parallel to the graphitic planes of HOPG are in good agreement with those of Taft and Philipp [2] determined by a Kramers-Kronig analysis. The value of $\varepsilon_1^{\perp}$ has been measured by several authors. Venghaus [7] has obtained $\varepsilon_1^{\perp} = 5.4$ by ELS, Zanini et al. [5] $\varepsilon_1^{\perp} = 3.2$ from reflection on the a-face, and Greenaway et al. [6] $\varepsilon_1^{\perp} = 2.34$ by a similar method as ours, but using visible light and no index matching geometry. Our value confirms that of Greenaway et al. Blinowski et al. [4] have deduced from their reflection data a carrier concentration similar to our value. They have not evaluated $\varepsilon_1^{\parallel}$ and $\varepsilon_2^{\parallel}$ explicitly but the agreement in N suggest also agreement in the dielectric function. The value of $\varepsilon_1^{\perp}$ and $\varepsilon_2^{\perp}$ have been determined in this paper for the first time for an intercalation compound. Note that $\varepsilon_1^{\perp}$ of stage 1 intercalated graphite is considerably smaller than that of pure HOPG, in good ageement with the decrease of the electrical conductivity along the c-axis.

ACKNOLEDGMENTS

We would like to thank A.W.Moore of Union Carbide for the HOPG used in this work. We are indebted to J.Greiner for developing the software and M.Schmid for technical assistance.

REFERENCES

1. M.S.Dresselhaus,G.Dresselhaus, Adv.Phys., 30 (1981) 139
2. E.A.Taft,H.R.Philipp, Phys.Rev., 138 (1965) A197
3. L.R.Hanlon,E.R.Falardeau,J.E.Fischer, Solid State Commun., 24 (1977) 377
4. J.Blinowski,N.Hau,C.Rigaux,J.Vieren,R.Le Toullec,G.Furdin,A.Herold,J.Melin, J.Physique, 41 (1980) 47
5. M.Zanini,D.Grubisic,J.E.Fischer, Phys.Status Solidi, B90 (1978) 151
6. L.Greenaway,G.Harbeke,F.Bassani,E.Tossati, Phys.Rev., 178 (1969) 1340
7. H.Venghaus, Phys.Status Solidi, B81 (1977) 221
8. D.Hoffman,R.Heinz,G.Doll,P.Eklund, Phys.Rev.B, 32 (1985) 1278

INTERCALANT STRUCTURE DETERMINATION OF ICl-GRAPHITE BY X-RAY ABSORPTION SPECTROSCOPY

W. Krone, G. Wortmann, and G. Kaindl

Fachbereich Physik, Freie Universität Berlin
D-1000 Berlin 33, Germany

R. Schlögl

Fritz-Haber-Institut der Max-Planck-Gesellschaft
D-1000 Berlin 33, Germany

ABSTRACT

A x-ray absorption (XA) study of the local structure of intercalated ICl molecules in graphite is reported. The near-edge (NEXAFS) as well as the extended XA fine structure (EXAFS) measured on the Cl-K and I-$L_{I,III}$ thresholds contain informations on the chemical form of the intercalated molecules and on their relative positions. We observed a significant increase of the intramolecular I-Cl bondlength and we propose an improved model for the intercalate structure.

INTRODUCTION

Graphite intercalation compounds (GICs) with halogen molecules (Br_2, ICl, and IBr) are considered as acceptor systems with Br_2-graphite as the most studied example.[1,2] The charge transfer mechanism is, however, not well understood since the intercalant is described in most studies as consisting of unreacted and unchanged halogen molecules without significant chemical reactions to negatively charged acceptor molecules like Br_3^-, as observed, for instance, in Br_2-doped polyacetylene.[3] The structure of the intercalated halogen molecules and their relationship to the graphite host lattice is still a matter of actual interest despite of numerous studies, especially on Br_2-graphite.[2] ICl-GICs were thought on geometrical arguments to be related in their structural properties closely to Br_2-GICs. However, Turnbull and Eeles[4] observed a considerably different in-plane structure may be due to the intrinsic dipole moment of the ICl molecule. The detailed arrangement of the intercalant superstructure, however, is not yet clarified.[4,5] In this study, we apply XA spectroscopy on the iodine $L_{I,III}$-thresholds and chlorine K-threshold to investigate the local intercalant structure of ICl-graphite. Information is obtained both from near-edge (EXAFS) as well as from the extended XA fine structure (EXAFS).

EXPERIMENTAL

ICl-GIC samples were prepared by exposing highly oriented pyrolytic graphite

(HOPG) to ICl vapour. X-ray diffraction patterns reveal (001)-reflections typical for stage-2 compounds (I_c = 10.45(5) Å) with a small admixture of higher stages. The chemical composition was found to be $C_{28}ICl$ with an I:Cl ratio of 1:1±0.05. XA spectra were recorded at HASYLAB/DESY (Hamburg). The linear polarization of the synchrotron radiation together with the highly-oriented, quasi-twodimensional sample enables the observation of polarization effects in the absorption processes, when the direction of the graphite c-axis is changed with respect to, *E*, the polarization vector of the synchrotron radiation. Similar polarization effects in XA spectra have been reported for various halogen-doped polyacetylene systems which have quasi-onedimensional character.[6]

EXPERIMENTAL RESULTS

NEXAFS on the I-L_I and Cl-K thresholds

The near-edge spectra of K- and L_I-thresholds probe transitions from 1s and 2s electrons, respectively, to unoccupied p-states. In the case of diatomic halogen molecules, these final-states are well defined σ- or π-orbitals. Since σ-orbitals are oriented in the direction of the molecular axis, *z*, and π-orbitals perpendicular to it, one expects a strong polarization dependence in XA spectra when the orientation of the molecules is changed in a defined way with respect to *E*. The Cl-K spectra of ICl-graphite (Fig. 1) show a dominant pre-edge resonance peak A representing a transition into the empty σ^*-states and reaching maximum amplitude for parallel orientation of *z* and *E*. The peak B shows the opposite polarization dependence, indicating a transition into unoccupied (Rydberg-like) π-orbitals. Similar results have been obtained from the I-L_I NEXAFS. The observed polarization dependences prove a parallel alignment of the ICl molecules with respect to the graphite planes.

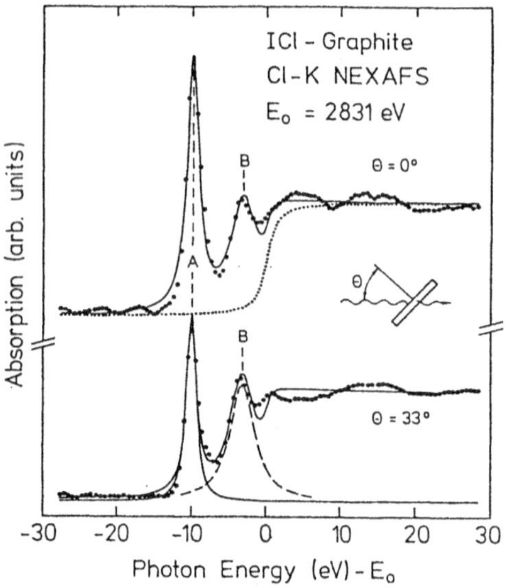

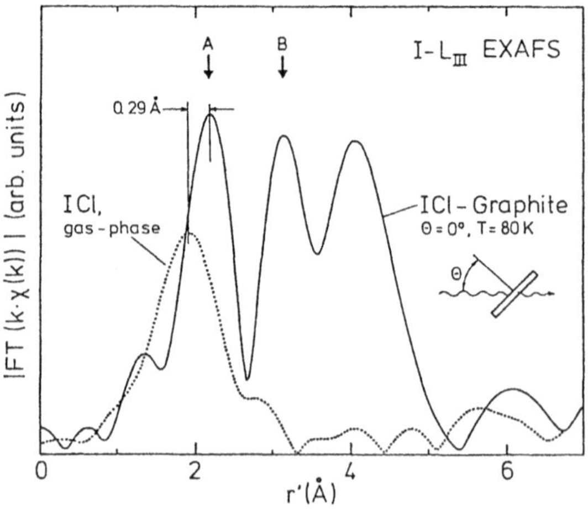

Fig. 1 (left figure). Cl-K NEXAFS spectra of ICl-graphite. The subspectra of the σ^* resonance (peak A) and π resonance (peak B) are indicated.

Fig. 2 (right figure). Fourier transforms of the I-L_{III} EXAFS of ICl-graphite at 80K with $\Theta=0^\circ$ (solid line). The dotted line is the corresponding Fourier transform of gaseous ICl demonstrating the increase of the intramolecular bondlength.

EXAFS on the I-L_{III} and Cl-K thresholds

The spectra of the extended XA fine structure probe radial distribution functions. The magnitude of the Fourier transform of the normalized EXAFS oscillations peaks at distances corresponding to nearest neighbor coordination shells[7] (see Fig. 2 as an example). The I-L_{III} EXAFS of ICl-graphite exhibit various peaks, with their relative intensities depending differently on temperature and orientation of the sample. The first peak (indicated by A in Fig. 2) is the prominent one in all spectra and is assigned to the intramolecular I-Cl distance d_i from comparison with I-L_{III} EXAFS data on gaseous ICl (see Fig. 2). With the usual analysis procedure,[7] one obtains a phase-shift corrected value of d_i = 2.59(2) Å. This points to a considerable increase of the intramolecular bondlength of intercalated ICl molecules with respect to the gaseous state (d_g = 2.302 Å taken from the literature[8]). One should note that the bondlength of the ICl molecule is already increased in the two solid modifications of ICl, when compared to the gaseous state. One finds d_s = 2.40 Å as average value in the α as well as in the β-phase.[8]

We assign the well-resolved second peak B from Fig. 2 to an intermolecular I-halogen distance. In contrast to peak A its amplitude is strong temperature dependent (EXAFS data not shown here). From an analysis of its intensity and from the known intermolecular I-I distances[4] we attributed peak B to a next-nearest I-Cl distance at 3.5 Å (phase-shift corrected). We hesitate to assign the other peaks in the Fourier transform to more distant neighbor shells because of the limited statistical accuracy of the present EXAFS data. The absence of any I-C distances in the Fourier transforms can be explained by the present geometrical arrangement in the XA measurements which favors contributions from neighboring in-plane atoms. It should be mentioned that EXAFS spectra on the Cl-K edge of ICl-GIC and gaseous ICl (not shown here) yield the same increase in bondlength as observed from the presented I-L_{III} data.

DISCUSSION

The most important finding of this study is the observation of an increased bondlength of the ICl molecule in the intercalated state. The interatomic distance in intercalated ICl is 0.29(2) Å larger with respect to the gas phase and still 0.19 Å larger with respect to the solid phase of ICl. This may originate from two principally different mechanisms. One mechanism is an intermolecular interaction between neighboring ICl molecules similar to that which probably causes the known chain-like structure in solid ICl.[8] The other mechanism is an interaction of the ICl valence orbitals with the π-system of the graphite resulting in a charge transfer from graphite to ICl, as required for an acceptor system.[9] An estimation of the charge transfer by comparing the bondlength increase with those of other halogen systems (e.g. I_2 and I_3^-) gives rather large values. Therefore, we suppose a superposition of both types of mechanisms to be responsible for the observed large I-Cl bondlength.

A combination of the results from XA spectroscopy (short-range) with those from electron and/or x-ray diffraction techniques (medium- and long-range order sensitive) can be used to determine the intercalate structure. In an early x-ray and electron diffraction study of ICl-graphite, Turnbull and Eeles proposed an arrangement of the ICl molecules within a rather simple pseudo-cell derived from a larger superlattice (Fig. 3a). In this pseudo-cell, the iodine positions have been accurately determined because of their large contributions to the scattering intensities. The chlorine positions, however, were rather ambigous and fixed by a trial and error method[4] assuming an intramolecular I-Cl distance similar to the solid phase. The present results with the increased value of d_i clearly rule out the proposed structure, since in this case the nearest intermolecular chlorine distance would be smaller than within the Cl_2 molecule. Using our values of intra- and intermolecular I-Cl distances and the iodine positions of Ref. 4, one can construct new in-plane arrangements. Fig. 3b gives, to our opinion, the most probable structure taking into account the suggested centrosymmetry of the pseudo-cell.[4] One should, however, regard this structure as preliminary, since a recent x-ray diffraction study modified this pseudo-cell.[5] Nevertheless, we want to emphasize that in both diffraction studies[4,5] the main parameters concerning the short-range order were found to be identical. This gives strong evidence that a subunit built up by 4 ICl molecules (square of iodines in Fig. 3) is contained in both proposed pseudo-cells. The arrangement shown in Fig. 3b allows for the possible occurence of chain-like ICl structures as they are known to exist in solid ICl.[8] Further details of this study are given in Ref. 10.

Acknowledgement. This work was supported by the BMFT, project number No. 05-313AX B/C3-06 and by the Sfb-337 (TP B2) of the DFG.

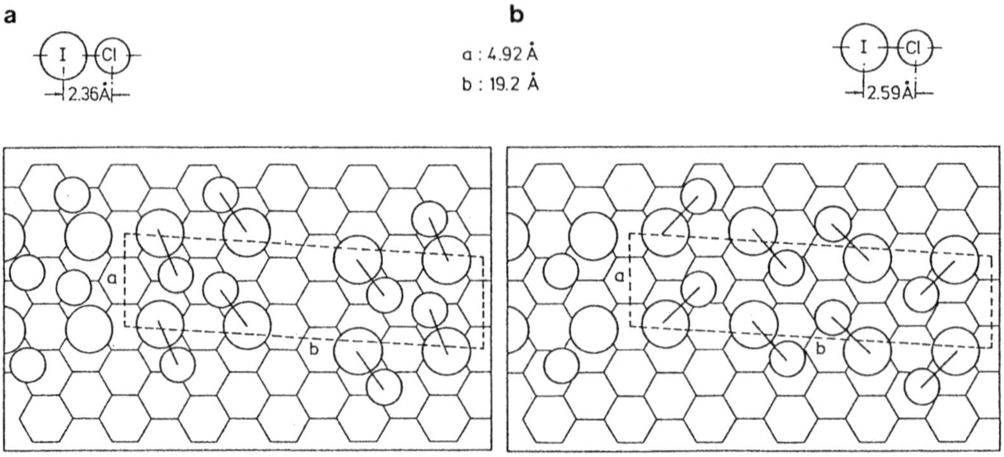

Fig. 3. (a) Pseudo-cell of the intercalate structure in ICl-graphite as proposed by Turnbull and Eeles.[4] (b) Improved in-plane arrangement of the ICl molecules using the intra- and intermolecular I-Cl distances determined in this study.

REFERENCES

1. M.S. Dresselhaus and G. Dresselhaus, Advances in Physics 30:139 (1981).
2. A.R. Kortan et al., Phys. Rev. Lett. 49:1427 (1982).
3. H. Oyanagi et al., J. Phys. Soc. Japan 53:4044 (1984); W. Krone et al., Solid State Comm. 52:253 (1984).
4. J.A. Turnbull and W.T. Eeles, Proc. 2nd Conf. Ind. Carbons and Graphite, 1965, Soc. Chem. Industry, London (1966) pp. 173.
5. R. Moret et al., Synth. Met. 7:289 (1983).
6. W. Krone et al., Synth. Met. 17:479 (1987).
7. E.A. Stern and S.M. Heald, ch. 10, in: "Handbook on Synchrotron Radiation, Vol. 1, Part B", E.E. Koch, ed., North-Holland, Amsterdam (1983).
8. A.J. Dones and C.J. Adams, "The Chemistry of Chlorine, Bromine, Iodine and Astatine", Pergamon Press, Oxford (1975).
9. M.S. Dresselhaus and G. Dresselhaus, Synth. Met. 12:79 (1985).
10. G. Wortmann et al., Proc. ISGIC-4, to appear in Synth. Met.

IN-SITU DIFFRACTION STUDY OF ELECTROCHEMICAL INTERCALATION OF POTASSIUM SOLVATED BY TETRAHYDROFURAN INTO GRAPHITE. OBSERVATION OF A FRACTIONAL STAGE n = 4/3

B. Marcus, J.L. Soubeyroux° and Ph. Touzain

Laboratoire de Science des Surfaces et Matériaux Carbonés, UA 413
ENSEEG, INPG,
B.P 75 Domaine Universitaire
38 402 , Saint-Martin d'Hères, (France)

°Institut Max Von Laue-Paul Langevin
B.P 156 X
38 042 Grenoble Cedex (France)

INTRODUCTION

Ternary graphite intercalation compounds, TGIC, with potassium and tetrahydrofuran (THF) can be obtained by two main synthesis methods : (i) by direct preparation from graphite plunged into a molar solution of a highly reducing agent as potassium naphtalenide into THF[1], and (ii) by ternarisation of a binary graphite intercalation compound, BGIC, plunged in liquid THF[2,3] or in a vapour-bath of THF with a vapour pressure of some Torrs[4,5].

With either of these synthesis ways, two different first stage compounds have been reported, differing in their structure and their composition. The first of them is characterized by an I_c parameter of 7.20 Å and the corresponding structure can be described by the mean plane of the organic molecules roughly parallel to the graphene layers. In the second, the I_c parameter equal to 8.9 Å is also attributed to the THF molecule size ; the molecules stay here perpendicular to the graphene layers. The first compound is noted "poor" because its composition is generally reported to be close to K-THF-C 1 : 1.6 : 24, while the second is called "rich" with a stoichiometry represented by K-THF-C 1 : 2.5 : 24. The ratios C/THF have been deduced of volumetric measurements of the organic liquid consumed by the reactions and then confirmed by intensity calculations , while the K/C ratios are generally assumed to be equal to the ratio in the binary compound : K/C = 1/24 in the two structures.

In order to follow the two kinds of transformations involved by the ternarisation of a second stage BGIC, is to say the stage 2-stage 1 transition and binary-ternary change, we have studied this reaction by neutron diffraction measurements in situ[5]. The structural observation revealed that the THF action on KC_{24} leads to the first stage TGIC via intermediary second stage ternary ones. With the two stages, we have observed the formation of the "lying" phase before the formation of the "standing up" one, as expected according to the amount of THF needed in each structure. Nevertheless, and for the first time, it has been observed the growing of the first stage binary compound KC_8 at the same time that the formation of the "lying" phase.

Such a mechanism has been explained by the alkali atom reorganisation inside the graphite layers : the bulky organic molecule penetration induces a back flow for some alkali atoms which induces the growth of the binary domains. This necessarily implies a ratio K/C weaker in the "poor" phase than in the initial binary compound, and so a ratio THF/K greater than 1.6. The vanishing of the diffraction lines of the binary KC_8 during the transformation between the "poor" and the "rich" phase showed furthermore that this ratio K/C is also different in the two first stage ternary compounds, which can respectively be represented by K-THF-C x_1 : 1.6 : 24 and K-THF-C x_2 : 2.5 : 24 with $x_1 < x_2$ and $x_1 < 1$.

Recent results of EXAFS measurements performed on these two phases reported only one distance K-O in the two structures, and the same oxygen number around the metal alkali atoms[6]. These data are consistent with our proposed mechanism of formation if we consider that at first, some THF molecules lying parallel to the gaphitic sheets are solvating the potassium ions and, when the THF pressure increases, these organic molecules straighten up and stay perpendicular to the host layers. Our results tend to a ratio THF/K close to 2 and in this hypothesis the phases would be represented by K-THF-C 1 : 2 : 30 and K-THF-C 1 : 2 : 20.

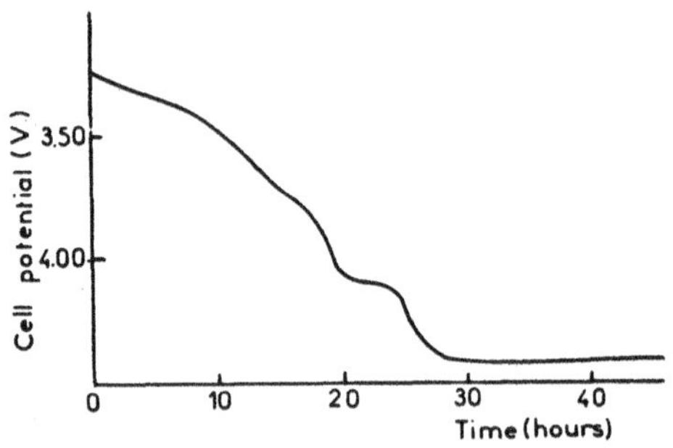

Figure 1 :Cell potential variation observed during the reduction of graphite by K-THF in 1 M potassium cyanotinethylborate THF solution.

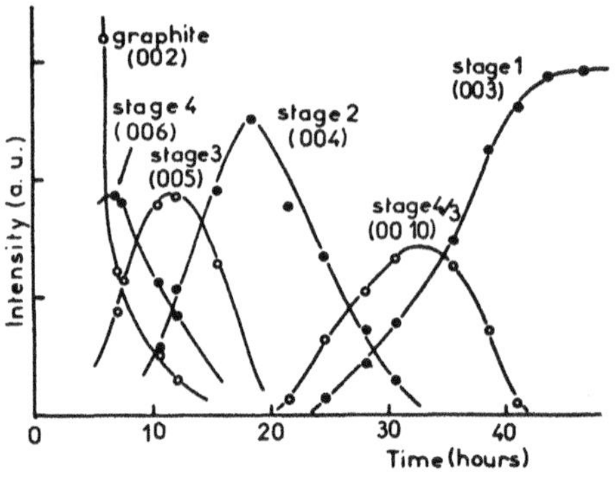

Figure 2 : Evolution of the (00n+2) peak intensity during the reduction of graphite by K-THF. Deuterated THF-d8 has been used for neutron scattering.

ELECTROCHEMICAL REDUCTION OF GRAPHITE

In order to examine this assumption, we have synthetized these compounds by a third way of preparation : the electrochemical synthesis. This electrochemical method has been used for the preparation of the ternary compounds K-DMSO-graphite[7]. Carried out with simultaneous diffraction measurements, this reaction of graphite reduction allowed us to determine the K/C ratio at any step of the intercalation processus and to relate the overall sample composition to the sample structure registered[8].

Moreover,it was interesting to follow the different stage stabilities with the potassium concentrations[9,10].

Figure 1 and 2 respectively show the variation of the cathode potential versus platinum, and the variation of the intensity for the (00n+2) Bragg peak of each phase recorded when the potassium concentration increases.

As it was observed during the electrochemical synthesis of the ternary graphite potassium-DMSO phases, the first stage appears only after the sequential formation of transient phases : the upper stages n = 4, 3 and 2. All these stages exhibit the same thickness of the intercalated layer : Δd = 5.55 Å, related to the THF molecule standing up between the graphene layers. Under our experimental conditions, the first stage was completed after 41 hours, corresponding to the expected ratio K/C equal to 1/20 close to total stoichiometry K-THF-C 1 : 2 : 20.

Contrary to our waiting we never observed the K-THF-C compounds with the organic molecules "lying" parallel to the host planes.

As it was also reported for the electrochemical synthesis of the K-DMSO-C, only the (00ℓ) diffraction lines of the first stage can be fitted with one gaussian function. The spectra recorded for the higher stages exhibit simultaneously two kinds of lines ; (i) the first kind was constituted of wide diffraction lines with a variable wavevector. the position of these Bragg peaks shifts from the position of the (00ℓ) stage n position to the (00ℓ') stage n-1 position when the cation concentration is increasing into the sample. This phenomenom is characteristic of diffraction by cristallographic domains of microscopic mixtures of stages, also called domains of solid solution of stages, and was first modeled by Hendricks and Teller[11]. (ii) the second kind was constituted of sharper peaks with a constant position expected for the pure stage domains. Figure 2 represents the integrated intensities of these narrow diffraction lines.

The existence of these different domains of pure and non pure stages can be explained by the advancement of the intercalated species inside the sample during the reduction of graphite. During the stage transitions, the inner part of the sample is made of pure stage n domains, the outer part of pure stage n-1 domains while the intermediary part is arranged in disordered staging resulting from the motion of some domains which are interpenetrating.

For the first time in the study of TGICs, a new phase corresponding to the fractional stage n = 4/3 has been observed between the second and the first stages (figure 2). The neutron diffraction spectra for the maximum intensity of this phase (t = 31.5 h) is represented in figure 3. The diffraction lines (003), (007) and (0010) respectively situated between the (001) line of the stage 2 and stage 1, between the (002) of the stage 1 and (003) of stage 2 and between the (004) of the stage 2 and the

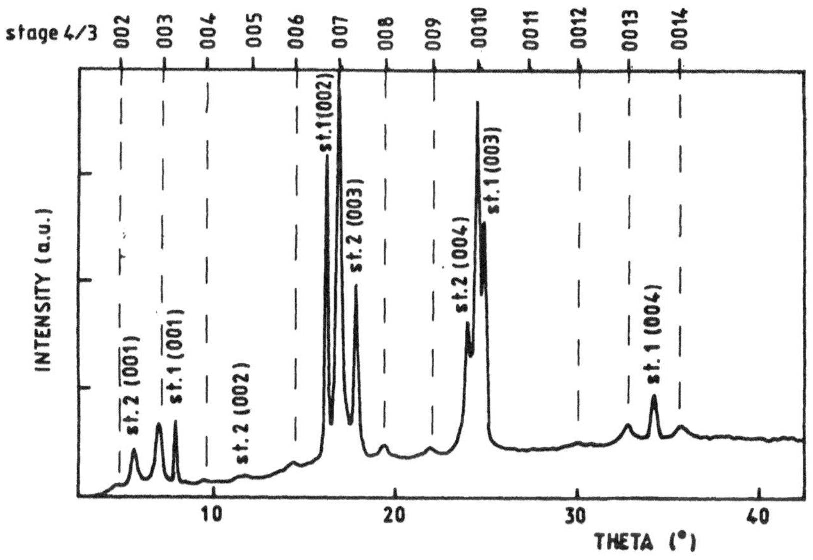

Figure 3 : Experimental neutron diffraction spectrum recorded for the maximum intensity of stage 4/3

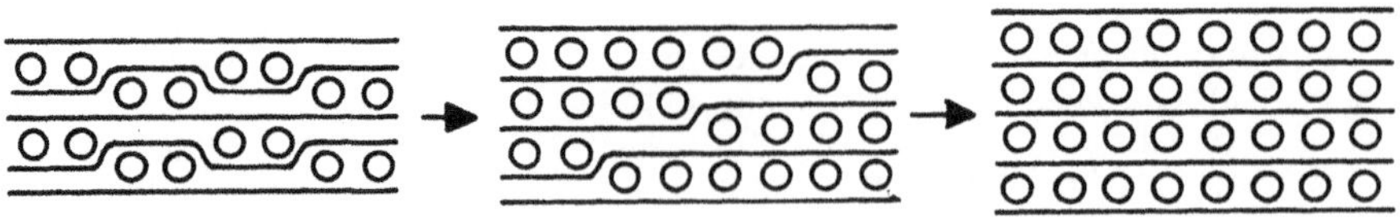

Figure 4 : A schematic representation of the stage 2-stage 1 transition.

(003), of the stage 1 show the higher intensities, but we have recorded eight other (00ℓ), which can also be indexed with the repeat distance d = 30.05 Å. This stage corresponds to domains consisting of an alternative array of two packages of stage 1 and one package of stage 2.

The full width at half heigth, a little more important in this fractional stage than in the two pure stages, indicates more numerous stacking faults in this intermediary stage than in the two pure ones, in agreement with a stage transition based on the Daumas-Herold model[12]. A simple mechanism of the transition from stage 2 to stage 1 is schematized in figure 4 according to this model. We can see that the intermediary state so represented exhibits the same density of the intercalated specie in each plane.

An other so ordered compound with a fractional stage n = 3/2 has been reported when hydrostatic pressure is applied on a binary compound KC_8[13]. It is given by a stage 1/stage 2 ratio equal to 1.

So fractional stage compounds are expected at the thermodynamic equilibrium for lamellar compounds in the theory of long-range interactions between the intercalated layers[9,10].

In this experiment, the formation of the 4/3 stage was observed while the intercalation reaction was proceding. To determine if this stage is also present at the thermodynamic equilibrium, we have to follow the variation of the diffraction line intensities with time when the reaction is stopped. It is yet noteworthy that with our experimental conditions, th (002) diffraction line of the graphite has fully disappeared after 12 hours indicating a penetration rate of about 10^{-5} cm/s. With these experimental conditions, the stage 2-stage 1 transition takes place during 22 hours, suggesting that the fractional stage formations is not due only to kinetic effects.

BIBLIOGRAPHY

1 - M. Nominé and L. Bonnetain, J. Chim. Phys., 66 : 1731 (1969)
2 - F. Beguin and R. Setton, Carbon, 13 : 293 (1975)
3 - F. Beguin, R. Setton, A. Hamwi and Ph. Touzain, Mat. Sci. Eng., 40 : 167 (1979)
4 - M.F. Quinton, H.P. Legrand and F. Beguin, Synth. Met., 14 : 179 (1986)
5 - B. Marcus, Ph. Touzain and A. Hamwi, Carbon, 24 : 203 (1986)
6 - J. Bouat, D. Donnin and F. Beguin, Proc. Intern. Conf. Carbon, Bordeaux, 318 (1984)
7 - J.O. Besenhard, Carbon, 14 : 111 (1976)
8 - B. Marcus and Ph. Touzain, 4th Intern. Conf. Carbon, Baden-Baden, 430 (1986)
9 - S.A. Safran, Phys. Rev. Lett., 44 : 937 (1980)
10 - S.E. Millman and G. Kirczenow, Phys. Rev. B, 28 : 3482 (1983)
11 - S. Hendricks and E. Teller, J. Chem. Phys., 10 : 147 (1942)
12 - N. Daumas and H. Herold, C.R. Acad. Sci. Paris, 268 : 373 (1969)
13 - C.D. Fuerst, J.E. Fischer, J.A. Axe, J.B. Hastings and D.M. Mc Whan, Phys. Rev. Lett., 50 : 357 (1983)

KINETICS OF ORDERING IN GRAPHITE $SbCl_5$ STUDIED BY ESR

S. Rolla, L.C. Walmsley and C. Rettori

Instituto de Física Gleb Wataghin - UNICAMP
Campinas - São Paulo - Brazil

ABSTRACT

Spin resonance experiments allowed us to study the kinetics of ordering in the quasi-2D order-disorder phase transition observed at Tc = (210 ± 2)K for $SbCl_5$-GIC stage 4. Quenching from the disorder phase (T $\gg$ Tc) reveals a time broadening of the resonance linewidth with scaling behavior ($\sim t^n$), which we attribute to the time growing of the average linear domains size of a weakly incommensurate superlattice.

INTRODUCTION

Antimony pentachloride graphites intercalated compounds ($SbCl_5$-GICs) have been recently a subject of many experimental studies[1-6] in consequence of the interesting properties originated in the quasi-2D phase transition experienced by their intercalant molecules at various temperatures. Since conduction carrier spin resonance (CCSR) seems to be an excellent tool to study the GICs, in this work we studied the kinetics of ordering in the phase transition observed at Tc = (210 ± 2)K in $SbCl_5$-GIC stage 4, by monitoring the time evolution of the CCSR linewidth around Tc.

EXPERIMENTS AND RESULTS

The handling, sample preparation, X-ray characterization and ESR experiments were described in detail in our previous work for $SbCl_5$-GICs[1].

Figure 1 shows the thermal hysteresis of the CCSR linewidth observed in $SbCl_5$-GIC 4 for different temperature change rates. The details about this hysteresis cycle can be found in ref. [1].

Figures 2a,b show the time dependence of the CCSR linewidth for $SbCl_5$-GIC 4 at various temperatures Tf within the irreversible region (180K < Tf < 240K) for Tf > Tc and Tf < Tc respectively. These data was taken after a quenching (approximately 100K/min) from room to low temperature (T ≃ 110K) and followed by a quick temperature rise to a final temperature Tf where the time evolution of the resonance linewidth was monitored during approximately five hours.

Figure 3 shows, for $SbCl_5$-GIC 4 at Tf > Tc and Tf < Tc, the scaling behavior observed in the time broadening of the CCSR linewidth, [$\Delta H(t)$-

$\Delta H(0)] \sim t^n$. The exponent n is obtained from the best fit of the experimental data in a $\ln[\Delta H(T)-\Delta H(0)] \times \ln t$ plot. Table 1 gives the exponents obtained for temperatures within the interval 180K < T < 235K. At T ≃ Tc we observed much faster kinetics with larger exponents than those given in Table 1, indicating probably the presence of critical phenomena near Tc. After approximately 50 minutes the kinetics become slower than initially, see figure 3.

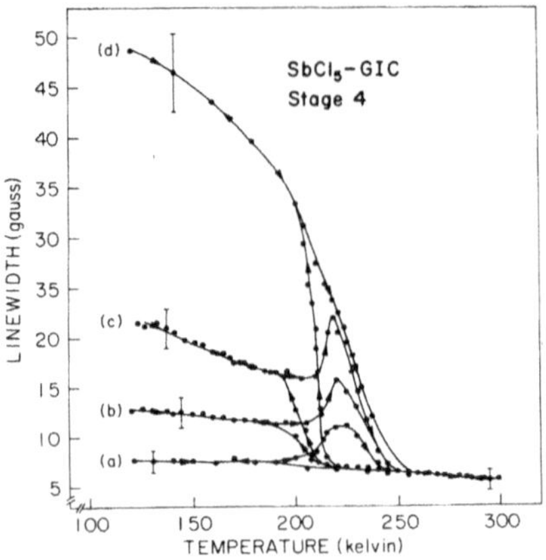

Fig. 1. Thermal hysteresis of the CCSR linewidth in $SbCl_5$-GIC 4.

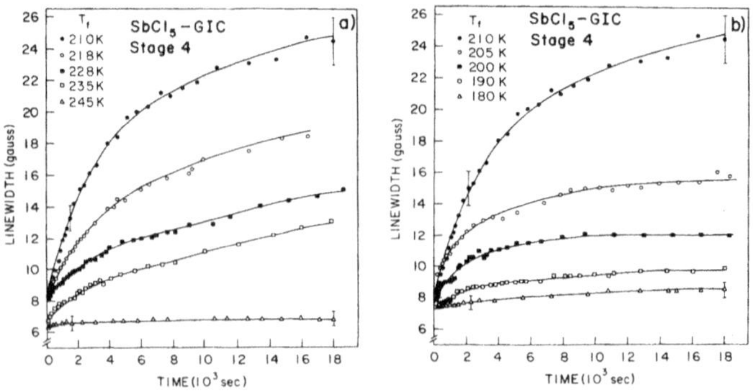

Figure 2. Time broadening of the CCSR linewidth in $SbCl_5$-GIC 4 at a) Tf > Tc and b) Tf < Tc.

Table 1. Quasi-2D Domain Growth Power Law Exponents

T (K) (±2K)	n (± 0.05) t < 50 min.	n (± 0.05) t > 50 min.
180	0.49	0.20
190	0.51	0.23
200	0.47	0.27
205	0.49	0.26
218	0.67	0.39
223	0.54	0.34
228	0.59	0.16
235	0.54	0.57

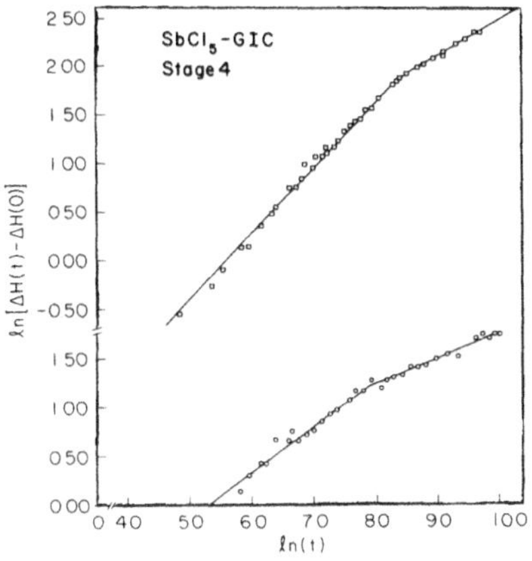

Fig. 3. Scaling behavior of the CCSR linewidth in $SbCl_5$-GIC 4 at Tf = (218±2)K (square) and Tf = (200±2)K (circles).

ANALYSIS AND DISCUSSION

According to Elliot theory[3] the main contribution to the CCSR linewidth is given by the spin lattice relaxation through a phonon scattering via conduction carrier spin-orbit coupling. Elliot's calculation leads to a linewidth which is proportional to the spin-orbit coupling parameter and to the inverse of the conduction carrier's mean free path. In our previous work[1,2], we attributed the observed hysteresis cycle in the CCSR linewidth to an order-disorder first order phase transition experienced by the $SbCl_3$ molecules within the graphite galleries, which via spin-orbit coupling can affect both the conduction carrier spin-flip scattering and g-value.

Recent in plane diffuse X-ray scattering experiments in $SbCl_5$-GIC by Homma and Clarke[4] reveals that $SbCl_3$ molecules, one of the coexisting intercalated molecular species, experience an "order-disorder" phase transition at 230K. They identified the low temperature "solid-like" phase as an almost commensurate (2% off) ($\sqrt{39} \times \sqrt{39}$) R(± 16.1°) superlattice. Ultrasonic experiments[5] and AC calorimetry[6] in $SbCl_5$-GICs, also atributed the observed transition at Tc = 230K to a "disorder-liquid" to "incommensurate solid" phase transformation experienced by the $SbCl_3$ molecules.

Following this ideas we suggest that $SbCl_5$-GIC 4 upon cooling through the first order phase transition at Tc = (210 ± 2)K the "supercooled-liquid like" phase may freeze in many small domains or islands of a weakly incommensurate ($\sqrt{39} \times \sqrt{39}$) structure of $SbCl_3$ molecules. Now, when the temperature is raised we claim that around Tc, but before that the "superheated solid" to "liquid-like" phase transformation takes place, the intercalant molecules may have enough thermal energy to migrate and the domains grow in time. It is possible, that the growth of this microdomains may allow the formation of regions, within the Daumas-Herold domains[7], with higher Sb/C concentration which via spin orbit coupling causes the shortening of the conduction carrier mean free path or either a distribution of g-shifts, leading to the observed time broadening of the CCSR linewidth. Therefore the CCSR linewidth would be measuring directly the time growing of the average linear domain size (L) and therefore $[\Delta H(t)-\Delta H(0)] \propto L$. The power law evolution of the experimental time broadening of the linewidth showed in figure 3 supports this idea.

Monte Carlo simulations, performed by Sahni et al[8], of the kinetics of the domain growth in 2D systems quenched from high temperatures ($T \gg Tc$) to low temperatures ($0.5Tc < T < Tc$), for both small and large Q degeneracies, show an universal or lattice independent power law behavior for all value of Q, but with exponents falling gradually from the classical value 0.5 (Q=2, Ising limit) to a constant value of 0.41 for large Q ($Q > 20$). Since no simulations of incommensurate structure have been done yet, our thermal broadening of the linewidth near Tc agrees, in terms of scaling, only qualitatively with Monte Carlo simulations for either triangular, square or even honeycomb lattices.

At low temperatures, the slow kinetics of ordering ($T < 180K$ for $SbCl_5$-GIC 4) could be an indication for the existence of a configuration of frozen-in microdomains in a quasi-2D metastable state.

The sluggish kinetics observed after an elapsed time of approximately 50 minutes (see figure 3) can be attributed to a pinning of the domain bounderies at the Daumas-Herold domain's walls, or at their next neighbour domain bounderies.

CONCLUSION

We believe that for $SbCl_5$-GICs our time broadening of the CCSR linewidth, the time dependence of the X-ray intensity[4] and the time dependence of the c-axis shear elastic constant[5], all have the same origin: the time growing of domains of weakly incommensurate superlattice of $SbCl_3$ molecules. For $T \ll Tc$ the microdomains may be forming a quasi-2D metastable state and the fast kinetics observed at $T \simeq Tc$ may suggest a critical behavior for the domains growth.

Finally we showed that CCSR experiments is a helpful technique to study the kinetics of quasi-2D domains ordering in GICs.

ACKNOWLEDGEMENTS

Partially supported by FAPESP and CNPq.

REFERENCES

1. S. Rolla, L. Walmsley, H. Suematsu, I. Torriani, C. Rettori and Y. Yosida, Solid State Comm. 58:333 (1986).
2. S. Rolla, L. Walmsley, H. Suematsu, C. Rettori and Y. Yosida, accept in Phys. Rev. Brief Report.
3. R.J. Elliot, Phys. Rev. 96:266 (1954).
4. H. Homma and R. Clarke, Phys. Rev. B31:5865 (1985).
5. D.M. Hwang and G. Nicolaides, Solid State Comm. 19:483 (1984).
6. D.N. Bittner and M. Bretz, Phys. Rev. B31:1060 (1985).
7. N. Daumas and A. Hérold, Bull. Soc. Chim. Fr. 5:1598 (1971).
8. P.S. Sahni, D.J. Srolovitz, G.S. Grest, M.P. Anderson and S.A. Safran, Phys. Rev. B28:2705 (1983).

SYNTHESIS OF NICKEL- AND IRON-HYDROXIDE GRAPHITE INTERCALATION COMPOUNDS AND THEIR APPLICATION TO ALKALINE SECONDARY BATTERY

Norio Iwashita and Michio Inagaki

Materials Science, Toyohashi University of Technology
Tempaku-cho, Toyohashi, 440 Japan

Recently one of authors developed new process to synthesize different graphite intercalation compounds(GICs) of metal chlorides by using their molten salts[1,2]. Ternary GICs of $NiCl_2$ and $FeCl_3$ were obtained from the molten salts of $NiCl_2$ and $FeCl_3$ with different ratios at 400°C after 3 days without any intentional addition of chlorine gas. The compounds were found to be stable in air and even in boiling water.

By using these ternary GICs, we studied the formation of GICs of nickel and iron hydroxides by electrolysis and also by charge-discharge repetition.

The ternary $NiCl_2$-$FeCl_3$-GICs were synthesized from natural graphite powders with average particle size of 400 μm in the molten salt of $NiCl_2$/ $FeCl_3$ in molar ratio of 50/50. After the reaction, excess molten salt was washed out by water until any chloride ions were not detected in rinsed water. The GIC synthesized was a mixture of 1st-stage(I_c=0.943 nm) and 2nd-stage(I_c=1.275 nm) compounds and homogeneous distribution of nickel and iron in flaks of the compounds was confirmed under X-ray microprobe analyzer. Chemical analysis showed the content of nickel of 8 wt%.

The electrochemical cell for electrolysis and charge-discharge cycles was constructed as follows; In the cathode the GIC powders are compressed in between two plastic plates with many small holes. The counter electrode is Pt wire for electrolysis and a plate of metallic zinc for charge-discharge cycles. The electrolyte is 7 N KOH. A Hg/HgO electrode in 7 N KOH is used as the reference electrode. Electrolysis was performed by 2 mA for 10 hours and charge-discharge by 2 mA was repeated 50 times. Detailed experimental procedure will be reported elsewhere[3].

The structural and compositional changes in the GICs were examined by using cyclic voltammetry, X-ray powder diffractometry and X-ray microprobe analysis.

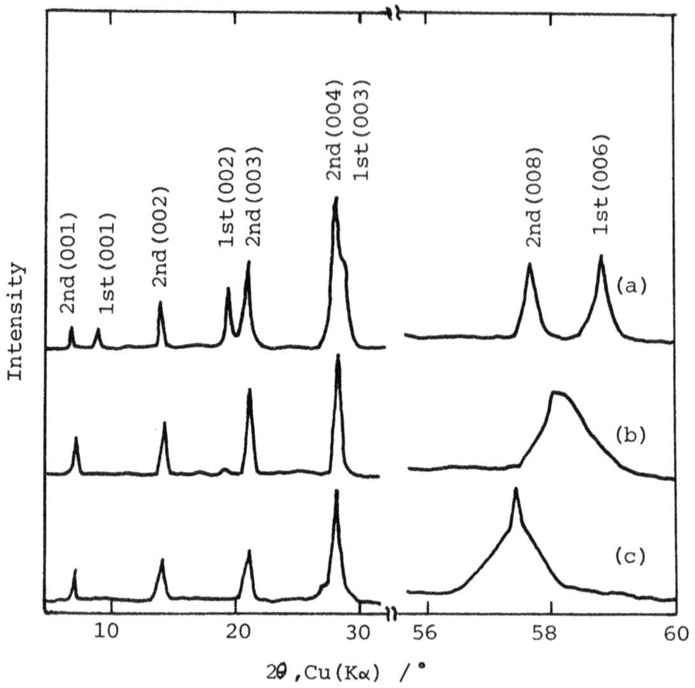

Fig. 1 X-ray powder diffraction patterns of GICs
(a):Starting material
(b):After electroltsis (2 mA,10 h)
(c):After 50 cycles

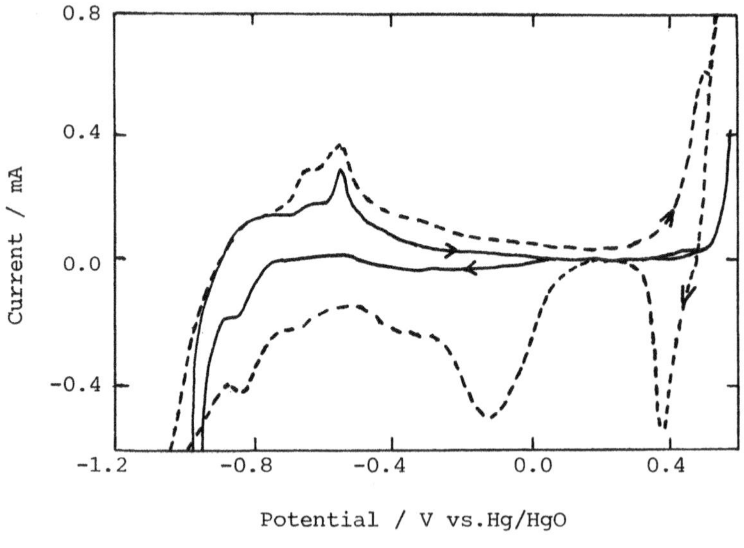

Fig. 2 Cyclic voltammograms of GICs in 7 N KOH solution, scan rate:1 mV/sec
——— Starting material
- - - - - - After electrolysis (2 mA,10 h)

X-ray powder patterns of the starting compound and the compounds obtained by electrolysis and after 50 cycles of charge-discharge are shown in Fig. 1. After electrolysis and after 50 cycles of charge-discharge, it is noteworthy that the 1st-stage structure in the starting compound disappeared and only 2nd-stage struture is remained, of which I_c-value is 1.28 nm, very similar to the starting compound. Energy spetra by XMA on the resulting compounds showed remarkable diminution of chlorine content, but only small change in nickel and iron contents. During charge-discharge cycles chlorine ions could be detected in KOH solution.

The cyclic voltamograms on the compounds electrolized and obtained by charge-discharge cycles in Fig. 2 show three cathodic peaks; the peak at 0.4 V can be attributed to reduction of NiOOH to $Ni(OH)_2$ and the one at -0.9 V to that of $Fe(OH)_3$ to $Fe(OH)_2$. The other peak around 0.0 V seems to be due to the reduction of oxygen adsorbed on the compound. The corresponding three anodic peaks are also observed.

Flandrois et al.[4] used $NiCl_2$-GIC as starting material for the cathode of alkaline secondary battery. They found gradual increase in battery capacity with charge-discharge cycles and attributed it to gradual change of intercalated nickel compound from chloride to hydroxide by electrochemical reduction in KOH solution. In the present work, similar gradual increase in capacity with charge-discharge cycle was also observed. X-ray powder pattern and XMA energy spectra on the compounds after different cycles showed that 1st-stage structure disappeared in the beginning of cycles and 2nd-stage structure was maintained through whole cycles. Gradual replacement of chloride by hydroxide was not detected from powder pattern because two GICs of chlorides and hydroxides have almost the same identity periods, as described above. However, the formation of hydroxide is confirmed from cyclic voltamogram, as shown in Fig. 2, and their occupancy in interlayer space of graphite is concluded from powder pattern.

By cathodic electrolysis in KOH aqueous solution, nickel- and iron-chlorides intercalated in graphite can be progressively converted to nickel- and iron- hydroxides, and consequently ternary graphite intercalation compounds of these hydroxides are obtained.

References

1. Inagaki, M., and Wang, Z. D., 1987, Synth. Met. 20:1.
2. Inagaki, M., Wang, Z. D., Okamoto, Y., and Ohira, M., 1987, Synth. Met. 20:9.
3. Inagaki, M., Iwashita, N., and Maeda, Y., Electrochem. Acta(to be published)
4. Flandrois, S., Masson, J. M., and Rouillon, J. C., 1981, Synth. Met. 3:195.

INFLUENCE OF THE ZIRCONIUM DISELENIDE STOICHIOMETRY

ON ITS BEHAVIOUR UPON LITHIUM INTERCALATION

Philippe Deniard[+], Luc Trichet[+] and Yves Chabre[++]

[+]Laboratoire de Chimie des Solides, associé au CNRS
Université de Nantes, 2 rue de la Houssinière, 44072 Nantes, F.

[++]Laboratoire de Spectrométrie Physique, associé au CNRS
Université Scientifique Technologique et Médicale de Grenoble
BP 87 38402 Saint Martin D'Heres, F.

INTRODUCTION

Zirconium diselenide belongs to the IVB group transition metal dichalcogenides which have lamellar structures and easily intercalate metals and molecules. From optical measurements (Lee et al.,1969) it appears as a semiconductor with a 1.2 eV energy gap. But from the Gleizes and Jeannin work (1970), it is known that zirconium diselenide does not exist as a stoichiometric compound but as $ZrSe_y$ with $1.85 < y < 1.95$.

In previous studies on chemically lithium intercalated powder compounds with y composition close to the upper limit (Berthier et al. 1981, 1984), we gave evidence for an intercalation induced semiconductor to metal transition for a lithium content close to x = .40. This transition is characterized mainly by an abrupt change in the ^{77}Se NMR shift while discontinuities in the a and c lattice parameter values are observed. In some works carried out at that time by other groups, either by conductivity measurements on chemically intercalated single crystals (Klipstein et al. 1983, 1984), or on electrochemically intercalated ones (Onuki et al. 1982), this transition was not observed. Later on, Onuki et al. (1984) did observe this intercalation induced transition as well on the lattice parameters, as we did, as by pointing out that the intercalated samples were superconducting below 1.5 K for lithium contents larger than x = .45.

Nevertheless in a structural study by X-ray diffraction in-situ in an electrochemical cell, Dahn et al. (1985) did not observe this intercalation induced transition. They suggested a lack of homogeneity of our chemically intercalated samples, precluding observation of resulting broad lines in the X-ray spectra, although our samples were annealed at 300°C for a few days. Moreover their voltage/composition curves were very dependent upon the liquid electrolyte composition, as was already noticed by Onuki et al. (1983). They thought that it was due to solvant co-intercalation effects rather than to stoichiometry effects.

With the idea that the discrepancies in the literature data could be due to difference in the properties of the pristine compounds, we undertook a study of the dependence of the physical properties of pure and lithium intercalated zirconium diselenide on the host lattice stoichiometry (Deniard et al.,1987). We have found that pure zirconium diselenide experiences a stoichiometry defect induced conductivity transition and that, on chemically intercalated samples, the structural behaviour upon intercalation also depends on the stoichiometry of the pristine material.

We present in this paper some of the results of this previous work and first results of an electrochemical intercalation study of the stoichiometry effects.

EXPERIMENTAL

The diselenides $ZrSe_y$ are prepared by direct synthesis from the elements. The mixture of selenium (Koch Light, 99,99%) and zirconium (Ventron, 99,9%) powders are heated under vacuum in sealed pyrex tubes from 300°C to 550°C in several steps. Then, after grinding, the compound is homogeneized at 850°C in sealed silica tubes. The composition is determined by roasting in air at 1000°C. Composition value indicated for one sample is the average of the nominal value and the chemical analysis one, which did never differ more than .01.

Single crystals have been grown by vapour transport of charges with y = 1.94, with either chlorine or hydrogen chloride as transport agent at 950°C and with or without an excess of selenium. Due to their small size the composition of the single crystals could not be determined chemically.

For the pristine powder compounds conductivity measurements have been performed with the van der Pauw method, on rods sintered at 700°C.

Chemical intercalation of the powder samples is performed by the n-buthylithium technique, at room temperature, under inert atmosphere. After a week the product is filtered, washed with hexane and dried under vacuum at 100°C. Then the compounds are annealed at 350°C for three days under vacuum. The lithium content is determined by atomic absorption.

Electrochemical intercalation is performed in sealed button-type cells, with polymer electrolyte to avoid any possible solvant co-intercalation. A detailed description of such cells is given in the paper of Yazami (1987) in this volume. We use poly-(ethylene oxide) complexed lithium salt (Armand 1979), mainly $P(EO)_8LiClO_4$ operated at 100°C in this work. The cathode is a mixed one (50% active material, 40% electrolyte and 10% very small grain size graphitic powder) in order to improve the ionic and electronic contacts of the $ZrSe_y$ powder grains and then to get 100% efficiency.

We study electrochemical intercalation doing either titration curves with intermitent discharge current (1h constant current, usually at capacity/30 rate, and 3h relaxation) or slow scan voltamperometry, scanning the voltage cell at -10 mV/h rate and recording the discharge current. In the extremely slow scanning limit, such a voltamperogram corresponds to the inverse derivative of the equilibrium titration curve, $f(x) = -dx/dV$. As originally proposed by Thompson (1978), this is a very efficient way to give evidence for plateaus and changes in the electrochemical potential in relation to phases limit, ordering ..etc. and this also hold for voltamperometry, if done slowly enough in order to probe the bulk active cathode material near equilibrium.

RESULTS

$ZrSe_y$ physical properties

We observed a dependence of the cell parameters on stoichiometry similar to that determined by Gleizes and Jeannin (1970), but we do not have determinations accurate and numerous enough to add information to their data. And even the c variation is so small that it cannot be used to control the stoichiometry of the powder compounds or determine that of the single crystals which cannot be done chemically, due to their small size.

As can be seen in fig. 1, the conductivity of powder samples shows an abrupt change of value and behaviour in the middle of the existence domain of the phase. The room temperature conductivity increases by more than six orders of magnitude when the stoichiometry decreases from y = 1.915 to y = 1.895. In addition the temperature dependence of the conductivity, observed in the 77K-300K range, changes from a semiconductor-type one, that is thermally activated, for y > 1.91 to a metal-like one, that is an inverse temperature dependence, for y = 1.88 and 1.895. For y = 1.865 and y = 1.85 the conductivity is almost temperature independent.

Similar conductivity behaviours have been found for single crystals, that is semiconductor-like or metal-like, depending on the batches, but as their stoichiometry are not known, that do not give more information on the critical stoichiometry value.

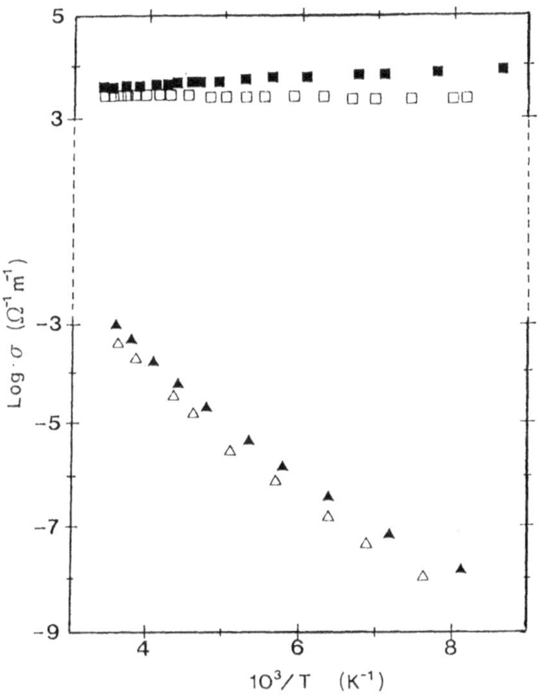

Fig. 1. Temperature dependences of the $ZrSe_y$ conductivity for y = 1.94 (△), 1.915 (▲) , 1.88 (■) and 1.865 (□).

Chemically intercalated Li_xZrSe_y

X-ray diffraction studies of $Li_xZrSe_{1.865}$ and $Li_xZrSe_{1.85}$ compounds show that the lattice parameters variations upon lithium intercalation depend on the stoichiometry of the pristine compound, as can be seen in fig.2. Values for the pure y = 1.865 and y = 1.85 hosts, a = 3.769(1) Å, c = 6.147(2) Å and a = 3.765(1) Å, c = 6.157(2) Å respectively, agree with the Gleizes and Jeannin (1970) determination for the low selenium contents. But instead of a plateau up to x =.40 as in the $Li_xZrSe_{1.94}$ (Berthier et al. 1981, 1984), the lattice parameters do not change significantly up to x =.15 only, then increase and, for x >.5, reach the same values as in $Li_xZrSe_{1.94}$.

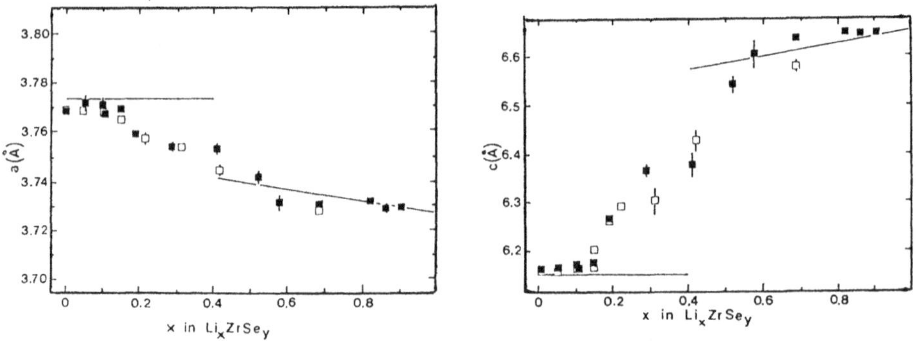

Fig. 2. Variations of the a and c lattice parameters vs. lithium content in Li_xZrSe_y, for y = 1.94 (—), 1.865 (□) and 1.85 (■).

Electrochemically intercalated Li_xZrSe_y

In fig. 3a are reported the first discharge slow scan voltamperograms recorded at -10mV/h, for y compositions 1.95, 1.915, 1.88 and 1.865. In fig. 3b we have reported the corresponding discharge curves, V = f(x), obtained from integration of the voltamperograms. One can see that total intercalation up to x =1 is achieved when the potential reaches 1.2 V. As the voltage scans usually start from 2.5 V, this is obtained within 130 h, which is the same order of magnitude as the duration of the intermittent intensiostatic titration curves we have also studied (from 80 to 180 h). The stoichiometry dependence of current peaks (or voltage plateaus of discharge curves) is clearly observed: for the most stoichiometric y = 1.95 compound, two well defined narrow peaks are observed at 2.25 V and 1.95 V whereas no intensity peaks occur above 1.90 V for the y = 1.865 compound and intermediate behaviours are observed for intermediate y values.

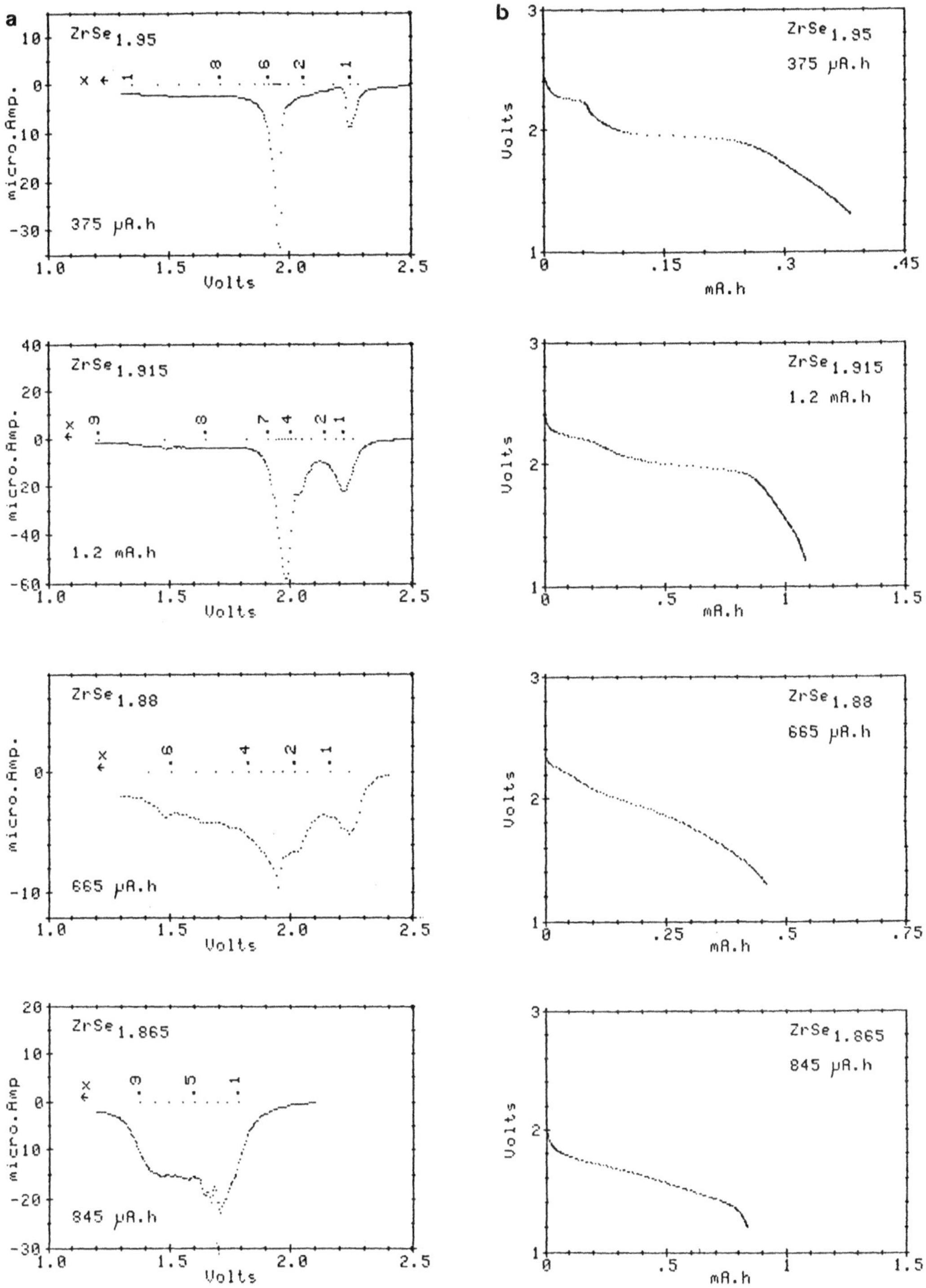

Fig. 3. a) First discharge voltamperograms recorded at -10 mV/h for various $ZrSe_y$ compounds with $P(EO)_8LiClO_4$ electrolyte operated at 100°C .

b) Corresponding discharge curves obtained from integration of the voltamperograms

First discharge intermittent intensiostatic curves also show a voltage plateau at 2.28 V at the early step of intercalation for $y \geqslant 1.88$. Then, for x values depending on the stoichiometry ($.1 < x < .2$), the voltage drops towards 2.05 V to a second plateau whose amplitude also depend on the soichiometry, before decreasing rapidly to about 1.2 V for $x = 1$. The slightly higher voltage values of the plateaus in the intermittent intensiostatic measurements come from that we are closer to equilibrium than for the voltamperometry studies, in which there are always ohmic and polarization potential drops.

For the lowest $y = 1.865$ value we studied, there are no voltage plateaus, neither at 2.25 V nor around 1.95 V, but a continuous decrease of the potential from 2 V to 1.2 V. Moreover the ohmic and polarization contributions appeared to be negligible for this compound and one had a perfect agreement of the intensiostatic intermitent titration curve with the discharge curve deduced from the voltamperogram. This confirm the qualitative change of the material for large departures from stoichiometry.

DISCUSSION

Our conductivity measurements on $ZrSe_y$ powders with various y values show the occurence of a stoichiometry induced semiconductor to metal transition for the composition $y = 1.905(10)$. The nature of the stoichiometry defects has been discussed by Gleizes and Jeannin (1970). From density and X-ray unit cell volume measurements they have determined the unit cell weight as a function of the departure from stoichiometry. They deduced that defects are selenium vacancies for the upper limit value $y = 1.945$ and that, when y decreases, the additional defects are excess zirconiums occupying the vacant selenium sites. These additional zirconiums should create impurity levels within the gap, close to the conduction band, and be responsible for the metallic behaviour when in a large enough concentration.

This scheme seems to be confirmed by our electrochemical measurements. We think that the first intensity peak (voltage plateau) at about 2.25 V is to be attributed to a first step for lithium intercalation, in which lithium ions occupy preferential sites bounded to selenium vacancies. When the departure from stoichiometry increases, this 2.25 V peak broadens, which can be related to the fact that the number of isolated selenium vacancies decreases, as excess zirconiums associate to some of them.

When the stoichiometry is close to the lower limit value (our $y = 1.865$ sample), the results are fondamentally different. Neither the 2.25 V state nor the 1.95 V one are observed, but we have a regular behaviour, with a discharge curve which agree with that obtained by Dahn et al. (1985) and which can be fitted with a lattice gas model with a pair interaction (Armand, 1978, 1980). Moreover the average potential is about .3 V lower and these effects are similar to the non-stoichiometry effect observed by Thompson (1979) for excess titanium in titanium disulfide.

That lead us to think that the 1.95 V peak (2.05 V plateau) for $y > 1.88$ is a two phase plateau related to the existence of an intercalated semiconducting phase, before occurence of the intercalation induced semiconductor to metal transition. This is discussed elsewhere (Chabre et al. 1987) as reversibility studies.

As to lattice expansion upon lithium intercalation our results on chemically intercalated $Li_xZrSe_{1.865}$ and $Li_xZrSe_{1.85}$ shows no lattice parameters changes up to x = .15, whereas for the most stoichiometric compound $Li_xZrSe_{1.94}$ this state extend up to x =.40. On another hand Dahn et al (1985) observed lattice expansion as soon as lithium intercalate.

As previously discussed (Deniard et al. 1987), we think that, when lithium associate to stoichiometry defects in the first step of the intercalation, that gives rise to lattice expansion as low as the one due to stoichiometry changes for the pristine material, d(c/a)/dy =.10. It is only when the compounds behave metallic, either due to lithium intercalation or to large enough non-stoichiometry of the pristine material, that large lattice expansion upon intercalation occurs. We then observed d(c/a)/dx =.32, in agrement with the Dahn et al (1985) results. We previously (Deniard et al. 1987) suggested that they could have missed this first intercalation step if the electrochemical intercalation was not totaly reversible, because they use average value of first recharge and second discharge curves to determine the lithium content of their X-ray cells. That could be true considering that the 2.25 V peak is not reversible , but, in view of our results on the y = 1.865 sample, we now think that their starting material should have been largely non-stoichiometric with a metallic character.

CONCLUSION

We have shown that zirconium diselenide, which does not exist as a stoichiometric compound but as $ZrSe_y$ with $1.85 < y < 1.945$ presents a conductivity transition at y = 1.905(10), from a semiconductor-type to a metallic one as the departure from stoichiometry increases. Analysis of lattice expansion upon chemical intercalation led us to suggest that for semiconducting host compound lithium ions associate to stoichiometry defects in the first step of intercalation. This is confirmed by electrochemical intercalation studies which also show that for large non-stoichiometry which gives it a metallic character, zirconium diselenide behave upon lithium intercalation as titanium disulfide with excess transition metal. These results confirm the Gleizes and Jeannin description of the stoichiometry defects.

Joint use of polymer electrolyte, which avoid any possible solvant co-intercalation as encountered in liquid electrolytes, and of very slow scan voltamperometry, appears to be very efficient for such studies of defects contribution to intercalation properties. Extension to in situ X-ray studies is currently in progress.

AKNOWLEDGMENTS

Electrochemical studies would not have been possible without the advices of M. Armand, close collaboration with R. Yazami who prepared the cells and the numerous discussions we had about electrochemistry.

REFERENCES

Armand M. (1978) Thesis, University of Grenoble.
(1979) in "Fast Ion Transport in Solids", P. Vashishta, J.N. Mundy and G.K.Shenoy eds., p.131, North-Holland, N.Y.
(1980) in "Materials for Advanced Batteries", Murphy D.W. Brodhead J. and Steele B.C. eds. p.145, Plenum Press, N.Y.

Berthier C., Chabre Y., Ségransan P., Chevalier P., Trichet L. and Le Méhauté A., (1981) Solid State Ionics, 5, 379.

Berthier C., Chabre Y., Ségransan P., Deniard P., Trichet L. and Rouxel J. (1984) in "Physics and Chemistry of Electrons and Ions in Condensed Matter" J.V. Acrivos, N.F. Mott and A.D. Yoffe eds. NATO ASI Series, C130, 561, D. Reidel, Dordretch.

Chabre Y., Deniard P. and Yazami R. (1987), presented at the sixth International Meeting on Solid State Ionics, Garmisch (D),1987 and to appear in Solid State Ionics.

Dahn J.R., Mc.Kinnon W.R. and Lévy-Clément C., (1985) Solid State Comm., 54, 245.

Deniard P., Chevalier P., Trichet L., Chabre Y. and Pannetier J., (1987) Sol. State Comm. , 64, 2, pp 175-180

Gleizes A. and Jeannin Y., (1970) J. Solid State Chem., 1, 180.

Klipstein P.C., (1983) Ph.D. Thesis, University of Cambridge.

Klipstein P.C., Pereira C.M. and Friend R.H., (1984) see Berthier (1984), NATO ASI Series, C130, 545.

Lee P.A.,Said G., Davis R. and Lim T.H. (1969) J. Phys. Chem. Sol., 30, 2719

Onuki Y., Inada R., Tanuma S., Yamanaka S. and Kamimura H., (1982) J. Phys. Soc. Japan, 51, 880.

Onuki Y., Inada R., Tanuma S., Yamanaka S. and Kamimura H., (1983) Solid State Ionics 8, 141.

Onuki Y., Hirai T., Shibutani K. and Komatsubara T., (1984) J. Inc. Phen.,2, 279.

Thompson A.H., (1978) Phys. Rev. Lett. 40, 1511.

Thompson A.H., (1979) in "Fast Ion Transport in Solids", P. Vashishta, J.N. Mundy and G.K.Shenoy eds., p.47, North-Holland, N.Y.

Yazami R., (1987), in this NATO ASI Series volume

^{77}Se NMR OBSERVATION AND MONTE-CARLO SIMULATION OF THE Li^+ ION DISTRIBUTION IN LITHIUM INTERCALATED TITANIUM DISELENIDE

Yves Chabre° and Philippe Deniard°°

° Laboratoire de Spectrométrie Physique, associé au CNRS (LA 08)
Université Scientifique Technologique et Médicale de Grenoble
BP 87 38402 Saint Martin D'Heres, F.
°° Laboratoire de Chimie des Solides, associé au CNRS (LA 279)
Université de Nantes, 2 Rue de la Houssinière, 44072 Nantes, F.

INTRODUCTION

Among the lithium intercalated IVB transition metal dichalcogenides most of the studies have been devoted to the Li_xTiS_2 system, due to its application as intercalation electrode in high energy density primary and secondary batteries (Whittingham, 1978).

To study the charge transfer from the intercalated alkali atom to the host compound and its effect on the electronic structure, the ^{7}Li NMR, and the ^{77}Se NMR for the diselenides, proved to be very usefull since the early work of Silbernagel (1975). In this work, undertaken on pure and fully intercalated disulfides and diselenides of group IVB and VB transition metals, it was shown that intercalation was acccompanied by the donation to the host of approximately one electron per lithium.

^{7}Li NMR studies of the Li_xTiS_2 system were carried out also to determine the lithium mobility (Kleinberg and Silbernagel, 1980) and to search for evidence of lithium ordering by ^{7}Li linewidth study in $Li_{.33}TiS_2$ (Kleinberg et al., 1982). No evidence of ion ordering was found in this way, although such an ordering, which has also been observed in similar systems, was shown to occur, using electrochemical spectroscopy (Thompson, 1978) .

Since the above mentionned NMR studies of Silbernagel (1975) lithium intercalation in $TiSe_2$ has been the subject of limited studies only. Murphy et al. (1976) studied punctual compositions obtained by n-buthyl-lithium intercalation and Thompson (1979) made precise studies of the voltage/composition relation upon electrochemical intercalation.

Some time ago Deniard (1984) has shown on chemically intercalated samples that the lattice parameter expansion upon intercalation is regular and similar to that observed for Li_xTiS_2 (Thompson and Symon, 1981). We undertook ^{7}Li and ^{77}Se NMR studies of this system and we have shown recently that the regularity of the intercalation is particularly well observed, from a perfect linear dependence of both the ^{7}Li quadrupolar coupling and the ^{77}Se NMR shift in the whole intercalation range, $0 \leqslant x \leqslant 1$ (Chabre and al., 1987).

In this paper we present the analysis of the low temperature ^{77}Se NMR spectra which are shown to contain information about the lithium distribution on the Van der Waals intercalation plane. Preliminary Monte-Carlo simulations of these spectra are presented which allow to determine the ion-ion interactions and show the tendency for local ordering for some fractionnal lithium contents.

SAMPLE PREPARATION AND EXPERIMENTAL CONDITIONS

The diselenide is prepared by direct synthesis from titanium (Ventron, 3N) and selenium powders (Koch Light, 4N). Intercalation is performed by the n-buthyllithium technique under inert atmosphere at room temperature. After one week the product is filtered, washed with hexane and dried under vacuum at 100°C. Then the compounds are heated at 300°C for three days for homogeneization.

NMR measurements were performed with a pulsed spectrometer at 17 MHz mainly, corresponding to a magnetic field around 21 kG for ^{77}Se. This frequency was choosen because the width of the ^{77}Se spectra is due to a hyperfine interaction distribution as in Li_xZrSe_2 (Chabre et al., 1983 a, b; Berthier et al.,1984) and at low temperature they already extend over more than 60 G at 21 kG. Lineshapes were recorded by analogic integration of dipolar echoes with a gated integrator and sweeping the magnetic field (Clark, 1964).

^{77}Se NMR RESULTS

The influence of the intercalated lithium ions on the ^{77}Se NMR spectra has been partly understood in our previous work on the Li_xZrSe_2 system (Chabre et al, 1983 a, b): there is an inhomogeneous charge redistribution in the sheets of the host layer compound in order to screen the potential of the ionized intercalant (see Mc. Kinnon, 1987, in these proceedings). This charge redistribution in the d-like conduction band creates on the selenium nuclei through the p-d hybridization a hyperfine field distribution which is time dependent as a result of the lithium ion mobility. At high temperature when the lithium mean jump frequency is larger than the hyperfine interaction distribution expressed in frequency units, it produces the usual motional narrowing of fine structures (Abragam, 1961, Chap.X). One then observes ^{77}Se homogeneous lineshapes which are characteristic of the powder distribution of an anisotropic hyperfine shift (Chabre and al., 1987).

At low temperature the motion of the lithium ions is frozen at the time scale of the hyperfine interaction betwen the ^{77}Se nuclei and the electronic screening of the surrounding lithium ions. One then observes ^{77}Se NMR spectra with several lines (fig. 1) which correspond to the various possible configurations of the lithium ions around each ^{77}Se nucleus. This was already observed in the Li_xZrSe_2 compound (Chabre et al. 1983 b) in which the lines were attributed to selenium nuclei with either 0, 1, 2 or 3 lithium ions in the three first neighbour octahedral sites of the Van der Waals intercalation plane.

The new interesting feature of these ^{77}Se spectra, as compared to the previous ones we get on the Li_xZrSe_2 system at higher freqency, is that, at least for low x values, we were able to determine the spin-spin relaxation of each component line. This was achieved through a 2D type NMR measurement: we recorded the spectra as Fourier transforms of dipolar echoes for increasing values of the echo delay. Then we analysed the time dependence of each component.

These determinations of the ^{77}Se spin-spin relaxation are reported in Table 1. For low lithium contents it appears that the larger are the shifts of the spectrum components from the position of the ^{77}Se line in the pristine compound the shorter are the spin-spin relaxation times. For x =.53 the only two lines observed have the same spin-spin relaxation times and for larger x values it seems there is a single relaxation time.

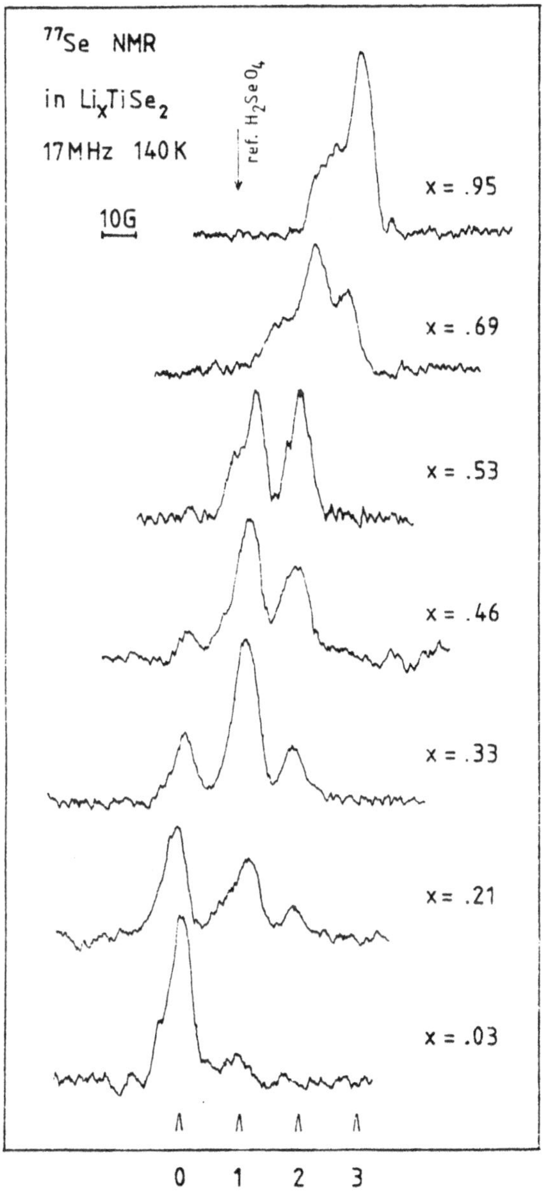

Fig. 1. Low temperature ^{77}Se NMR spectra in Li_xTiSe_2, which are Fourier transforms of the dipolar echo at 500 μs. Labels (0), (1), (2) and (3) refer to ^{77}Se resonance lines from selenium nuclei with 0, 1, 2, or 3 lithium ions in the three nearest nieghbour octahedral sites of the Van der Waals intercalation plane.

Table 1. Spin-spin relaxation times (ms) of the ^{77}Se low temperature NMR spectra lines for some Li_xTiSe_2 samples. Relaxation is not simply exponential. The characteristic time is taken at 1/e of the initial amplitude of the studied component.

line		0	1	2	3
	0	6.5 (5)	-	-	-
	.21	4.2 (2)	3.5 (2)	3.0 (2)	-
x =	.33	4.0 (2)	3.5 (2)	3.0 (2)	-
	.53	-	2.0 (1)	2.0 (1)	-
	.95	-	-	1.05 (10)	

This distribution of spin-spin relaxation times within the ^{77}Se spectra for low x values was overlooked in our previous study of the Li_xZrSe_2 compounds (Chabre et al., 1983 b) where $^{77}T_2$ was determined by measuring the maximum amplitude of the dipolar echoes. New measurements on this system with the Fourier transform method gave results similar to the above ones on Li_xTiSe_2. Thus the sharp drop in $^{77}T_2^{-1}$ previously observed at 205 K in $Li_{.29}ZrSe_2$ could not be the signature of an order-disorder transition as we suggested but an NMR artefact at the temperature for which lithium moves slowly enough compared to the individual line splittings, so that the spectrum becomes inhomogeneous.

In a Van Vleck second moment calculation with contributions of like and unlike spins (Abragam, 1961, Chap.IV) one can estimate that since the natural abundance of the ^{77}Se is 7.5 % only and that of the ^{47}Ti and ^{49}Ti are 7.5 % and 5.5 % with low gyromagnetic ratios, the ^{77}Se-^{7}Li contribution to the dipolar magnetic interaction is the dominant one as soon as the lithium content is of the order of 10 %.

An exact calculation of the spin-spin relaxation of selenium nuclei with either 0, 1, 2 or 3 lithium ions in the surrounding sites cannot be done. But these spin-spin relaxation measurements confirm our previous result on the Li_xZrSe_2 system: the charge redistribution which screens the Li^+ ion potential is localised to the close vicinity of the ion. This gives rise to an hyperfine shift of the ^{77}Se resonance which is almost proportional to the number of neighbouring ions, while the spin-spin relaxation rate also is an increasing function of this number.

Thus the relative intensities of the component lines of the spectrum for a given x give the relative numbers of selenium nuclei surrounded by either 0, 1, 2, or 3 lithium ions. We thus have a short range observation of the lithium distribution in the Van der Waals intercalation plane as seen from the selenium nuclei.

THE EFFECT OF ION ORDERING

Let us consider what can be expected for x =.33. One can easily calculate that if the lithium ions were randomly distributed on the octahedral sites of the Van der Waals intercalation plane the proportion of seleniums with 0 to 3 lithiums in the nearest-neighbour sites, that is the relative intensities of the '0' to '3' lines, would be 29.6 %, 44.5 %, 22.2 % and 3.7 % respectively. On the other hand, if lithium were perfectly long range ordered along a $a\sqrt{3} \times a\sqrt{3}$ superlattice each selenium would have only one lithium in the first-neighbour sites and one would observe the '1' line only. Analysis of the low temperature spectrum reported in fig. 1 for x = .33 gives 18 %, 61 %, 21 % and 0 % for the relative intensities of the four lines. Thus this ^{77}Se NMR spectrum gives clear evidence for some ordering of the ions on the 2D triangular intercalation lattice due to ion-ion interaction.

To describe the lithium-lithium interaction we have assumed that the energy of one lithium on one octahedral site was an increasing function of the number of occupied first-neighbour lithium sites, as Berlinsky et al (1979) have done previously when studying lithium ordering in Li_xTiS_2 from the electrochemical titration curve measurements of Thompson (1978). Then we have done 2D Monte-Carlo simulations of the lithium distribution on the intercalation plane, varying the site energies in order to fit the relative intensities of the low temperature ^{77}Se lines for a given x. We have calculated also the room temperature voltage/composition curve for comparison with Thompson's (1979) determination.

MONTE-CARLO SIMULATIONS

Preliminary simulations have been performed on a n x n triangular lattice with periodic boundary conditions. For a given concentration we start from a random distribution and then simulate the lithium redistribution in the following way: we choose randomly an ion; then if this ion jumps to one of the first neighbour vacant sites its site energy and that of the ions first and second neighbours of its initial position will change. So for every possible jump to the first neighbour sites we calculate the energy changes of a 5 X 5 area centered on the initial position of the ion. The jump then occurs in one of the vacant site with a probability proportional to the Boltzman factor with the appropriate energy difference and temperature. Whenever on the average the ions have jumped once, i.e. after a time equal to the jump correlation time, the lithium distribution around the selenium is examined and the intensities of the spectrum components are calculated. Numerical results converge rather rapidly, within a few jumps per ion. Simultaneously the tendency for local ordering can be observed on the n x n lattice.

Results of such simulations are shown on fig. 2 for fractional compositions for which long range order can develop. Fig. 2b shows the lithium distributions on an 20 x 20 lattice after 8 jumps per ions, from the corresponding initial random distributions represented on fig. 2a. For x = 1/3 the tendency for local ordering along an $a\sqrt{3}$ x $a\sqrt{3}$ superlattice appears clearly. For x = 1/2 one can see the tendency for forming alternatively full and empty parallel rows of lithium sites which correspond to a rectangular a x $a\sqrt{3}$ superlattice. For x = 2/3 one observes the construction of the hexagonal superlattice complementary to the triangular $a\sqrt{3}$ x $a\sqrt{3}$ one, obtained for x = 1/3.

These figures have been obtained with lithium site energies depending quadratically on the number n of first-neighbour occupied lithium sites ($0 \leqslant n \leqslant 6$), $E(n) = E(0) + n^2 . E(1)$, and E(1) = 5 meV. E(0) is obtained from attempts to fit the experimental voltage/composition curve (Thompson 1979) with that of a lattice-gas of interacting particles calculating the average site energy $\langle E(n) \rangle$ and taking into account the configurational entropy term, that is :

$$V(x) = \langle E(n) \rangle / e - (kT/e) \log(x/(1-x))$$

where e is the electron charge.

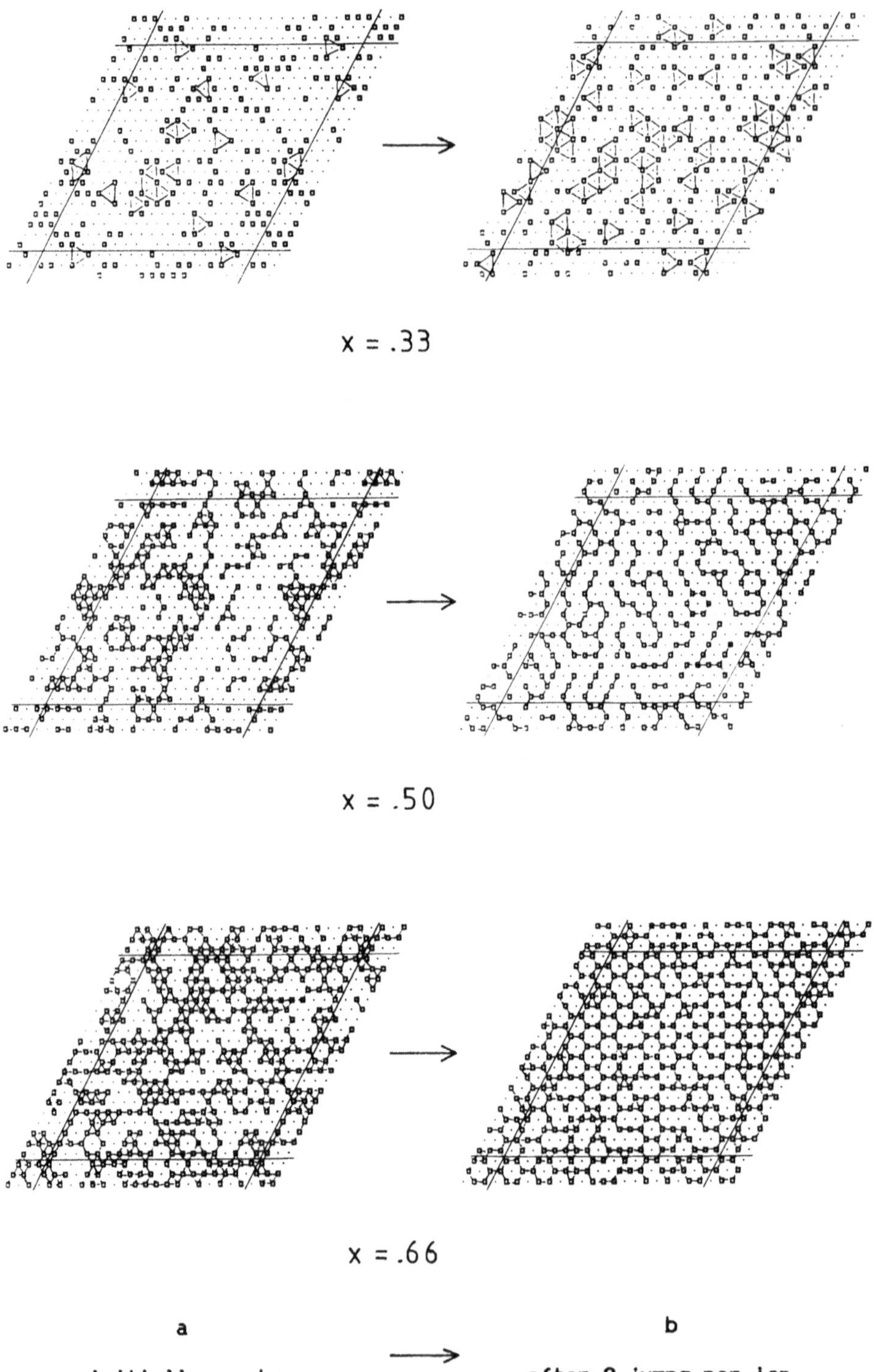

Fig. 2. Results of the Monte-Carlo simulations on a 20 x 20 lattice with periodic boundary conditions, at 140 K and E(1) = 5 meV. Starting from a random distribution one see the tendency for local ordering for these fractional x lithium contents.

Table 2. 140 K experimental and calculated intensity percentage of the ^{77}Se spectra components for some x lithium contents. Rand. is the result of a random distribution and M.C. that of the Monte-Carlo simulation at 140 K with E(1) = 5 meV.

x	Line :	0	1	2	3
.33	Exp.	18	61	21	-
	Rand.	29.6	44.5	22.2	3.7
	M.C.	18	64	18	-
.53	Exp.	-	52	48	-
	Rand.	10.4	35.1	39.6	14.9
	M.C.	1.8	47.8	48.8	1.5
.69	Exp.	-	20	48	32
	Rand.	3	19.9	44.2	32.9
	M.C.	0	10	80	10

Thompson (1979) experimental data and Monte-Carlo simulation results with E(0) = 2.05 eV and E(1) = 5 meV are reported in fig. 3. Agreement is good for $x \leqslant .5$. Attempt to get a better agreement for larger x values increasing E(1) to 8 meV is also reported, but with this value one gets poorer fits for the low lithium content ^{77}Se NMR spectra.

Experimental and calculated relative intensities of the lines of the selenium spectra for samples compositions close to the above represented fractionnal ones are reported in Table 2. Agreement between experimental determinations and results from the Monte-Carlo simulations appears to be fairly good for the x = .33 and .53 samples. On the contrary for the x = .69 the Monte-Carlo simulation with the same set of parameters gives values strongly differing from the observed relative intensities which are very close to what can be calculated for a random distribution.

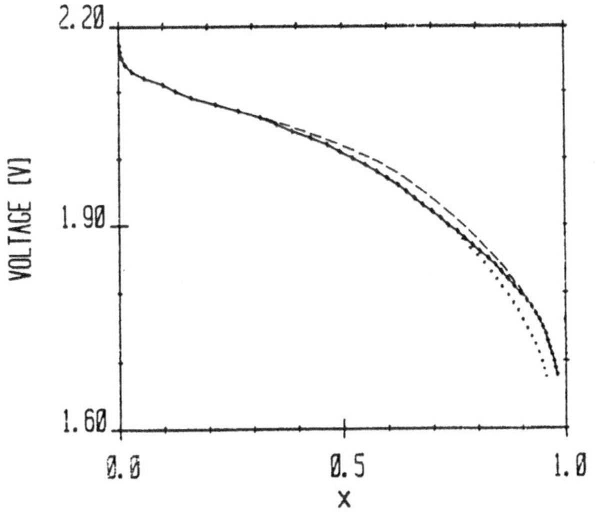

Fig. 3. Li_xTiSe_2 voltage/composition curves. (----) Thompson (1979) and results of the Monte-Carlo simulations with E(0) = 2.05 eV and E(1) = 5 meV (-----) or 8 meV (·····).

This cannnot be attributed to a qualitative change in the electron screening since $TiSe_2$ appears to behave very regularly upon intercalation as can bee seen from a perfect linear dependence of both the ^{7}Li quadrupolar coupling and the ^{77}Se NMR shift on the whole intercalation range, (Chabre et al. 1987). Thus one has to introduce more parameters than a site energy depending quadratically upon the number of first neighbour occupied lithium sites. Further investigation could be in two directions: (i) one could introduce next nearest neighbour contributions which become important for large lithium contents and (ii) one should also look for possible elastic contributions which may play a role at the beginning of intercalation as shown by Dahn et al. (1982) for TiS_2.

CONCLUSION

With this work, the ^{77}Se NMR is confirmed to be a very powerfull tool for studying intercalation in lamellar diselenides. Evidence is given for the very short range screening of the Li^+ by the electrons transfered to the host during the intercalation and it is demonstrated that the low temperature ^{77}Se spectra give access to the details of the lithium distribution on the sites of the Van der Waals intercalation plane.

By attempting to simulate the NMR spectra one has a way to determine the ion-ion interaction, in addition to the simulation of the voltage/composition curve. Moreover as the NMR lineshapes correspond to a local observation of the lithium distribution the NMR spectra simulation which should give the relative intensities of four lines for each lithium content thus should allow a more detailed description of the ion-ion interaction than can achieve the simulation of the voltage/composition curve.

REFERENCES

Abragam A., 1961, "Principles of Nuclear Magnetism", Clarendon Press, Oxford (1978)

Berlinsky A.J., Unruh W.G., Mc.Kinnon W.R. and Hearing R.R., 1979, Solid State Comm., 31, 135

Berthier C., Chabre Y., Ségransan P., Deniard P., Trichet L. and Rouxel J. 1984, in "Physics and Chemistry of Electrons and Ions in Condensed Matter", J.V. Acrivos, N.F. Mott and A.D. Yoffe eds. NATO ASI Series C 130, 561, D. Riedel, Dordrecht.

Chabre Y., Berthier C. and Ségransan P., 1983 a J. Phys. Lettres, 44, L619

Chabre Y., Berthier C., Ségransan P. Trichet L. and Deniard P., 1983 b, Solid State Ionics, 9-10, 467

Chabre Y., Deniard P. and Trichet L., 1987, submitted to Solid State Comm.

Clark W.G., 1964, Rev. Sci. Inst., 35 316.

Deniard P., 1984, Thesis, Université de Nantes, unpublished.

Dahn J.R., Dahn D.C. and Hearing R.R., 1982, Solid State Comm., 42, 179

Kleinberg R.L. and Silbernagel B.G., 1980, Solid State Comm., 36, 345

Kleinberg R.L., Silbernagel B.G. and Thompson A.H., 1982 Solid State Comm., 41, 401

Mc.Kinnon W.R., 1987, in this NATO ASI Series volume.

Murphy D.W.,Di Salvo F.J., Hull G.W.Jr. and Waszczak J.V., 1976 Inorg. Chem. 15, 17

Silbernagel B.G., 1975, Solid State Comm., 17, 361

Thompson A.H., 1978, Phys. Rev. Lett., 40, 1511

Thompson A.H., 1979, in "Fast Ion Transport in Solids", P. Vashista, J.N. Mundy and G.K.Shenoy eds., p. 47, North-Holland, Amsterdam.

Thompson A.H. and Symon C.R., 1981, Solid State Ionics, 3/4, 175

Whittingham M.S., 1978, Progress in Solid State Chem., 12, 41

POLYVALENCY IN POST TRANSITION METAL INTERCALATION COMPOUNDS OF $2H\text{-}TaS_2$: A TDPAC STUDY

T. Butz*, U. Klapp* and A. Lerf**

* Physics Department of TUM, 8046 Garching, FRG

** Walther-Meissner-Institut, 8046 Garching, FRG

ABSTRACT

We report on the in situ measurement of the nuclear quadrupole interaction (NQI) of radioactive ^{181}Ta labels in $2H\text{-}TaS_2$ during thermal intercalation of the post transition metals In, Ga, Sn and Cd. At sufficiently high temperatures all intercalates diffuse rapidly enough to yield a motionally averaged NQI. Based on a previously determined NQI-frequency versus charge transfer relationship we obtain for a formal uptake of $x = 1$ (intercalate/TaS_2) a formal average valence state of $\langle v \rangle = +1$ for In and Ga and $\langle v \rangle \sim +2/3$ for Cd and Sn, suggesting the formation of an intercalate conduction band in the latter two cases. During intercalation metastable intermediate phases with high charge transfer show up for Ga and Sn. Moreover, for In_xTaS_2 we observe a continuous decrease of $\langle v \rangle$ from $+1.75$ towards $+1.0$ when x increases from ~ 0.33 towards ~ 1.00. Both observations suggest the formation of higher oxidation states (Sn^{4+}, In^{3+}, Ga^{3+}) at low intercalate concentrations.

INTRODUCTION

The layered compound $2H\text{-}TaS_2$ consists of hexagonal layers with S- Ta-S... S-Ta-S ...stacking and readily intercalates cations between adjacent sulfur layers, i.e. into the so-called van der Waals gap /1/. Upon intercalation the lattice expands in order to accommodate the guest species. The required elastic deformation energy has to be overcompensated by an electronic energy gain due to charge transfer from the intercalate to the host. It, therefore, seems plausible that polyvalent metals should intercalate with their maximum oxidation state. However, the mutual Coulomb repulsion between highly charged intercalated ions is adverse to a high packing density. Therefore, the maximum uptake of $x = 1$ (M_xTaS_2), which is sterically possible, is likely to be obtained with the intercalate in lower oxidation states, i.e. there is partial charge transfer only. This is to be expected anyway for poor electron donors.

Thermal intercalation of post transition metals into $2H\text{-}TaS_2$ was reported in 1973 by Di Salvo et al. /2/. For In, Ga, Cd and Sn a maximum uptake of $x = 1$ was reported. Since it is generally assumed that one e^- can be accommodated in the conduction band of $2H\text{-}TaS_2$ only, it was concluded that these metals do not intercalate in the maximum oxidation state (In^{3+}, Ga^{3+}, Cd^{2+}, Sn^{4+}) but transfer charge partially only. For In and Ga a value of v = +1 was proposed, in good agreement with the lattice expansion data for larger monovalent metal radii. Since Cd and Sn do not adopt a valence v = 1, a mixture of Cd^{2+} and Cd^{o} and Sn^{2+} and Sn^{o}, or - in other words - the formation of an intercalate conduction band is required to reconcile the large uptake x = 1 with the maximum possible charge transfer n = 1.

In a series of experiments we have shown that the NQI at ^{181}Ta in $2H\text{-}TaS_2$ depends very sensitively on the charge transfer /3/, essentially independent of the guest species. This NQI can be easily measured by perturbed angular correlation (TDPAC) of γ-rays emitted by ^{181}Ta decaying from excited states. We therefore monitored the thermal intercalation of In, Ga, Sn and Cd into $2H\text{-}TaS_2$ in situ in order to determine the charge transfer n and to look especially for indications of polyvalency. Some results on In and Ga were briefly reported previously /4/.

TIME DIFFERENTIAL PERTURBED ANGULAR CORRELATION (TDPAC)

A TDPAC experiment requires a nucleus in an excited state which decays via the successive emission of two γ-rays. In the present case we use the 133 keV - 482 keV γ-γ-cascade in ^{181}Ta, which is fed via β-decay from ^{181}Hf ($T_{1/2}$ = 42 d). Therefore, all $2H\text{-}TaS_2$ samples were doped with $\lesssim$ 200 ppm neutron activated Hf. If the intermediate nuclear state of the cascade has a lifetime τ_N a time resolved (differential) coincidence measurement of both quanta yields:

$$W(t) \sim \exp(-t/\tau_N) \qquad (1)$$

As a consequence of conservation of angular momentum there is a correlation between the emission probability of each of the γ-quanta and the nuclear spin orientation. Hence, the emission of γ_2 depends on the emission direction of γ_1. An angle resolved coincidence measurement then yields an anisotropic countrate:

$$W(t,\Theta) \sim \exp(-t/\tau_N)(1 + A_{22}P_2(\cos\Theta)), \qquad (2)$$

where A_{22} denotes the anisotropy and $P_2(\cos\Theta)$ is a Legendre polynomial; Θ is the angle between the two detectors for γ_1 and γ_2. This angular correlation can be perturbed by extranuclear fields, e.g. via the interaction of the nuclear quadrupole moment with electric field gradients (EFG), which in turn arise from the l = 2 multipole component of the extranuclear charge distribution. This perturbation causes the nuclear spin to precess around the quantization axis. Because the emission probability pattern for γ_2 simultaneously precesses with the nuclear spin, a periodic modulation of the coincidence countrate results:

$$W(t,\Theta) \sim \exp(-t/\tau_N)(1 + A_{22}G_2(t)P_2(\cos\Theta)) \qquad (3)$$

The perturbation function $G_2(t)$ contains all information on the fate of the nucleus while in the intermediate excited state. For the present case of a spin I = 5/2 and a pure axially symmetric NQI we have

$$G_2(t) = 0.2 + 0.371\cos\omega t + 0.286\cos 2\omega t + 0.143\cos 3\omega t \qquad (4)$$

Each of the three cosine-terms corresponds to the energy difference of the hyperfine-split I = 5/2 level:

$$\hbar\omega = E_{\pm 1/2} - E_{\pm 1/2};\ 2\hbar\omega = E_{\pm 5/2} - E_{\pm 3/2};\ 3\hbar\omega = E_{\pm 5/2} - E_{\pm 1/2} \qquad (5)$$

For I = 5/2 the quadrupole frequency ν_Q is related to the precession frequency ω via

$$\nu_q = e^2qQ/h = 10\omega/3\pi \qquad (6)$$

with eq denoting the largest component of the EFG-tensor, Q being the nuclear quadrupole moment of the intermediate state, and h is Planck's constant.

Our setup consists of four detectors in a plane at 90° intervals using NaI(Tl) sczintillators on RCA 8850 photomulipliers. The details of this spectrometer are described elsewhere /5/. From coincidence spectra of the type described in eq (3) the perturbation factor muliplied by A_{22} is extracted and then Fourier transformed. A single Ta site would thus show up as a series of peaks located at o, ω, 2ω and 3ω with relative intensities of 0.2, 0.371, 0.286 and 0.143. Since the finite time resolution of the spectrometer (∿ 2 nsec FWHM) reduces the sensitivity at higher precession frequencies we often loose the highest 3ω signal. The instrumental line width depends on the time window (typically 80 nsec) for the Fourier transform as well as on the window function used to reduce parasitic sidelobes and is roughly 75 Mrad/s (FWHM).

RESULTS

Indium

A series of experiments with a metal/TaS_2 ratio x = 0.31, 0.55 and 1.2 was performed with host particle diameters ranging from 100 μm up to about 1 mm. The sealed and evacuated quartz ampoules were heated in a furnace placed between the γ-detectors. The reaction started at temperatures of about 628 - 670 K (for x = o.55 and x = 1.2) and about 750 K (for x = 0.31), as indicated by the loss of the signal of the empty host (ω = 820 Mrad/s). It is interesting to note that this is substantially above the melting point of In (429 K). The results for x = 0.55 during

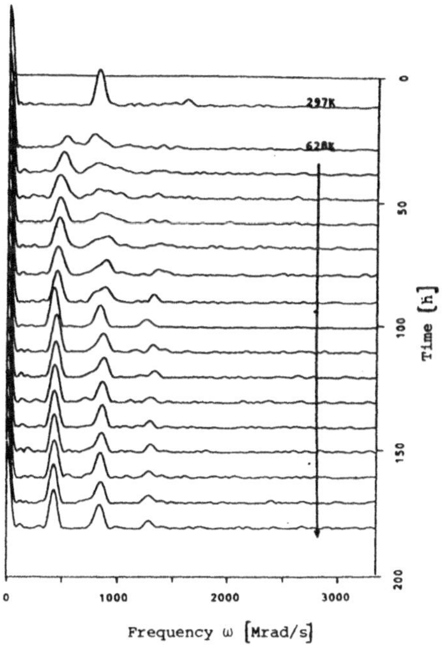

Fig. 1. Stacked plot of Fourier transformed TDPAC data during thermal intercalation of In with x = 0.55.

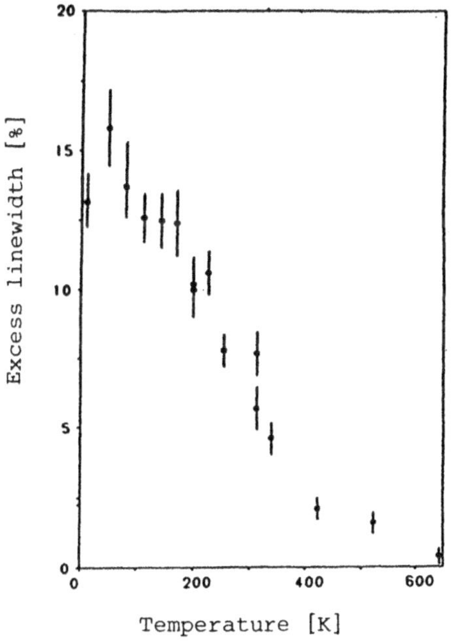

Fig. 2. Excess line-broadening during cooldown of In_xTaS_2 (x = 1.2) indicating slowing down of intercalate mobility.

isothermal reaction at 628 K are shown in Fig. 1. A similar figure for $x = 1.2$ is given in /4/. It is clear that there is a superposition of unreacted 2H-TaS_2 and a partially intercalated phase at the onset of the reaction. As soon as no further decrease of the frequency took place the reaction was assumed to be complete. Increasing the temperature further did not yield lower frequencies. On the contrary, shortly before the TDPAC spectra were irreversibly lost /4/, possibly due to ^{181}Hf probe diffusion, we observed a slight increase of ω, which we interpret as being due to the onset of thermal deintercalation. The maximum observed frequency, the derived charge transfer n based on the previously determined relationship /3,6/

$$\omega = 820 - 754\ n + 293\ n^2\ \text{(Mrad/s)} \qquad (7)$$

and the deduced average valence <v> are summerized in table I.

In all cases the temperature variation of ω was followed down to at least room temperature and found to be small. Hence, the error introduced by comparing observed frequencies at temperatures between 750 K and 950 K, where ω_{min} was obtained, with eq. (7) based on room temperature data, is small. Upon cooling, the linewidth increased for the sample with $x = 1.2$ (see Fig. 2).

Table 1

composition x	ω_{min} (Mrad/s)	n (e^-/TaS_2)	<v>
0.31	520	∿ 0.50	∿ 1.6
0.55	425	∿ 0.75	∿ 1.4
1.0*	390	∿ 0.88	∿ 1.0

*Despite a metal/TaS_2 ratio $x = 1.2$ we assume that intercalation took place up to $x = 1$ only.

This is interpreted to arise from the slowing down of the intercalate motion. At 300 K, a mean residence time of about 50 nsec can be estimated from this line-broadening. For lower uptakes we have evidence for segregation into phases with higher and lower In concentrations. Interestingly enough, the average frequency $\langle\omega\rangle$ for these two-phase spectra migrated considerably during thermal homogenization. Provided the total uptake does not change during this process (we are still far below the thermal deintercalation temperature), this calls for changes in the average valence <v>. In fact, the valence state should be a strongly nonlinear function of the Indium packing density, as already suggested by the relatively weak increase of <v> with lower x (see Tab. 1). Hence, there is clear evidence that the intercalation starts with a relatively dilute phase of In^{3+} - which requires less lattice expansion - and then turns over to more densely packed layers with the averarge valence <v> rapidly approaching +1, in agreement with the proposed valence state for $x = 1$ /2/.

Gallium

Here we performed a single measurement only with $x = 1.4$, a 2H-TaS_2 particle diameter of 315 - 400 µm and a reaction temperature of 750 K - 850 K. (Note, that again the reaction starts well above the melting point of Ga, which is 303 K.) Contrary to the case of In we observed a very low metastable frequency component around $\omega = 100$ Mrad/s (see Fig. 3).

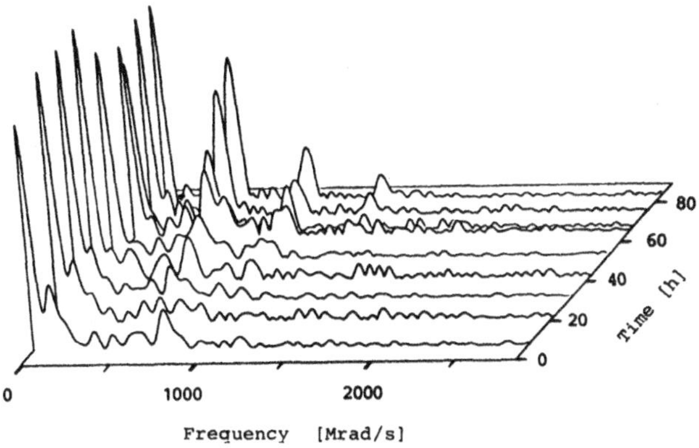

Fig. 3. Perspective view of Fourier transformed TDPAC spectra during intercalation of Ga into $2H\text{-}TaS_2$ ($x = 1.2$). The reaction temperature was slowly increased from 750 - 850 K within 80 hours.

The lowest observed frequency was ω_{min} = 385 Mrad/s at 850 K, in striking agreement with In ($x = 1.2$). Hence, we confirm the proposed valence state of Ga^{1+} /2/. The metastable low frequency component is clear evidence for polyvalency in this system, too. Here, Ga^{3+} is supposed to prevail at low concentrations, as was the case for In. Moreover, Ga-cluster formation is suggested even at $x = 1$ because we observed a dramatic splitting of the single site TDPAC spectrum into a three component spectrum when cooling the sample below about 600 K.

Cadmium

A single experiment with $x = 1$, a $2H\text{-}TaS_2$ particle diameter of 315 - 450 µm, and a reaction temperature of 620 - 700 K was performed. A perspective view of the observed frequencies is shown in Fig. 4.

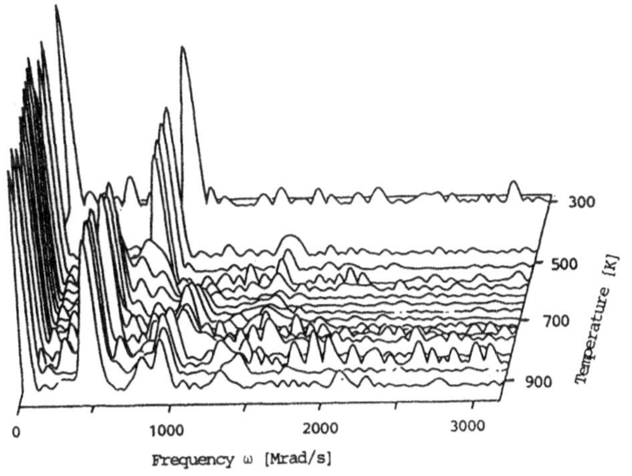

Fig. 4. Perspective view of Fourier transformed TDPAC spectra during intercalation of Cd into $2H\text{-}TaS_2$.

The onset of reaction is around 640 K, still above the melting point of Cd (594 K), despite the significantly higher vapor pressure (∿ 1 Torr at 665 K, compared to 10^{-6} and 10^{-5} Torr at 1065 K and 878 K for Ga and In, respectively). The minimum observed frequency was ω = 465 Mrad/s at 923 K, corresponding to n = 0.55 - 0.65. This could imply an average valence of $\langle v \rangle = +2/3$, which means that an intercalate conduction band is formed, as already suggested for $Sn_{1.0}TaS_2$ and $Pb_{1.0}TaS_2$ on the basis of Sn and Pb NMR Knight shift data /7/. This agrees with the lack of any line-broadening upon cooling to room temperature, where the intercalate mobility should be low enough to prevent motional averaging. Here, the site inequivalencies are strongly screened by the conduction electrons within the intercalate layer. A static snapshot could possibly consist of 1/3 Cd^{2+} on a $\sqrt{3}a \times \sqrt{3}a$ superlattice embedded into the remaining 2/3 Cd^{0} in order to guarantee $\langle v \rangle = +2/3$.

Tin

Two experiments with x = 1 and x = 3, a $2H\text{-}TaS_2$ particle diameter of <100 µm and <200 µm, respectively, and reaction temperatures of 673 - 937 K were performed. A stacked plot of the Fourier transformed TDPAC data for x = 1 is shown in Fig. 5.

The most striking observation is the appearance of a metastable frequency component at ω = 350 Mrad/s. Since it can safely be assumed that the Sn uptake is a monotonous function of time we interpret the occurrence of this metastable phase as due to a low concentration phase with Sn^{4+}. We can rule out thermal deintercalation at temperatures above 900 K as a possible reason for the disappearance of the ω=350 Mrad/s phase because we repeated the intercalation with a massive excess of Sn (x = 3) with identical observations. The minimum frequency was ω_{min} = 480 Mrad/s at 1020 K, very

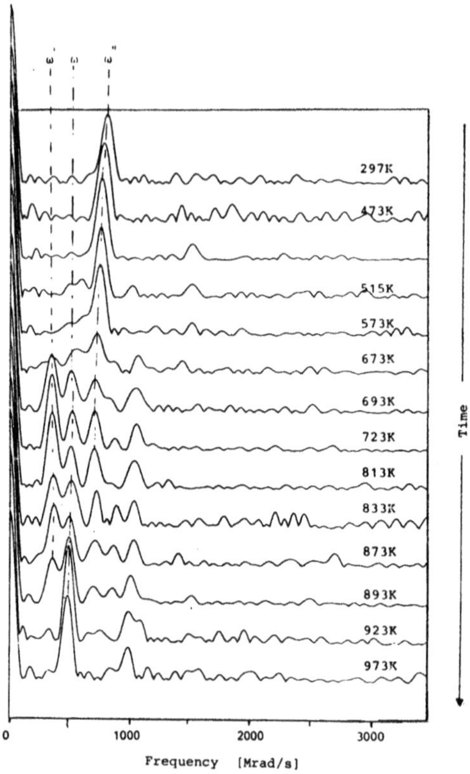

Fig. 5. Stacking plot of Fourier transformed TDPAC spectra during intercalation of Sn (x = 1) into $2H\text{-}TaS_2$.

close to that observed for Cd. Hence, we have $\langle v \rangle = 2/3$ and we propose the formation of an intercalate conduction band, as was concluded before on the basis of NMR Knight shift data /7/ and Sn Mössbauer experiments /8/. This is in keeping with the lack of any line-broadening upon cool-down to room temperature.

SUMMARY

By measuring the NQI at ^{181}Ta in intercalation compounds of 2H-TaS_2 with In, Ga, Cd and Sn we where able to deduce the formal average valence state $\langle v \rangle$ and found that partial charge transfer takes place only for the maximum uptake $x = 1$. While at $x = 1$ we have In^{1+} and Ga^{1+} there is clear evidence for the formation of an intercalate conduction band for Cd and Sn where $\langle v \rangle \approx +2/3$. Clear evidence for higher oxidation states at low concentrations stems from the observation of metastable low frequency phases (Ga, Sn), from the increase of $\langle v \rangle$ for In at low concentrations, and from a migration of $\langle \omega \rangle$ in low uptake In-systems during thermal homogenization. The average valence state adopted by these polyvalent post transition metals, which are poor electron donors, seem to be governed by the balance between electronic energy due to charge transfer to the host and the mutual Coulomb repulsion between intercalate ions, whether "unscreened" (In, Ga) or "screened" (Cd, Sn).

ACKNOWLEDGEMENT

The continuous interest and support of Professors G.M. Kalvius and K. Andres is gratefully acknowledged. This work was funded by the German Federal Minister of Research and Technology (BMFT) under the contract number 03-KA1TUM-4 and the DFG.

REFERENCES

1. R. Schöllhorn, Angew. Chem. Int. Ed. Engl. 19 (1980) 983
2. F.J. Di Salvo, G.W. Hull, L.H. Schwartz, J.M. Vorhoeve, J.W. Wasczak, J. Chem. Phys. 59 (1973) 1922
3. T. Butz, A. Lerf, Rev. Chim. Min. 19 (1982) 496
4. T. Butz, A. Lerf, Ber. Bunsenges. Phys. Chem. 90 (1986) 638
5. A. Lerf, T. Butz, Angew. Chem. Int. Ed. Engl. 26 (1987) 110
6. T. Butz, A. Lerf, J.O. Besenhard, Rev. Chim. Min. 21 (1984) 556
7. A.C. Gossard, F.J. Di Salvo, M. Yasuoka, Phys. Rev. B9 (1974) 3965
8. R.H. Herber, F.J. Di Salvo, R.B. Frankel, Inorg. Chem. 19 (1980) 3135

LAMELLAR FERRIC OXYCHLORIDE PROPERTIES

FROM INTERCALATION TO TOPOCHEMICAL REACTION

B. Bujoli*, P. Janvier*, J. Villieras*, P. Palvadeau**
and J. Rouxel**

* Laboratoire de synthèse organique
et sélective et matériaux
** Laboratoire de chimie des solides
Universite de Nantes
2, rue de la Houssiniere
44072 Nantes Cedex 03 France

The layered structural type of FeOCl[1] has often been reported for its ability to be intercalated by lithium and various molecules[2,3,4]. The lattice is built up with two alternate metal oxygen-layers sandwiched by a double layer of chlorine atoms (Fig. 1).

But an additional feature allows interesting developments : irreversible "topochemical" reactions are possible, involving the partial or complete replacement of the external layers of the sheets (chlorine atoms) by other ions layers such as OH^-, NH_2^- ... leaving only the inner layer unchanged. The first step of this reaction process is an intercalation of the molecule in FeOCl. The electrons transferred from guest to host are accepted on discrete atomic levels. Magnetic measurements and Mössbauer studies of intercalated FeOCl clearly show a reduction of Fe^{3+} to Fe^{2+} in agreement with the transfer that can be expected from the stoichiometry of samples[5]. In such localized systems the ability of intercalating depends basically on three factors which are the stability of the +3 oxidation state, the ionization potential or related property of the guest, the rigidity of the FeOCl lattice (Fig. 2).

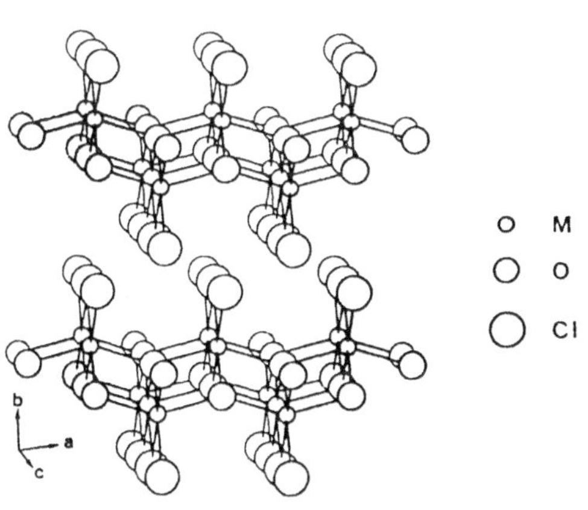

Fig 1: The FeOCl structural type.

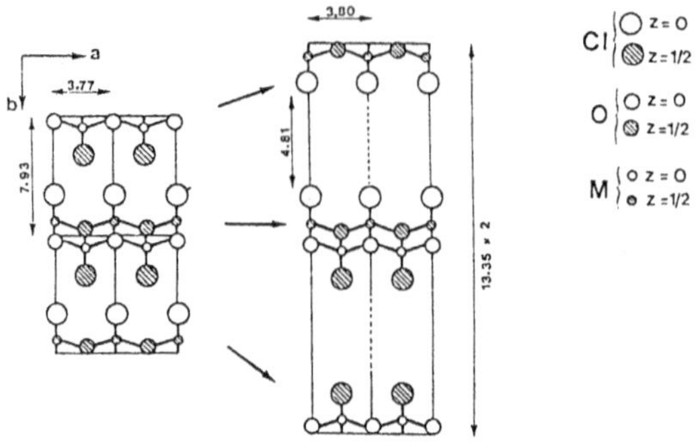

Fig 2: Separation and relative displacement of the FeOCl layers during intercalation.

Many substitution attempts in FeOCl using organic molecules possessing <u>one or two</u> nucleophilic groups will be described.

- In the first case simple grafting reactions occur that lead to the substitution of a simple layer. They are carried out in accordance with the following process.

$$RK + FeOCl \longrightarrow KCl + FeOR$$

R is an organic group. We have used the aliphatic hydrocarbon potassium salts from CH_3OK up to $C_9H_{19}CO_2K$. The rate of reaction and the degree of substitution depends on the lenght of the organic molecule. Prolonged reactions lead to a progressive amorphisation of the network. For the small molecules CH_3OK and CH_3CO_2K we observe different structural results. $FeOOCH_3$ prepared from potassium methylate has been synthesized previously by KOIZUMI et al.[6]. The interlayer parameter is found to be 9.97 Å[7] (Fig. 3) ; $FeOOCOCH_3$ possesses a smaller b parameter of 7.56 Å. This corresponds to a pseudo tetragonal cell that requires an interpenetration and not a superposition of the organic groups (Fig. 4).

The IR spectra of $FeOOCH_3$ carried out between 500 - 50 cm^{-1} on a F.T. IR BRUKER IFS 113 show the disappearance of the sharp band at 113 cm^{-1} attributed to the Fe-Cl vibration[8,9].

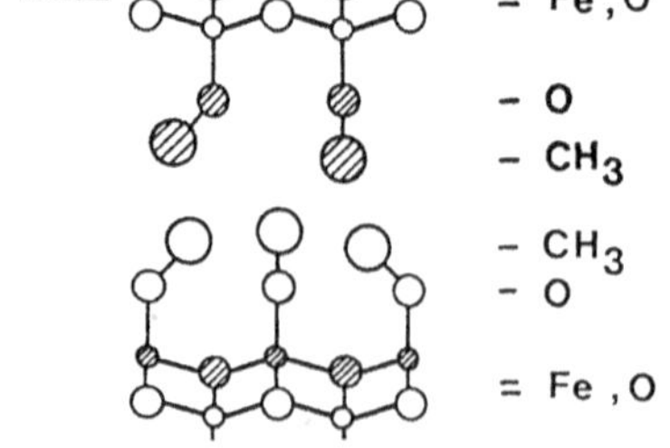

Fig 3: Structure of $FeOOCH_3$ (from ref.7).

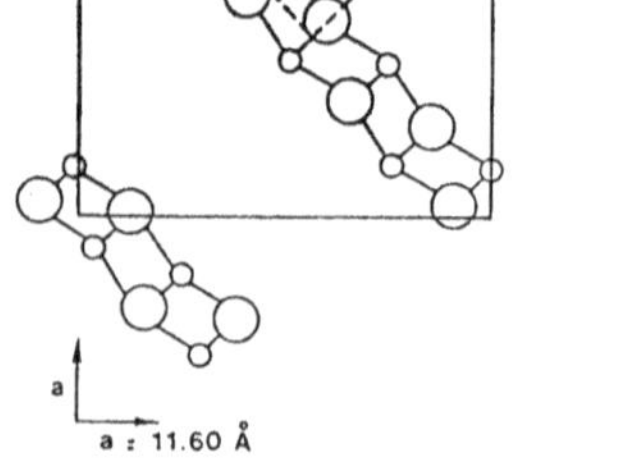

Fig 4: $FeOOCOCH_3$ related to the cell of FeOCl.

The 3000-100 cm^{-1} range allows one to classically characterize the valence and deformation vibrations of the CH_3 and CO groups[10].
The solvent plays a considerable role in the kinetic of the reactions and polar ones such as acetone, acetonitrile, DMF or methanol are the most efficient. Other attempts are planned to study the behaviour of silyl compounds such as $SiMe_3CN$ and $CH \equiv CSiMe_3$ to FeOCl that might lead in this case to $SiMe_3Cl$ as a volatile coproduct more easy to remove than potassium chloride.

- In the second case (molecules bearing two nucleophilic groups) we realize double grafting or pillaring reactions based on the scheme :

$$\text{K-R-K} + 2\ \text{FeOCl} \longrightarrow 2\ \text{KCl} + \text{FeO-R-OFe}$$

While this reaction occurs it does not systematically lead to the formation of bridges between the sheets. Just as in the work of KOIZUMI, KANAMARU et al.[11] we have found that the b parameter, which is in the direction normal to the sheets, may not correspond to the formation of intersheet bridges but to a bending and folding of a molecule doubly fixed to the same sheet (Fig. 5).

This is the case with potassium oxalate $(CO_2K)_2$. To avoid this phenomenon we therefore choose more rigid organic molecules with either multiple bonds or rings. Such features are present in the potassium salt of 4-hydroxybenzoic acid KO-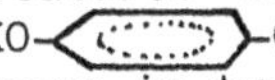-CO_2K for example. In this case the reaction is rapid and occurs in two stages.

- Intercalation with increase of the b parameter from 7.90 Å to more than 13 Å.
- Substitution of chlorine atoms with a decrease of the b parameter to 11.3 Å, in good agreement with the bulk of the molecule. As mentioned previously, we will investigate silyl compounds such as $Me_3Si-C \equiv C-SiMe_3$ (application of pillared compounds as molecular sieves or complexing agents).

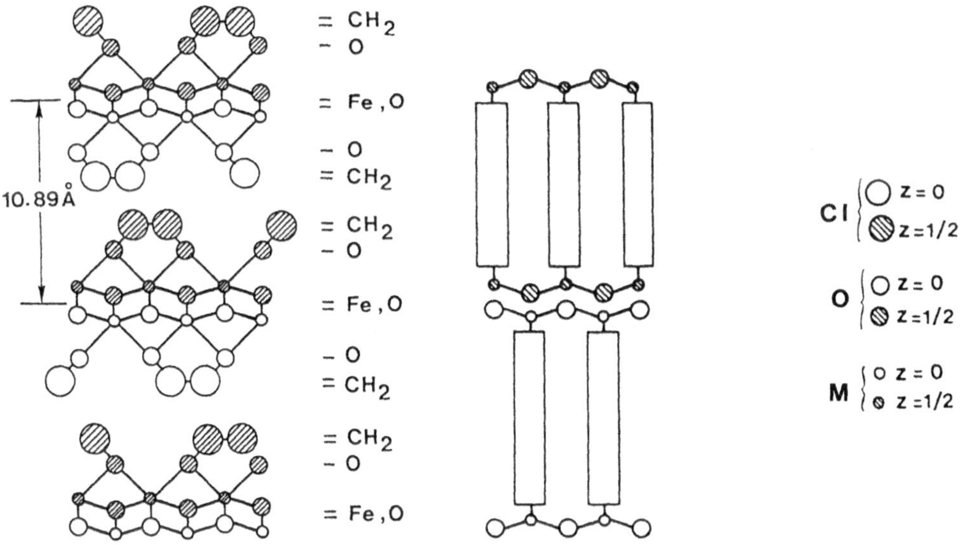

Fig 5: Left view: Double grafting on the same sheet $(CO_2K)_2$.
Right view: Double grafting reaction: pillaring.

Finally, the action of anionic polyhedra containing transition metals (carbonyl metals, Zintl phases, clusters) has been considered. The first attempt was made using disodium tetracarbonylferrate in order to get a metallic iron scattering (catalytic properties) inside the host structure by decarbonylation of the intercalated complex. Although the reaction seems to get on as expected (X-ray powder spectra show an interlayer separation increase from 7.9 Å to 11.2 Å along with a progressive attenuation of the FeOCl pattern and NaCl formation) it turns out that the phenomenon is more complex. Even if a ν -CO is present on IR spectra, its frequency is very low (1620 cm^{-1}) compared with carbonylferrates (1800 cm^{-1})[12] and its intensity is relatively weak in agreement with elemental analyses that point out the low percentage of carbon in the samples after reaction. MÖSSBAUER spectra were recorded (4.2 K, 300 K) but no metallic iron (0 oxidation state) has been characterized until now. However a chemical process occurs owing to the significant formation of sodium chloride while the structure becomes increasingly amorphous. The difficulty of the NaCl elimination and the extremely hydrophilic nature of the products limit their overall interest. The work currently in progress on other reactions of this type, ought to enable to be free of these constraints.

FeOCl appears to be a highly flexible structure capable of a large number of topochemical reactions. New mechanisms of "chimie douce" have also been discovered. Nevertheless, beyond questions of fundamental interest, the implications for the domain of application which one might have been able to deduce from such a flexibility are often compromised by the low thermal and chemical stability of these phases.

REFERENCES

1. S. Goldstaub. Bull. Soc. Chim. Fran. Minér., 56:49 (1935).
2. F. Kanamaru, M. Shimada, and M. Koizumi, J. Solid State Chem., 7:297 (1973).
3. M. Armand, L. Coic, P. Palvadeau, and J. Rouxel, J. of Power Source, 3:137 (1978).
4. T. R. Herbert, and J. Scanlon, Mat. Res. Bull., 13:221 (1979).
5. P. Palvadeau, L. Coic, J. Rouxel, F. Menil, and L. Fournes, Mat. Res. Bull., 16:1055 (1981).
6. S. Kikkawa, F. Kanamaru, and M. Koizumi, Inorg. Chem., 15:2195 (1976).
7. J. Villieras, R. Chiron, P. Palvadeau, and J.P. Venien, Rev. Chim. Minér., 22:209 (1985).
8. J. S. Avery, C.D. Burbridge, and D.M.L. Goodgame, Spectrochim. Acta, 24A:1721 (1968).
9. G. Guillot, J. Cerisier, P. Palvadeau, O. Repka, and S. Lefrant, Solid State Comm., to be published.
10. R. H. Herbert, and A. J. Rein, J. Chem. Phys., 73(2):6345 (1980).
11. S. Kikkawa, F. Kanamaru, and M. Koizumi, Inorg. Chem., 19:259 (1980).
12. K. Nakamoto, "Infrared and Raman spectra of inorganic and coordination compounds, "John Wiley, New York (1978).

INTERCALATION OF TETRASELENAFULVALENE (TSF) AND TETRAMETHYLTETRASELENAFULVALENE (TMTSF) INTO IRON OXYCHLORIDE (FeOCl)

J. F. Bringley[1], J. M. Fabre[2] and B. A. Averill[1]

1 Department of Chemistry, University of Virginia Charlottesville, Virginia 22901 U.S.A.

2 Laboratoire de Chimie Structurale Organique U.S.T.L. 34060 Montpellier Cedex France

Conducting organic materials (1) have attracted the interest of scientists for about two decades. Radical cation salts derived from tetrathia- and tetraselenafulvalenes have been much more studied because some of them have been found to exhibit superconductivity at low temperatures (2). From accumulated data it has been established structural factors are crucial in producing electrically conductive materials (3). Thus, as in TTF-TCNQ (4), the formation of regular segregated stacks (5) is considered as a condition necessary for the conducting network to be formed. But such a formation is very often unpredictable (6). That is why we decided to enforce a segregated structure upon organic π-donors using lamellar materials as macroanionic electron π-acceptors.

At first, we investigated the intercalation of tetrathiafulvalene (TTF) and other tetrathiolenes into the layered host iron oxychloride (FeOCl) (7). Unfortunately, even whether the desired perpendicular host-guest orientation was observed (8), the reactions proceed with complete charge transfer (9). Furthermore, in the case of TTF, sulphur-sulphur contacts were found at distances greater than the sum of van der Waals radii (10) resulting in poor molecular overlap. Trying to avoid such drawbacks, we planned to use the selenium analogs of TTF and TMTTF as new intercalates because:

first, the selenium π-donors might provide partially oxidized guest stacks within FeOCl, according to their poorer ability to give up one electron compare to the sulphur analogs (TTF : 0.33 V ; TSF : 0.48 V).

second, the larger size of selenium might result in greater molecular overlap between adjacent guest species.

We report here the synthesis and the characteritics of intercalated materials involving selenium derivatives of the TTF molecule :

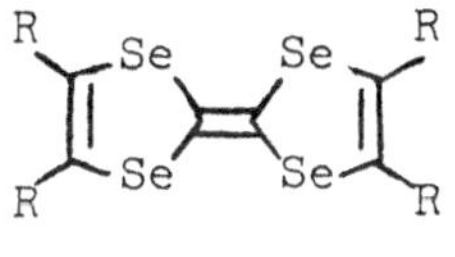

- TSF (tetraselenafulvalene) :
  R = H
- TMTSF (tetramethyltetraselenafulvalene) :
  R = CH_3

1. - Synthesis ; characterization

The selenium π-donors were synthesized as schemed below:
- TSF (procedure adapted from litterature data) (11).

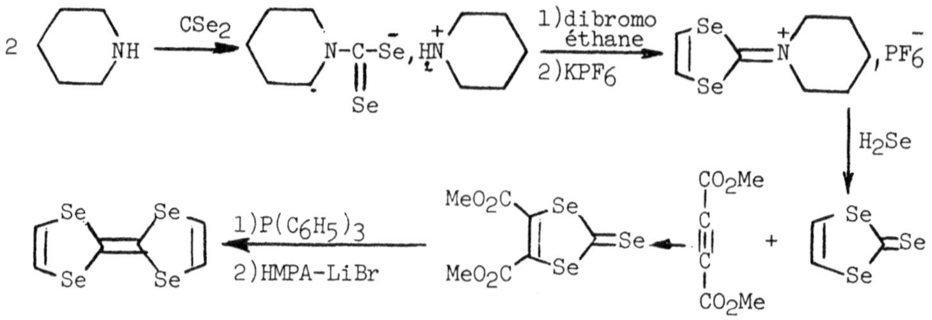

- TMTSF (12) or commercial (ALDRICH 27,440-2).

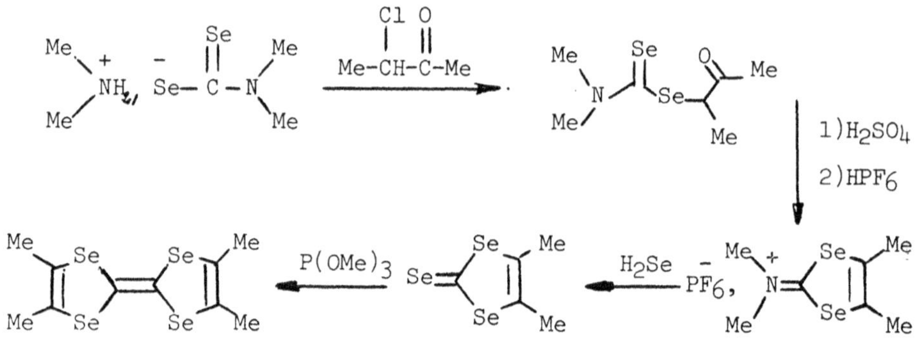

The intercalates were prepared by treating FeOCl (13) with equimolecular solutions of the appropriate tetraselenolene in anhydrous dimethoxyethane (DME). The resulting black solids were then washed using the same solvent and dried in vacuo.

Elemental analyses (Fe, Cl, H, and Se) are consistent with stoichiometries FeOCl (TSF)1/8.5 and FeOCl (TMTSF)1/10.

Intercalation was confirmed for all materials by powder X-ray diffraction. The absence of the characteristic 7.917 A interlayer reflection for pristine FeOCl (14) was used to verify complete intercalation.

2. - Guest orientation ; degree of charge transfer

Upon intercalation, the host undergoes an interlayer expansion corresponding to the size of the guest molecules. These expansions are only 1.3 Å and 1 Å less than the width of TSF and TMTSF respectively (see table 1). Such a result is consistent with orientation of the planes of the guest molecules perpendicular to the host layers but slightly "nested" in the corrugated chloride layers of FeOCl.

The maximum stoichiometries calculated for perpendicularly oriented close-packed guest species are FeOCl(TSF)1/8 and FeOCl(TMTSF)1/9. Thus the guest molecules are very nearly close-packed within the interlayer space.

The FTPIR spectra of the intercalates are consistent with the presence of guest radical cations due to complete charge transfer making each guest molecule completely oxidized. FeOCl(TMTSF)1/10 shows a strong absorption at 1302 cm^{-1}, which can be assigned to the ag 4 mode of the radical cation TMTSF(15). Similar results are noted with FeOCl(TSF)1/8.5. These data suggest that in each case the guests are present as the radical cation within the host lattice.

As reported in the table I, the conductivities of the intercalates are observed to increase by 10^4 - 10^5 $\Omega^{-1}.cm^{-1}$ over that of pristine FeOCl.

TABLE I - Structural and physical properties of FeOCl and the intercalates.

	FeOCl	FeOCl(TSF)1/8.5	FeOCl(TMTSF)1/10
Interlayer distance (Å)	7.917	13.56	14.24
Interlayer expansion (Å)		5.6	6.2
Nesting (Å)		1.3	1.0
Charge transfer		1 e^-	1 e^-
RT $(\Omega.cm)^{-1}$	10^{-7}	4.7×10^{-2}	1.3×10^{-3}

CONCLUSION

As it can be seen in the table I, even with the use of selenium analogs of the TTF which are poorer donors than their sulphur analogs, we observed a complete charge transfer upon intercalation into FeOCl making the donor stacks non-conductive. The increasing in conductivity being probably due to the electrons added to the host lattice upon intercalation. To improve the conductivity of such a material, it is now necessary to partially reduce the donor stacks in order to produce an uncomplete oxidation of the intercalated TSF or TMTSF molecules. To attain such a result, we are currently examining the cointercalation of alkali metals within FeOCl(TSF)1/8.5 and FeOCl(TMTSF)1/10.

REFERENCES

(1) a- D.O. COWAN and F.M. WIYGUL ; chemical and engineering July 21, 1986, p. 28.

b- K. BECHGAARD, ANDERSEN J.R. ; "Physics and chemistry of low dimensional solids" L. Alcacer ; ed. D. Reidel. Press Holland 1980, 247-263.

(2) a- D. JEROME, A. MAZAUD, M. RIBAULT and K. BECHGAARD. C.R. Acad. Sci. Paris, 1980, t-290, B-27.

b- J.M. WILLIAMS, M.A. BENO ; H.H. WANG ; P.C. LEUNG, T.J. EMGE; V. GEISER ; K.D. KARLSON ; Acc. Chem. Res. 1985, 18, 261.

(3) a- P. DELHAES ; C. COULON ; J. AMIELL ; S. FLANDROIS ; E. TORREILLES ; J.M. FABRE ; L. GIRAL. Mol. Cryst. Liq. Cryst., 1979, 50, 43.

(3) b- J.M. FABRE ; E. TORREILLES ; M. VIGROUX ; L. GIRAL ; J. Chem. Res. (M), 1980, 4564.
(4) T.E. PHILIPS ; R.J. KISTENMACHER ; J.P. FERRARIS ; D.O. COWAN; J.C.S. Chem. Commun ; 1973, 471.
(5) K. BECHGAARD ; C.S. JACOBSEN ; K. MORTENSEN ; H.T. PEDERSEN; N. THORUP ; Solid State Commun ; 1980, 33, 1119.
(6) F. WUDL ; Acc. Chem. Res., 1984, 17, 227.
(7) B.A. AVERILL ; S.M. KAUZLARICH ; M.R. ANTONIO ; J. de Physique (Paris), 1983, 44, C_3-1973.
(8) S.M. KAUZLARICH ; B.K. TEO ; B.A. AVERILL ; inorg. Chem. 1986, 25, 28.
(9) B.A. AVERILL ; S.M. KAUZLARICH ; B.K. TEO ; J. FABER. Jr ; Mol. Cryst. Liq. Cryst., 1985, 120, 259.
(10) S.M. KAUZLARICH ; J.L. STANTON ; J. FABER Jr; B.A. AVERILL ; J. Amer. Chem. Soc., 1986, 108, 7946.
(11) M.V. LAKSHMIKANTAN and M.P. CAVA ; J.C.S. Chem. Commun, 1975, 882.
(12) K. BECHGAARD ; D.O. COWAN ; A.N. BLOCH ; J.C.S. Chem. Commun, 1974, 937.
(13) Reaction conditions : TSF ; 14-17d. at 70°C ; 0.03 M TMTSF, 4-6 week at 85°C. After work up, the intercalates were treated again with guest solutions if intercalation was incomplete.
(14) M.D. LIND ; Acta. Cryst., 1972, B-26, 1058.
(15) M. MENEGHETTI ; R. BOZIO ; I. ZANON ; C. PECILE ; C. RICOTTA; M. ZANETTI ; J. Chem. Phys., 1984, 80, 6210.

MPS_3 LAYER MATERIALS AS POLYNUCLEAR COORDINATION COMPOUNDS : WHEN THE GUEST ($NiPS_3$) REACTS WITH THE HOST (DIBENZENE-CHROMIUM)

Martine Doeuff and René Clement

Laboratoire de Spectrochimie des Eléments de Transition (UA 420), Batiment 420, Université Paris-Sud, 91405 Orsay Cedex(France) and LURE, CNRS and Université Paris-Sud Orsay (France)

ABSTRACT

This paper describes the results of a structural and electronic study of amorphous materials $NiPS_3Cr_x(arene)_y$, ($x \sim y$), which reveal new aspects of the reactivity of $NiPS_3$ with organometallics.

INTRODUCTION

Earlier work on the MPS_3 layer compounds had led one of the authors to introduce the idea that these materials, initially considered as more analogs of the layered transition metal dichalcogenides, were in fact more properly described as coordination compounds (1,2,3), each MPS_3 layer consisting of M^{2+} ions coordinated to $P_2S_6^{4-}$ ligands. The basic reason for that was the ability exhibited by several MPS_3 to exchange M^{2+} ions thanks to an heterogeneous equilibrium $MPS_3(s) \rightleftharpoons M^{2+} + 1/2(P_2S_6^{4-})$, giving rise eventually to an unusual intercalation mechanism (1,3). This point of view is also supported by the fact that lithium insertion in MPS_3 cathodes involves the reduction of M^{2+} ions to the zero oxydation state (4,5). We present here evidence that a vast amount of further unexpected coordination chemistry involving $NiPS_3$ can occur in toluene or benzene medium.

It had been shown earlier (6) that $FePS_3$ and $NiPS_3$ were able to react with a dibenzenechromium/toluene (or benzene) solution around 125°C to yield intercalation compounds $MPS_3(CrBz_2)_{0.3}$. It had been reported later (7,8), in the case of $FePS_3$, that the above reaction proceeded far beyond simple intercalation, if the reaction temperature was slightly higher (150°C). The compounds $FePS_3Cr_xC_yH_z$ obtained in the latter case had a much higher conductivity than $FePS_3$, but this result couldn't be interpreted because of a lack of electronic and structural characterization. It was nevertheless obvious that unusual chemistry was involved and therefore we decided to undertake a very detailed study in order to obtain at least a clear understanding of the nature of these materials.

We wish to emphasize here that characterization at the atomic scale of such amorphous materials, containing six different elements, is a very difficult task. The strategy followed was to combine usual analytical techniques with a thorough EXAFS and XANES study of the environment of all elements available in the materials. For technical reasons, we have decided to carry out this study with $NiPS_3$ (rather than $FePS_3$) as the starting material.

II - EXPERIMENTAL AND RESULTS

Samples of about 300mg of $NiPS_3$ (powder or platelets) were treated with various amounts of a benzene or toluene solution of $CrBz_2$ in pyrex ampoules sealed under vacuum. After two days at $\sim$ 150°C, the solution had become colorless, except those where the initial $CrBz_2/NiPS_3$ ratio was larger than $\sim$ 2.5. After washing (toluene then methanol) and vac. drying, a black solid was obtained. According to elemental analysis, all samples have a " global " composition $NiPS_3Cr_xC_yH_z$, where x,y and z appear to vary continuously. No attempts were made below $x < 0.5$. The maximum value of x lies between 2.5 and 3 and in all samples we observed $x \sim y \sim z$. We note that such a high Chromium content and the presence of an organic residue imply that these compounds are very different from usual metal intercalates such as Cr_xTaS_2 (9). The organic residue is strongly bound to the solid. It has been removed by heating at $\sim$ 350°C, then condensed as a liquid at room temperature and finally identified by NMR as toluene (or benzene, if benzene is used for the synthesis). We can therefore write the simpler formulation $NiPS_3Cr_x(arene)_y$ ($x \sim y$) for the obtained samples.

The X-ray pattern of $NiPS_3Cr_x(arene)_y$ for $x \sim 2.5$ (noted hereafter (1)) only shows a broad reflexion at 11.5 Å. Inspection of a sample of (1) by scanning electron spectroscopy (SEM) reveals an " exfoliated " morphology. These experiments rule out the possibility of Chromium being present in (1) under the form of metallic particles. Indeed, a sample containing 40% of metallic chromium would give rise to X-ray reflections ; in addition, some aggregates would be seen by SEM.

The X-ray pattern of $NiPS_3Cr(arene)_y$ materials for intermediate values of x show the existence of a diphasic mixture of (1) and of $NiPS_3$, a conclusion confirmed by SEM which shows that each " microcrystallite " has the morphology of a stack of " ordered " and " exfoliated " domains.

As already noted for $FePS_3$, the electrical properties of (1) drastically differ from those of $NiPS_3$. Whereas the latter is essentially insulating, ($\sigma = 10^{-9}\ \Omega^{-1}\ cm^{-1}$ at 25°C), the conductivity of (1), measured using gold electrodes evaporated on a platelet, is much larger ($\sigma = 10^{-1}\ \Omega^{-1}\ cm^{-1}$) and is no longer thermally activated. We note that this behavior recalls that of Li_xNiPS_3 (10).

The IR spectra (between 4000 cm^{-1} and 400 cm^{-1}) of the $NiPS_3Cr_x(arene)_y$ materials show a progressive weakening, as x increases, of the intensity of the $\nu(PS_3)$ stretching mode at 570 cm^{-1}. The band even completely disappears for $x \sim 2.5$, demonstrating that the P-S bonds are affected by the reaction. A weak band at 530 cm^{-1} is present in the spectrum of (1), whose attribution is not clear. We also note that the opacity of the samples strongly increases as x increases, and it has not been possible to obtain a UV/visible spectrum.

As discussed above, a very detailed EXAFS and XANES study of a "saturated" $NiPS_3Cr_x(arene)_y$ ($x \sim y \sim 2.5$) has been achieved at the K edges of Ni, Cr, P and S. The experiments have been carried out at LURE (The French Synchrotron Laboratory), using spectrometers connected to both DCI and ACO storage rings. Because of the limited available space, it is only possible to give here a short account of the results which will be published in detail elsewhere :

Cr edge : EXAFS at the Cr edge reveals the existence in (1) of Chromium-Sulfur bonds, with d (Cr-S) = 2.39 Å. This bond length suggests that such Chromium ions are in the (+III) oxidation state. However, comparison of the XANES spectrum of (1) with the spectra of other Cr(+III)materials (Fig. 1) shows the presence for (I) of a strong shoulder on the

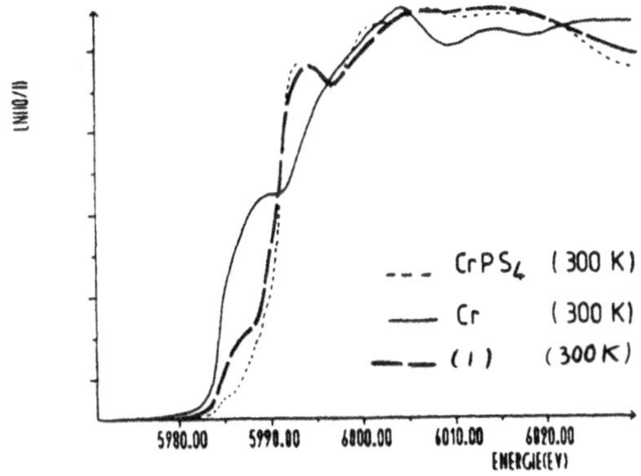

Fig. 1 : Comparison of the XANES spectra of (1), $CrPS_4$ and metallic Chromium

low energy side (5985 eV) of the edge. Interesting enough, this shoulder corresponds to an intense absorption region of metallic chromium. These features reveal the existence in (1) of a second kind of Chromium atoms in a much lower oxidation state, close to the metallic state. Keeping in mind thè conclusions drawn from the X ray and SEM studies, these " neutral " Cr atoms must be dispersed, individually or in small clusters throughout the host structure. These considerations account for other experimental facts, such as the termally non activated conductivity of (1) and they also strongly recall the observation of metal atoms in the zero oxidation, state in the Li_xNiPS_3 materials (4,5).

Ni edge : EXAFS data at the nickel edge are best fit by assuming two kinds of neighbours : Sulfur (Ni-S=2.33 Å) and Carbon (Ni-C=2.10 Å). The latter suggests that some Ni ions are linked to arene rings (for instance Ni-C=2.10 Å in nickelocene), which in turn implies that such Ni atoms no longer occupy intralamellar sites. The bonding of the arene rings to paramagnetic centers is supported by proton NMR measurements. We note that the Ni-S bond in (1) appears significantly shorter than in $NiPS_3$ (2.46 Å).

P and S edges : The XANES spectrum of (1) at the phosphorus K edge is shown in Fig. 2 and compared to that of $NiPS_3$. The absorption edge of (1) exhibits a strong shoulder on the low energy side which is not present in the case of $NiPS_3$. This shoulder occurs at the same energy as the edge of " red " phosphorus (also shown in Fig. 2 for comparison), where P is at the zero oxidation state. These results therefore unambiguously show that the phosphorus atoms are strongly reduced upon the reaction of dibenzenechromium with $NiPS_3$.

Fig. 2: Comparison of the XANES spectra of (1), $NiPS_3$ and red Phosphorus

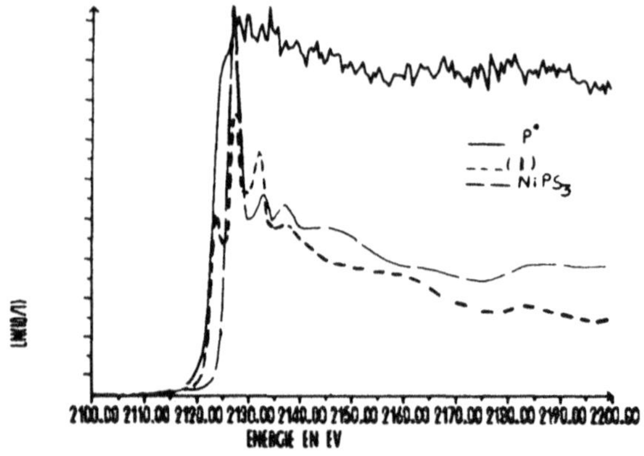

In addition, the EXAFS spectra recorded at both P and S edges show the existence of P-S bonds 2.20 Å in length, much longer than in $NiPS_3$ (2.02 Å). This clearly supports the existence of reduced phosphorus entities in (1).

III - CONCLUSION

A detailed discussion of these results will be published elsewhere. Nevertheless the following conclusions concerning the reactivity of $NiPS_3$ may be reasonably drawn :

i) - The reaction of $CrBz_2$ with $NiPS_3$ involves the reduction of the phosphorus P(IV) ions and a correlative oxidation of a fraction of the Chromium atoms : Cr(0)⟶Cr(+III).

ii) - These Cr(+III) ions are bound to Sulfur atoms and they have probably expelled some Ni(+II) ions from their intralamellar sites.

iii) - Dispersed " quasi metallic " chromium entities (Cr(0) ?) are present, probably in insertion in interlamellar sites, being stabilized by the host structure as in the Li_xNiPS_3 materials.

Therefore, the kind of chemistry encountered in this work is consistent with the idea that the MPS_3 materials react, even in toluene or benzene medium, as coordination or organometallic compounds (ligand exchange and transformation, stabilization of low oxidation states, redox reactions).

Acknowledgements: The authors thank all the staff of LURE, in particular P. LAGARDE and A. M. FLANK.

REFERENCES

(1) Clément, R. J. Chem. Soc., Chem. Commun. 1980, 647.
(2) Clément, R.; Michalowicz, A. Rev. Chim. Miner. 1984, 21, 426.
(3) Clément, R.; Garnier, O.; Jegoudez, J. Inorg. Chem. 1986, 25(9) 1404.
(4) Rouxel, J.; Brec, R. Ann. Rev. Mater. Sci. 1986, 16, 137.
(5) Brec, R. Solid State Ionics 1986, 22, 3.
(6) Clément, R.; Green, M.L.H. J. Chem. Soc., Dalton Trans. 1979, 1566

(7) Clément, R.; Garnier, O.; Mercier, H.; Audière, J.P.; Michalowicz A. Rousseau, B.; Setton, R. J. Chem. Soc., Chem. Commun. 1984, 1354.
(8) Clément, R.; Mercier, H.; Michalowicz, A.; Rev. Chim. Miner. 1985, 22, 135.
(9) Beal, A.R., Physics and Chemistry of Materials with layer Structure : Intercalated layer Materials ; Levy Ed. (Reidel, Dordrecht, 1979).
(10)Foot, P.J.S.; Nevett, B.A. Solid State Ionics 1983, 169.

MAGNETIC RESONANCE AND ELECTRONIC SPECTROSCOPY OF $M_2P_2S_6$ INTERCALATION COMPOUNDS

Efrat Lifshitz and Anthony H. Francis

Department of Chemistry
University of Michigan
Ann Arbor, MI 48109-1055

INTRODUCTION

The transition metal chalcogeno-phosphates, $M_2P_2X_6$ (M=Mn, Cd, Zn, Fe; X=S, Se) belong to a family of broad-band semiconductors with two dimensional, layered crystal structures. The basic structural repeat unit of these compounds is a sandwich of three covalently bonded atomic layers: a central layer of metal atoms and P-P pairs (in a ratio 2:1), between two layers of chalcogen atoms. These units are held together by weak van der Waals (VDW) bonds to form at the lamellar crystal lattice. The stacking arrangement produces a monoclinic or trigonal lattice at ambient temperature. Under appropriate conditions guest molecules can diffuse into the VDW region causing changes in the structural and physical properties of the host materials.

The structural and electrical properties of intercalated transition metal chalcogeno-phosphates have been only partially revealed by previous experimental work. Many of the most widely employed experimental techniques (X-ray diffraction, TGA, magnetic susceptibility, neutron diffraction) are sensitive to the average properties of the sample and are of limited value for the study of atomic-scale structure in the disordered or heterogeneous samples frequently encountered in intercalation chemistry. In the following sections, we describe several experiments utilizing spectroscopic techniques that are sensitive to molecular or atomic scale structure (ESR, NMR, photoluminescence, and IR).

ESR OF INTRA-LAMELLAR MN^{+2} 'SPIN-PROBES'

ESR spectra were obtained from doped single crystals of $Cd_2P_2S_6$, $Zn_2P_2S_6$, $Mg_2P_2Se_6$ and $Cd_2P_2Se_6$ containing 0.1 - 1.0 % Mn^{+2}. The Mn^{+2} impurity serves as a dilute magnetic 'spin-probe' which is sensitive to local structural and chemical changes. At ambient temperatures, the doped host lattices exhibit a 30 line ESR spectrum typical of Mn^{+2} in a well ordered environment. A representative spectrum of Mn^{+2} impurity and $Cd_2P_2S_6$, obtained with the external magnetic field parallel to c* is shown in figure 1.

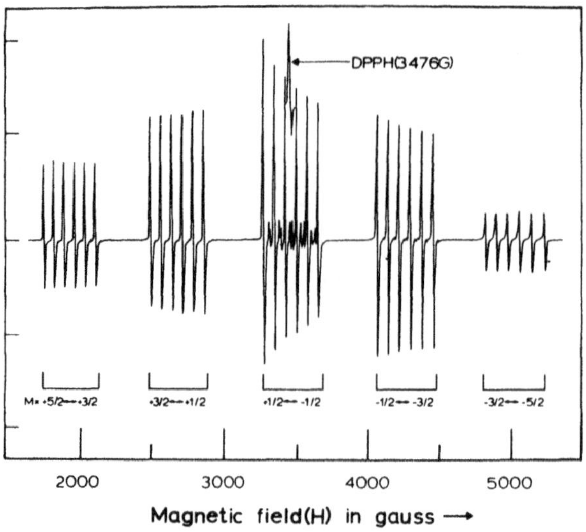

Fig. 1. ESR Spectrum of Intra-Lamellar Mn^{+2} in $Cd_2P_2S_6$

It consists of five fine-structure transitions, +-5/2<-->+-3/2, +-3/2<--> +-1/2 and +1/2<-->-1/2. Each fine structure transition is split into hyperfine sextet by the I=5/2 nuclear spin of the Mn^{+2}. The ESR spectrum indicates the oxidation state of the metal ion and provides detailed information concerning the metal-ligand interactions. Detailed analysis of these spectra confirm that the Mn^{+2} ions substitutionally replace lattice metal in trigonally distorted octahedral sites [1]. The effect of the trigonal distortion on the local crystal field may be obtained from the parameters of the spin Hamiltonian. Additionally, the Mn^{+2} magnetic probes were used to observe phase transition behavior in the nominally pure lattice, and perturbation of the host lattice structure caused by intercalation.

FIRST-ORDER STRUCTURAL PHASE TRANSITIONS IN $M_2P_2S_6$ LATTICES

ESR spectra of intra-lamellar Mn^{+2} in $Cd_2P_2S_6$ were recorded as a function of temperature in the range 180-400°K. Below about 220 K a second set of resonance lines appear which gradually become dominant at the lower temperature. These lines correspond to Mn^{+2} ions in a different lattice structure and the appearance of a second spectrum indicates the onset of a structural phase transition. Moreover, a plot of the mole fraction of the high-temperature phase as a function of temperature (figure 2) shows substantial hysteresis, characteristic of a first-order structural phase transition. The phase transition is from high-temperature monoclinic to a low temperature orthorhombic phase as a result of the sliding motion of the layers one against the other [2]. The displacement of adjacent chalcogen atom planes between the two phases is 2A.

INTERCALATION INDUCED LATTICE DISORDER

Changes in the host lattice crystal structure induced by intercalation were followed by observing the ESR spectra of intra-lamellar Mn^{+2} 'spin probes'. The ESR lines (figure 1) were found to broaden dramatically upon intercalation until only the (nearly) magnetically isotropic +1/2 <-->-1/2 fine-structure transition could be observed. This result indicates both static and dynamic disorder sensed by the 'spin probe' in the transition metal plane [3]. Evidently, the pyridine intercalate interacts with the chalcogen layers to induce a change in the local coordination of

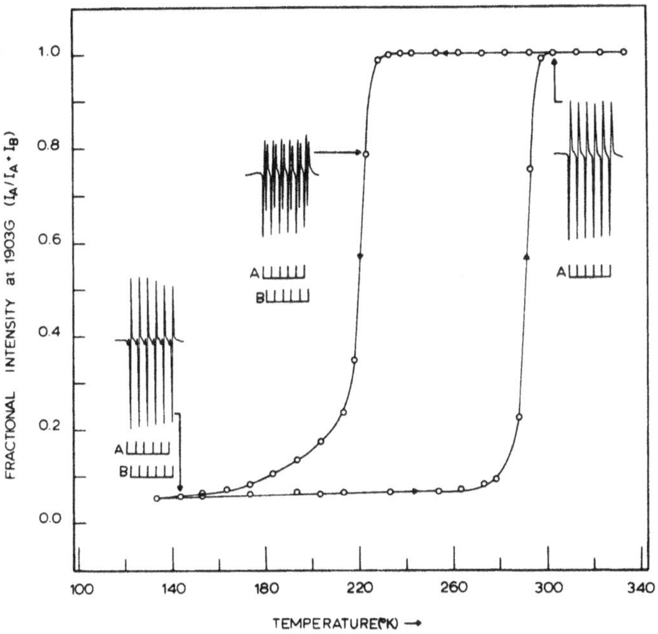

Fig. 2. A First-Order Phase Transition in $Cd_2P_2S_6$

the ESR line. Molecular motion in the intercalate layer causes a modulation of the crystal field at the intra-lamellar metal sites, which gives rise to an additional homogeneous broadening of the ESR signal. ESR spectra of pyridine intercalated $Cd_2P_2S_6$ exhibit a strong temperature and frequency (e.g. L-band, X-band, Q-band) dependence, indicating a thermally activated motion of the pyridine intercalate. Analysis yields a correlation time for the pyridine motion of $8x10^{-12}$ seconds, characteristic of a collision time in a liquid, and an activation energy of 1083 cm^{-1}.

The pyridine intercalated $Cd_2P_2S_6$ (1% Mn) crystals were deintercalated by heating in situ and the effect upon the Mn^{+2} ESR spectrum was observed. Upon complete deintercalation, the ESR spectrum of intra-lamellar Mn^{+2} was regained in very nearly its form prior to intercalation. A slight residual broadening indicates irreversible disorder in the crystal introduced by the intercalation-deintercalation cycle.

CATION EXCHANGE INTERCALATION REACTIONS IN $M_2P_2S_6$ LATTICES

When crystalline $Cd_2P_2S_6$ is treated with pyridine containing small amounts of a transition metal-pyridine complex, the transition metal ion is found to co-intercalate with pyridine solvent [4]. Intercalation is accompanied by the loss of an equivalent amount of Cd^{+2} lattice metal. Chemical analysis can reveal the loss of lattice metal and the uptake of intercalate ion by the host lattice, but does not establish the structure of the intercalation compound. When the intercalate is a paragmatic species and the host lattice is diamagnetic, it is possible to follow the intercalation reaction using ESR spectroscopy. We have examined the intercalation of paramagnetic Mn^{+2} ions complexed by pyridine into $Cd_2P_2S_6$.

The spectrum consists of the partially resolved manganese nuclear hyperfine (I=5/2) structure associated with the +1/2<-->-1/2 fine structure transition and is typical of Mn^{+2} in a disordered environment. The spectrum is isotropic and is essentially temperature independent between room temperature and 123 K. The strongly anisotropic structure associated with the 5/2<-->3/2 and 3/2<-->1/2 fine structure transitions, clearly evident in figure 1, is absent. This demonstrates that the spin-axes of the complexed Mn^{+2} ions randomly oriented within the inter-lamellar space.

The intercalated crystal was heated to successively higher temperatures resulting in thermal deintercalation. As the amount of pyridine remaining in the lattice decreases, the ESR spectrum becomes progressively more detailed and lines associated with the anisotropic fine structure reappear. The re-emergency of the 5/2<-->3/2 and 3/2<-->1/2 anisotropic fine structure transitions can only be interpreted in terms of long range, periodic ordering associated with translational site equivalence and alignment of the Mn^{+2} spin axes. As more pyridine is thermally intercalated, the ESR spectrum gradually approaches that of the intra-lamellar (Mn^{+2}) Cd center shown in figure 1. The spectrum linewidth never becomes as narrow as that shown in figure 1, indicating that the crystalline perfection has been irreversibly effected by the intercalation process. Heating above about 970K results in the deterioration of the spectrum due to decomposition of the host lattice.

The experimental results are consistent with a cation exchange mechanism for the intercalation of the Mn^{+2}/pyridine/water complex into the $Cd_2P_2S_6$ host lattice. The ability of the $M_2P_2S_6$ lattices to exchange intra-lamellar cations with intercalation cationic species is unique in the chemistry of layered materials. This analysis supports the mechanism suggested by Clement [5] for cation exchange.

DEUTERIUM NMR STUDY OF PYRIDINE INTERCALATED $Cd_2P_2S_6$

Deuterium NMR spectra of the intercalation complexes of lamellar $Cd_2P_2S_6$ with various deuterated pyridine molecules were examined in order to develop a more detailed picture of structure and dynamics within the intercalate layer. Conclusions concerning the static equilibrium structure and high-temperature dynamical behavior of the intercalate layer were derived from the magnitude and orientation dependence of the deuterium quadrupole splitting. Each deuteron in an ordered sample exhibits a two-line spectrum with splitting $D=3/2\{e^2qQ/h <1/2(3\cos^2\phi-1)>\}$, where e^2qQ/h is the principal component of the quadrupole interaction, ϕ is the instantaneous angle between the C-D bond and the magnetic field, and the angular brackets indicate averaging over fast motions (on the NMR time scale).

The deuterium NMR measurements demonstrate that the pyridine intercalate lies within the van der Waals gap with the molecular plane parallel to the chalogen layers. Above a temperature of about 213 K, the molecules exhibit a rapid reorientational motion (on the NMR time scale) about an axis perpendicular to their molecular plane [6].

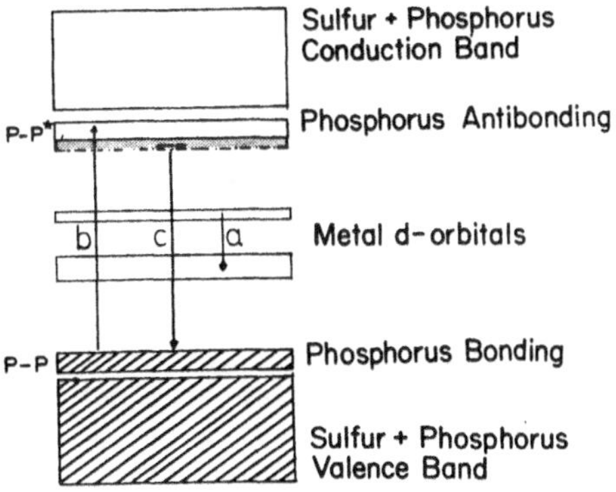

Fig. 3. Energy Levels Structure of $M_2P_2S_6$ materials.

PHOTOLUMINESCENCE OF $M_2P_2S_6$ AND THEIR PYRIDINE INTERCALATES

Photo-luminescence spectroscopy preferentially samples states near the conduction and valence band edges, which are particularly sensitive to the perturbations caused by intercalation. Ultraviolet photoexcitation of $Cd_2P_2S_6$ and $Zn_2P_2S_6$ (d^{10} electronic configuration), single crystals produces a characteristic photoluminescence between 2.5 and 1.5 eV [7]. $Mn_2P_2S_6$ (d^5 electronic configuration) exhibits only low energy emission due to transitions between localized d-orbital states [8] (labeled a in figure 3). These results are consistent with the traditional electronic band structure model for the $M_2P_2S_6$ compound shown schematically in figure 3, which predicts highly localized states associated with the transition metal d-orbitals and delocalized, band-like states derived principally from sulfur and phosphorus orbitals [9]. Analysis of the ultraviolet photoexcitation spectra of several $M_2P_2S_6$ compounds indicates that all exhibit a narrow (FWHM < 0.5 eV) band <--> transition at about 3 eV on the absorption edge of the main conduction-to-valence band transition (labeled b in figure 3). This band may be assigned to the P-P pair antibonding-to-bonding transition that is predicted by the traditional electronic band model. More recent calculations [10] suggest that the P-P sigma-antibonding orbital is considerably higher in energy and not available for this assignment.

Finally, both excitation and photoluminescence spectroscopy of pyridine intercalted $Cd_2P_2S_6$ indicate that small amounts of water, when present during the intercalation process, result in profound and irreversible changes to the host lattice electronic band structure. Pyridine/H_2O intercalation of these layered lattices intensifies the photoluminescence and results in a 0.4 eV shift of the spectrum to lower energy (labeled c in figure 3). The behavior is consistent with either a broadening of the conduction band edge into the band gap or the activation of forbidden indicrect band-gap transitions.

MID-IR ELECTRONIC ABSORPTION OF $M_2P_2S_6$ LATTICES

Because the coordination of the transition metal ion in the $M_2P_2S_6$ lattices is trigonally distorted (D_{3d}), the t_{2g} orbital set is split into an a_{1g} and a e_g orbital pair, resulting in a low energy electronic transition in the mid-IR spectrum. The magnetic properties depend upon the energy levels of those states populated within a given temperature range and also of unpopulated levels close enough to the ground-state to mix significantly under the magnetic moment operator. Therefore, the trigonal field splitting may have consequences for the magnetic properties of the $M_2P_2S_6$ materials and should be sensitive to changes in both the metal coordination geometry and in the nature of the metal-chalcogen bond. Thus, changes in the trigonal field splitting which accompany lattice intercalation may be helpful in the interpretation of electronic and structural changes caused by the intercalate.

We have examined the mid-IR absorption spectrum of single crystals of several $M_2P_2S_6$ materials and have observed directly the trigonal field splitting of the ground state. The mid-IR spectra of $Fe_2P_2S_6$ is illustrated in figure 4. The majority of the features correspond to vibrational transitions which have been discussed previously and the band positions agree closely with those reported for P-2 and M-2 deformations. The spectrum also exhibits a weak, broad (FWHM=1350 cm^{-1}) absorption band centered at 1885 cm^{-1} with a peak absorption coefficient of about 8 cm^{-1}. The band energy and intensity are consistent with its assignment as an electronic transition between the $^5A_{1g}$ and the ^5E_g trigonal field components of the ground state $^5T_{1g}$ term. The trigonal field splitting of the Fe^{+2} ground term is found to be large, in agreement with the predictions from previous Mossbauer studies. The observed splitting (1885 cm^{-1}) is intermediate between the higher values found in many minerals in which iron is oxygen coordinated (2000-3000 cm^{-1}) and in more ionic lattices such as $MgCO_3$ (760 cm^{-1}).

The FTIR spectrum of $Co_2P_2S_6$, like the spectrum of $Fe_2P_2S_6$, exhibits a broad absorption feature (1224 cm^{-1}, FWHM=830 cm^{-1}, k_{max}=8 cm^{-1}), characteristic of a low-lying electronic excitation. The free-ion ^4F electronic ground state of Co^{+2} is split by the octahedral field into $^4A_{2g}$, $^4T_{2g}$ and $^4T_{1g}$ states. These are further split by the trigonal field.

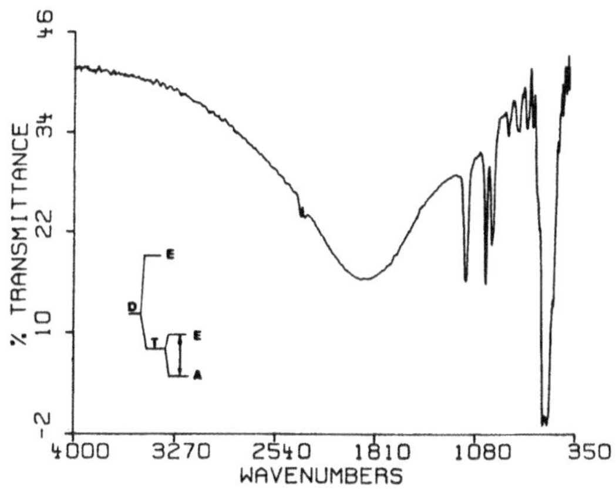

Fig. 4. Mid-IR Spectrum of $Fe_2P_2S_6$.

REFERENCES

1. E. Lifshitz, A. H. Francis, J. Phys. Chem. 86 (1982) 4714.
2. E. Lifshitz, A. H. Francis, R. Clarke, Solid State Commun. 45 (1983) 273.
3. E. Lifshitz, A. Gentry, A. H. Francis, J. Phys. Chem. 88 (1984) 3038.
4. D. Cleary, J. Groh, E. Lifshitz, A. H. Francis (J. Phys. Chem., to be published).
5. R. Clement, O. Garneir, J. Jegoudez, Inorg. Chem. 25 (1985) 1404.
6. E. Lifshitz, S. Vega, Z. Luz, A. H. Francis, H. Zimmerman, J. Phys. Chem. Solid State 47 (1986) 1045.
7. D. A. Cleary, A. H. Francis, E. Lifshitz, J. Lumin. 35 (1986) 163.
8. J. Boerio-Goates, E. Lifshitz, A. H. Francis, Inorg. Chem. 20 (1981) 3019.
9. F. S. Khumalo, H. P. Hughes, Phys. Rev. B. 23 (1981) 5375; M. Piacentini, F. S. Khumalo, C. G. Olson, J. W. Anderegg, D. W. Lynch, Chem. Phys. 65 (1982) 289.
10. M. H. Whangbo, R. Brec, G. Ouvrard, J. Rouxel, Inorg. Chem. 24 (1985) 2459.

OPTICAL STUDIES OF LITHIUM INTERCALATED INDIUM SELENIDE

C.Julien and M.Jouanne

Laboratoire de Physique des Solides, associé au CNRS
Université Pierre et Marie Curie
4, place Jussieu 75252 Paris Cedex 05, France

Optical properties of Li_xInSe are presented as a function of the lithium concentration and temperature. Investigations of the photoluminescence and absorption measurements show that the presence of alternating sheets of lithium ions induces changes in the electronic band structure. New radiative recombination centres due to the lithium insertion are also observed. The Raman spectrum does not change appreciatively, unless it is modified by the presence of excitons near certain resonance conditions.

INTRODUCTION

Layered materials such as the MX compounds of the III-VI group (M=Ga, In and X=S, Se), with large structural anisotropy, seem to be a promising family of materials suitable for intercalation applications, mainly as electrodes in solid state batteries[1,2].

There are two main effects in intercalation when we introduce foreign atoms between the sheets of a layered compound[3]. First of all, an intercalation reaction can be described in a simple scheme as a process in which a defined matrix of the solid remains unaltered with respect to composition, and in which small variations like super-lattice formations can appear. Here we refer mainly to electronic properties such as band structure and optical properties. The second effect is the transfer of charge (electrons) from the intercalated atom to the host material, the extent of this charge transfer varying with the particular system. In general one now replaces the relatively weak Van der Waals forces with stronger Coulomb interactions and the lattice does not expand very much along the c-axis normal to the layer. The system can now be considered to be more three dimensional.

The intercalated substance Li_xInSe is generally a mixed electronic and ionic conductor. The performance of a lithium battery using intercalation cathodes will depend on the intercalation capacity of the cathode material, which is in turn related to its electronic structure. Insertion of Li into a layered compound such as InSe produces structural changes observable through the lattice dynamics and the electronic band structure modifications of the host.

ELECTRONIC BAND STRUCTURE

Indium selenide is a layered semiconductor, crystallized into stacked hexagonal layers. Each layer consists of four two-dimensional sheets of atoms in the sequence Se-In-In-Se with mainly strong covalent binding between the sheets, while the complete four fold layers are bound together by Van der Waals type forces, forming different crystalline structures (polytypes) with different stacking sequences. The γ-polytype has only one layer in the elementary cell. The rhombohedral unit cell contains one M_2X_2 molecular unit. Three sites are available for guest atoms, one octahedral site (O) and two tetrahedral sites (T_1) and (T_2) in the Van der Waals plane of the lamellar compound. The electronic band structure of InSe is almost flat in all directions within the layer plane. The uppermost valence bands are derived mainly from the selenium 4p state and the bottom conduction band is antibonding and formed mainly from the indium s(In) state. The Se-In bond is tilted at only 29° out of the layer plane and the In-In bond is normal to the layer indicates that the Se $4p_x$ and $4p_y$ electrons have much greater Coulomb interaction with In cations than the Se $4p_z$ electron do. Because of such anisotropy of crystalline field, Se $4p_x$ and $4p_y$ states lie well below Se $4p_z$ states[5].

RESULTS AND DISCUSSION

In figure 1 are shown the emission spectra of pure and Li intercalated InSe. The Li intercalation was performed by direct reaction with n-butyl lithium dissolved in hexane forming a 1.6 M solution. We observed two sets of emission bands the first in the near infrared region with several bands and a peak at 1.352 eV and the second with only a broad peak at 2.53 eV. The two peaks at 1.352 eV and 2.53 eV are due to the two excitonic states associated with the two direct gaps between the two valence bands and the conduction band[6]. Here we observe a significant difference between the spectrum for pure material and that of an intercalated sample. After intercalation the excitonic spectrum is drastically modified. The first exciton at

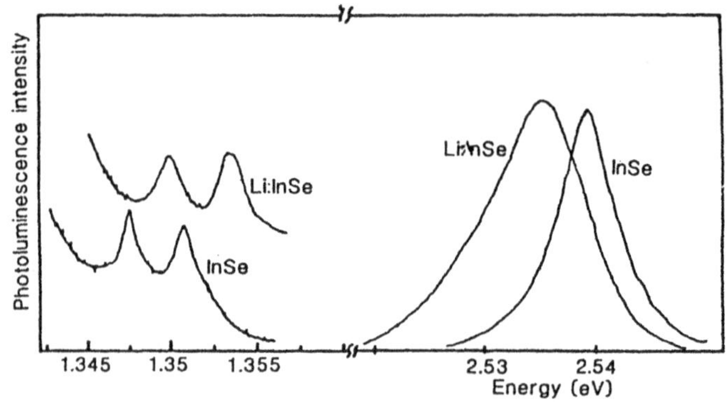

Fig.1 Emission spectra of pure and Li intercalated InSe.

1.352 eV shifts towards higher energies while the second exciton at 2.53 eV shifts towards lower energies with increasing Li dose (figure 2.).

In the lower energy region of the first set of emission bands we also observe the appearance of new bands at lower energies upon lithium intercalation. The position of the band at 1.2815 eV shifts towards higher values with increasing temperature up to 25 K and then decreases linearly as temperature increases. In contrast, the second band situated at 1.198 eV at 5 K has an energy position which steadily shifts towards higher energies with increasing temperature. The analysis of the dependences of these peaks on temperature and incident excitation power leads to the conclusion that the band at 1.2815 eV is due to plastic deformation and the band at 1.198 eV is presumably related to optical transition involving Li donor centres.

In figure 3. is shown the luminescence intensity as function of inverse temperature of the two new bands appeared upon intercalation. The thermal dissociation energies of these bands are found to be 17.8 meV and 55 meV respectively.

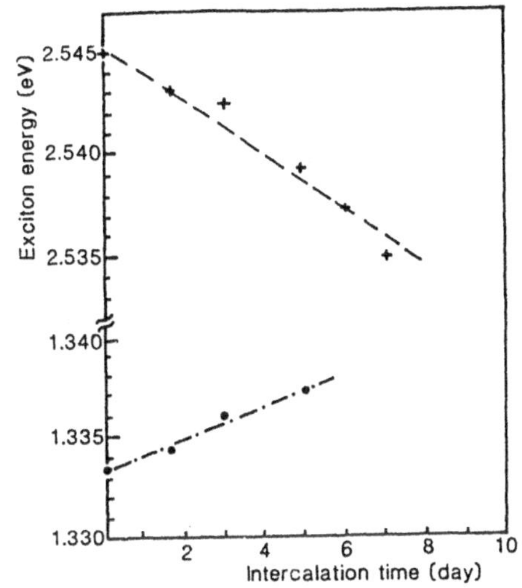

Fig.2 Energy position of excitonic peaks versus Li concentration

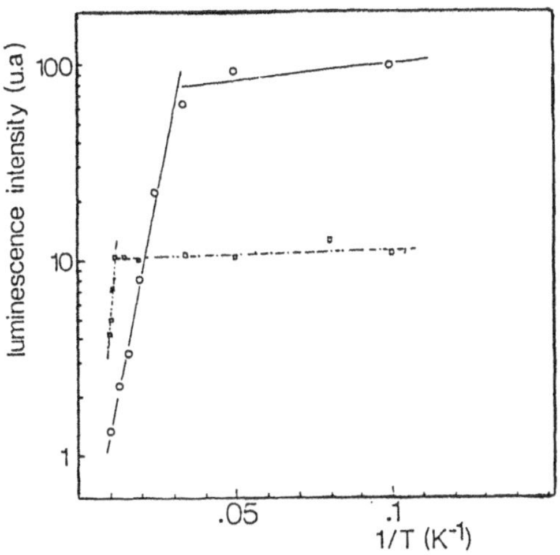

Fig.3 Luminescence intensity versus inverse temperature of new peaks appeared upon intercalation

To investigate more deeply the electronic band structure modifications, absorption measurements were carried out in the region of the first edge at 1.35 eV. Similar beheviours to the photoluminescence transitions are observed. The excitonic peak also shifts towards higher energies and the shift is of the same order as the one observed in emission spectrum. Figure 4. represents the temperature dependence of the energy position of the free excitonic peak of pure and Li intercalated InSe. It is often convenient to approximate the temperature dependence of the excitonic position with the simple analytical expression[7] :

$$E_T = E_o - B \left(\exp (\hbar\omega_p/kT) - 1 \right)^{-1}$$

where T is the sample temperature and $\hbar\omega_p$ is the phonon energy which interacts. Taking one phonon energy of 118 cm^{-1} related to the lowest A_2'' optical mode the fitting gave the parameters E_o=1334.6 meV, B=74.7 meV and E_o=1339 meV B=82.2 meV for pure and Li intercalated material respectively. Intercalation can act to noticiably alter the nature and the dispersion of some of the levels of the host material. The Γ_4^- transitions consist of antisymmetric LCAO's involving p_z orbitals while the deeper valence band Γ_9^- involves p_x and p_y. It is indicated that Se p_x and p_y electrons have much greater Coulomb attraction with In cations than Se p_x electrons do. Consequently the strongest interaction will take place between the Li 2s electrons and the outer electrons of the Se.

Figure 5. shows the Raman spectra of pure and Li intercalated InSe at room temperature. The Li intercalaton was performed by the electrochemical technique using the galvanostatic method and the Li dose is then x=0.2 mole. All spectra are recorded using a vacuum cell, with low laser excitation (50 mW). The resonant Raman spectra show that the peaks at 116 cm^{-1}, 199 cm^{-1} and 228 cm^{-1} i.e vibrational modes along the z-axis, are strongly in resonance when the incident photon energy is in coincidence with the E_1' exciton energy at 2.53 eV. These observations are in good agreement with the work of Kuroda et al. at 77 K^6.

After intercalation, the exciton energy decreases and, as a consequence, we observe a resonant magnification of the above phonon modes when the laser excitation at 530.8 nm is used. Measurements at low temperatures confirm these results.

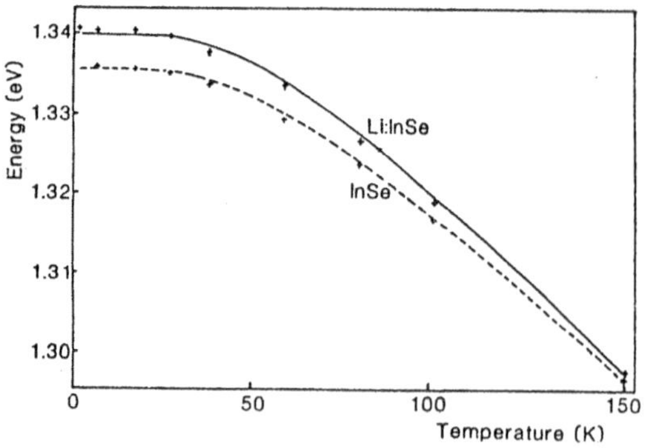

Fig.4 Energy position of the exciton versus temperature of pure and Li intercalated InSe.

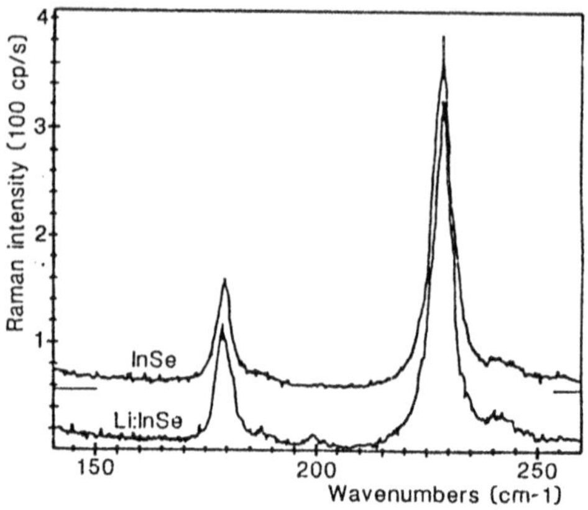

Fig.5 Raman spectra at RT.

CONCLUSION

Raman spectroscopy confirms the photoluminescence measurements regarding the E_1' exciton energy shifts with respect to the lithium intercalation dose. No new modes appear in the Raman spectrum.

These two observations agree well with the hypothesis that the intercalated lithium atoms are situated into the interlayer space. These atoms do not disturb the intralayer vibrational properties. However, the electronic states are modified mainly on the uppermost valence bands. We have to consider not only the Li intercalation with the host, but also the effect of the increase of the z-axis. The decrease in splitting of the energies of the orbitals probably indicates a decrease of anisotropy. We therefore suggest that at least two mechanisms act at the same time : (i) one of crystallographic origin, due to the increase of the c-axis lattice parameter upon intercalation, (ii) one of electronic origin due to the Li-InSe interaction. As these mechanisms are of different physical origin, they may act in a coherent or a counter-balancing way.

REFERENCES

1. See for example : Intercalation in layered materials, M.S.Dresselhaus Ed., NATO-ASI Series, B148, Plenum Press, N.Y.(1986) and references therein.
2. C.Julien, E.Hatzikraniotis, K.M.Paraskevopoulos, A.Chevy and M.Balkanski, Solid State Ionics, 18-19:859 (1986)
3. C.Julien, K.M.Paraskevopoulos, M.Jouanne and M.Balkanski, Ext.Abstr. of 3rd Int.Meeting on Lithium Batteries (Kyoto, Japan, 1986)p.307.
4. C.Julien and E.Hatzikraniotis, Materials Letters, 5:134 (1987)
5. N.Kuroda, I.Munakata and Y.Nishina, Solid State Commun., 33:687 (1980)
6. N.Kuroda and Y.Nishina, Solid State Commun., 34:481 (1981)
7. J.Camassel, P.Merle, H.Mathieu and A.Chevy, Phys.Rev., B17:4718 (1978)

ELECTROCHEMICAL LITHIUM INCORPORATION IN InSe THIN FILMS

I.Samaras, M.Tsakiri and C.Julien

Laboratoire de Physique des Solides, associé au CNRS
Université Pierre et Marie Curie
4, place Jussieu 75252 Paris Cedex 05, France

Indium selenide films prepared by flash evaporation were studied as possible cathodes in lithium batteries. Electrical and electrochemical properties are presented for films obtained under different conditions. Intercalation processes have also been investigated using the potential sweep voltametry and the influence of the host structure has been observed.

INTRODUCTION

All-solid microbatteries are currently being investigated for use in microelectronic devices[1]; recent studies have shown that the use of indium selenide as the cathode material is possible[2]. During the discharge process, this material serves as an ionic conductor for the cation Li^+; this property along with the high values of electrical conductivity, lithium diffusivity and electrochemical potential relative to lithium, make InSe a quite favourable material for battery operation. Furthermore, lithium intercalation in InSe has proved to be reversible over many charge-discharge cycles in laboratory experiments.

In a previous paper[3], we reported the structural and optical properties of In_xSe_{1-x} thin films. It has been established that the flash evaporation technique leads to good InSe stoichiometry with acceptable reproducibility. This is an important factor for solar energy conversion, and is connected with lithium micro-battery operation using the same material.

In this work, we present the electrical properties and the electrochemical behaviour of InSe thin films prepared under different conditions, i.e. as a function of substrate temperature (T_s) and thermal annealing, the results being used as a feedback to optimize the conditions of their preparation.

ELECTRICAL PROPERTIES

Electrical resistivity and Hall measurements were performed using the five-probe technique. Ohmic contacts were provided by evaporation of In spots on the films. The polycrystalline films were formed by subsequent thermal annealings with T_a ranging between 300K and 600K, in a purified flowing argon atmosphere. All flash evaporation films were initially n-type semiconductors.

Figure 1 shows the influence of annealing temperature (T_a) on the Hall mobility which increases consequently with T_a. This behaviour is mainly due to the crystallite size formation in the film. Temperatures of about 600K

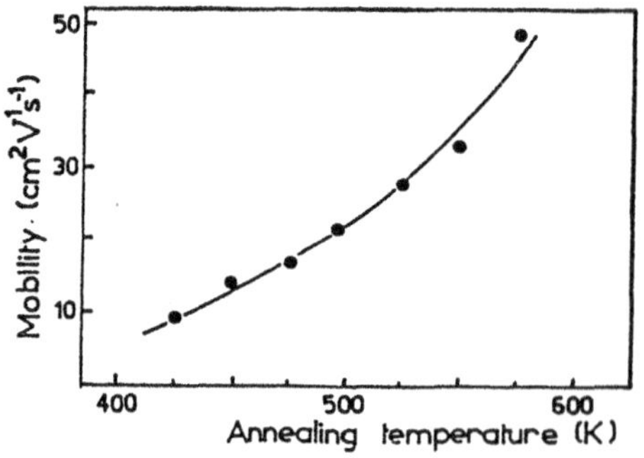

Fig.1 Hall mobility versus annealing temperature for InSe thin films prepared at T_s=433K.

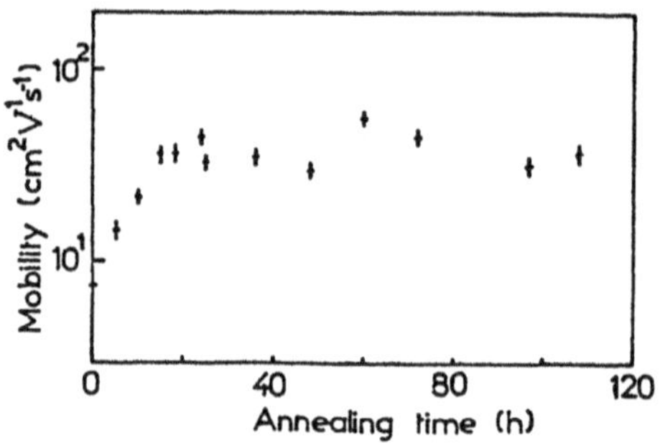

Fig. 2 Hall mobility versus annealing time for InSe thin films prepared at T_s=433K.

seem to be an upper limit, because higher annealing temperatures induce chemical reactions. The apparent decomposition can be attributed to the evaporation of the Se. The film's Hall mobility remains weak compared to the crystal's, which can reach 1000 cm^2/Vs. Figure 2 represents the variation of the Hall mobility with the annealing time: the optimum value appears after few hours. After 15 hours of annealing, the Hall mobility remains stable and constant at 50 cm^2/Vs, while the carrier concentration does not change notably.

Figure 3 represents the conductivity of InSe thin films as a function of the inverse absolute temperature, for different annealing treatments. The electrical conductivity shows an exponential behaviour as a function of temperature and the activation energy decreases as annealing temperature increases. The room temperature conductivity increases by three orders of magnitude when the annealing temperature rises from 300 up to 573K; for T_a>573K the electrical conductivity remains constant.

The electrical transport properties of InSe films are governed by carrier trapping at the grain boundaries[4]. To simplify the model, we assume that the films are composed of identical crystallites having a grain size L, and that only one type of totally ionized impurity is present. The energy band diagram in the crystallites, is calculated using an abrupt depletion approximation. For a higher potential barrier height the values of the theoretical Petritz mobility[4] are in good agreement with the experimental values, because the average grain size, measured by electron microscopy, is comparable to the depletion zone width, which is of the order of 20-80 nm, depending on the thermal treatment.

ELECTROCHEMICAL PROPERTIES

The InSe films evaporated on pyrex are electrochemically active, giving higher initial EMF than that of the bulk InSe. Galvaniç cells with the following geometry: Li // $LiClO_4$ in PC // InSe, were constructed. Each cell, containing three different cathodes with the same Li reference electrode, gave initial EMF values:

a. film InSe prepared at T_s=433K: EMF=3.3 V

b. film InSe prepared at T_s=RT : EMF=3.4 V

c. bulk InSe cathode: EMF=2.9 V

The insertion of Li into the InSe films was studied first by a galvanostatic electrochemical method using a three-electrode cell; lithium metal was used for both counter and reference electrodes. An Apple IIe computer was used to control the current pulse and to monitor both timing and cell voltage. As a cathode for this experiment, we used an InSe film prepared at T_s=433K and annealed for 50 h at 473K. Each step was consisted of a current pulse I_p=1 μA during 435 min, which corresponds to x=0.05 mole of Li

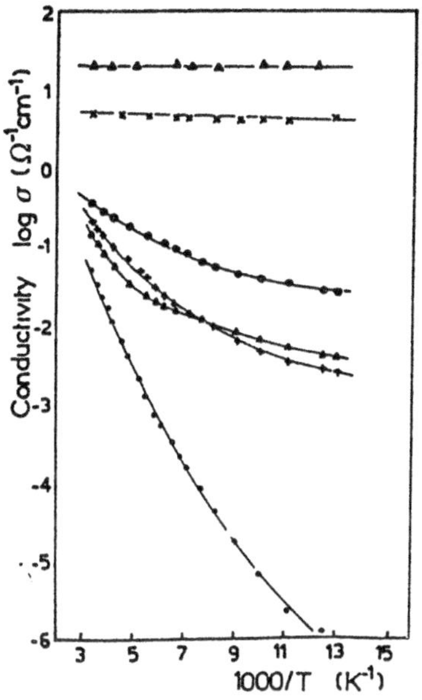

Fig.3 Electrical conductivity versus absolute inverse temperature for films annealed at:(•) none, (+) 493K,(Δ) 513K,(o) 533K, (×) 553K,(Δ) 573K.

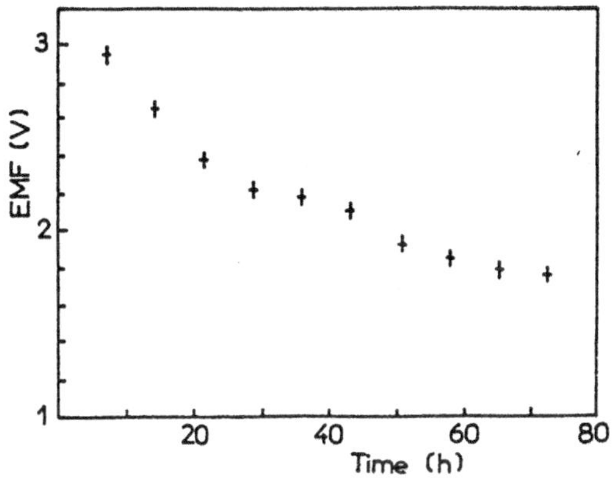

Fig.4 EMF versus time for a film InSe cathode prepared at T_s=433K and annealed for 50 h at 473K.

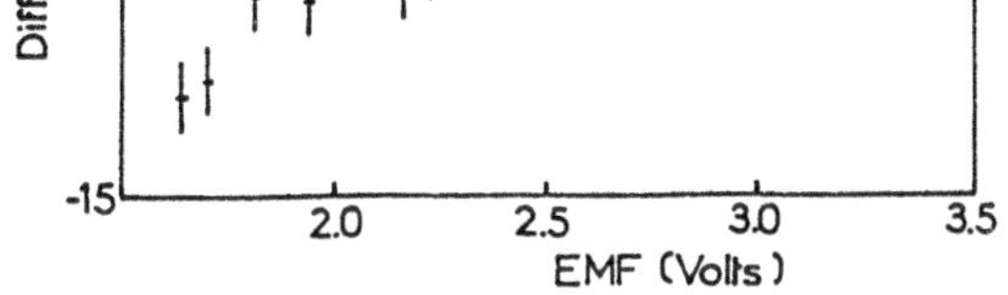

Fig.5 Chemical diffusion coefficient versus cell's EMF.

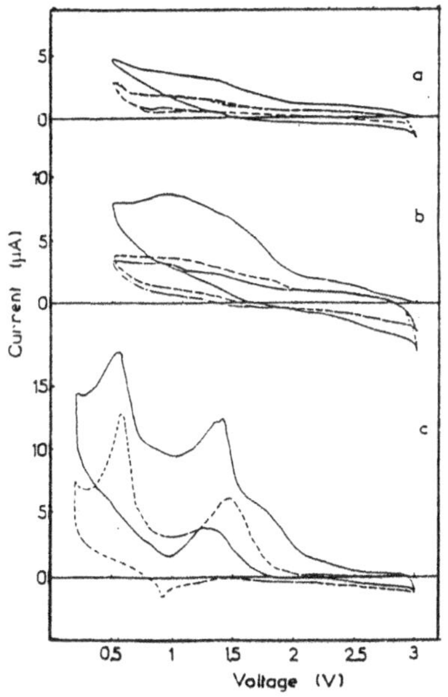

Fig.6 Linear sweep voltagrams for cathode cells: (a) a-film (b) b-film, (c) bulk InSe for (—) 1st cycle, (--) 2nd cycle and (-•-) 3rd cycle.

for the estimated cathode mass. When the cell's voltage became 0.5 V lower than the last EMF value, the current pulse was switched off for 5 min. Finally the predifined charge (x=0.05) was transferred to the cathode and the cell was left for relaxation, while the computer was recording the cell's voltage up to the point where the slope in OCV became less than 1 mV/h. The EMF of this cell for subsequent steps is presented in fig. 4. The open circuit voltage versus lithium curve, shows that a plateau at 1.7 V appears when the discharge depth reaches the value 0.5. The cathodes appeared unchanged after electrochemical cycling and no cracking or exfoliation was observed. Although no long term cycling studies were performed, the thin film InSe material was charged and discharged several times with no apparent lack of capacity.

In order to investigate the cause of the cell's polarization, the diffusivity of lithium within InSe was determined as a function of lithium composition. Using the constant current pulse method[5], a plot of overvoltage versus $1/\sqrt{t}$ allows a calculation of the chemical diffusion of Li. At the end of each step we applied a current pulse of 1 μA lasting 10s and we recorded the relaxation voltage of the cell. Figure 5 represents the Li diffusivity in InSe film annealed at 433K for 50 h as a function of open circuit voltage. The highest $\tilde{D}$ is 4 orders lower than the corresponding value of bulk InSe.

For comparison between the InSe films prepared at T_s=RT and T_s=433K, we used the linear sweep voltametry (LSV) technique[6]. In fig. 6 we observe that the films present a behaviour similar to the bulk InSe, with current peaks which are weak due to the higher cathode resistivity. We also observe that the diffusion coefficient of the films prepared at T_s=RT is lower than of those prepared at T_s=433K. On the other hand, the behaviour of the films at T_s=RT upon LSV cycling seems to be better regarding the difference in current level from the first to the second and the third LSV cycle.

CONCLUSION

InSe films prepared by flash evaporation were examined as possible cathodes for room temperature secondary micro-batteries. The electrical and electrochemical properties of the films were used as a feedback to optimize the conditions of their preparation.

The electrical measurements suggest that the InSe films which are prepared at T_s=433K, become "better" cathodes compared to those prepared at T_s=RT, because they have two orders higher conductivity; they also present an improved electronic mobility, when they are subjected to an annealing treatment at T_a=473K up to 10 h. The films' grain sizes obtained here are small, predominately due to the scattering processes associated with the microcrystalline nature of the films.

The electrochemical investigations suggest that the InSe films are electrochemically active with an EMF higher than that of the bulk InSe. However, the Li diffusivity is much lower (4 orders of magnitude) and it is still diminiched during intercalation. When the films are prepared at T_s=433K (a-samples), they seem again to be "better" cathodes, over those with T_s=RT; we observe that the a-samples have higher current in the LSV measurements, higher lithium diffusivity and appearence of weak current peaks, similar to those of the bulk material.

REFERENCES

1. K. Kanehori, K. Matsumoto, K. Miyanchi and T. Kudo, Solid State Ionics, 9-10:1469 1983).
2. C. Julien and E. Hatzikraniotis, Materials Letters, 5:134 (1987).
3. J. P. Guesdon, C. Julien, M. Balkanski and A. Chevy, Phys. Stat. Sol. (a), 101 (1987) in press.
4. R. L. Petritz, Phys. Rev., 104:1508 (1956).

5. S. Basu and W. L. Worrell, in "Fast Ion Transport in Solids", P. Vashishta, J. N. Mundy and G. K. Shenoy ed., North-Holland, Amsterdam (1979), p. 149.
6. J. R. Dahn and R. R. Haering, Solid State Ionics, 2:19 (1981).

ON THE LITHIUM UPTAKE BY THE n-BUTYL-LITHIUM INTERCALATION METHOD

E. Hatzikraniotis, K. Chrissafis and N.A. Economou

Solid State Physics Dept. (313-1)
University of Thessaloniki
54006 Thessaloniki - Greece

The direct reaction with n-butyl lithiumhas been proved as an effective and easy to use way of Li intercalation to a variety of layered host structures (1). This method has been succesfully applied to most of the members of the transition metal dichalcogenides (TX_2), as well as to the MPX_3 (2) and III-VI family compounds (3,4). However this potential method has a major disadvantage. The total amount of the inserted Li is not directly controlled.

One of the foundamental aspects of intercalation is the swelling of the layer separation while there is a negligable (if any) change in the interatomic distances whithin the layer sandwich. This is the basis of the justification that the quasi 2-D band structure of the host material (InSe) is preserved (Rigit Band Approximation (RBA) (5).

Electrical properties however, are drastically affected due to the varriation of the number of the free carriers. We have examined the conductivity of the In-Se family compounds (3,4). We have observed an enhancement of the n-type character and have concluded that the inserted Li act as mobile donnors. In the present work we shall combine the change of conductivity upon lithiation by butyl-lithium to the Li uptake. Our discusion will be based on the classical Diffusion Theory.

The samples used in this work were prepared by splitting the InSe single crystals grown in our laboratory by the Bridgman-Stockvarger method from direct fusion of high purity elements using 5% excess of In as proposed by Chevy (6). Depending on the growth conditions we were able to obtain a variety of n-type samples with different initial free carrier concentrations.

The Li intercalation of InSe was effected by immersing large samples in 1.6M solution of butyl-lithium in hexane. During the intercalation process the conductivity of the samples was monitored as described elsewhere (3,4). It is interesting to note that despite the initial differences all the samples examined end up with approximately the same value of conductivity (fig. 1).

The room temperature (RT) mobility before intercalation, was of about the same value ($\mu = 205$ cm^2/Vs) for the three different samples pressented, while the free carriers were varied from $5 \cdot 10^{14}$ to $2 \cdot 10^{15}$ cm^{-3}. After intercalation, the mobility slightly decreased ($\mu = 180$ cm^2/Vs. We thus conclude that the S type variation of the conductivity upon time, is mainly due to the increase of the free carrier concentration.

According to the RBA model we are not expecting any drastical changes on the band scheme of InSe upon intercalation. The lowest conduction band at the Γ point of the Brillouin zone, (Γ3) is partially occupied in the pure material. As the excess carriers are filling up the Γ3 band, the

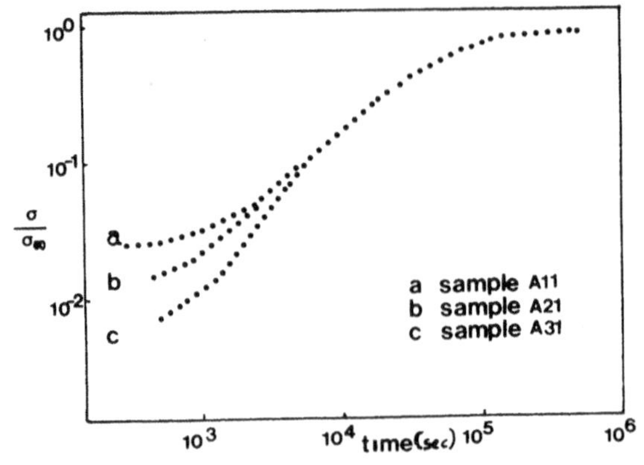

Fig. 1. Time evolution of the normalized conductivity ($\sigma(t)/\sigma(t\to\infty)$) of different samples upon the lithiation by butyl lithium.

increase of their effective number (and hence the increase of the conductivity) is directly related to the total amount of the inserted Li.

Our samples were initially free of lithium. As the lithiation reaction is spontaneous, samples will uptake Li which will be intitially accumulated at the close to the interface region. Lithium uptake is dominated by the Li kinetics into the host. As the time passes, the concentration profile, will decay into a uniform Li distribution all over the bulk of the sample. Denote Co, to be the equilibrium Li concentration.

The rate of Li exchange at any instant, depends on the concentration of the solution in the vincinity of the sample, and the Li concentration at the sample/solution interface. The simplest reasonable assumption is that the rate of Li exchange is directly proportional to the difference between the actual Li concentration, Cs, at the interface at any time, and the equilibrium concentration C_∞. The boundary condition of the 2-D diffusion equation at the sample-solution interface is

$$-D\ (dC/dr) = \alpha\ (Cs - Co) \tag{1}$$

where r is the radial coordinate and D is the diffusion constant. The constant of proportionality α expresses the rate of exchange and mainly depends on the concentration of the solution and on the electronic properties at interfacial level. Within the framework of the classical diffusion theory the total amount of diffusing substance Mt, entering the sample over the time period (0-t), expressed as a fraction of M_∞ (i.e. the amount related to the equilibrium concentration Co), is given by (7).

$$\frac{Mt}{M_\infty} = 1 - \frac{4\alpha^2}{b^2\ (b^2 + \alpha^2)} \exp\ (-b^2 Dt/R^2) \tag{2}$$

where b are the roots of

$$b\ J_1(b) - f\ J_o(b) = 0 \tag{3}$$

where $J_o(b)$, $J_1(b)$ are the Bessel functions of the first kind and zero and first order respectively and the factor f is related to the radial dimension of the sample (R) and to the diffusion constant ($f = \alpha R/D$). In figure 2 we present the calculated uptake fraction vs. the dimensionless parameter $\tau = Dt/R^2$ for different values of the factor f.

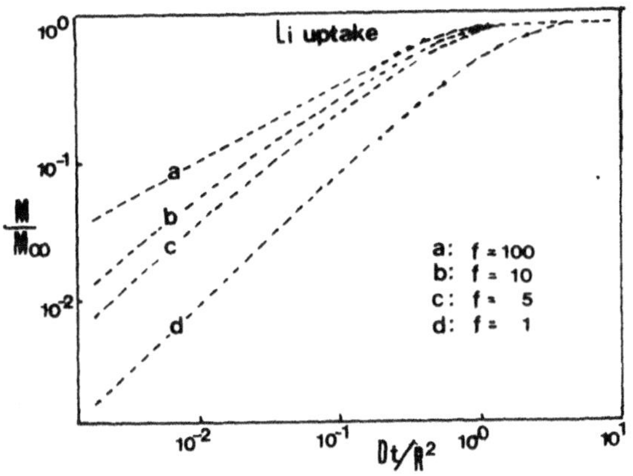

Fig. 2. Calculated Li uptake (eq. 3) for different values of the factor f.

The shape of the time evolution of the conductivity $\sigma(t)$ is very close to the calculated lithium uptake even to our rough approximation ($f = 1$). The excact fitting of the experimental data to equation 3 and hence the direct estimation of the degree of lithiation, consists of a NLLS fit of the two parameters of interest (f,D). When the value of the diffusivity D is known (which is generally the case), and alternative way is to rescale the actual time values to the characteristic "time" $\tau = Dt/R^2$. The NLLS problem is then reduced in finding the value of the factor f that fits the experimental data (fig. 3).

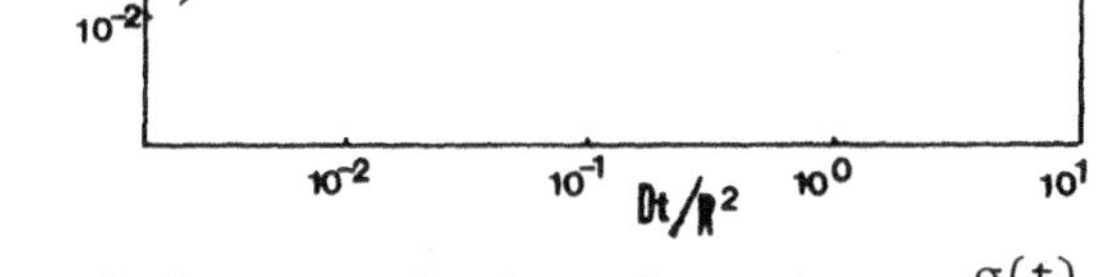

Fig. 3. Evolution of the normalized conductivity $\dfrac{\sigma(t) - \sigma(t=0)}{\sigma(t\to\infty) - \sigma(t=0)}$. Solid line represents the theoretical Li uptake (eq. 2,3). The actual experimental time has been rescaled to Dt/R^2, taken into account the measured value of the diffusivity from ref. (10).

Concluding this pressentation, we have shown that insitu conductivity measurements can be easily combined to the classical Diffusion Theory to estimate the degree of the lithium uptake. This method can be applied when the inserted Li significantly alters the free carrier concentration and has a minor effect to the carrier mobility at least at the temperature of application (usually the room temperature). The method is destructionless and offers the simplicity and the ease of application as compared to optical (4,8) or elaborated radioactive tracer (9) methods which are also sensitive to the degree of lithiation.

REFERENCIES

1. M.S. Withingham et. al., Electrochem. Soc. Meeting, Dallas Texas (1975).
2. D.W. Murphy, F.J. DiSalvo, G.W. Hull and J.V. Waszczak, Inorg. Chem. 15, 17 (1976).
3. C. Julien, E. Hatzikraniotis, A. Chevy and K. Kambas, Matt. Res. Bull, 20, 287 (1985).
4. C. Julien, E. Hatzikraniotis, K.M. Paraskeyvopoulos, A. Chevy and M. Balkanski, Solid State Ionics, 18-19, 859 (1986).
5. W.Y. Liang in "Intercalation in Layered Materials", Eds. M.S. Dresselhaus, Plenum Press, 1986.
6. A. Chevy, PhD.Thesis, University of Paris V, (1981).
7. J. Crank, "The Mathematics of Diffusion", Oxford Univ. Press. (1967), p.p. 72.
8. K.M. Paraskevopoulos and E. Hatzikraniotis in "Intercalation in Layered Materials", Eds. M.S. Dresselhaus, Plenum Press, 1986, p. 161
9. R. Muhlbewrger, T. Butz and A. Lerf. Physica, 105b, 218 (1981).
10. J. Samaras, C. Julien and M. Balkanski, in this volume.

IN SITU X-RAY DIFFRACTION STUDIES OF INTERCALATION BATTERIES

C. Levy-Clement

CNRS, Laboratoire de Physique des Solides
1, Place Aristide-Briand
92195 Meudon, Principal Cedex, France

INTRODUCTION

Beside their technological usefulness as high energy density batteries, intercalation batteries can be used to study the process of intercalation itself. They can be used in in situ experiments such as in situ X-ray diffractometry. In this technique a specially constructed electrochemical cell is charged or discharged while an X-ray diffraction profile is taken. When high resolution measurements of the voltage of the cell versus the amount of the intercalated species are done simultaneously with the X-ray diffraction experiments, correlation between the two different sets of results permits the physics and chemistry of the intercalation process to be studied.

ELECTROCHEMICAL INTERCALATION

The principle of a lithium intercalation cell is very simple and is shown in Fig. 1. The cell is composed of an intercalation cathode made of the Li_x Host compound and an anode made of Li metal immersed in an organic electrolyte in which a Li Salt is disolved.

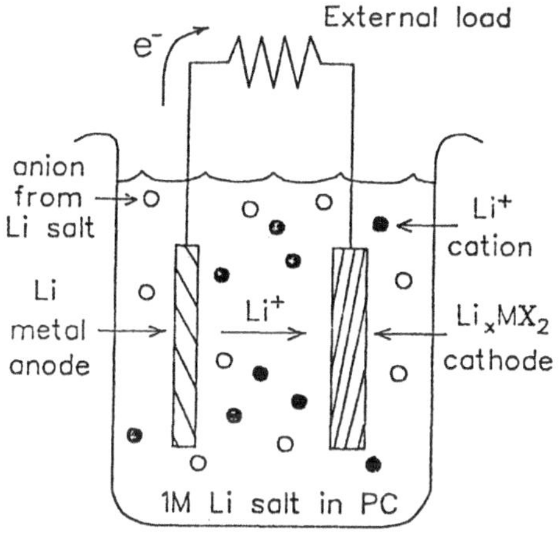

Fig. 1. The electrochemical cell.

Under discharge or charge the electrochemical reactions at the electrodes are the following:

At the anode $Li \Rightarrow Li^+ + e^-$
At the cathode $xLi + e^- + host \Rightarrow Li_x\ Host$

The Li^+ ions migrate through the electrolyte and electrons through the external circuit. The voltage difference between the two electrodes is given by

$$v(x) = 1/e\ (\mu_a - \mu_c\ (x)\ )$$

where μ_a and μ_c (x) are the chemical potentials of the lithium atoms in the lithium metal anode and the Li_xHost cathode respectively, e is the magnitude of the electronic charge. Because μ_a is constant the voltage of the cell directly measures the chemical potential of the lithium in the Host and thus permits to study the process of intercalation itself [1-4].

Gross features of the variation of the voltage with lithium concentration x, in reversible intercalation batteries v(x) give already much information on the mechanisms of intercalation in a particular sytem, such as the filling by the guest species of different cristallographic sites or the existence of two phase mixtures. However to observe smaller coexistence ranges it may be necessary to look at the derivative $- \delta_x\ /\delta_v$. From theoretical analysis it may be possible to correlate the shape of $-\delta_x/\ \delta_v$ with one kind of intercalation mechanism. For example a sharp peak at a certain voltage and composition may indicate the presence of an order disorder transition of the intercalant [5-6]. In all cases measurements of δ_x/δ_v serve to characterize the intercalation system more precisely than V(x) and allow a fine structure in v(x) to be seen more easily, as it has been shown for the Li/Li_xTiS_2 system (Fig.2) and other systems [7-10].

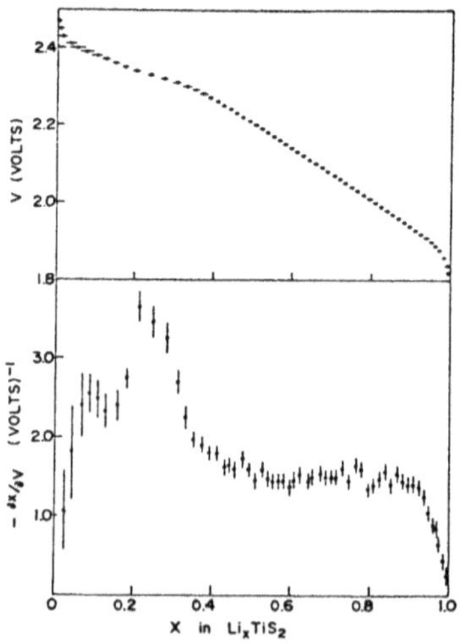

Fig. 2. The V(x) and $-\Delta x/\Delta V$ behaviour of Li_xTiS_2 cells.

Derivative constant current chronopotentiometry, in which the cell voltage is measured as a function of time at constant charge or discharge current and then numerically differentiated, has appeared as a valuable technique to obtain - δ_x/δ_v [7].

IN SITU X-RAY DIFFRACTION

To be interpreted, electrochemical measurements must be supported by other experimental studies. X-ray diffraction studies provide information on the crystal structure of the Host intercalation lattice. This information is of great importance, since the structure determines the number and nature of available sites where the guest species may be situated. It is also of great interest to know what changes in the Host lattice occur upon intercalation, such as expansion or contraction of the lattice and changes in the symmetry of the Host lattice. Many difficulties, that arise when classical X-ray diffraction is used, are avoided with the In situ X-ray diffractrometry technique. An unique electrochemical cell which incorporates a beryllium window makes possible simultaneous measurements of high resolution voltametry and X-ray diffraction during the charge and discharge processes. The main advantage is that quantitatively comparative studies may be done as all the measurements are performed on the same sample. The variation of the width of a given bragg peak upon intercalation can be measured, as the apparition of Bragg peaks with a weak intensity due to a new cristalline phase may be appreciated. Also formation of poorly crystalline phases due to intercalation may be detected because original peaks decrease and brooden out.

The first In situ X-ray measurements were done by CHIANELLI et al to study Li_xTiS_2 [11]. Later on a well design X-ray diffraction electrochemical cell where rapid equilibration of the entire intercalation electrode can be obtained was published by DAHN and HAERING [12]. The schema of this cell is shown in Fig.3.

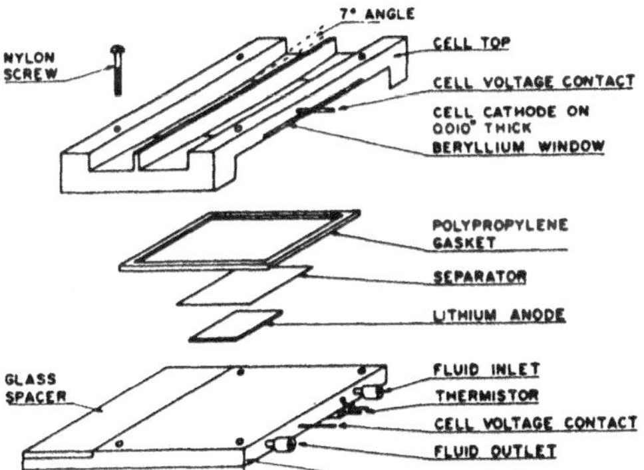

Fig. 3. "Exploded" view of the electrochemical cell used for in situ X-ray diffraction studies of lithium intercalation compounds.

The base of the cell is made of stainless steel and the cell top holds a 0.25 mm thick beryllium foil which is transparent to X-rays. The electrochemical cell is constructed in the region between the beryllium top and the cell base and is sealed by a 0.40 mm polypropylene gasket. The anode is a lithium foil. The intercalation compound is electrically connected to the beryllium window. Electron transfert between the electrodes inside the cell is prevented by the separator between the two electrodes. Before used this separator is soaked in the electrolyte which is an organic solvant with a lithium salt disolved. Voltage contacts are made on top and base of the cell. All cell assembly and lithium handling have to be carried out in a dry glove box. The X-ray diffraction electrochemical cell constructed in this way is totally air tight.

The problems associated with this type of cell include contamination of the diffraction profile with unwanted peaks (Be, Li...), prefered orientation of the cathode particles and cell alignment. These three problems are discussed in details in ref. [12]. This paper also gives several experimental results obtained with Li/Li_xTiSe_2 cells, which illustrate the uses of the in situ X-ray diffraction technique. These results include lattice changes in Li_xTiS_2 upon intercalation, the study of dynamic effects such as solid state diffusion and electrolyte co-intercalation [12].

RESULTS

Several other intercalation systems have been investigated using this in situ X-ray diffraction technique [9,13-18].

Structural changes observed by lithium or sodium intercalation using in-situ X-ray diffraction measurements have been observed for the following systems : Li $/Li_xMoS_2$ [13], Li $/Li_xNbSe_2$ [9], Li/Li_xZrS_2 [14], $Li/Li_xMo_6Se_8$ [15-16], $Li/Li_xRu_zMo_{6-z}Se_8$ [16-17] and $Na_xMo_6X_{8-y}I_y$ (x = S,Se and y = 0 to 2) [18]. The anomalies observed in the electrochemical curves (v (x) and- $\delta x/d_v$ vs v) correlate with the observed structural changes.

Electrolyte co-intercalation corresponds to the simultaneous intercalation in the Host of the alcali-ion and of its solvation cloud. Such a phenomenon can be observed for propylene carbonate based electrolyte in Li_xMX_2 systems.

Co-intercalation leads to a two phase region and this phenomenum is observed as a plateau in the voltage curve [19]. Co-intercalation has been detected by in situ X-ray diffraction studies of intercalation of lithium in TiS2, ZrS_2, $ZrSe_2$, TaS_2 [20,19, 21, 10]. Fig.4b shows a portion of a profile of X-ray spectrum of Li/Li_xZrS_2, in 1M $LiAsF_6$ propylene carbonate, during the first discharge. The variation of the voltage is represented in the Fig. 4a. In that case co-intercalation arises and a slow growth of a set of Bragg peaks associated with the electrolyte co-intercalated material is noticed on X-ray patterns, while the structure of the Host compound becomes amorphous. No shift or broading with time of the Bragg peaks of the Host lattice is observed.

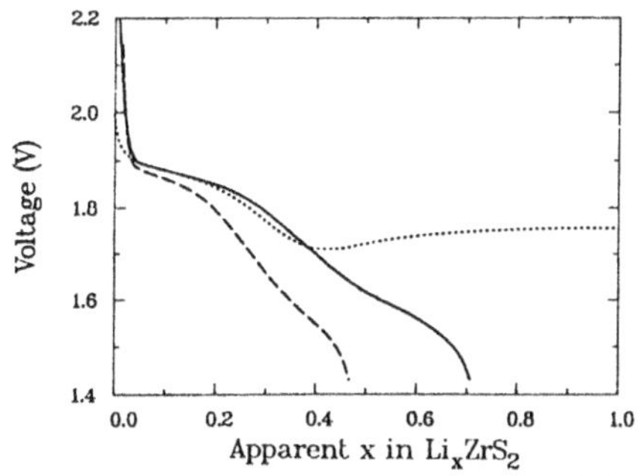

Cell voltage V vs. the apparent value of x in Li_xZrS_2 for the first discharge of Li/Li_xZrS_2 cells. (· · · ·): 1M $LiAsF_6$/PC electrolyte, 25h rate. (———): Saturated $LiAsF_6$/PC electrolyte, 50h rate. (– – – –): Saturated $LiAsF_6$/PC electrolyte, 18h rate. The apparent value of x was calculated only from the current and the mass of ZrS_2. The rate quoted corresponds to the time to change the apparent x by $\Delta x = 1$.

Fig. 4a.

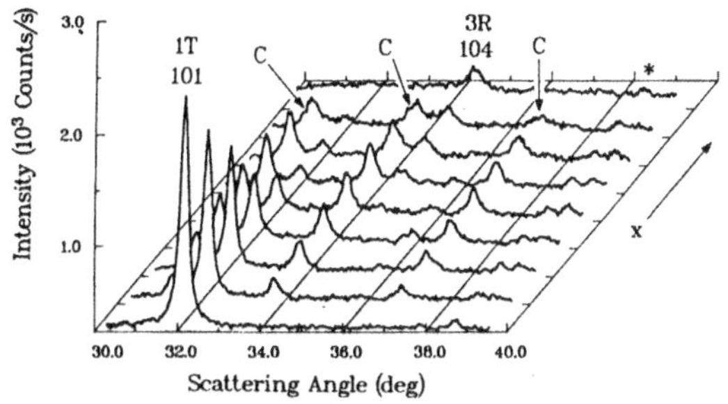

As in Fig.5,. except for cells with 1M $LiAsF_6$/PC electrolyte at a 40h rate (The last scan is from a different cell than the other eight). Starting from the front, the apparent x is 0.07, 0.12, 0.18, 0.24, 0.30, 0.35, 0.41, 0.51, 2.04. In addition to the peaks in Fig.5, are peaks from cointercalated ZrS_2 (labeled as C).

Fig. 4b.

In this case co-intercalation is an irreversible process. Use of a Li salt saturated electrolyte, as saturated $LiAsF_6$ propylene carbonate reduces co-intercalation. In (Fig.5b) is shown portion of a profile of X-ray spectrum of Li/Li_xZrS_2 where co-intercalation does not occur. A phase transition is then observed between ZrS_2 (1T structure) and Li_xZrS_2 (3R structure). Coexistence of the two phases 1T(ZrS_2) and 3R($Li_{0.23}ZrS_2$) for $0<x<0.23$ corresponds to the plateau in the voltage curve in Fig. 5a.

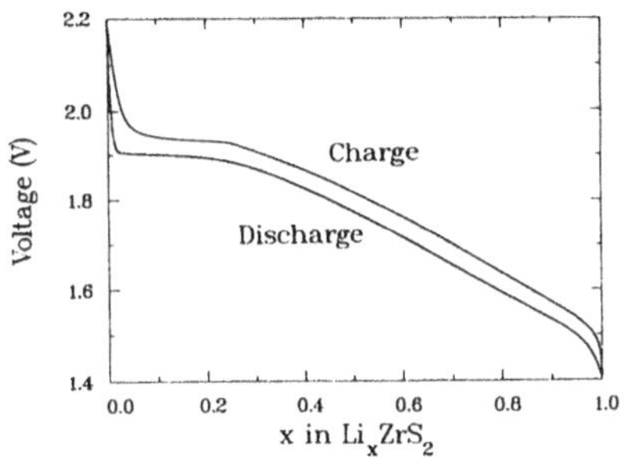

V(x) for Li/Li_xZrS_2 cells

Fig. 5a.

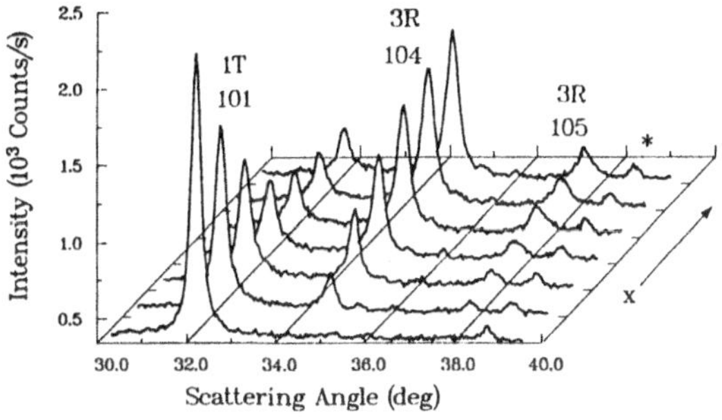

X-ray diffraction spectra for ZrS_2 during the first discharge in a Li/Li_xZrS_2 cell with saturated $LiAsF_6$/PC electrolyte. The cell was discharged at constant current at a 25h rate. The apparent x for the scans, starting from the front, is 0.03, 0.12, 0.21, 0.29, 0.38, and 0.47. The 101 peaks from 1T-ZrS_2 and the 104 peak of 3R-Li_xZrS_2 are shown, together with a peak from some component of the cell (*).

Fig. 5b.

Recent improvements to the in situ X-ray diffraction technique has allowed **high precision measurements of lattice parameters changes** in lithium intercalation compounds. The method proposed by DAHN and McKINNON has been experimented to study the Li/2H-Li_xTaS_2 system (22).

With classical X-ray diffraction, which involve the use of separate samples for each composition of lithium, the precision attained in the measurements of variation of lattice parameters vs lithium concentration is about 1 per cent (largely due to uncertainty in the lithium composition).

With in situ X-ray diffraction technique when measurements are done on homogeneously intercalated samples prepared by fixing the cell voltage v and waiting for equilibrium, the precision within the variation of lattice parameters vs lithium is one part in 10^3.

Improvements are obtained when position, intensity and width of selected Bragg peaks are recorded while the cell is charged or discharged at constant current. A second computer monitors the cell voltage so that structural changes can be correlated with the features in v(x). Bragg peaks under study are "tracked" as intercalation proceeds. The X-ray data are collecting by measuring X-ray intensity versus scattering angle at n points in the vicinity of several Bragg peaks. Near the maximum of the peaks, the variation of intensity with angle is parabolic. The position, height and width of each peak are determined from the parameters of a least squares fit of a quadratic polynomial to the m data points closest to the maximum of each peak. The position of each peak is then used as the center of the next n point scan over that peak. This method leads to a relative precision of one part in 10^5 for the variation of the lattice parameters.

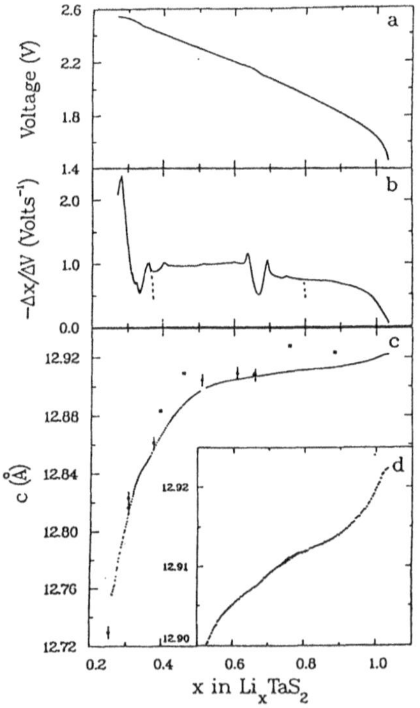

Fig. 6. a) Voltage and b) -dx/dV of Li/2H-Li_xTaS_2 cells and c) c(x) of 2H-Li_xTaS_2 vs x. d) shows an enlargement of some of the data in c). The data were collected during 3 partial charges of the cell at an 89 hour rate. The dashed portions in b) were measured before the cell reached steady state after the current started.

Experiments on 2H-Li_xTaS_2 demonstrate the sensitivity and reproducibility of the method. The results are reported in Fig. 6 which shows v(x), -dx/dv and C(x) for 2H-Li_xTaS_2 measured simultaneously on the same cell. Anomalies observed in the three curves near x = 1/3 and x = 2/3 are presumably due to order-disorder transitions in the intercalated lithium. At x = 1/3 and x = 2/3 the lithium is ordered and occupies one or two of the three V3a x V3a triangular superlattices of octaedral sites, respectively, within each Van der Waals gap [7-10-23]. The anomalies of c(x) may be related to the stacking of these occupied superlattices ; the stacking sequence is not known and has to be clarified.

Reaction of lithium with the Host lattice may occur instead of intercalation. Voltametric and in situ X-ray diffractometry studies have shown that this is the case of Li/InSe system [24]. The first plateau in v(x) between 0<x<2 (Fig. 7a) corresponds to the transfer

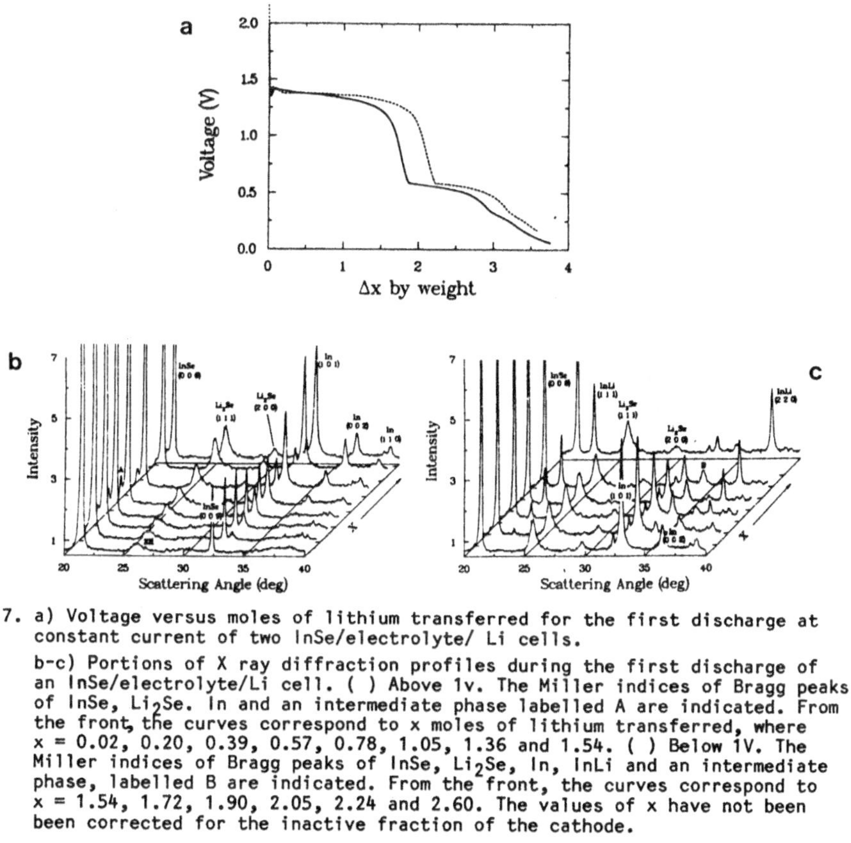

Fig. 7. a) Voltage versus moles of lithium transferred for the first discharge at constant current of two InSe/electrolyte/ Li cells.
b-c) Portions of X ray diffraction profiles during the first discharge of an InSe/electrolyte/Li cell. ( ) Above 1v. The Miller indices of Bragg peaks of InSe, Li_2Se. In and an intermediate phase labelled A are indicated. From the front, the curves correspond to x moles of lithium transferred, where x = 0.02, 0.20, 0.39, 0.57, 0.78, 1.05, 1.36 and 1.54. ( ) Below 1V. The Miller indices of Bragg peaks of InSe, Li_2Se, In, InLi and an intermediate phase, labelled B are indicated. From the front, the curves correspond to x = 1.54, 1.72, 1.90, 2.05, 2.24 and 2.60. The values of x have not been been corrected for the inactive fraction of the cathode.

of 2 lithiums to the InSe cathode per mole of InSe. X-ray spectra show, that Li_2Se and In are formed, and, also the formation of a phase Li_xInSe that is either metastable or is an intermediate step in the decomposition reaction (Fig. 7b). The second plateau at 0.5 v corresponds to the reaction of Li with the In to form InLi. During the first discharge an intermediate phase (labelled B) is observed which may be some In-Li alloy (Fig. 7c).

CONCLUSION

Intercalation of guest species into Host compounds may proceed by means of electrochemical reactions using test cells in which the electrodes are made of the Host compound and the guest species. The cell voltage V (x) measured as a function of time, while the cell is charging or discharging at constant current rate, gives us direct thermodynamic data from which interesting structural and electronic information can be deduced. When the voltametric technique is combined with in situ X-ray diffractometry, an extremely powerful experimental method for the investigation of phase diagrams of intercalation compound is obtained. The crucial factor in a sucessful application of the technique is a good electrochemical cell design. In this paper, the principle and sensitivity of the technique has been described. Experimental results when the guest species are Lithium or Sodium has been presented to illustrate the uses of the in situ X-ray diffraction, for studies like high precision measurement of the variation of lattice parameters, electrolyte cointercalation, staging, phase diagram investigations.

REFERENCES

[1] M.S. Whittingham, Prog.Solid.State Chem. 12 : 41 (1978)
[2] M. Armand, Thèse d'Etat, Grenoble (1978)
[3] A. Le Mehaute, Thèse d'Etat, Paris (1979)
[4] W.R. McKinnon and R.R. Haering in "Modern Aspects of Electrochemistry", ed. Plenum, New-York, 15 : 235 (1983)
[5] A.J. Berlinsky, G.W. Uruh, W.R. McKinnon and R.R. Haering, Solid State Commun., 31 : 175 (1979)
[6] R. Osorio and L.M. Falicov, J.Phys.Chem.Solids, 43 : 73 (1982)
[7] J.R. Dahn and W.R. McKinnon, J.of Electrochem.Soc., 131 : 1823 (1984)
[8] A.H. Trompson, Phys.Rev.Lett., 40 : 1511 (1978)
[9] D.C. Dahn and R.R. Haering, Solid State Commun., 44 : 29 (1982)
[10] J.R. Dahn and W.R. McKinnon, J.Phys.C Solid State Phys., 17 : 4231 (1984)
[11] R.R. Chianelli, J.C. Scanlon and B.M.L. Raoj, J.Electrochem.Soc. 125 : 1563 (1978)
[12] J.R. Dahn, M.A. Py and R.R. Haering, Can, J.Phys. 60 :307 (1982)
[13] M.A. Py and R.R. Haering, Can, J.Phys., 61 : 76 (1983)
[14] W.R. McKinnon, J.R. Dahn and C. Levy-Clement, Solid State Commun. 50 : 101 (1984)
[15] J.R. Dahn, W.R. McKinnon and Coleman, Phys.Rev. B 31 : 484 (1985)
[16] W.R. McKinnon (*in this volume*)
[17] W.R. McKinnon and L.S. Selwyn, Physics Rev. B 35 : 7275 (1987)
[18] J.M. Tarascon, G.W. Hull, P. Marsh and L. Ter Haar, J.of Solid State Chemistry (1986)

[19] W.R. McKinnon and J.R. Dahn, J.Electrochem.Soc. 132 : 364 (1985)
[20] J.R. Dahn, Thesis of University British Colombia, Vancouver B.C. (1981)
[21] J.R. Dahn, W.R. McKinnon and C. Levy-Clement, Solid State Commun. 54 : 245 (1985)
[22] J.R. Dahn and W.R. McKinnon, Solid State Ionics (1985)
[23] W.R. McKinnon and J.R. Dahn, Solid State Comm. 48 : 43 (1983)
[24] C. Levy-Clement, J. Rioux, J.R. Dahn and W.R. McKinnon, Mat.Res.Bull., 19 : 1629 (1984)

NEW TRENDS IN ALKALI METALS ELECTROCHEMICAL INTERCALATION INTO GRAPHITE AND G.I.C's IN SOLID STATE CELLS

Rachid Yazami

Laboratoire d'Ionique et d'Electrochimie du Solide
(UA 1213 CNRS) E.N.S.E.E.G.,
B.P. 75, 38402 Saint Martin-d'Hères (France)

INTRODUCTION

Electrochemistry is one of the most important fields in which graphite intercalation compounds (GIC's) have reached an industrial scale of applications. In the year 1985, more than 50 million primary lithium batteries using poly-(graphite fluoride) $(CF)_n$ as cathode active material were manufactured in Japan by Matsushita Battery Industrial Co. and commercialized around the world[1]. The market is constantly increasing and must respond to an higher demand, particularly for small consumer items (LCD calculators, quartz watches, cameras, ...)

Besides this spectacular success in the electrical energy storage field, electrochemistry provides a powerful tool for more basic studies concerning ions electrochemical intercalation (I.E.I) into graphite (G) and GIC's. Many techniques based on voltage and current measurements allow the determination of crucial data concerning the host electrode material behavior towards intercalation (i.e. reaction thermodynamics : ΔG, ΔH, ΔS, stage formation..., kinetics: chemical diffusion coefficient D_A , interface properties, ...). These data are of great importance for battery performance knowledge.

The purpose of this article is to give some basic aspects about I.E.I. into G and GIC's and present some recent results concerning sodium and lithium intercalation in solid state galvanic cells using polymer electrolyte. This will concern two domains of applications : new synthesis and energy storage.

BASICS ON ION ELECTROCHEMICAL INTERCALATION (I.E.I)

Mechanism of I.E.I.

A simple picture showing the different steps of an I.E.I. into graphite is depicted in figure 1: at the electrode / electrolyte interface an electrostatic double layer (here // to the c-axis) is formed as a consequence of adsorbed ions which are electrically balanced by charges from graphite planes. Taking the graphite band structure[2] into account, the charges are electrons e^- for adsorbed cations A^+(donor-type), and holes h^+ for anions X^- (acceptor-type).

The electrode potential E_o (Volt) can be related to the molar chemical potential μ_i of the adsorbed ions which expresses the tendancy to give a reduced form for A^+ and an oxidized one for X^- following the equilibrium reactions :

$$A^+ + e^- \underset{\text{oxidation}}{\overset{\text{reduction}}{\rightleftharpoons}} (A^+, e^-) \qquad (1)$$

or

$$X^- + h^+ \underset{\text{reduction}}{\overset{\text{oxidation}}{\rightleftharpoons}} (X^-, h^+) \qquad (2)$$

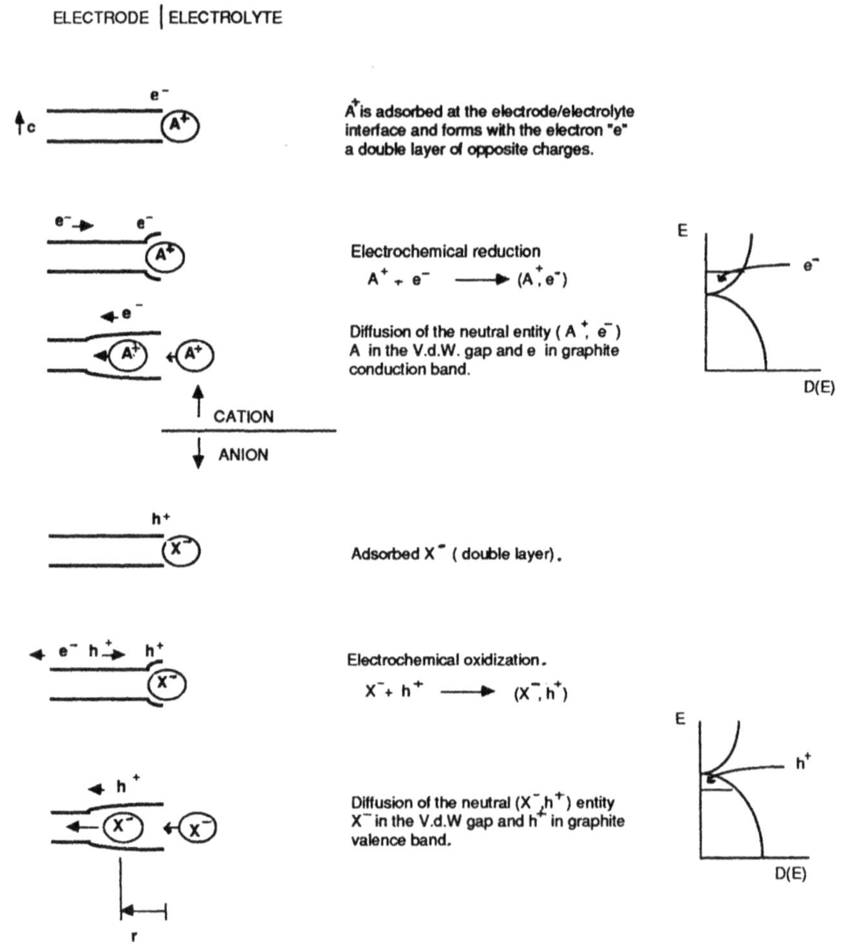

Fig. 1. Schematic representation of ion electrochemical intercalation into graphite (the charge transfer on the graphite electronic bands is illustrated on the right).

We have the relation :

$$E_o = - \mu_i / F \quad + \quad \text{constant} \qquad (F = \text{Faraday constant}) \qquad (3)$$

Switching the electrode potential to lower values for A^+ or higher values for X^- leads to a charge transfer at the interface and thus to the formation of the neutral entities (A^+, e^-) or (X^-, h^+).

Ideally, the diffusion of (A^+, e^-) or (X^-, h^+) becomes effective if the two following stability conditions are fulfilled : a) No internal recombination through a reduction or oxidation reaction for (A^+, e^-) or (X^-, h^+) respectively. This supposes that (A^+, e^-) and (X^-, h^+) do not lead to other more stable reduced or oxidized forms respectively. The choice of cations and anions should then be limited to highly reduced forms as alkali metals or tetra-alkylammonium[3-5] or highly oxidized and delocalized anions as ClO_4^-, HSO_4^-, PF_6^-, BF_4^- ... [6] . b) No reaction with the electrolyte which must have a sufficient electrochemical stability domain upon reduction for (A^+, e^-) and oxidization for (X^-, h^+). Polar organic electrolyte such as DMSO (dimethyl sulfoxide), DME (1,2-dimethoxy ethane) or PC (propylene carbonate) [3-5] show relative good stability for A^+ intercalation, whereas very concentrated acid solutions can be used for ClO_4^- or HSO_4^- [7].

The recent emergence of polyethers solid electrolytes (POE-like : PEO = poly-(ethylene-oxide)) particularly developed for their use in solid state batteries[8,9] , should bring a new tool for I.E.I. into graphite and derivatives and this for three main reasons i) ion move free from any solvent molecules and so one avoids electrolyte co-intercalation as generally observed in liquid media. Binary GIC's with lithium[10] and potassium[11] has been synthetized in PEO based solid state cells whereas in DMSO media, Besenhard [4] has obtained ternary alkali metal-DMSO-GIC's. ii) the wide electrochemical stability domain (0 to 4 V vs Li^+ / Li), so PEO is particularly suitable for alkali metals intercalation. iii) As elastomer, PEO is adapted when intercalation induces a volume variation (i.e. electrode swelling), thus the electrode / electrolyte interface mechanical properties are conserved during the intercalation.

The last step of I.E.I. is the neutral species (A^+, e^-) or (X^-, h^+) diffusion within the graphite structure which offers available paths for ions in the Van der Waals gap with concomitant c-axis expansion, and electrons or holes in the graphite conduction or valence bands respectively. This mixed conductivity (ions + electrons or holes) is a key property for graphite and GIC's and gives them good I.E.I. electrode properties [12,13].

Another important feature of these materials is the remarkable stability of the carbon atoms hexagonal array towards important charge transfer which corresponds to their strong reduction in $C_6^- Li^+$ [10] and oxidation in $C_{24}^+HSO_4^- (H_2SO_4)_y$ [14]. It should be noticed however, that over-oxidation can lead to irreversible formation of C-O bonds with a modification of the hybridization nature of carbon atoms ($sp^2 \longrightarrow sp^3$). As a consequence, a decrease of the electronic conductivity is observed (i.e. graphite oxide formation [15]).

Thermodynamics

As shown above, cation and anion electrochemical intercalation proceeds in a symmetrical way. Therefore this section will be limited only to alkali cations , A^+.

A typical electrochemical cell contains three electrodes, including a reference one to which the working intercalation electrode potential is compared. However as a first approximation, we shall consider the activity of the alkali metal electrode as invariant during the cell operation and so it constitutes a reference electrode in the electrochemical chain of the type :

$$(-)\ A\ /A^+, X^-\ \text{Electrolyte}\ / A_x H\ (+) \quad (H = \text{host structure : G or GIC})$$

The open circuit voltage E_{th} (OCV) measured between the extreme electrodes is related to the difference of the chemical potential of the alkali metal A (μ_A) in the pure metal A and when intercalated in H to form $A_x H$:

$$E_{th} = - \Delta \mu_A / F = - (\mu_A(A_x H) - \mu_A(A)) / F \qquad (4)$$

In $A_x H$, x denotes the relative part of the maximum amount of intercalate A in the host structure, following the cathode reaction :

$$x A^+ + xe^- + H \longrightarrow A_x H \qquad (0 < x < 1) . \qquad (5)$$

It is then tempting to relate μ_A (H) to x for E_{th} determination.

μ_A(H) is the sum of an ionic contribution $\mu_A{}^+$ (H) and an electronic one $\mu_e{}^-$ (H) = $F\Phi$ (Φ = inner electrode potential).

$$\mu_A(H) = \mu_A{}^+ (H) + F\Phi \qquad (6)$$

$\mu_A{}^+$(H) can be related to the chemical activity $a_A{}^+$(H):

$$\mu_A{}^+(H) = \mu^o{}_A{}^+(H) + RT\, ln\ a_A{}^+(H) \qquad (7)$$

In a solid state solution model, Armand [16] has shown that E_{th} has three terms : a standard term E_o, a configuration term (entropic), and an interaction term (ion-ion and ion-lattice) which is assumed to be proportional to x :

$$E_{th} = E_o - \frac{RT}{F}\, ln\, \frac{x}{1 - x} - f x . \qquad (8)$$

The validity of this latter equation has been confirmed by our measurements in lithium cells using transition metal chlorides- GIC's [17] and graphitic oxide as cathode material [18].

However, from Gibbs' phase rule, one should expect that potential / composition curves present one or several plateaus either when cation intercalation is followed by a displacement reaction consistent with the irreversible apparition of a new phase (i.e. Li / CF [19]) or when intercalation proceeds by formation of successive stages as found in Li / LiC_6 [10] and K / KC_8 [11] systems.

From E_{th} temperature dependence, it is possible to determine the electrode entropy variation ΔS corresponding to precise x value using the relation:

$$\Delta S = F\, \partial E_{th} / \partial T \qquad (9)$$

ΔS is related to possible site occupancies of both ions and electrons and is very sensitive to ordered phase formation during the intercalation process.

Energetics

The maximum theoretical energy W_{th} that the battery can provide to the external circuit is given by the relation :

$$W_{th} = \int E_{th}\, F\, dx \qquad (10)$$

Generally W_{th} is related to the total mass or volume of active material (anode+ cathode) during the battery operation and one thus defines the specific energy and volumic energy densities (in $Wh.kg^{-1}$ or $Wh.dm^{-3}$) respectively . In practical use, a loss of energy is observed as a consequence of electrolyte ohmic drop and electrodes polarizations as well as of lower coulombic yield ($x < 1$). It is customary to define the energy yield as the ratio of practical energy W_{pr} to W_{th} . It should be pointed out that covalent type GIC's such as graphitic oxide of general formula $C_nO_xH_y$ and fluoride $(CF)_x$ constitute very performent cathode materials in organic liquid lithium batteries and energy densities as high as 1800 $Wh.kg^{-1}$has been experimentally reached[19,20].

Kinetics

An important characteristics of electrochemical methods is the direct determination of the intercalation reaction kinetics by measurement of the current i . Assuming that reaction (5) is the sole cathodic process, we have the relation:

$$i = -F\, dx / dt \qquad (11)$$

Current density is related to the charged particles flux and is obtained from the ratio of i to the electrode / electrolyte active surface area (here $\perp$ to c- axis). The characterization of this latter interface geometry is of in importance in the kinetics of intercalation.

As mentioned above, mass transport from the electrode / electrolyte interface to the bulk of the electrode is governed by the diffusion of the neutral specie (A^+, e^-) under the chemical potential gradient wich can be deduced by derivation of eq. (6) and (7) :

$$\text{grad}\, \mu_A(H) = \text{grad}\, \mu_{A^+}(H) + F\, \text{grad}\, \Phi = RT\, \text{grad}\, \ln a_{A^+}(H) + F\, \text{grad}\, \Phi \qquad (12)$$

Thus, in the ion transport process two terms should be distinguished : a) the chemical activity gradient related to the ion concentration $[A^+]$ variation at the interface electrode / electrolyte and to the activity constant γ ($a_{A^+} = \gamma [A^+]$) b) the internal electrical field (grad Φ) mainly due to the inhomogeneous density of the charge transfer in the electrode. For sufficiently electron conducting materials such as donor or acceptor-type GIC's, grad Φ is small ($< 10^{-4}$ V. cm^{-1}) and can be neglected.

The chemical diffusion coefficient D_{A^+} can be determined by several electrochemical techniques i.e. for the most classical : galvanostatic or potentiostatic intermittent titration as first developped by Weppner et al.[21], or by impedance spectroscopy [22, 23]. Concerning Li^+ and K^+ intercalation into graphite, we have found $D_{Li^+} \sim 10^{-8}$ to 10^{-7} [10] and $D_{K^+} \sim 10^{-7}$ to 10^{-6} cm^2.s^{-1} [11] using galvanostatic intermittent titration technique (GITT).

EXPERIMENTAL

Electrochemical cell

The experimental cell is represented in figure 2. It consists of a button-type battery, sealed in argon glove box with less than 2 v.p.m. $H_2O + O_2$ content. High purity alkali metal is cut into a disk of approximatly 0.5 mm thickness and 17 mm diameter, then pressed on a stainless steel disk of the same diameter which constitutes the electron collector.

The cathode is obtained by spreading a composite suspension of a very fine powder of graphite or $(CF_y)_n$ (~ 3 µm in diameter) and P(EO)-AX in acetonitrile on a stainless steel disk of 20 mm diameter. Once the solvent is evaporated (under primary vacuum at 80°C), the composite cathode contains 50 % vol. of P(EO)-AX ; it is approximatly 10-20 µm thick and 5 -10 mg in weight.

The solid electrolyte, $P(EO)_n$ AX, is a very thin film (~ 100µm) cut in a disk of 21 mm in diameter and pressed in between the alkali metal and the composite electrode to insure an ionic contact between them. The cell is operated at a constant temperature (80°C ± 1°C) under argon atmosphere.

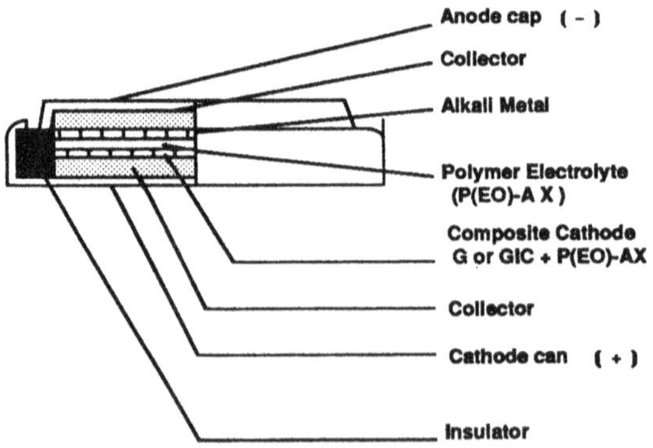

Fig. 2. Experimental solid state electrochemical cell schematic drawing.

Sodium electrochemical intercalation

Although it is well known that most alkali metals form rich first stage GIC's (i.e. LiC_6 , MC_8 : M = K, Rb, Cs), sodium shows surprisingly low reactivity during its chemical intercalation into graphite. Metrot et al.[24] have demonstrated that richest compounds (mixture of 4 th to 6 th stages) are obtained when the reaction temperature is lowered to 150°C with extended duration (many months). At higher temperature (400° C) , only the 8th stage of composition NaC_{64} is obtained.

The electrochemical intercalation of sodium by cathodic graphite reduction has been performed in liquide DMSO electrolyte by Besenhard [4] and led to the formation of a ternary Na-DMSO-GIC. OCV measurements has been performed by Aronson et al.[25] at 300°C on a liquid sodium / Na^+ glass electrolyte / $C_{64}Na$ cell. This has permitted the determination of some thermodynamical data concerning sodium intercalation reaction.

In this preliminary study, we will describe the graphite cathodic behaviour during sodium intercalation in the electrochemical chain :

$$(-)\ \ Na / P(OE)_{12}\, NaCF_3SO_3 / C\ \ (+)$$

following the reaction :

$$xNa^+ + xe^- + C \longrightarrow Na_xC \qquad (\text{ or } NaC_n\ (n = 1/x)\,) \qquad (13)$$

A constant current i ~ 2 μA is maintained during τ = 1 hour, then released for 3 hours. The cell voltage E is recorded during the two phases and in figure 3 we have plotted $E_i(\tau)$ (curve a) and $E_0(4\tau)$ (curve b) corresponding respectively to the E values

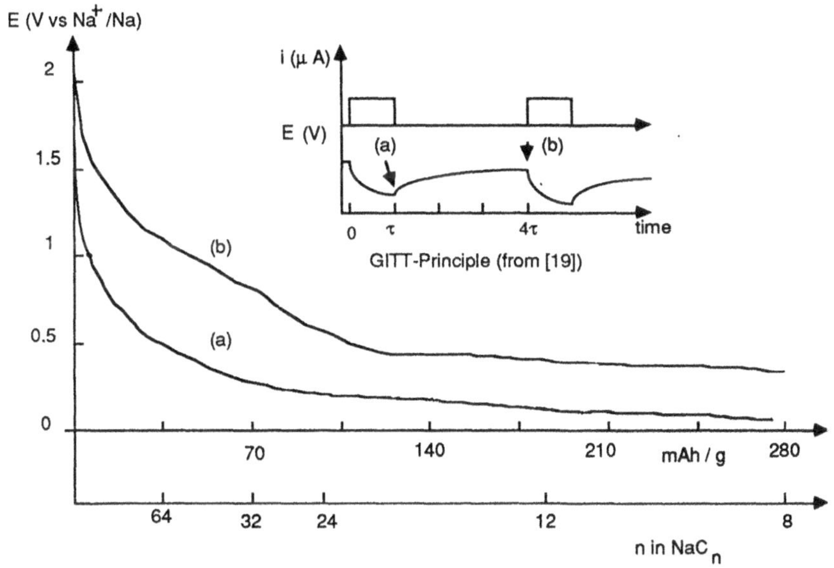

Fig. 3. $Na/P(EO)_{12}NaCF_3SO_3/Na\ C_n$ cell GITT-curves, (a) undercurrent flux (b) after 3h of relaxation.

reached at the end of each phase as functions of the total specific charge transfered to the graphite cathode (in $mAh.g^{-1}$). It should be noticed that $E_0(4\tau)$ is not an equilibrium OCV, since steady state is reached only after 48 hours. This is indicative of a very slow diffusivity of Na into graphite. However, it shows small curvature variations at n = 32 and n ~ 16 which could be the signature of 4th and 2nd stage formation.

From transient ($0 < t < \tau$) and relaxation ($\tau < t < 4\tau$) curves analysis, we have found some important differences in the electrode process when compared with Li [10] and K [11] : sodium intercalation proceeds at least in two steps which could be attributed to the diffusion phenomena (with $D_{Na^+} \sim 10^{-11} cm^2 s^{-1}$) followed by a phase nucleation (slower process). Note that D_{Na^+} (80° C) and D_{K^+} (59° C) measured by the same electrochemical technique differ by four orders of magnitude ($D_{Na^+} << D_{K^+}$). This emphasizes the role of kinetics in the alkali metals intercalation into graphite.

We have also studied the temperature dependence of the steady state cell voltage E_{th} (T) when n = 8 in NaC_n. As shown in figure 4, a linear dependence is found between the temperature limits of sodium melting (~98° C) and PEO vitreous transition (~ 65° C). The slope is - 5 $mV.K^{-1}$ and corresponds, from eqn (9) to a constant entropy variation of NaC_8 : ΔS = - 480 $J.K^{-1}$ (Na-g. atom)$^{-1}$.

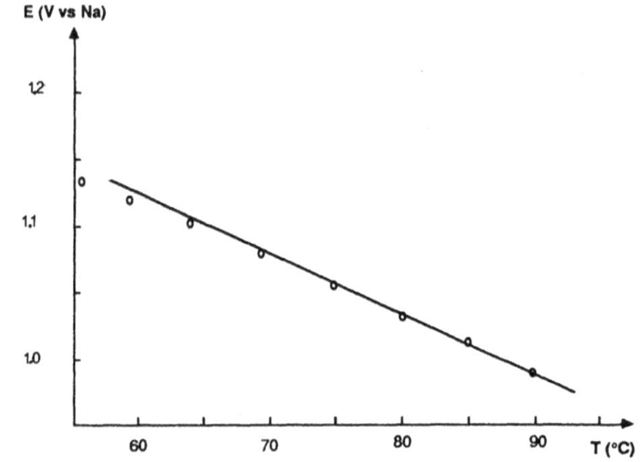

Fig. 4. Na/P(EO) $NaCF_3SO_3/C_8Na$ cell OCV temperature dependence, $\partial S/\partial T$ = 5mVK^{-1}.

Evidence for sodium intercalation is obtained from the X-ray diffraction analysis performed on the final composite electrode in which the theoretical cathode formula was NaC_8 . It shows the total disappearance of the prestine graphite characteristic 002 line at 3.35Å. However, no evidence of any stage formation has been found; some of the room temperature $P(OE)_{12}NaCF_3SO_3$ cristalline phase X-ray diffraction lines interfer with those of the cathode. Further investigations concerning the electrode process and cristallography are under way.

Lithium / R.T.graphite fluoride solid state cell

A new poly-(graphite fluoride) compound has been recently synthetized by Hamwi et al. [26] at room temperature (R.T.) using a preliminary graphite oxidation before fluorination by a gaseous catalyser. GIC's of this new family have the general formula : $CF_y(catalyser)_z$ with $y \sim 0.7$ and $z \sim 10^{-2}$. Unlike the high temperature $(CF)'_x$ which has a strong covalent character, R.T.-CF'_y shows a iono-covalent behaviour which explains the remaining planar arrangement of the carbon hexagons (sp^2 hybridization). The composite electrode conductance was enhanced by adding vol. 10 % of a thin powder graphite (~3 µm grain size) before the spreading operation.

The galvanostatic discharge curve (i = 35 µA, total discharge in 40 h) of the electrochemical cell : **Li / $P(OE)_8$ $LiClO_4$ / R.T.CF_y** is depicted in figure 5 for y = 0.66. The cathode reaction involves lithium intercalation following the equation

$$xLi^+ + x\,e^- + CF_y \longrightarrow Li_xCF_y \qquad (14)$$

It shows a remarkably high potential ($E_i > 2.9$ V) and coulombic efficiency ($x / y > 90$ %). This leads to a practical energy density of 1600 $Whkg^{-1}$ which is one of the highest ever reached in solid state PEO-based lithium batteries [9].

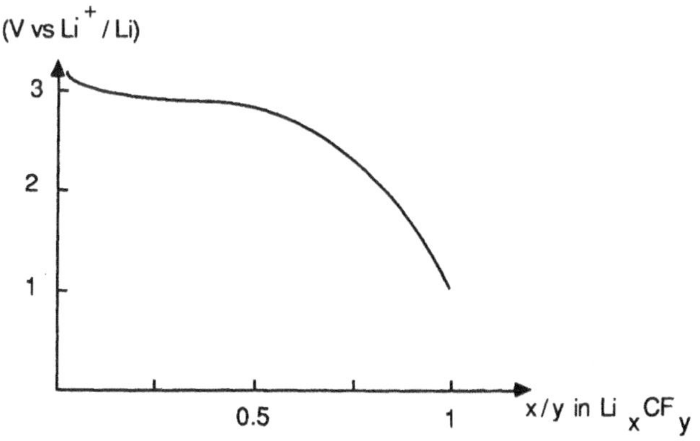

Fig. 5. Galvanostatic discharge curve. i=35µA. Cathode capacity - 1420 µAh composed of ~2 mg RT-$CF_{0.66}$+0.38 mg graphite + 1.24 PEO-$LiClO_4$.

CONCLUSION

We summarize here the most important characteristics of the electrochemical methods for studying ion intercalation into host lattices :

1) Determination of the amount x of intercalate ion.
2) Direct measurement of the rate of the electrode reaction (current measurement).
3) Possibility of observation of phase formation(OCV measurement).
4) Measurement of the thermodynamics and kinetics of the intercalation reaction.
5) Characterization of the electrolyte / electrode interface (a.c. measurement).
6) Synthesis of new compounds (or phases) which can not be obtained by other chemical methods.
7) Study of the reversibility (deintercalation).

The use of new and stable electrolytes such as solid state polyethers (i.e. PEO), and thin composite electrodes will widen the domain of electrochemical intercalation studies since the neutral electrolyte molecules co-intercalation is avoided This allows a more precise knowledge of the nature of intercalate species and the characterization of the host structure.

Our preliminary study concerning the preparation of binary sodium-graphite compounds at lower temperature (~80°C), shows the particular behaviour of this alkali metal when compared with Li and K, in both intercalation process and its kinetics.

For solid state lithium batteries applications, the new R.T.-CF_y has good cathode material characteristics, and because of its easier and cheaper preparation, it will constitute a good candidate for larger scale utilization.

AKNOWLEDGEMENTS

The author is indebted to Dr A. Hamwi from the Laboratoire de Chimie du Solide (Clermont-Ferrand, France) for kindly providing of R.T.-CF_y samples and for stimulating discussions.

REFERENCES

1. H. Ogawa, in "Proceeding of 3rd Int. Meeting on Lithium batteries", 27-30 May 1986, Kyoto (Japan), p. 19.
2. J. E. Fisher in "Physic and Chemistry of Materials with Layered Structures", F.A. Levy, ed. D. Reidel Publish. Co., Dordrech (Holland), 6 : 481 (1979).
3. J.O. Besenhard and H.P. Fritz, Electroanal. Chem.53: 329 (1974)
4. J.O. Besenhard . Carbon 14:111 (1976)
5. J.O. Besenhard, H. Möhwald, J.J.Nickl, Carbon18:399(1980)
6. see for exemple F. Beck, H. Junge and H. Krohn Electochem: acta 26:799 (1981), and for a more detailed review J.O. Besenhard and H.P. Fritz, Angew. Chem. int. ed. Engl.22: 590 (1983).
7. L.C.F. Blackman, J.F. Mathiews and A.R. Ubbelohde. Proc.R.Soc.A 258: 329 (1960)
8. M.B. Armand in "Solid State Batteries" ed. C.A.C. Sequeira and A. Hooper Martinus Nijhoff. Pub.63 (1985).
9. M. Gauthier, D. Fauteux, G. Vassart, A. Beranger, M. Duval, P. Ricoux, J.M. Chabagno, D. Muller, P. Rigaud, M.B. Armand and D. Deroo. J. Electrochem. Soc. 132: 1333 (1985).
10. R. Yazami and Ph. Touzain, J. of Power Sources, 9 : 365 (1983).
11. R. Yazami and Ph. Touzain, Solid State Ionics 9-10 : 489 (1983).
12. M. Armand and Ph. Touzain. Mater.Sci. Eng. 31: 319 (1977).
13. S. Flandrois. Synthetic metals 4: 255 (1982)
14. J.O. Besenhard and H.P. Fritz, Z. Naturforsch. B. 26: 1125 (1971)
15. J.O. Besenhard and H.P. Fritz, Z. Anorg. Allg. Chem. 416: 106 (1975)
16. M.B. Armand, in "Electrode Materials for Advanced Batteries", ed. Nato Adv. Studies, Plenum Press, New-York, 145 (1980).

17. Ph. Touzain and R. Yazami, in "Proceeding of Journées d'Electrochimie 1985", 28-31 Mai 1985, Florence (Italy), Abstr. n° 7-15 and Ph Touzain in "Colloque Franco-japonais sur les Composés d'Insertion du Graphite", Paris, 8-11 October 1985, Ann. Phys., 11 : 23 (1986).
18. M. Mermoux, R. Yazami and Ph. Touzain, J. of Power Sources (1987) under press.
19. N. Watanabe, in "Proceeding of 16th Biennel Conference on Carbon", July 18-22 , 1983, San Diego, p. 250. and H. Touhara, H. Fujimoto, N. Watanabe and A. Troussaud, Solid State Ionics 14: 163 (1984)
20. R. Yazami and Ph. Touzain, Synthetic Metals, 12:499 (1985).
21. W. Weppner and R.A. Huggins, J. Electrochem. Soc., 124 : 1569 (1977).
22. W. L.Arsher and R.D. Armstrong, Specialis Periodical Reports, The Chemical Society London, "Electrochemistry 7 ", 157 (1980).
23. L. Dawson and D.G. John, J. Electroanal. Chem. 110: 37 (1980).
24. A. Metrot, D. Guerard, D. Billaud and A. Herold, Synthetic Metals, 1: 363 (1979-1980).
25. S. Aronson and F.J. Salzano. Nucl. Sci. Eng. 38: 187 (1969).
26. A. Hamwi, M. Daoud and J.C. Cousseins (to be published).

NEGATIVES FOR SECONDARY Li-BATTERIES: Li-ALLOYS OR METALLIC Li ?

J.O. Besenhard, J. Gürtler and P. Komenda

Department of Inorganic Chemistry
University of Münster
Wilhelm-Klemm-Str. 8, D-4400 Münster

INTRODUCTION

Metallic lithium anodes as well as Li-alloy anodes are currently used in commercial rechargeable organic electrolyte batteries. The disadvantage of Li-alloys is of course low energy density and power density compared to metallic lithium; for some applications this disadvantage may be overbalanced by better cycling and storage behaviour of the alloys.
Metallic lithium is thermodynamically unstable in any organic electrolyte; the same is true for Li-rich Li-alloys. Both types of anodes can only be operated in organic electrolytes because of protective surface layers, consisting of inorganic and organic electrolyte decomposition products [1,2]. These layers have to be permeable to Li^+-ions and should be as impermeable as possible to all the other electrolyte components. Because unsolvated Li^+-ions are unique with respect to their extremly small diameter, the condition of selective permeability is not too difficult to fulfil - Li and Li-alloy anodes are fairly "stable" in various organic electrolytes.
Nevertheless, the bottleneck of rechargeable lithium batteries is still the poor cycling and storage behaviour of the negative electrodes. This may be surprising because i) Li-anodes in primary batteries can be stored for 10 years and more without substantial capacity losses and ii) large cycle numbers have been demonstrated to be possible with metallic Li and Li-alloy anodes.
The point is that the morphology of Li and Li-alloy anodes changes drastically after the first full capacity discharge/charge cycle. For various reasons compact starting materials are divided into smaller particles. Each individual particle gets covered with a Li^+-permeable but electronically insulating surface layer. Therefore, the electrical and mechanical contact between the particles is poor. Further cycling of the anode creates new reactive surfaces, i.e., the surface filming process doesn't come to an end and electrolyte decomposition products are accumulated on the anode. Because of these complications the storage behaviour of already used secondary Li-batteries is not well predictable. Large cycle numbers achieved in short-term experiments do not give much evidence for durability of a Li-battery.
There are two main routes to get around the problems of particle size reduction and surface area increase:

- o improvement of electrolytes to get less dendritic, compact depositions of lithium during charging,
- o "dimensionally stable" Li-alloy electrodes which do not change significantly their morphology with Li-content.

IMPROVED ORGANIC ELECTROLYTES FOR NONDENDRITIC Li-DEPOSITION

There are many papers dealing with the dependence of the morphology of electrodeposited lithium on the kind of electrolyte [2,3-6]. As a rule, electrolytes made up from low viscosity and low dielectric constant, i.e., from "low polarity" solvents such as ethers yield fairly compact Li-depositions. This compact morphology is paralleled by good cycling efficiency. On the other hand, "high polarity" and hence high viscosity and high dielectric constant solvents (typical example: propylene carbonate) usually cause dendritic Li-depositions and poor cycling efficiency.
There are some exceptions to the rule that Li^+-electrolytes based on high polarity solvents yield dendritic Li-depositions, e.g., N.N-dimethylformamide- and N,N-dimethylacetamide-propylene carbonate mixed electrolytes [6,7]. These electrolytes, however, do not form protective layers on lithium, i.e., lithium cannot be stored therein. The storage behaviour of lithium in ether electrolytes is also disappointing - in spite of the above mentioned excellent cycling efficiency.
We believe that dendritic deposition of lithium is strongly related with immediate formation of an almost completely covering surface layer during the deposition process. Nonuniformities of this layer "trigger" a stepwise onset of Li-deposition; this finally leads to dendrite growth. Organic Li^+-electrolytes which only slowly or not at all do form "electrolyte-tight" surface layers on lithium favour more compact Li-depositions - on the expense of poor long-term storage behaviour. On the other hand fast and effective surface protection, as it is provided by most of the "high polarity" solvents, favours dendrite formation.
This idea of dendrite formation mechanism is illustrated by Figs. 1a and 1b. Fig. 1a shows a relatively smooth deposition obtained in a $LiClO_4$/tetrahydrofuran(THF)-propylene carbonate(PC) electrolyte. Doping of this electrolyte with some SO_2 - which forms $Li_2S_2O_4$ surface layers in a fast reaction - drastically changes the morphology of deposited lithium from smooth to dendritic.
One possibility to reduce dendrite growth in "high polarity" solvents such as propylene carbonate is the use of chemically stable surface additives which are - reversibly - adsorbed on the lithium surface and <u>slow down</u> the - irreversible - formation of surface layers caused by electrolyte decomposition. However, as the adsorption energies are low compared to the energy of the reaction between electrolyte and lithium, this surface reaction is not finally prevented.
In organic electrolytes, hydrocarbons are significantly surface active [8,9]. Doping of $LiClO_4$/PC electrolytes with saturated hydrocarbons, especially with alicyclic hydrocarbons, has been shown to improve very much the morphology of electrodeposited lithium [6,10]. Due to the more compact morphology of the Li-depositions, the cycling behaviour of lithium electrodes in presence of hydrocarbon surfactants is significantly improved. This is demonstrated by Fig. 2, showing the anodic Li-recovery*) in dependence of cycle number during cycling of Li on a stainless steel substrate.

---

*) Li-recoveries were determined at a stainless steel grid hanging in a large reservoir of electrolyte. In close-packed configurations the Li-efficiencies are much better [6,11].

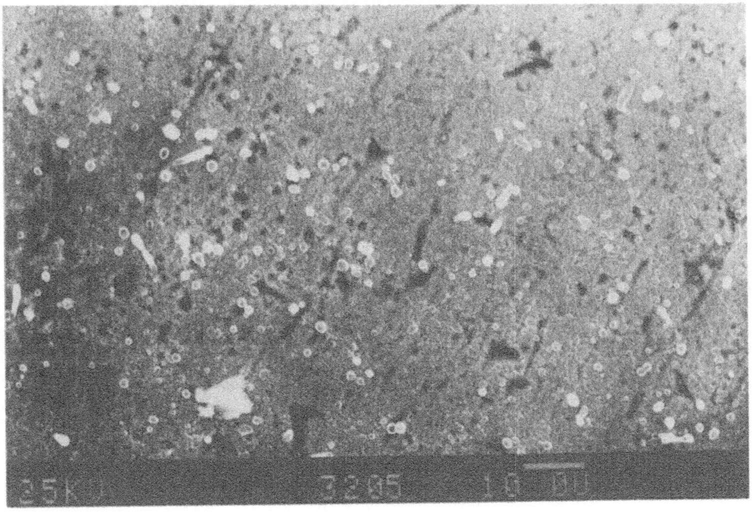

Fig. 1a. SEM micrograph of Li deposited on a Ni-substrate from 0.5 M $LiClO_4$/PC-THF (50-50), i = 2 mA/cm^2, Q = 3.6 C/cm^2.

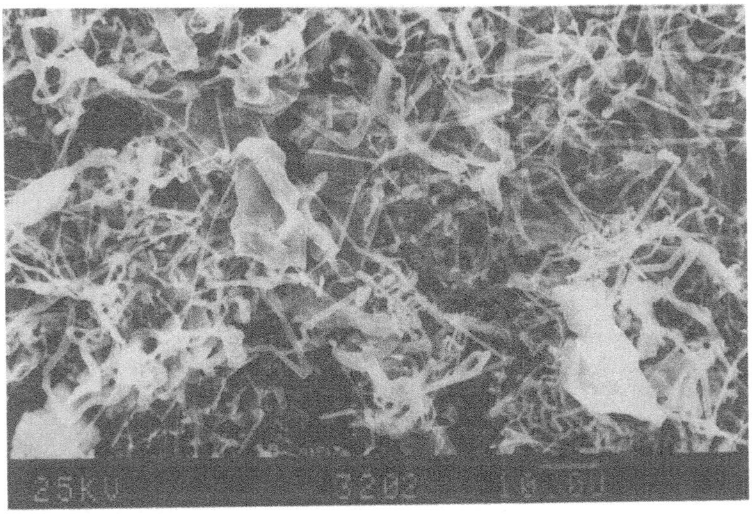

Fig. 1b. SEM micrograph of Li deposited on a Ni-substrate from 0.5 M $LiClO_4$/PC-THF (50-50) + 0.1 M SO_2, i = 2 mA/cm^2, Q = 3.6 C/cm^2.

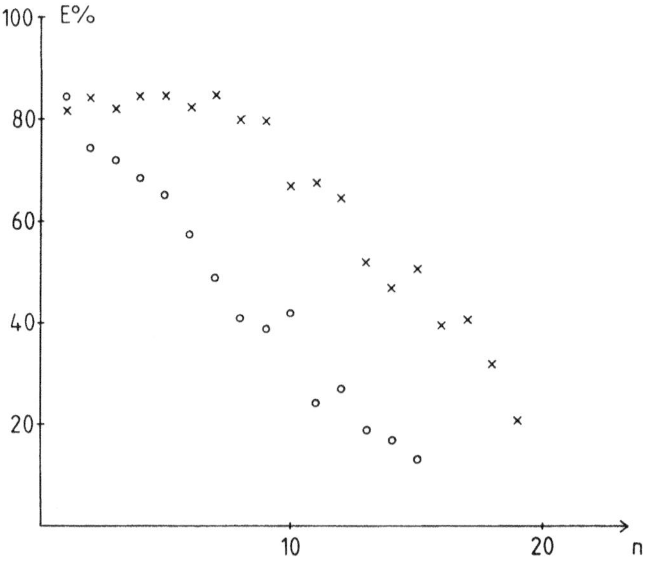

Fig. 2. Anodic Li-recovery in dependence of cycle number,
Li on stainless steel,
o 0.5 M $LiClO_4$/PC,
x 0.5 M $LiClO_4$/PC + 1 wt% tricyclo(5.2.1.0.^{2}.^{6})-decane,
$i_{ox} = i_{red} = 1$ mA/cm^2, Q = 2.7 C/cm^2,
cut-off = +1 V vs Li/Li^+.

There is also some beneficial effect of hydrocarbon additives on the storage behaviour of Li-depositions. This seems to be mainly related with their more compact morphology, although there may be some additional corrosion inhibition effect, as it has been observed for Li_xIn-alloy electrodes [6]. Fig. 3 shows the dependence of Li-recovery on storage time for lithium deposited from $LiClO_4$/PC in presence and absence of decalin.
The beneficial effect of the decalin additive becomes evident after about 3 days of storage. As already mentioned at Fig. 2, the values refer to an electrode hanging unprotected in electrolyte; close-packed configurations behave much better.
Not any hydrocarbon which is strongly adsorbed on the lithium electrode has a significant effect on the morphology of deposited lithium. Protection of a dynamic electrode surface (1 mA/cm^2 corresponds to deposition of 5 monolayers of lithium per second) requires immediate availability of the surfactant. Therefore only well soluble (> 1 wt%) and, nevertheless, highly surface active hydrocarbons are effective.

Li-ALLOY ELECTRODES

There are various metals - especially soft ones with large atomic radius such as Al, In, Sn, Pb, Zn, Cd, Mg, Bi and Tl [8,12-15] - which readily form Li-alloys at room temperature. Li^+-mobility is very high in these alloys; chemical diffusion coefficients of Li^+ at room temperature are in

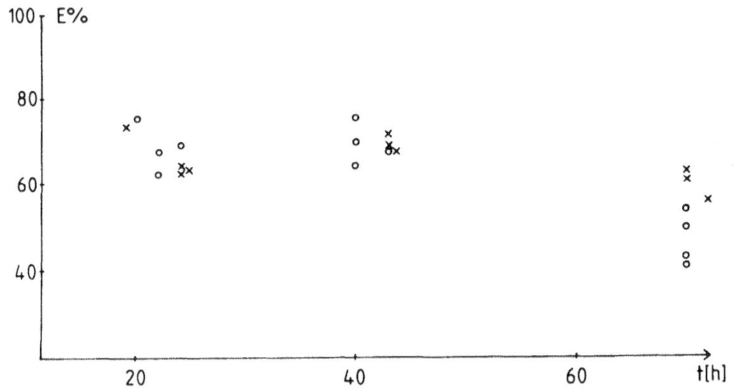

Fig. 3. Li-recovery after storage of lithium deposited on stainless steel,
o 0.5 M $LiClO_4$/PC,
x 0.5 M $LiClO_4$/PC + 1 wt% cis/trans-decalin (perhydronaphtalene),
$i_{ox} = i_{red} = 1$ mA/cm^2, Q = 2.7 C/cm^2.

the order of 10^{-8} cm^2/s for LiAl or $Li_{0.7}Sn$, respectively [16-18]. Li-alloy electrodes avoid the problem of dendrite formation during charging, however, because of the volume differences between the alloy Li_xM and the metallic "host matrix" M, they go powdery after prolonged cycling.
Despite their less negative standard potentials compared to pure lithium, Li-alloys are also chemically unstable in organic electrolytes. Like lithium, these alloys are protected by Li^+-permeable surface layers. Alloy/electrolyte combinations which do not form protective layers (e.g. LiIn/$LiClO_4$ in propylene carbonate) react completely within a few days. This fast corrosion can, however, be prevented by addition of hydrocarbon surfactants [8,10]. For corrosion protection of the "static" LiIn-electrode, fast availability of the hydrocarbon molecules is not an important condition. All saturated hydrocarbons tested by us so far (aliphatic and alicyclic hydrocarbons from 10-16 C-atoms) were effective.
In simple binary Li_xM-alloy electrodes volume changes between charged and discharged state of the electrode are unavoidable. The rate of powderization of an alloy electrode is related with these volume changes. Typical material losses per cycle are in the order of e.g., ca. 5 % for Al/β-LiAl (molar volume ratio 51.2/100 [19]) and 1.0-1.5 % for In/LiIn (molar volume ratio 66.5/100 [19]).
As the phase ranges in intermetallic phases Li_xM are fairly narrow, there does not seem to be a chance for a single phase "solid solution" Li-alloy electrode with reasonable energy density. Nevertheless, the overall geometry of a Li-alloy electrode can be stabilized by nonreactive "binder materials" in a ceramic type multicomponent electrode. A multicomponent Li-alloy electrode, using an intermetallic Li^+-ion conductor (Li_xSn) as binder material for Li_xSi as active material was already presented in 1981 by Huggins et al. [20]. As - in a considerable range of x - the standard potentials of phases Li_xSi are more negative than those of phases Li_xSn, the Sn-phases can be used as a nonreactive binder matrix for Li_xSi as the active material. The weak point of the Li_xSn matrix is that it is destroyed on overdischarge.
Multicomponent Li-alloy electrodes are good for very high cycle numbers, so long the changes of Li-content, i.e., changes of morphology are limited.

"Woods' alloy" (Bi-Pb-Sn-Cd) and related alloys with low melting point have been shown to be excellent "Li-absorbing" matrix materials [21]. 1000 cycles and a life expectance of 10 years are claimed for commercial cells using negatives of this type [22].
We have tried to make up dimensionally stable multicomponent alloy electrodes with an at least practically nonreactive binder component. For this purpose a mixture of ß-LiAl powder (ca. 80 wt%) and Cu-powder (ca. 20 wt%) has been sintered for ca. 60 minutes at ca. 1000 °C in a sealed vacuum container. Under these conditions copper forms intermetallic phases with aluminum and some of the lithium is set free. Approximately 80 % of the originally used lithium are still electrochemically active after the sintering process; this fraction of lithium is available for cycling.
Pressed powder electrodes made from the sintered ß-LiAl/Cu-mixture keep the coppery and shiny appearance of their surface irrespective of the state of charge. We believe that the Cu-Al-phases resulting from the sintering process form a kind of fine-meshed 3-dimensional network, keeping the electrochemically active ß-LiAl particles together and improving the electronic contact between them. This view is supported by X-ray diffraction, showing ß-LiAl in the charged and Al in the discharged electrodes. The complicated line structure resulting from various Cu-Al-phases remains essentially unchanged after discharge. Electrochemical experiments also verify that Cu-rich Cu-Al alloys are practically unable to absorb lithium at room temperature.
Sintering of the ß-LiAl/Cu-mixture onto a Ni grid yields electrodes which can be cycled at full capacity (see Fig. 4). Similar electrodes prepared by simply pressing a previously sintered ß-LiAl/Cu-mixture onto the same grid material behave much worse during cycling. The better performance of the "sintered onto" electrodes is obviously related with the formation of intermetallic phases Ni-Al and Ni-Cu at the contact grid/active material [8].
Copper is probably not the best choice for a metallic additive M to ß-LiAl to get M-Al alloys which are supposed to be chemically inert vs lithium. The chemical stability of Cu vs Li is doubtful [14], although there are no Cu-Li-phases that have been identified. On the other hand, there are various other metals M which do not form intermetallic phases with lithium (e.g. Ti, Mo, Ni) but readily react with Al.

CONCLUSIONS

Metallic lithium negatives as well as Li-alloy negatives have interesting fields of application in secondary ambient temperature lithium batteries.

- Metallic lithium is indispensable for high energy density - in this field, batteries with Li-alloy anodes could hardly compete with established aqueous electrolyte systems such as silver/zinc.

- Li-alloy negatives are very promising in the field of long lifetime and high reliability but low energy and power density.

ACKNOWLEDGEMENT

We are obliged to Deutsche Forschungsgemeinschaft and Fonds der Chemischen Industrie for financial support.

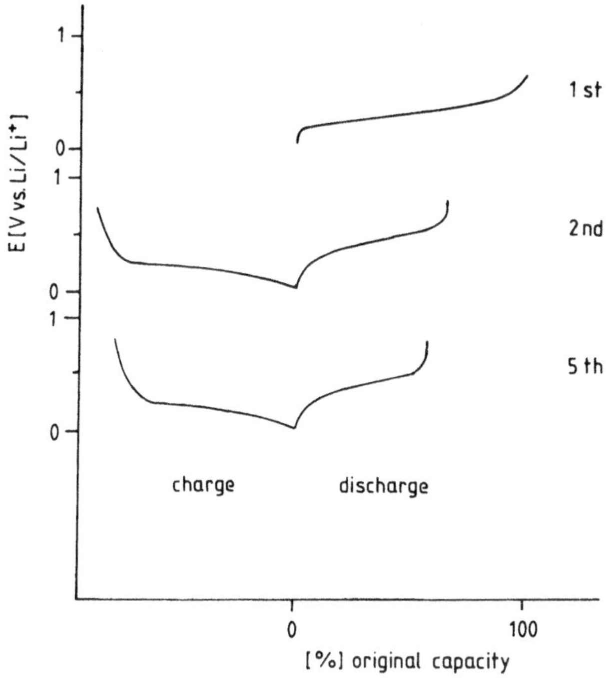

Fig. 4. Full capacity cycling of ß-LiAl (85-15 wt%) sintered onto Ni, 0.5 M $LiClO_4$/PC, i_{ox} = 1 mA/cm^2, i_{red} = 0.3 mA/cm^2, original capacity = 4.8 mAh.

REFERENCES

1) I. Epelboin, M. Froment, M. Garreau, J. Thevenin and D. Warin, J. Electrochem. Soc., 127 (1980) 2100.
2) A.N. Dey, Thin Solid Films, 43 (1977) 131.
3) V.R. Koch, J.L. Goldman, C.J. Mattos and M. Mulvaney, J. Electrochem. Soc., 129 (1982) 1.
4) S.-I. Tobishima and T. Okada, Electrochim. Acta, 30 (1985) 1715.
5) M. Garreau, J. Thevenin and D. Warin, Progress in Batteries & Solar Cells, 2 (1979) 54.
6) J.O. Besenhard, J. Gürtler, P. Komenda, DECHEMA Monographien, in press.
7) J. Gürtler, Dissertation, Techn. Univ. Munich, 1987.
8) J.O. Besenhard, P. Komenda, A. Paxinos and E. Wudy, Solid State Ionics, 18&19 (1986) 823.
9) R. Artes, J. O. Besenhard and H.P. Fritz, J. Electroanal. Chem., 81 (1977) 205.
10) J.O. Besenhard, J. Gürtler, P. Komenda and A. Paxinos, 3rd International Meeting on Lithium Batteries, Kyoto 1986, Extended Abstracts p. 346.
11) K. Brandt, J. Power Sources, 18 (1986) 117.
12) J.R. van Beek and P.J. Rommers, in J. Thompson (ed.), Power Sources 7, Academic Press, London, 1979, p. 595.
13) J.O. Besenhard, J. Electroanal. Chem., 94 (1978) 77.
14) A.N. Dey, J. Electrochem. Soc., 118 (1971) 1547.
15) J.O. Besenhard and H.P. Fritz, Electrochim. Acta, 20 (1975) 513.
16) T.R. Jow and C.C. Liang, J. Electrochem. Soc., 129 (1982) 1429.
17) A.S. Baranski and W.R. Fawcett, J. Electrochem. Soc., 129 (1982) 901.
18) J. Wang, I.D. Raistrick and R.A. Huggins, J. Electrochem. Soc., 133 (1986) 457.
19) K. Kurijama, S. Saito and K. Iwamura, J. Phys. Chem. Solids, 40 (1979) 457.
20) B.A. Boukamp, G.C. Lesh and R.A. Huggins, J. Electrochem. Soc., 128 (1981) 725.
21) Y. Toyoguchi, T. Matsui, J.Yamaura and T. Iijima, 3rd International Meeting on Lithium Batteries, Kyoto 1986, Extended Abstracts p. 113.
22) Matsushita - Panasonic, data sheet.

DIMENSIONALLY STABLE VITREUOS OXIDE INTERCALATION CATHODES FOR Li-BATTERIES

J.O. Besenhard, P. Komenda and M. Schwake

Department of Inorganic Chemistry
University of Münster
Wilhelm-Klemm-Str. 8, D-4400 Münster

N. Misailidis

Department of Physical Chemistry
Aristoteles University
Thessaloniki, Greece

INTRODUCTION

Many of the crystalline oxide intercalation cathodes go through one or more phase transitions until they are completely discharged. In secondary batteries repetitive phase transitions during cycling cause problems. Apart from fatal irreversible transitions, e.g. due to overdischarge, ever highly reversible phase transitions are undesired, because they are related with dimensional changes and strains and - in the long run - may result in a "powderisation" of the cathode material.
One way to get around these problems is the use of amorphous vitreous intercalation cathodes, stabilized by a "network former" such as P_4O_{10} (in the order of 5-10 wt%) are sufficient to get glasses such as V_2O_5-P_4O_{10} [1,2] or MoO_3-P_4O_{10} [3,4].
We report on electrochemical Li-intercalation into electronically conductive MoO_{3-x}-P_4O_{10} glasses which have been prepared by reduction of MoO_3 with red phosphorus.

EXPERIMENTAL

MoO_{3-x}-P_4O_{10} glasses have been prepared by melting MoO_3 + P_4O_{10} + P_{red} in sealed quartz ampoules at ca. 900 °C. After ca. 1 h the reaction was completed and the ampoules could be opened. Electrodes were prepared by dipping Pt-wires into the molten glass; the thickness of the glass layers obtained by this procedure was in the order of 50 μm.

RESULTS

Electronic Conductivity

The electronic conductivity of glasses MoO_{3-x}-P_4O_{10} depends mainly on x,

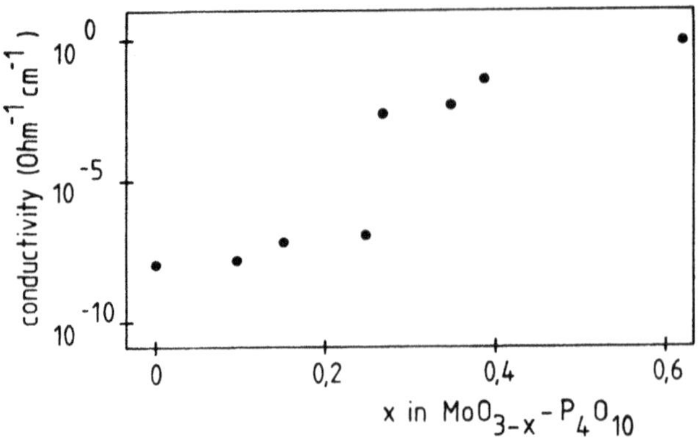

Fig. 1. Electronic conductivities of glasses MoO_{3-x}-P_4O_{10} (87.5 - 12.5 wt%)

i.e. on the average oxidation state of Mo. Samples with about constant content of P_4O_{10} (ca. 12.5 wt%)*) have been prepared; their typical room temperature conductivities (DC, 2-point) are shown in Fig. 1.

Discharge characteristics in $LiClO_4$/propylene carbonate

In contrast to crystalline MoO_3, MoO_3-P_4O_{10} and MoO_{3-x}-P_4O_{10} glasses show a "quasi single phase" discharge behaviour, i.e., a capacitor-like slopy discharge characteristic. There are, however, some reproducible irregularities in the OCV vs composition plots (see Fig. 2a), indicating a non-equivalence of the available Li-sites.
As expected from the Mo oxidation states, the OCVs of Li-intercalated MoO_3-P_4O_{10} glasses are slightly higher than those of MoO_{3-x}-P_4O_{10} glasses with the same content of lithium. On the other hand, under load the MoO_{3-x} glasses behave much better than the MoO_3 glasses (see Fig. 2b). This is only partly due to the better electronic conductivity of the oxygen deficient glasses**). An additional significant contribution to this effect is from the higher mobility of intercalated lithium (see Fig. 2c). Surprisingly, the maximum uptake of lithium is not much dependent on the oxidation state of Mo. For x in MoO_{3-x} = 0.6, i.e. average oxidation state of Mo = 4.8, still about 2 Li per Mo can be accommodated.
Based on the OCV vs composition diagramm, the theoretical energy density of a MoO_3-P_4O_{10} (87.5 - 12.5 wt%) glass is ca. 700 Wh/kg (for 1.5 V cut-off); this is about the same value as that of crystalline MoO_3 (ca. 650 Wh/kg). Unfortunately, the Li^+-mobility in molybdenum oxide glasses is about three orders of magnitude lower than that in crystalline MoO_3.

---

*) Calculated compositions, assuming oxygen transfer from MoO_3 to P_{red}, yielding MoO_{3-x}-P_4O_{10}.

**) Conductivity of MoO_3-P_4O_{10} glasses is much improved by Li-intercalation. Nevertheless, the IR-drop in intercalated MoO_{3-x}-P_4O_{10} glass electrodes is always significantly lower and fairly independent of content and distribution of the intercalated lithium.

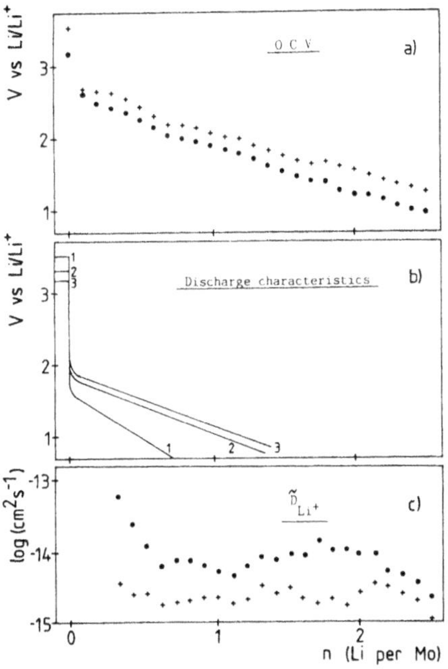

Fig. 2 a) Open cell voltage vs composition plot
+: MoO_3-P_4O_{10} (87.5 - 12.5 wt%)
•: $MoO_{2.4}$-P_4O_{10} (87.5 - 12.5 wt%)

b) Discharge characteristics of MoO_{3-x}-P_4O_{10} glasses in 0.5 M $LiClO_4$/propylene carbonate; i = 200 $\mu A/cm^2$
1: MoO_3-P_4O_{10} (87.5 - 12.5 wt%)
2: $MoO_{2.7}$-P_4O_{10} (87.5 - 12.5 wt%)
3: $MoO_{2.4}$-P_4O_{10} (87.5 - 12.5 wt%)

c) Chemical diffusion coefficient $\tilde{D}_{Li^+}$ vs composition plot*)
+: MoO_3-P_4O_{10} (87.5 - 12.5 wt%)
•: $MoO_{2.4}$-P_4O_{10} (87.5 - 12.5 wt%)

Reversibility of Li^+-intercalation

Roughly, not more than 50-60 % of the lithium intercalated into "virgin" glasses MoO_3-P_4O_{10} or MoO_{3-x}-P_4O_{10} (cut-off compositions: 1 Li per Mo) could be deintercalated by charging them at current densities in the range 10-20 $\mu A/cm^2$. This percentage was not much improved by lowering the charging current. Similar findings have been reported for V_2O_5-P_4O_{10} glasses [6]. Apparently a considerable fraction of the intercalated lithium occupies sites from which it is difficult to remove. On the other hand, there are also indications for limitations in the charging process due to insufficient electronic conductivity in parts of the electrode.
MoO_3- and MoO_{3-x}-P_4O_{10} glasses are dimensionally stable during lithium intercalation/deintercalation. For lithium contents up to 1.2 - 1.4 Li per Mo there is not any observable change of the glassy and shiny appearance of the electrode surface. Higher lithium contents cause cracks and finally an irreversible surface roughening.

ACKNOWLEDGEMENT

We are obliged to Fonds der chemischen Industrie for financial support.

REFERENCES

1) T. Pagnier, M. Fouletier, J.L. Souquet, Solid State Ionics, 9+10 (1983) 649.
2) Y. Sakurai, J. Yamaki, J. Electrochem. Soc., 132 (1985) 512.
3) I.J. Schultz, Z. Anorg. Allgem. Chem., 99 (1955) 281.
4) B. Bridge, N.D. Patel, J. Mater. Sci., 21 (1986) 1187.
5) B.B. Scholtens, Mat. Res. Bull., 11 (1976) 1533.
6) Y. Sakurai, S. Okada, H. Ohtsuka, J. Yamaki, T. Okada, 3rd International Meeting on Lithium Batteries, Kyoto 1986, Extended Abstracts p. 283.

---

*) Chemical diffusion coefficients were determined by the galvanostatic pulse method [5].

FORMATION OF TiO_2(B) BY PROTON EXCHANGE AND THERMOLYSIS OF SEVERAL LAYERED ALKALI METAL TITANATE STRUCTURES

T.P. Feist[1], S.J. Mocarski[1], P.K. Davies[1], A.J. Jacobson[2], and J.T. Lewandowski[2]

[1]Department of Materials Science and Engineering, University of Pennsylvania, Philadelphia, PA 19104, USA
[2]Exxon Research and Engineering Corp., Annandale, NJ 08801, USA

INTRODUCTION

Titanates of the formula $A_2Ti_nO_{2n+1}$ ($3 \leq n \leq 5$) consist of corrugated sheets of corner-sharing and edge-sharing TiO_6 octahedra with monovalent cations occupying the interlayer spaces. The materials undergo rapid ion-exchange under relatively mild conditions, and thermal treatment of acid-exchanged materials has been shown to lead to the formation of new metastable forms of titanium dioxide (1).

The ratio of edge to corner sharing octahedra in layered titanates is determined by the stoichiometry of the compound. (Figure 1) The sheets are essentially infinite in the b-direction, and stacking along the a-axis follows either an AAA pattern, as in $Na_2Ti_3O_7$, or an ABA pattern, as in $Cs_2Ti_5O_{11}$.

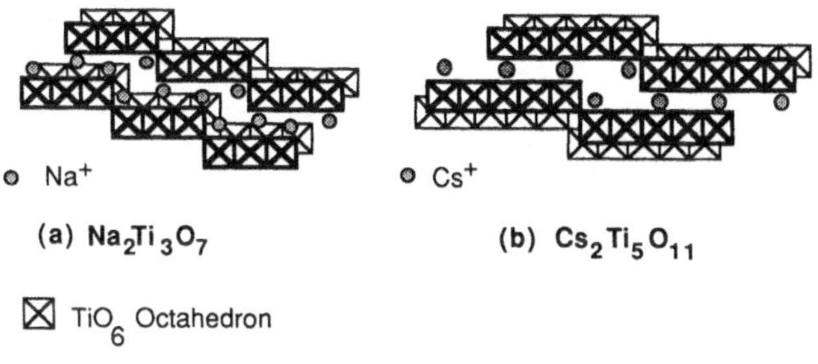

Figure 1. The structures of (a) $Na_2Ti_3O_7$ and (b) $Cs_2Ti_5O_{11}$.

Acid exchange reactions have been used to prepare new compounds from layered titanates. Marchand et al (1) reported that acid-exchange of $K_2Ti_4O_9$ followed by dehydration of the resulting compound, $H_2Ti_4O_9 \cdot H_2O$, led to the formation of a previously unobserved metastable form of titania, TiO_2(B). The authors suggested that TiO_2(B) formed during a multistep topotactic dehydration of the exchange product (2), but no detailed mechanism was presented. In this work, we have examined the intercalation chemistry and dehydration reactions of acid-exchanged $Na_2Ti_3O_7$ and $Cs_2Ti_5O_{11}$. In both systems, the powder x-ray diffraction patterns of the final dehydration products are identical to that of TiO_2(B).

EXPERIMENTAL

$Na_2Ti_3O_7$ was prepared by conventional solid state reaction of intimately mixed powders, in a method similar to that of Andersson and Wadsley (3). Na_2CO_3 and TiO_2(anatase) were mixed in a 1:3 ratio, pressed into a pellet, and fired at 850°C for 24 hours. The sintered pellet was then ground, repressed, and fired at 850°C for another 24 hours. The product was a white powder, with an x-ray diffraction pattern that matched the reported data for $Na_2Ti_3O_7$ (3). $Cs_2Ti_5O_{11}$ was prepared by an analogous method, using $CsNO_3$ and TiO_2(anatase) in a 2:5 ratio. Both products appeared to be single phase, with no evidence of residual anatase.

Ion-exchange reactions were carried out by stirring the powders in a 1M HCl solution (~200 fold excess) at 60°C for one to three days. For $Na_2Ti_3O_7$, the product after three days is $H_2Ti_3O_7$, and $Cs_2Ti_5O_{11}$ yields $H_2Ti_5O_{11} \cdot 3H_2O$ after 24 hours in the exchange solution. Residual Na and Cs, as measured by Inductively Coupled Plasma (ICP) analysis, was less than 2%. More complete exchange can be obtained by repeating the process using a fresh HCl solution.

The dehydration behavior was investigated using Thermogravimetric Analysis (TGA), which showed several, distinct water-loss steps for each compound. Hydrogenated samples were annealed at temperatures corresponding to each dehydration step. After heat-treatment, the products were examined by powder x-ray diffraction.

RESULTS AND DISCUSSION

The results of thermogravimetric analysis of $H_2Ti_3O_7$ are presented in Figure 2. There are two, or possibly three, distinct weight loss regions. Each weight loss is thought to be accompanied by significant structural rearrangement and formation of stable intermediates. The x-ray patterns obtained after heating a sample at various temperatures for several days (Figure 3) confirm this hypothesis. The heat treatment temperatures are somewhat lower than the temperatures suggested by the TGA plot since the TGA scan, performed at 5°C/min, does not represent equilibrium heating. The final product of the dehydration is TiO_2(B), and the lattice parameters are: a=12.186Å, b=3.745Å, c=6.53Å, ß=106.9°. TiO_2(B) transforms slowly to anatase when heated at temperatures above approximately 350°C.

The exact structures of the intermediate compounds are not yet known, but the diffraction data for the first intermediate structure strongly resemble the calculated pattern for $H_2Ti_6O_{13}$, a compound formed by the condensation of $H_2Ti_3O_7$ such that the unshared corners of TiO_6 octahedra from each sheet connect, forming tunnels which run through the structure. Water loss from the intermediate compound was one half a mole of H_2O compared to the end product ($3TiO_2$), implying a formula of $H_2Ti_6O_{13}$. The second intermediate structure, formed by the loss of approximately 0.75 moles of H_2O from the starting material, seems closely related to TiO_2(B).

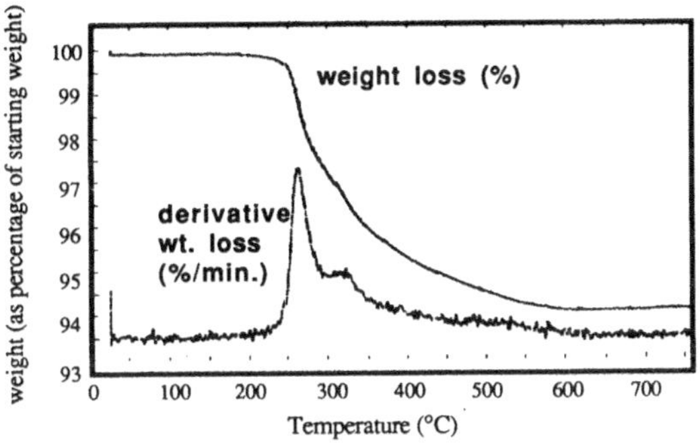

Figure 2. Weight loss behavior of $H_2Ti_3O_7$.

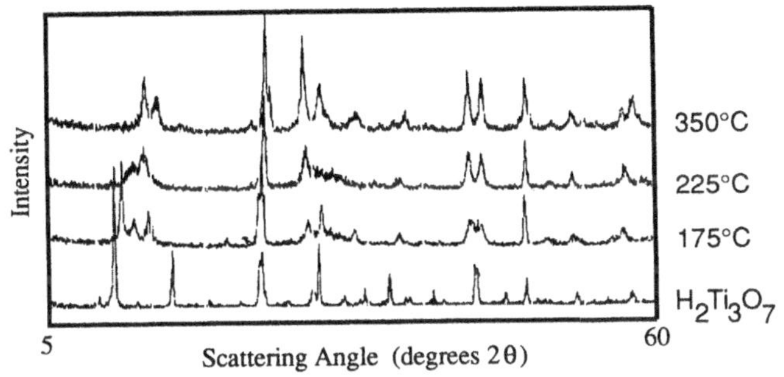

Figure 3. Powder X-ray diffraction patterns of $H_2Ti_3O_7$ heated at the temperatures indicated.

$Cs_2Ti_5O_{11}$ undergoes a similar set of reactions which also lead to the formation of TiO_2(B). The intermediate products are less stable than those from $H_2Ti_3O_7$ (and hence less crystalline), and anatase forms more easily than TiO_2(B) as the end product.

These results indicate that the formation of TiO_2(B) is not completely dependent on the specific starting titanate structure. Further investigations using x-ray, electron, and neutron diffraction are currently being undertaken to further elucidate the details of the mechanism of TiO_2(B) formation.

ACKNOWLEDGEMENTS

This research was supported by NSF DMR831669 (S.J.M. and P.K.D.) and Exxon Research and Engineering Corp. (T.P.F.).

REFERENCES

1. R. Marchand, L. Brohan, and M. Tournoux, *Mat. Res. Bull.*, **15**, 1129 (1980).

2. M. Tournoux, R. Marchand, and L. Brohan, *Progress in Sol. State Chem.*, **17**(1), 33 (1986).

3. S. Andersson and A. Wadsley, *Acta Cryst.*, **14**, 1245 (1961).

MAGNETIC ORDERING IN NONTRONITE PILLARED WITH Al-POLYOXO CATIONS

N.H.Gangas[†], T.Bakas[+], A.Moukarika[+], D.Petrides[++] and A.Simopoulos[++]

+ Dept.of Physics, Univ.of Ioannina,GR-453 32 Ioannina, Greece
++ Inst.of Mat.Sc., NRCPS "Democritos" GR-153 41 Athens, Greece
† Kifissias Av., 108, GR-115 26 Athens, Greece

In this short communication we present preliminary results from our current investigation by ^{57}Fe Moessbauer Spectroscopy of the effects of pillaring dioctahedral smectites with aluminium polyoxo cations. In particular we report here the marked effect found on the magnetic ordering of iron in Nontronite after pillaring it with the above cations.

The mineral used in this study, SWa-1 Nontronite*, is an iron rich dioctahedral 2:1 sheet silicate with the chemical composition:

$$< Mg_{0.22} \; K_{0.03} > [Si_{3.52} \; F^{3+}_{0.13} \; Al_{0.34}] \{Fe^{3+}_{1.71} \; Al_{0.29}\} \; O_{10}(OH)_2$$

Bracket <> denotes interlayer cations, while the brackets [ ] and { } denote cations in four and six coordination in silicate lattice respectively.

On the basis of Moessbauer, magnetic susceptibility and magnetic neutron diffraction data[1,2] it was found that in the SWa-1 Nontronite there is no sign of long magnetic ordering down to 2K although the paramagnetic Curie temperature is θ = -20K. Moessbauer spectra showing magnetic splitting appear only at about 1.6 K. This behaviour is markedly distinct from that of another iron rich member of the 2:1 dioctahedral sheet silicate group, the Ferripyrophyllite, for which Coey et al[1] found that its $\theta = -22$ K and it orders antiferromagneticaly with a Neel temperature 18 K.

Recently, Gangas et al[3] obtained at 10 K magneticaly splitted Moessbauer spectra for the SWa-1 Nontronite after pillaring it with iron oxides.

---

* Source Clay Minerals Repository. Dept. of Geology, University of Missouri. Columbia, Missouri 65211, U.S.A

The pillared Nontronite was prepared by calcining the iron-polyoxo-Nontronite precursor at 825K for 28h. The similarly calcined Na^+-Nontronite has shown no magnetic splitting at 10K.

In order to differentiate the effects of the pillaring process *per se* on the lattice and on the magnetic structure of Nontronite, we have intercalated it with Moessbauer silent cations. To this end the Na^+-saturated clay fraction of the mineral reacted with a base-hydrolysed water solution of $AlCl_3$ (OH/Al ratio = 2.0). The intercalated clay product, after washing and drying, showed an interlayer spacing d_{001} = 18 Å after calcination of the material at 500° C for 12h.

Figure 1(a-e) displays Moessbauer spectra obtained at 4.2 K of the as prepared material (a) and of samples thereof which have been calcined for 5 hours at: 300°C (b), 400°C (c) and 500°C (d). The spectrum 1(e) corresponds to the Na^+-saturated form of SWa-1 Nontronite after calcining it at 500°C for 5 h.

These spectra show that magnetically splitted as well as magnetically relaxing spectral components dominate only in the spectra of Al-polyoxo - Nontronite calcined at 400°C and 500°C. This fact points out that the onset of magnetic interactions is not a direct consequence of the modification of the silicate lattice because of its normal dehydroxylation upon calcination. This modification has been detected by Moessbauer spectra at a higher temperature (25K) - not shown here - where magnetic ordering has not set up for any of the samples. For the Na^+-Nontronite as well as for the Al-polyoxo-Nontronite a drastic increase in the quadrupole splitting is detected for calcination temperature above 300°C.

Infra-red spectra obtained in the region 625-4000 cm^{-1} confirm that the normal dehydroxylation of the Nontronite lattice starts at 400°C and is practically completed at 500°C for both, the original and the pillared clay. Such low dexydroxylation temperatures are typical for Nontronites.

Recently, on the basis of magic angle spinning NMR and IR data the occurence of new modifications in the Al-polyoxo-smectite lattice have been detected upon heating an Al-polyoxo-Montmorillonite[4] precursor above 400°C and an Al-polyoxo-Beidellite[5] precursor at 400°C. The SWa-1 Nontronite has its negative layer charge at the tetrahedral sheets and in this respect resembles to Beidellite. Two quite distinct structural models[4,5] have been proposed for these modifications which are related to the cross-linking of the pillars to the clay lattice. The Moessbauer spectra presented here could *prima facia* be regarded as confirming the cross-linking of the pillars to the octahedral sheet via an O^{2-}-bridge between the two partners as a result of the condensation of the OH^{-} belonging to each of them[4]. Such a bridge could make the magnetic interactions among the iron ions in

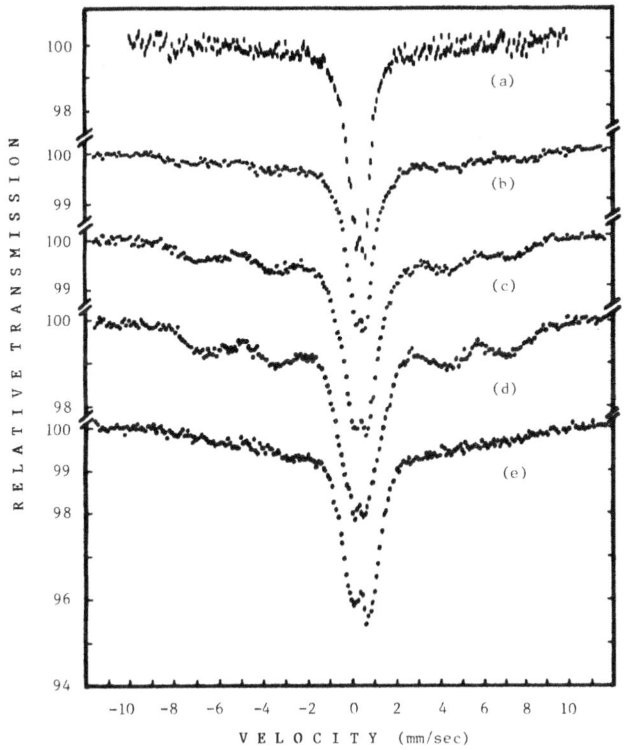

Fig. 1. Moessbauer spectra at 4.2 K. (a) Intercalated sample; (b) Intercalated and calcined for 5 hours at 300°C; (c) Intercalated and calcined for 5 hours at 400°C; (d) Intercalated and calcined for 5 hours at 500°C; (e) Na^{+}-saturated sample SWa-1 Nontronite calcined for 5 hours at 500°C.

the Nontronite lattice sensitive to the magnetic character of the cations forming the pillars.

The extend of the magnetic ordering for the lattice iron found in this work at 4.2 K for Al-polyoxo-Nontronite is definitely smaller than that found[3] at 10 K in Fe-polyoxo-Nontronite. In the latter case all lattice iron has been ordered at 10 K. This different behaviour between the Al- and Fe-pillared forms of the same clay may depict the above mentioned sensitivity and therefore hints that the model proposing for Al-pillared Beidellite cross-linking of the tetrahedral sheet to the pillars[5] may be not appropriate in the case of pillared Nontronite.

Evidently, this evaluation is not final in view of the fact that our investigation is still in progress. A full report on the present work will be published elsewhere.

In conclusion the present data point our that the promotion of the magnetic interactions among the iron ions in the Al-pillared Nontronite lattice is related to a structural modification of the lattice different from that due to the usual dehydroxylation of the original clay. This modification occurs upon calcining the Al-polyoxo-Nontronite at temperatures between 300°C and 400°C.

References

1. J.M. D. Coey, F. V. Chukhrov, and B. B. Zvyagin, Cation distribution, Moessbauer spectra, and magnetic properties of Ferripyrophyllite, Clays and Clay Minerals 32:198 (1984).
2. O. Ballet and J.M. D. Coey, Magnetic properties of sheet silicates; 2:1 layer minerals, Physics and Chemistry of Minerals 8:218 (1982).
3. N.H. J. Gangas, J. van Wonterghem, S. Morup, and C.J. W. Koch, Magnetic bridging in nontronite by intercalated iron, Journal of Physics C 18:L1011 (1985).
4. D. Tillak, B. Tennakoon, W. Jones, and J. M. Thomas, Structural Aspects of Metal-oxide-pillared sheet silicates, Journal of the Chemical Society, Faraday Trans.1 82:3081 (1986).
5. D. Plee, F. Borg, L. Gatineau, and J. J. Fripiat, High-Resolution Solid-State ^{27}Al and ^{29}Si Nuclear Magnetic Resonance Study of Pillared Clays, Journal of the American Chemical Society 107:2362 (1985).

NEW ROUTES TO THE PREPARATION OF PILLARED MONTMORILLONITE CATALYSTS BY INTERCALATION OF NIOBIUM COORDINATED COMPLEXES

E.M. Farfan Torres, P. Grange and B. Delmon

Laboratoire de Catalyse et de Chimie des Matériaux Divisé
Place Croix du Sud 1
1348 Louvain-la-Neuve, Belgium

ABSTRACT

A new synthetic method to obtain pillared clays by intercalation of niobium coordinated compounds is reported. Niobium pillared montmorillonites were prepared by treating dispersed Na-montmorillonite with an ethanolic solution of niobium polymerics derivatives of the type $Nb_4O_7Cl_4(OC_2H_5)_2(dipy)_2.2H_2O$. The solids were calcined at different temperatures. At 500°C the basal spacing of the Nb intercalated clay is 16.0 Å and its specific surface area is 180 m^2g^{-2}.

Keywords: Montmorillonite/Pillared Clays/Interlayering/Intercalation/Niobium.

INTRODUCTION

Smectite clay minerals consist of two-dimensional silicate layers which are bonded to each other by means of Van der Waals and electrostatic bonding arises as a consequence of substitution of Si and Al cations in the sheet framework, resulting in an anionic charge on the sheet, normally compensated by hydrated exchangeable cations such as Na^+, K^+, Ca^{++}, etc. Intercalation between the anionic silicious sheets of these minerals is possible by exchanging these compensation cations with different kinds of cationic species that can be considered as pillars. This causes the mineral to swell and expose the interlayer pore space to a possible adsorption. The phenomena of developing thermally stable clay structures by using pillars of organic or inorganic cations is known as "pillaring of clays" (1) or "cross-linking of clays" (2). But as these materials are obtained by grafting of complexe cations in the bidimensional structure of the clay resulting in a tridimensional network of pores, these materials can be considered as "bidimensional zeolites" too (3).

There are different approaches for the preparation of pillared clays. The best known one consists in the use of specially prepared polyhydroxy water-soluble polymers. The polymers can be prepared either by basic titration of a metallic salt or by hydrolysis of an appropriate precursor. These inorganic pillared clays, derived from polyhydroxy cations, are stable at quite high temperatures. Al (4-6), Zr (7-9), Si (10), Ni (11), Cr (11), Fe (13) etc. cations have already been reported. The other way of preparation consists in using organic pillars as tetralquilamonium (1). But, in this case, the solids do not present high thermal stability. Nevertheless, in the recent literature, several authors have reported the use of clusters and organometallic complexes as pillaring agents for cations that would not develop polyhydroxylic structures. Endo et al. (10,14) synthetized a silica pillared montmorillonite via the hydrolysis in the interlayer of tri (acetyl acetonate) silicon (iv) cation and Pinnavaia et al. (15) have prepared niobium and molybdenum pillared montmorillonites from clusters of the type $M_6Cl_{12}^{n+}$ (n=2,3; m=Nb, Ta, Mo).

In this paper we report preliminary studies dealing with the preparation of a niobium-pillared montmorillonite by means of polymeric complexe compounds like oxo-hydroxo-ethoxo-2,21-dipyridyl) Niobium (v), synthetized by in situ hydrolysis in ethanolic solutions of the chlorine complexes.

EXPERIMENTAL

Preparation of the sample. A Wyoming Montmorillonite, supplied by ECC International Ltd was used. The initial clay was completely saturated with Na^+, by several exchanges with a 10 N NaCl solution, with subsequent washing by centrifugation and dialysis to remove excess chloride. Only the fraction ≤2μ was used. The cation exchange capacity (CEC) was determined by saturation of the clay with amonium acetate 0.5N, followed by successive washing with technical methanol up to complete elimination of NH_4^+excess. Then the tenor of residual NH_4^+ was determined by the micro Kjehldal method on 50 mg of sample. The CEC, determined on the basis of sample weight after drying at 110°C, was 86 meq/100 g of clay.

Niobium Polymeric Complex (PI)

$Nb_4O_7Cl_4(OC_2H_5)_2(dipy)_2.2H_2O$. The polymeric complex was synthetized from Niobium pentachloride according to the procedure reported by Djordjevic et al. (16). The complex was purified by disolution in 98% ethanol and subsequent reprecipitation with a chloroform-carbon tetrachloride mixture (1:3). Finally, the precipitate was filtered, washed with chloroform and then dried at room temperature.

Niobium Pillared Montmorillonite (Nb Montmorillonite)

The niobium pillared clay was prepared by in situ hydrolysis of polymeric complex PI. For this, 2 gr of polymer PI were fully dispersed in an ethanolic aqueous suspension of Na-Montmorillonite. After standing several hours at room temperature with occasional shaking, the dispersed montmorillonite was washed and freeze-dried. The samples were then calcined, stepwise, in air up to 600°C.

Physico-chemical Analysis

X-ray powder diffraction (XRD) patterns were measured by use of a Kristalloflex 805 Siemens diffractometer with Ni filtered Cu-Kα radiation. Infrared (IR) spectra of samples prepared in the form of KBr pellets were obtained in a Brucker IFS88 FT-IR spectrometer. Thermogravimetric Analyses (TGA) and Differential Thermal Analysis (DTA) were carried out in a Setaram equipment. Nitrogen adsorption-desorption isotherms at liquid nitrogen temperature were performed with a Micromeritics Asap 2400 sorbtometer. Prior to the analysis, all samples were outgassed at 120°C overnight.

RESULTS

In Figure 1, a typical X-ray diffractogram showing the d001 peaks for a Na^+-montmorillonite before and after pillaring with niobium polymer is given. The basal spacings together with the BET surface areas are shown in Figure 2, as a function of the heat treatment temperature. These results indicate that the pillared clay is thermally stable up to 500°C. Although the spacing slightly decreases on heating, a value of about 16 Å is maintained upto this temperature. Heating at higher temperatures caused a modification of the interlayered structure denoted by the absence of the d001 diffraction line.

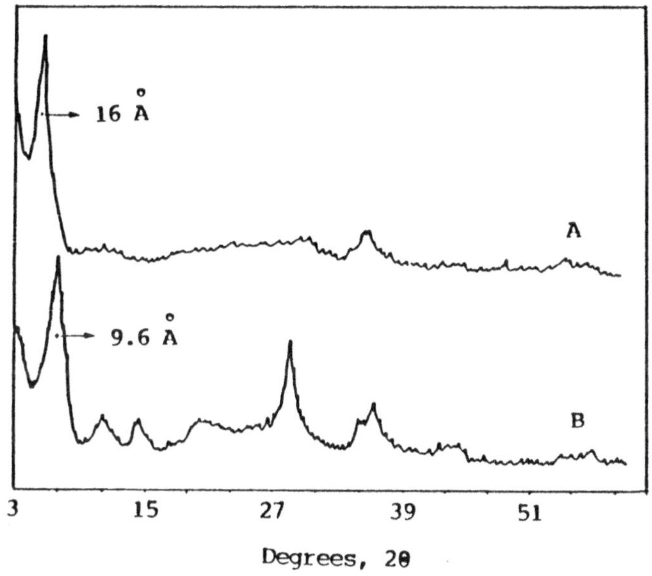

Fig. 1: X-ray diffraction patterns of Na^+-Montmorillonite (A) after and (B) before pillaring with niobium.

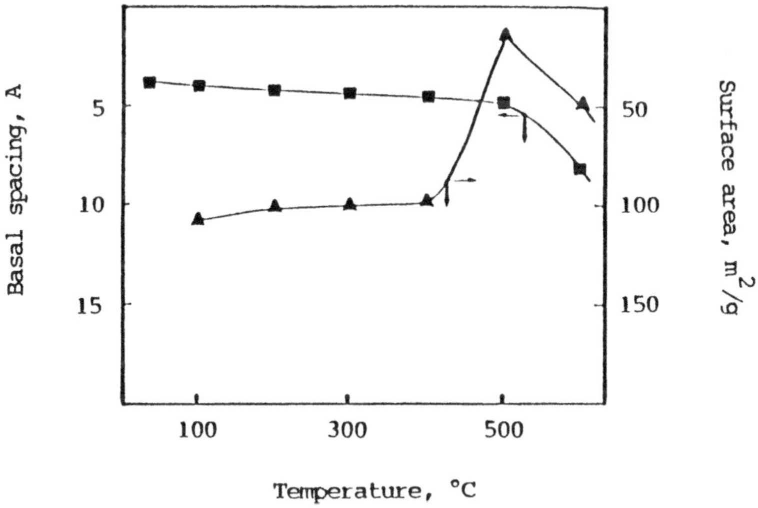

Fig. 2: Basal spacing and BET surface areas of Nb-Montmorillonites as a function of heat treatment temperature.

Figure 3 illustrates the middle IR-spectrum in the region between 1700 and 400 cm^{-1} of Nb-Montmorillonite (A) in comparison with the spectra of Na^+-Montmorillonite (B) and Niobium Complex PI (C). Infrared spectrum of niobium polymeric complex shows, near 767 cm^{-1}, a characteristic absorption band corresponding to an oxygen bridge polymeric structure, of the -(Nb-O-Nb)- type, as it was previously reported by Djordjevic et al. (16). This band is also present in the clay polymer complex spectrum.

Figure 4 illustrate the DTA curves for Nb and Na^+-Montmorillonite . The strong exotherm peak near 500°C of the Nb-Montmorillonite is very close to the decomposition temperature of the complex.

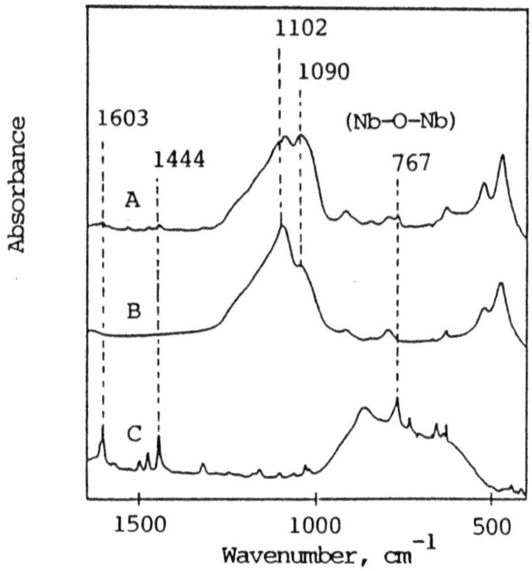

Fig. 3: Infrared spectra at 100°C of Nb-Montmorillonite (A), Na^+-Montmorillonite (B), Complex I (C).

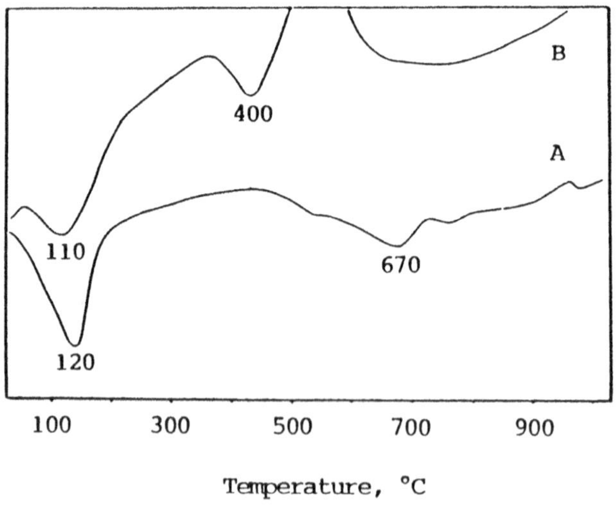

Fig. 4: DTA profiles of (A) Na^+-Montmorillonite and (B) Nb-Montmorillonite.

Figure 5 shows the percentage weight loss of Nb and Na^+ clay samples after heating at temperatures ranging between 25 and 1000°C. The whole process of thermal decomposition for both samples can be divided into two different stages. The first between room temperatures and 110°C associated with dehydration of water molecules loosely adsorbed in the interlayer spaces and on external surfaces, in both samples. For Nb-Montmorillonite, the second stage between 300 and 550°C, corresponds to the exothermic peak detected by DTA. For Na^+-Montmorillonite the second stage begins at 500°C and corresponds to a decomposition of the clay layers.

The BET surface areas are reported in Fig. 2, together with the basal spacing values of the solid. Surface area values showed a maximum of 500°C.

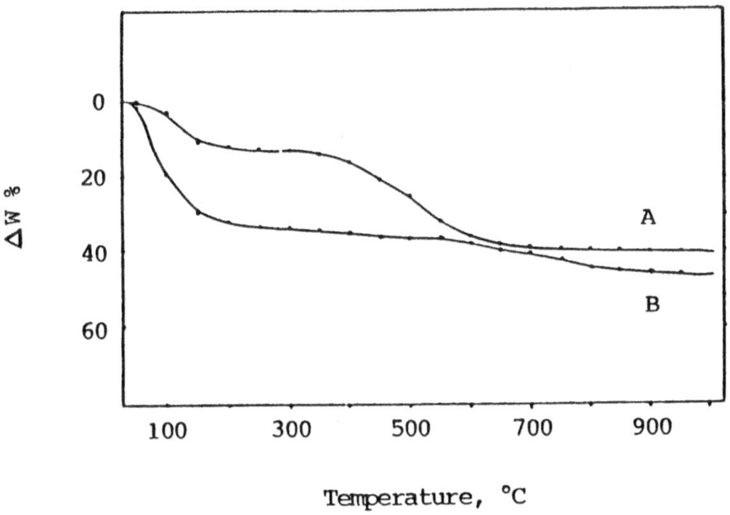

Fig. 5: Weight change of (A) Nb-Montmorillonite and (B) Na^+-Montmorillonite when heated at successively high temperatures.

DISCUSSION

The XRD and adsorption data clearly show that a new niobium pillared clay is obtained by the reaction of montmorillonite with niobium polymeric complex PI. Early works on complex PI show that Nb(v) forms polymeric species with a polyhedron chain amorphous type structure, wity a co-ordination number 6 for niobium, as shown below:

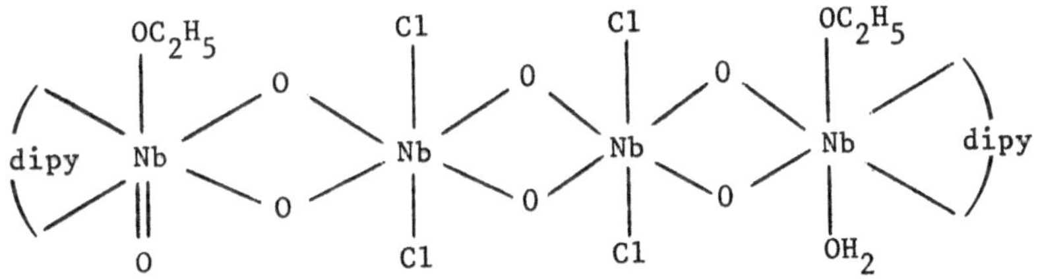

On dissolving these polymers in ethanolic solution, all chloro ligands are hydrolysed and replaced by OH groups. Complex cationic polymeric Nb(v) species can then be obtained, and intercalation between layers in montmorillonite occurs by displacement of interlayer Na^+.

IR data for the intercalated clays also confirms the formation of a complex clay-niobium polymer, as is shown by the presence of characteristic bridging metal-oxygen stretching at 767 cm^{-1} and also the characteristic bands of coordinated dipyridyl at 1610, 1600, 1440 cm^{-1} in the Nb-montmorillonite spectrum (Fig. 3). Pillared clay samples show basal spacing to be thermally stable in air at about 580°C; at higher temperatures an amorphous structure is detected. This is in agreement with the results observed in thermal analyses. The exothermic peak between 480 and 600°C, coincides with the weight-loss observed in this temperature range. This indicates a transition from complex to oxide type structure for niobium. In addition, from adsorption data, we observed that surface areas remain nearly constant for temperatures under 500°C and increase suddenly to give a maximum value at 500°C, when the oxide structure occurs. This indicates that the structure of the complex remains stable inside the layers up to the temperature of decomposition, maintaining the interlayer distance and the specific surface area almost constant up to 500°C. When the niobium oxide is formed, the specific surface area of the solid drastically increases without modification of the basal distance. However, this parameter decreases at temperatures higher than 500°C and the d(001) diffraction band disappears at 600°C. The surface area also progressively decreases.

This behaviour may be explained in two different ways; either by migration to the surface or a sintering of the niobium oxide. Assuming that sinterized niobium oxide particles, prepared by decomposition and calcination of alcoxides, may have sizes of the order of 100 Å, the second hypothesis may be realistic. Under these conditions at very high temperatures the regular edge to edge pillared clay would be transformed into edge to face delaminated clay. Experiments are in progress in order to confirm this hypothesis.

CONCLUSIONS

In this work we present a new simple way of preparing niobium pillared clays. A polymeric complex $(Nb_4O_7Cl_4(OC_2H_5)_2(dipy)_2.2H_2O$ is hydrolyzed giving Nb(v) species. At 500°C this complex is decomposed into niobium oxide. The solid obtained is stable and presents at this temperature a basal spacing of 16 Å and a specific surface area of 180 m^2g^{-1}. At higher temperatures both a decrease of the surface area and disappearance of the basal spacing seem to indicate the formation of delaminated clays.

ACKNOWLEDGMENTS

We greatly acknowledge Les Services de Programmation de la Politique Scientifique (Belgium) for the financial support. E.M. Farfan thanks Le Command Général aux Relations Internationals for the grant accorded to her. We would also like to thank Micromeritics for the textural measurements and Dr M. Martinez Lara and Dr Laureno Moreno for DTA TDG analysis.

REFERENCES

1. R.M. Barrer and D.M. Mac Leod, Trans.Faraday Soc., 51, 1290 (1955).
2. N. Lahav, V. Shani and J. Shabtai, Clays Clay Miner., 26, 107 (1978).
3. D. Plee, F. Borg, L. Gatineau and J.J. Frippiat, J. Am. Chem. Soc., 107, 2362 (1985).
4. R.H. Loeppert, M.M. Mortland and T.J. Pinnavaia, Clays Clay Miner., 27, 201 (1979).
5. M.F. Knudson, J.L. Mc Atee, Clays Clay Miner., 21, 19 (1973).
6. G.W. Brindey, R.E. Sempela, Clays Clay Miner., 12, 229 (1977).
7. S. Yamanka, G.W. Brindey, Clays Clay Miner., 27, 119 (1979).
8. R. Burch and C.I. Warburton, J. Catalysis, 97, 503 (1986).
9. R. Burch and C.I. Warburton, J. Catalysis, 97, 511 (1986).
10. T. Endo, M.M. Mirtland and T.J. Pinnavaia, Clays Clay Miner., 28, 105 (1980).
11. S. Yamanka and G.W. Brindley, Clays Clay Miner., 28, 281 (1980).
12. G.W. Brindley and S. Yamanka, Amer. Miner., 64, 830 (1979).
13. S. Yamanka, T. Doi, S. Saka, M. Hattori, Mat. Res. Bull., 19, 161 (1984).
14. T. Endo, M.M. Mortland and T.J. Pinnavaia, Clays Clay Miner., 29, 153 (1985).
15. S.P. Christiano, J. Wang and T.J. Pinnavaia, Inorg. Chem., 24, 1222 (1985).
16. C. Djordjevic and V. Katovic, J. Less. Common Metals, 21, 325 (1970).

LAYER RIGIDITY OF CLAY INTERCALATION COMPOUND: $[Me_4N^+]_{1-x}[Me_3NH^+]_x$-V

S. Lee[+], H. Kim[*], S.A. Solin[+] and T.J. Pinnavaia[*]

Center for Fundamental Materials Research
Michigan State University
East Lansing, MI 48824-1116

Vermiculite is an Alumino-Silicate clays which is one of the layered materials that can form intercalation compounds [1-3]. The host layer of Vermiculite is classified as a 2:1 layered silicate due to its structure [2]. This 2:1 layered silicate is responsible for several of the distinctive properties of Vermiculite. First, unlike graphite layers which are charge neutral, 2:1 layered silicate has a negative layer charge. To compensate this the galleries of Vermiculite are occupied by cations and in turn this makes the intercalation process in CIC's (Clay Intercalation Compounds) an ion exchange process. Second, because the 2:1 silicate layers are composed of multiple, cross-linked planes of atoms, one can expect the clay layers to be relatively rigid to transverse distortions. The rigidity of the silicate layer is important not only for the study of fundamental physical properties of quasi-two dimensional systems but also for practical applications such as catalysis [4,5]. In any case, it is very important to know how rigid the layer is and what factors affect layer rigidity.

To answer the above questions we have used the following strategy. First, we prepared the ternary CIC's: $A_{1-x}B_x$-V, $0 \leq x \leq 1$ with two different monovalent intercalants where the ionic radious r_A is larger than r_B. Ions A and B are randomly distributed in the gallery and the gallery height is different from that of the pristine host material. Second, we used x-ray diffraction to measure the c-axis repeat distance between two successive host layers (basal spacing). The basal spacing is in general a non-linear function of the intercalant composition x and is governed by the layer rigidity, intercalant dimensions, the difference in size and compressibility of the intercalants and the effects of inter-layer correlation.

For this study $[Me_4N^+]_{1-x}[Me_3NH^+]_x$-V(ermiculite) was made from Mg-Vermiculite using an ion exchange method [6], and the result was compared with the previous study of $[Cs^+]_{1-x}[Rb^+]_x$-V [7]. The x-ray diffraction experiment was performed with a computer controlled Huber 4-circle diffractometer coupled to a Rigaku 12Kw rotating anode Mokα source through a vertically bent graphite monochrometer.

[+]Department of Physics and Astronomy
[*]Department of Chemistry

The (OOL) x-ray diffraction patterns for different compositions (x) are shown in Fig. 1 together with an insert showing a rocking curve of the (004) reflection. Vermiculite sample films are the mopological analog of HOPG and this rocking curve indicates a mosaic spread of ~5.25° which is comparable to that of graphoil [8]. From the x-ray diffraction pattern one can see many orders of peaks and these are used to determine basal spacing from a Q-plot.

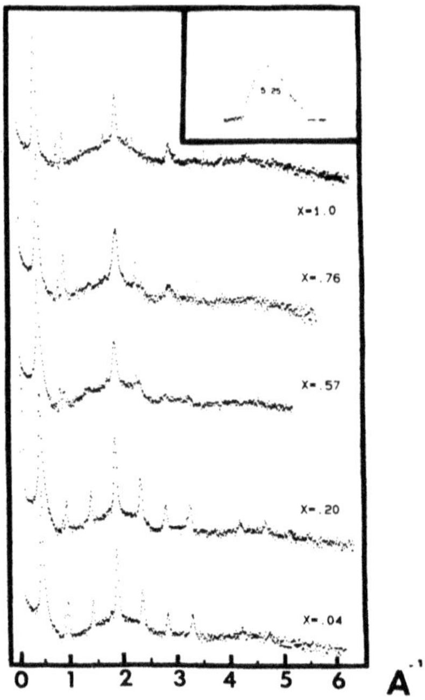

Fig. 1. X-Ray diffraction patterns of $[Me_4N^+]_{1-x}[Me_3NH^+]_x$-V

The following discussion of the behavior of basal spacing versus composition is based on the assumption of a mixed heteroionic system, meaning ions A and B are randomly decorating a 2D Kagomé lattice in each gallery. On the other hand, it is well known that clays tend to be interstratificated which means ions A and B can segregate into different galleries in a sequence units which may be randomly stacked [1,9,10]. To make it clear which is the case, a comparison between the experimental x-ray diffraction patterns and the computer generated patterns is made. In generating x-ray diffraction patterns, the Lorentz-polarization factor and the layer scattering factor are considered with two different interference functions, a Bragg function and a Hendricks-Teller (HT) function [1,11]. For the Bragg interference function, 25 layers and for the HT intereference function infinite number layers are assumed. The resultant computer generated peaks based on Bragg function give a superior agreement with experimental data compared to those based on the HT function. Also no sign of basal spacings on a scale greater than 20Å, characteristic of e.g. A/B/A/B stacking, has been observed. Hence one can conclude safely that the intercalants are indeed randomly distributed in the galleries.

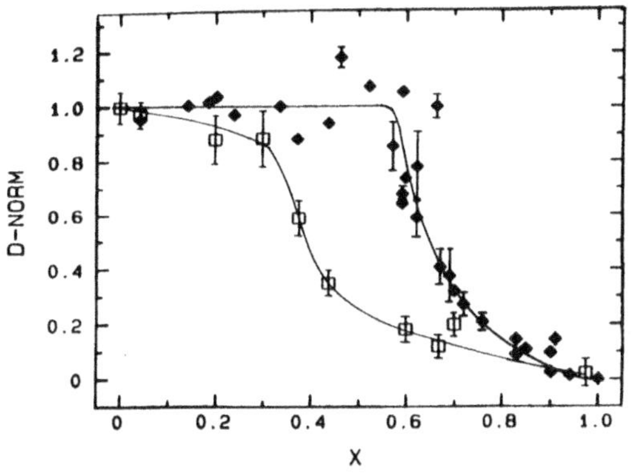

◆ : $[Me_4N^+]_{1-x}[Me_3NH^+]_x$-V

□ : $[Cs^+]_{1-x}[Rb^+]_x$-V

Fig. 2. Normalized basal spacing of CIC's

To compare the composition dependence of basal spacing between two different CIC's a normalized basal spacing defined as $d(x)_{norm}=[d(x)_{obs}-d_{min}]/[d_{max}-d_{min}]$ is used [7]. Here $d_{max}=d(0)_{obs}$ and $d_{min}=d(1)_{obs}$. The normalized basal spacings of $[Me_4N^+]_{1-x}[Me_3NH^+]_x$-V are shown in Fig. 3 with that of $[Cs^+]_{1-x}[Rb^+]_x$-V for comparison. Both systems exhibit a composition dependent normalized basal spacing which does not follow a Vegard's law but rather has a step-like character. Qualitatively the sharp drop in d_{norm} means silicate layers are quite rigid [12] but the different locations, x=.4 for $[Cs^+,Rb^+]$ system and x=.6 for $[Me_4N^+,Me_3NH^+]$ system suggests that intercalant ions are playing an important role in determining basal spacing.

To address the intercalant issue, the encapsulement parameter δ which is a measure of the degree to which the cation is enveloped by the bounding host layers [7] and the sagging limit Δh_{max} which quantifies the finite rigidity of the host layer are introduced. In Table 1, relevant parameters are listed. As one can see from the first and second columns of the table, $d(M^+\text{-}V)_{obs}$ increases with the height of cation M but there is a certain descrepancy between $[Cs^+,Rb^+]$ ions and $[Me_4N^+,Me_3NH^+]$ ions which is more vividly expressed in terms of δ in the table. In CIC's intercalants are sitting on the hexagonal pocket of an oxygene kagomé lattice and it is well known that these oxygene planes are capable of performing torsional distortions [1,7,13]. These hexagonal pockets have an undistorted size of 2.67Å and for the cations which have lateral dimensions that are less than or comparable to the pocket size these oxygene layers have a torsional distortion which increases δ. But if the cation becomes too large, the host layers cannot provide sufficient torsional distortion to accomodate them. Hence the basal spacing change from that of the pristine sample is far less sensitive to the ion size with the smaller ion set than with the larger ion set.

For finite rigidity layers we can express the sagging of the host layer from its maximum height as $\Delta h(x)=\Delta h_{max}[1-\exp(1-L(x)/a)]$ where Δh_{max} is the sagging limit due to finite layer rigidity (i.e. for infinitely rigid layer Δh_{max} is zero), L (x) is the separation between two nearest big ions with given composition x and a is the unit length of the kagomé lattice cell. For x=0 limite (all big ion limit) we have L(0)=a and Δh(x) becomes zero, for high x limit L>>a and Δh becomes Δh_{max}. Now compare Δh(x) with Δr, the difference of ion radii. If Δh(x) is smaller

TABLE 1. Relevant Parameters

	d(Å)	H(Å)	D(Å)	δ(Å)	δ/H(%)
$[Me_4N^+]$	13.34	4.2	4.8	0.20	4.76
$[Me_3NH^+]$	12.70	3.2	4.0	<0	
Cs^+	10.57	3.38	3.38	2.15	63.6
Rb^+	10.23	2.96	2.96	2.07	69.9
Na^+(*)	9.75	1.96	1.96	1.55	79.1

*:ref. 14

d is the basal spacing of M^+-V, H and D are height and diameter of cation, and encapsulement parameter is defined as $\delta=(T+H)-d$ where T=9.34(Å) is the thickness of the silicate layer.
cf. $D(Li^+)$=1.36(Å) and $D(Mg^{+2})$=1.30(Å)

than Δr, the smaller ion does not contribute to basal spacing which will then be determined by d_{max} and $\Delta h(x)$. If $\Delta h(x)$ is larger than Δr, the layer sagging is limited by the smaller ions and the basal spacing will be affected by the presence of the smaller ions. Because $\Delta h(x)$ increases as x increases, one can expect to observe the contribution from the smaller ion at smaller x for the system which has smaller Δr. As shown in Table 1 $\Delta r[Cs^+,Rb^+]$=.42Å and $\Delta r[Me_4N^+,Me_3NH^+]$=1.0Å, in addition the torsional distortion of the Kagomé lattice reduces the effective $\Delta r[Cs^+,Rb^+]$ even less than .42Å as explained in the encapsulement discussion. This is the reasons why sharp drops in basal spacing were observed near x=.4 for $[Cs^+,Rb^+]$ system and near x=.6 for $[Me_4N^+,Me_3NH^+]$ system, respectively.

ACKNOWLEDGEMENTS

Useful discussions with S. Mahanti and W. Jin are gratefully acknowledged. This work was supported by NSF-MRG Grant DMR 85-141545 and in part by the MSU Center for Fundamental Materials Research.

REFERENCES

1. G.W. Brindly and G. Brown (eds.): Crystal Structure of Clay Minerals and Their X-Ray Identification (Minerological Society, London, 1980).
2. R.E. Grim: Clay Minerology (McGraw-Hill, New York, 1968).
3. S.A. Solin: J. Molec. Catalysis 27, 293 (1984).
4. T.J. Pinnavaia: Science 220, 365 (1983)
5. M.S. Wittingham and A.J. Jacobson (eds.): Intercalation Chemistry Article by J.M. Thomas (Academic Press, New York, 1982).
6. H. Kim and T.J. Pinnavaia: unpublished.
7. B.R. York, S.A. Solin, N. Wada, R.H. Raythatha, I.D. Johnson, and T.J. Pinnavaia: Solid State Comm. 54 475 (1985).
8. M.S. Dresselhaus and G. Dresselhaus: Adv. Phys. 30, 139 (1981).
9. D.H. Fink, F.S. Nakayama, and B.L. McNeal: Soil Sci. Soc. Amer. Proc. 35, 552 (1971).
10. M.B. McBride and M.M. Mortland: Clays and Clay Minerals 21, 323 (1973).
11. S.B. Hendricks and E. Teller: J. Chem. Phys. 10, 147 (1942).
12. M.S. Dresselhaus (ed.): Intercalation in Layered Materials, NATO ASI Series 148, Article by S.A. Solin (Plenum Press, New York, 1986).
13. M. Ishii, T. Shimanouchi, and M. Nakahirs: Inorganica Chimica Acta /1:3/, 387 (1967).
14. D.R. Hines, N. Wada, and M. Suzuki: Bull. Am. Phys. Soc. 32, 559 (1987).

THEORETICAL CHARACTERIZATION OF THE ELECTRONIC STRUCTURE OF PHTHALOCYANINE AND RELATED MACROCYCLES

E. Ortí [1a] and J.L. Brédas [1b]

Laboratoire de Chimie Théorique Appliquée
Centre de Recherches sur les Matériaux Avancés
Facultés Universitaires Notre-Dame de la Paix
B-5000 Namur (Belgium)

ABSTRACT: Valence Effective Hamiltonian (VEH) calculations are used to characterize the electronic structure of phthalocyanine-type macrocycles which give rise to "molecular metals" after doping. The theoretical estimates for photoelectron binding energies and optical transitions in the valence region are compared with available data, showing good overall agreement. The main features of those electronic properties of importance with regard to conductivity are discussed in terms of structural changes.

Molecular crystals of phthalocyanine-based compounds exhibit electrical conductivities in the insulating or semiconducting range but become conductors after treatment with oxidant agents like iodine [2]. In the case of free base phthalocyanine, the crystals resulting from the intercalation process with iodine yield conductivities 13 orders of magnitude higher than that of the undoped compound. The temperature dependence of the conductivity is metallic with a conductivity maximun of about 4000 S/cm at 15 K [3]. As a first step towards understanding the properties of these and related "molecular metals", the knowledge of the electronic properties of the building blocks is of major interest.

Our main goal is to characterize from a theoretical point of view the electronic structure of free base phthalocyanine (Pc). However, we have extended the study to the related macrocyclic systems, tetraazaporphyrin (Tap), tetrabenzoporphyrin (Tbp), and porphyrin (P) which also give rise to conducting molecular solids [4]. All valence electron calculations have been carried out using the Valence Effective Hamiltonian (VEH) method [5]. This method provides monoelectronic energies of ab initio double zeta quality and has been succesfully applied to the calculation of the optical and electronic properties of conjugated systems containing the pyrrole moiety [6].

As can be seen in Figure 1, all the molecules studied here have the same basic structure but differ in the units linking the pyrrole moities and in the extension of the π

system depending on the presence of pyrrole or isoindole rings. The atomic coordinates used for Pc and P are taken from crystallographic data [7-9] and the geometries employed for Tap and Tbp are based on those for Pc and P. In all cases, a few geometrical adjustments are made to comply with D_{2h} symmetries.

Table I collects the orbital symmetries and energies for the upper occupied and lower unoccupied molecular orbitals of the four molecules studied. All these orbitals have π symmetry indicating that there are at least six occupied π orbitals which are less stable than the highest occupied lone pair orbital. As a rough illustration of the character of the highest two occupied orbitals, their percentage compositions are reported in Table II.

Table I.– VEH energy levels ($-\epsilon_i$, in eV) obtained for Pc, Tap, P, and Tbp. The highest six occupied and the lowest three unoccupied levels are included. All orbitals are of π nature.

Pc [a]		Pc [b]		Tap		P		Tbp	
a_u*	3.60	a_u*	3.60	a_u*	4.18	a_u*	3.14	a_u*	3.00
b_{3g}*	4.73	b_{3g}*	4.83	b_{3g}*	5.36	b_{2g}*	4.48	b_{2g}*	4.32
b_{2g}*	4.90	b_{2g}*	4.86	b_{2g}*	5.55	b_{3g}*	4.75	b_{3g}*	4.36
a_u	6.15	a_u	6.14	a_u	7.12	a_u	7.26	a_u	6.26
b_{1u}	9.13	b_{1u}	9.23	b_{1u}	9.62	b_{1u}	7.77	b_{1u}	7.84
b_{2g}	9.18	b_{2g}	9.27	b_{1u}	9.64	b_{3g}	9.79	b_{3g}	8.96
b_{1u}	9.19	b_{3g}	9.28	b_{3g}	9.67	b_{1u}	9.79	b_{2g}	8.97
b_{3g}	9.20	b_{1u}	9.29	b_{2g}	9.68	b_{2g}	9.93	b_{1u}	9.21
a_u	9.50	a_u	9.42	b_{1u}	10.61	b_{1u}	9.97	a_u	9.23

a,b) Geometries from crystallographic data of Pc and CoPc, respectively [9,10].

Table II.– Percentage composition (square of the LCAO coefficients) of the highest two occupied levels. N_p denotes pyrrole nitrogens, C_α and C_β are the alpha and beta pyrrole carbons, C_o and C_m are the benzo carbons *ortho* and *meta* to the pyrrole attachment, and N_b (C_b) refers to the bridging atoms.

		N_p	C_α	C_β	C_o	C_m	$N_b(C_b)$
a_u	Pc	0.0	56.2	14.6	14.4	14.8	0.0
	Tap	0.0	62.0	37.4	-	-	0.6
	P	0.0	63.6	36.4	-	-	0.0
	Tbp	0.0	57.2	13.3	15.1	14.4	0.0
b_{1u}	Pc	6.0	0.1	44.0	0.0	45.7	4.2
	Tap	13.9	0.1	79.6	-	-	6.4
	P	19.7	4.8	18.6	-	-	56.9
	Tbp	22.8	5.0	11.4	1.0	8.3	51.5

Pc and P systems have been more extensively studied than Tap and Tbp and represent the extreme structural cases in our molecular set. Indeed, Tap and Tbp can be viewed as intermediate structures between Pc and P. For Pc, VEH calculations place the highest occupied molecular orbital (HOMO) at –6.15 eV, almost 3 eV above the following set of four (almost degenerate) π orbitals situated around –9.15 eV. These values correlate very well with the experimental ionization energies obtained from gas-phase photoelectron spectra which show a sharp peak at 6.41 eV followed by a rather broad band centered at 8.75 eV [10]. It must be noted from Table I, that these values are almost independent from the geometry employed. Thus, the use of crystallographic data from Pc [7] or CoPc [8] does not introduce any significant changes in the occupied orbitals distribution and only the relative positions of the lowest unoccupied orbitals are affected as will be discussed later.

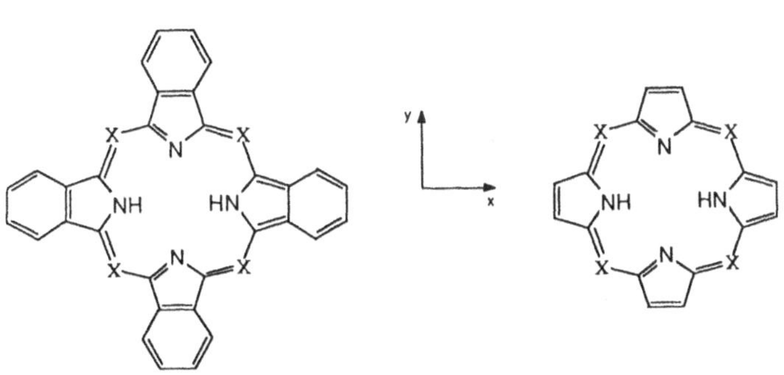

Figure 1.– Structures of the macrocyclic molecules phthalocyanine (Pc), tetraazaporphyrin (Tap), tetrabenzoporphyrin (Tbp), and porphyrin (P).

The energy level distribution of P shows important differences in relation to that obtained for Pc. In P, the HOMO is located at –7.26 eV (i.e., it is 1.1 eV more stable than in Pc) and no important energy gap is observed between this orbital and the b_{1u} orbital at –7.77 eV. The next occupied orbital is separated by a gap of 2 eV. These results are in total agreement with photoelectron spectra which exhibit a narrow band at 6.9 eV with a shoulder at 7.2 eV followed by a broad band centered at 8.8 eV [11]. The theoretical ab initio calculations reported for P [12-15] show some controversy in the assigment of the HOMO (a_u or b_{1u}) but all of them corroborate the trends we obtain for the relative energetic positions of the levels. Particularly, our results are in very good accord with the orbital distribution reported by Christoffersen [13] (–6.8 eV (a_u), –7.2 eV (b_{1u}), and –9.2 eV (b_{3g})) and also with the trends obtained at double zeta level for the FeP complex by Rohmer [15].

The closure of the energy gap between the a_u and b_{1u} occupied orbitals in going from Pc to P should be viewed as a consequence of replacing bridging nitrogen atoms by CH units. As can be seen in Table II, the atomic orbital composition is basically the same for the a_u HOMO in all cases, but differs completely for the b_{1u} level. While for Pc the b_{1u} level has almost no contribution from the linking units, the major contribution comes from those units for P. Looking at the nature of all orbitals, the b_{1u} level of P at –7.77 eV must be correlated with the significantly more stable b_{1u} level of Pc located at –10.61 eV.

The energy level distributions obtained for Tap and Tbp systems corroborate the effects discussed above. Tap, which only differs from Pc by the absence of benzenic rings, exhibits the same orbital distribution than Pc: (i) an energetic gap of 2.5 eV is observed between the HOMO at –7.12 eV and the following b_{1u} level; (ii) the b_{1u} level resulting from contributions of the bridging nitrogen atoms is situated lower in energy at –10.61 eV. On the contrary, for Tbp, which has linking CH units, a b_{1u} level appears at –7.84 eV splitting the gap between the highest occupied orbitals.

The effects of the two main structural differences found in the molecular set studied can be easily discerned. The presence of benzenic rings fused to the pyrrole rings produces a general destabilization of the energy levels of Pc and Tbp with respect to those of Tap and P, respectively. This destabilization is especially important for the HOMO level (≃1 eV). On the other hand, the presence of CH units linking the pyrrole rings causes the appearance of a b_{1u} level around –7.8 eV in the orbital distribution of Tbp and P which is not present in Pc and Tap.

From Table I, it can be inferred that the electronic and optical properties of the systems studied here as well as the crystals and polymers derived from them will be mainly determined by the highest occupied and the lowest two unoccupied levels. In this sense, some theoretical attempts have been carried out to explain the optical spectra of Pc in terms of those levels [16-18].

Taking into account that the VEH method generally provides good estimates for $\pi \rightarrow \pi^*$ energy gaps [6], our results for Pc predict electronic transitions of 1.25 and 1.42 eV for the $a_u \rightarrow b_{2g}^*$ and $a_u \rightarrow b_{3g}^*$ excitations, respectively. These values underestimate the absorption energy of the Q band observed in the vapor phase absorption spectra but reproduce quite well the splitting that shows this band (1.81-2.00 eV) [19]. In relation to other theoretical studies, our predictions agree with the results derived from EHT [16] and PPP-CI [17] calculations but not with those obtained from nonempirical Hartree-Fock-Slater calculations [18] since in the later case the b_{1u} level is assigned as the HOMO.

In going from Pc to Tap, Tbp and P, the energy gap between the HOMO and LUMO increases to 1.57, 1.90, and 2.51 eV, respectively. These values are in very good agreement with the experimental Q band energy only for Tbp (1.86 eV) [16], since they underestimate that energy for Tap (1.99 eV) and overestimate it for P (2.02 eV) [14, 16]. Finally, it must be noted that the relative energy ordering of b_{2g}^* and b_{3g}^* levels depends a geat deal on the geometry adopted. As can be observed in Table I, these orbitals become nearly degenerate when the crystallographic structure of CoPc complex is used instead of that of the free base compound. A more detailed analysis of those problems will be given elsewhere [20].

Our calculations indicate the adequacy of the VEH method for reproducing the main electronic and optical features of phthalocyanine-type macrocycles. They suggest the VEH method as an appropriate theoretical tool for predicting the electronic band structure of polymers and crystals derived from those systems. On the other hand, the similarity of the molecular orbital distribution obtained for Pc and Tap illustrates the suitability of Tap to serve as a simplified model to represent Pc systems.

ACKNOWLEDGEMENTS: One of us (E. O.) thanks the Facultés Universitaires de Namur for hospitality and is indebted to the Consellería de Cultura, Educación y Ciencia de la Generalitat Valenciana for a Postdoctoral grant. The authors are grateful to the Belgian National Fund for Scientific Research (FNRS), IBM Belgium, and the Facultés Universitaires Notre-Dame de la Paix for the use of the Namur Scientific Computing Facility (SCF).

References

[1] (a) Permanent address: Departamento de Química Física, Facultad de Química, Universidad de Valencia, Burjassot (Valencia), Spain. (b) Chercheur Qualifié of the Belgian National Fund for Scientific Research (FNRS).
[2] For recent reviews see: (a) J. Simon, and J.-J. André, "Molecular Semiconductors", J.M. Lehn, and Ch.W. Rees, eds., Springer-Verlag, Berlin, 1985; (b) T.J. Marks, Science, **227**, 881 (1985).
[3] T. Inabe, T.J. Marks, R.L. Burton, J.W. Lyding, W.J. McCarthy, C.R. Kannewurf, G.M. Reisner, and F.H. Herbstein, Solid State Commun., **54**, 501 (1985).
[4] B.M. Hoffman, and J. Ibers, Acc. Chem. Res., **16**, 15 (1983).
[5] G. Nicolas, and Ph. Durand, J. Chem. Phys., **70**, 2020 (1979); **72**, 453 (1980).
[6] J.L. Brédas, B. Thémans, and J.M. André, J. Chem. Phys., **78**, 6137 (1983).
[7] B.F. Hoskins, S.A. Mason, and J.C.B. White, J. Chem. Soc., Chem. Commun., 554 (1969).
[8] G.A. Williams, R. Mason, S.A. Mason, and P.E. Fielding, J. Chem. Soc., Dalton Trans., 1688 (1980).
[9] B.M.L. Chen, and A. Tulinsky, J. Am. Chem. Soc., **94**, 4144 (1972).
[10] J. Berkowitz, J. Chem. Phys., **70**, 2819 (1979).
[11] P. Dupuis, R. Roberge, and C. Sandorfy, Chem. Phys. Lett., **75**, 434 (1980).
[12] J. Almlöf, Int. J. Quantum Chem., **8**, 915 (1974).
[13] R.E. Christoffersen, Int. J. Quantum Chem., **16**, 573 (1979).
[14] U. Nagashima, T. Takada, and K. Ohno, J. Chem. Phys., **85**, 4524 (1986).
[15] M.-M. Rohmer, in "Quantum Chemistry: The Challenge of Transition Metals and Coordination Chemistry", A. Veillard, ed., D. Reidel, 1986.
[16] A.M. Schaffer, and M. Gouterman, Theor. Chim. Acta, **25**, 62 (1972).
[17] A. Henriksson, and M. Sundbom, Theor. Chim. Acta, **27**, 213 (1972).
[18] Z. Berkovitch-Yellin, and D.Z. Ellis, J. Am. Chem. Soc., **103**, 6066 (1981).
[19] L. Edwards, and M. Gouterman, J. Mol. Spectrosc., **33**, 292 (1970).
[20] E. Ortí, and J.L. Brédas, to be published.

SULFUR INTERCALATED INTO BARIUM-COPPER-RARE EARTH SULFIDES

R. Ithnin, D.J. Gilbert, S. Arnold and J.V. Acrivos

San Jose State University
San Jose CA 95192, USA

INTRODUCTION

We have prepared the sulfide analogues of the new warm (≈ 90 K) superconducting oxides ($M_1M'_2Cu_3X_{8-\delta}$, M = Y, Eu, Gd, M'=Ba, X=S and δ= 0 to 1.5). These are referred to in the literature by the stiochiometry **e.g.,** (1,2,3,7.1) is one of the superconducting phases reported in the literature. The purpose of this work is to find out if the sulfide analogues of the above oxides can be prepared by high temperature reactions in a sulphur atmosphere, the composition of the metallic (superconducting?) phases if present and the structure of the unit cell.

Murugesan **et.al.**[1] have prepared and characterized analagous ternary rare earth transition metal sulfides, $LnMS_3$ (Ln = La, Nd, and Gd, M = V and Cr as well as Ln = La and M = Mn, Fe, Co, Ni). However these metals were not found to be superconducting, but were metallic. Also they did not prepare and characterize the copper or titanium ternary compounds.

The new warm superconducting oxides[2], may be considered as layers of oxygen intercalated in between alternating layers of BaO and rare earth oxide. The oxide phase is usually prepared by chemical coprecipitation of the component oxides[3], followed by heating the samples at high temperatures in an oxygen rich atmosphere. The best annealing temperature has not been determined for the oxides or the sulfur analogues. In this work we wish to report the parameters found for the sulfide analogues of the compounds, annealed in a sulfur rich atmosphere.

EXPERIMENTAL

The starting mixture stoichiometry is on the chalcogen rich side of the superconducting oxides. The sulfur analogues have been prepared in our laboratory, starting with the sulphides and excess sulfur in quartz ampoules. A black metallic phase grows on the surface of the quartz tube and quartz plates mixed with the bulk at T> 800°C. Films appear to be formed starting from a seed of a single crystal. This suggests to us that by vapor transport one may be able to grow single crystals at a future date. Pellets to measure the conductivity as well as the X-ray measurements can be obtained by scraping the films from the sides of the tubes. Leads can be attached to these films to measure resistance of less than one ohm/cm.

The films were analyzed using ESCA and it was determined that 75-80% of the copper is in the plus one oxidation state to within the accuracy of the measurement. ESCA measurements have detected very little rare earth in our samples. It also shows that the stoichiometry for the europium sample is (0.3,3,4,9) and for the gadolinium sample it is (0.1,2,4,9). X-ray diffraction studies were carried out by preparing slides of powder scraped from the films using a DIANO XRD 6D/S system. X-ray diffraction studies were also carried out using Debeye-Scherrer X-ray powder techniques.

Chemical analyses were made on the metallicfilms. These were oxidized in concentrated nitric acid, then the weight of barium sulphate and the rare earth sulphate were determined from the yellow white films left on the quartz. The barium can also be determined by DC plasma and the copper can be determined by atomic absorption. The sulfur can be determined by difference.

DISCUSSION

The X-ray powder measurements indicate that the sulfide films, described above, are not in a single phase. We have observed strong diffraction lines with 2θ near 32 degrees (Cu K α) which is comparable to the strong diffractions observed for superconducting oxide samples.[4,5,6] We have compared our observed diffraction spectra for the sulfide with the (1,2,3,7)[7] and the (3,3,6,14)[4,6] perovskite theoretical models. The calculations have been adjusted to account for the sulfide using Moore's tables[8]. We can only index seven of the intense lines to the above structures. This comparison suggests that our unit cell is not based on these perovskite structures. This is to be expected since we know that the copper is in the plus one oxidation state. In this state the copper must be tetrahedrally coordinated instead of the octahedral coordination required be the perovskite structure.

CONCLUSIONS

We have prepared sulfide analogues of the new warm superconducting oxides, but a considerable amount of work remains to be done, **e.g.,** varying the chalcogen composition, the annealing atmosphere and temperature.

ACKNOWLEDGEMENTS

This work was supported by NSF Grant DMR 8612904. We would like to thank R.J. Lipari for ESCA analysis. RI wishes to thank the Malaysian Government for a fellowship to do graduate work at SJSU. DG wishes to thank the ARCS Foundation fellowship and others for financial support during her undergraduate work at SJSU.

REFERENCES

1. T. Murugesan, S. Ramesh, J. Gopalakrishnan, and C.N.R. Rao, J. Solid State Chem. **38**, 165(1981).

2. K.A. Mller, M. Takashige, and J.G. Bednorz, *Phys. Rev. Lett.* **58**, 1143(1987).

3. P.H. Hor, R.L. Meng, Y.Q. Wang, L. Gao, Z.J. Huang, J. Bechtold, K. Forster, and C.W. Chu, *Phys. Rev. Lett.* **58**, 1891(1987).

4. L. Er-Rakho, C. Michel, J. Provost, and B. Raveau, J. Solid State Chem. **37**, 151(1981).

5. J.M. Tarascon, L.H. Greene, W.R. McKinnon, and G.W. Hull, Phys. Rev. B **35**, 7115(1987).

6. S.B. Qadri, L.E. Toth, M. Osofsky, S.Lawrence, D.U. Gubser, and S.A. Wolf, Phys. Rev. B **35**, 7235(1987).

7. R.J. Cava, B. Batlogg, R.B. van Dover, D.W. Murphy, S. Sunshine, T. Siegrist, J.P. Remeika, E.A. Rietman, S. Zahurak, and G.P. Espinosa, *Phys. Rev. Lett.*, **58**, 1676(1987).

8. F.H. Moore, Acta Cryst. **16**, 1169(1963).

CONTRIBUTORS

INDEX

GPSR Compliance

The European Union's (EU) General Product Safety Regulation (GPSR) is a set of rules that requires consumer products to be safe and our obligations to ensure this.

If you have any concerns about our products, you can contact us on ProductSafety@springernature.com

In case Publisher is established outside the EU, the EU authorized representative is:

Springer Nature Customer Service Center GmbH
Europaplatz 3
69115 Heidelberg, Germany

Batch number: 10755801

Printed by Printforce, the Netherlands